# Current Progress in Bioengineering

# Current Progress in Bioengineering

Edited by Gretchen Kenney

Syrawood Publishing House
New York

Published by Syrawood Publishing House,
750 Third Avenue, 9th Floor,
New York, NY 10017, USA
www.syrawoodpublishinghouse.com

**Current Progress in Bioengineering**
Edited by Gretchen Kenney



International Standard Book Number: 978-1-68286-513-2 (Hardback)



**Cataloging-in-Publication Data**

Current progress in bioengineering / edited by Gretchen Kenney.
p. cm.
Includes bibliographical references and index.
ISBN 978-1-68286-513-2
1. Bioengineering. 2. Biotechnology. I. Kenney, Gretchen.
TA164.B56 C87 2018
660.6--dc23

Printed in China.

# TABLE OF CONTENTS

# PREFACE

Over the recent decade, advancements and applications have progressed exponentially. This has led to the increased interest in this field and projects are being conducted to enhance knowledge. The main objective of this book is to present some of the critical challenges and provide insights into possible solutions. This book will answer the varied questions that arise in the field and also provide an increased scope for furthering studies.

Bioengineering is the field of engineering that applies principles of engineering to biological systems. It is a vast field that seeks to design and develop tools and technology for the advancement of life sciences in general. Prospects in this field include manufacturing and manipulating living tissue and matter with the help of technological aids such as computational modeling, materials synthesis and machine learning. This book is meant for students who are looking for an elaborate reference text on this field. It will be of great help to students, researchers, academicians and scientists in the fields of biosystems engineering, biomechanics and tissue and metabolic engineering.

I hope that this book, with its visionary approach, will be a valuable addition and will promote interest among readers. Each of the authors has provided their extraordinary competence in their specific fields by providing different perspectives as they come from diverse nations and regions. I thank them for their contributions.

**Editor**

1

# The Molecular Basis for the Broad Substrate Specificity of Human Sulfotransferase 1A1

**Ilana Berger[1]◑, Chen Guttman[1]◑, Dotan Amar[1], Raz Zarivach[1,2]*, Amir Aharoni[1,2]***

**1** Department of Life Sciences, Ben-Gurion University of the Negev, Be'er Sheva, Israel, **2** National Institute for Biotechnology in the Negev, Ben-Gurion University of the Negev, Be'er Sheva, Israel

## Abstract

Cytosolic sulfotransferases (SULTs) are mammalian enzymes that detoxify a wide variety of chemicals through the addition of a sulfate group. Despite extensive research, the molecular basis for the broad specificity of SULTs is still not understood. Here, structural, protein engineering and kinetic approaches were employed to obtain deep understanding of the molecular basis for the broad specificity, catalytic activity and substrate inhibition of SULT1A1. We have determined five new structures of SULT1A1 in complex with different acceptors, and utilized a directed evolution approach to generate SULT1A1 mutants with enhanced thermostability and increased catalytic activity. We found that active site plasticity enables binding of different acceptors and identified dramatic structural changes in the SULT1A1 active site leading to the binding of a second acceptor molecule in a conserved yet non-productive manner. Our combined approach highlights the dominant role of SULT1A1 structural flexibility in controlling the specificity and activity of this enzyme.

**Editor:** Niall James Haslam, University College Dublin, Ireland

**Funding:** This work was funded by a grant from the European Research Council (ERC) 'Ideas Program', the Israel Science Foundation (ISF), and the Deutsch-Israelische Projektkooperation (DIP) program. The funders had no role in study design, data collection and analysis, decision to publish, or preparation of the manuscript.

**Competing Interests:** The authors have declared that no competing interests exist.

* E-mail: zarivach@bgu.ac.il (RZ); aaharoni@bgu.ac.il (AA)

◑ These authors contributed equally to this work.

## Introduction

The cytosolic sulfotransferase (SULTs) family catalyzes the transfer of a sulfate group from the universal 3′-phosphoadenosine 5′-phosphosulfate (PAPS) donor to a wide variety of acceptor molecules bearing a hydroxyl or an amine group [1,2,3]. Sulfonation results in inactivation of the majority of acceptors, including neurotransmitters, steroid hormones and drugs, thus modulating their biological activity and rendering the product more soluble and readily excretable. Sulfonation thus provides a line of defense by enabling the detoxification of a variety of chemicals. In some cases, however, sulfonation results in the metabolic activation of carcinogens and mutagens, while natural polymorphism in SULT genes has been shown to be associated with an increased risk for cancer [4].

In humans, 13 different SULT genes have been identified and can be divided into four families, including SULT1, SULT2, SULT4 and SULT6 [5,6]. Of these, the SULT1 family is the largest and is responsible for the sulfonation of phenol, thyroid and steroid hormones, as well as a variety of xenobiotics and known drugs [7]. The SULT1 family can be further divided into an additional 5 sub-families, each displaying distinct substrate preference and overlapping substrate specificity. For example, SULT1A1 [8] displays substrate preference for small phenolic compounds, while SULT1E1 shows a preference for estrogen acceptors [9]. Despite differences in substrate preference, a considerable degree of structural homology exists between the SULTs, highlighting the difficulties in understanding the molecular basis for the broad specificity of these enzymes [7].

A prevalent phenomenon in sulfotransferase activity is partial inhibition at high substrate concentrations. The X-ray structure of SULT1A1 in complex with *p*-nitrophenol (*p*NP) revealed two *p*NP molecules bound to the enzyme active site, suggesting that substrate inhibition is due to a decreased rate of catalysis when both *p*NP molecules are bound [10]. However, to understand the structural changes leading to binding of a second acceptor molecule to SULT1A1 in a non-catalytic mode, much more detailed structural data is needed.

In this study, we examined the molecular basis for the broad specificity and substrate inhibition of SULT1A1 using structural, protein engineering and kinetic approaches. We determined five new X-ray structures of SULT1A1, including that of SULT1A1 in complex with the 3′-phosphoadenosine 5′-phosphate (PAP) donor product and 3-cyano-7-hydroxycumarin (3CyC) or 2-naphthol (2NAP) acceptors. These structures demonstrate the high plasticity of the SULT1A1 active site in accommodating different types of acceptor molecule. Our SULT1A1 structures in complex with one or two 3CyC molecules provide new insights into the structural basis of SULT1A1 substrate inhibition. Comparison between the two structures shows a dramatic movement of Phe247 allowing the exposure of a conserved binding site for the second 3CyC molecule. Directed evolution of SULT1A1 for enhanced thermostability and catalytic activity allowed the identification of residues found on the protein surface and in the vicinity of SULT1A1 active site, respectively, that play crucial roles in controlling these SULT1A1 properties. Finally, we determined the structure of the SULT1A1 D249G mutant that confers increased activity towards *p*NP and 3CyC, to highlight the importance of structural flexibility in tuning SULT1A1 transfer activity.

## Results

To obtain detailed understanding of the molecular basis for the broad substrate specificity of SULT1A1, we solved four crystal structures of wild type (WT) SULT1A1 in complex with different acceptors. We determined the structure of SULT1A1 in complex with PAP alone (SULT1A1-PAP), with PAP and 3CyC (SULT1A1-PAP-3CyC1 and SULT1A1-PAP-3CyC2) and with PAP and 2NAP (SULT1A1-PAP-2NAP) to resolutions of 2.0–2.7 Å (**Table S1** and **S2**). Our overall determined structures resemble the typical α/β core fold of SULT1A1, comprising four β-strands framed by two α-helices on either side [7,9,10] (**Fig. 1A**). In all determined structures, as well as in previously described SULT1A1 structures, the PAP moiety is located at the same position in a catalytically competent manner with all binding site residues oriented to minimize conformational freedom of the PAP molecule (**Fig. 1B**) [10,11].

### Acceptor binding site

The SULT1A1-PAP binary complex structure, in the absence of acceptor, reveals a large L-shaped empty binding site proximate to the PAP donor product (**Fig. 1C**). This empty site is very similar to the binding site that is occupied by the different acceptors (**Fig. 1** and below), indicating that PAP binding leads to the pre-formation of the acceptor binding pocket in SULT1A1. The acceptor cavity can be divided into two compartments, referred to as pocket-1 and pocket-2, formed by Phe24, Phe81, Lys106, His108, Val148 and Phe247 (colored red mesh in **Fig. 1C**) and by Phe76, Phe84, Ile89, Tyr240, and Phe247 (colored blue mesh in **Fig. 1C**), respectively. We found that SULT1A1 crystals formed only in the presence of PAP, suggesting that PAP binding leads to dramatic conformational changes in the protein, as observed in the case of the SULT1A3 structure [12,13].

The refined 3D structure of SULT1A1-PAP-2NAP revealed a single 2NAP molecule within pocket-1, positioned in a catalytically competent manner in which the hydroxyl group is positioned 2.34 Å from the catalytic amine group of His108 and 3.33 Å from Lys106 (see **Fig. 1D**). The position of 2NAP is stabilized by hydrogen bonding with a nearby water molecule, as well as by stacking hydrophobic interactions with pocket-1 hydrophobic residues.

We determined SULT1A1 structures in complex with PAP and one or two molecules of the 3CyC acceptor (SULT1A1-PAP-3CyC1 and SULT1A1-PAP-3CyC2, respectively). The SULT1A1-

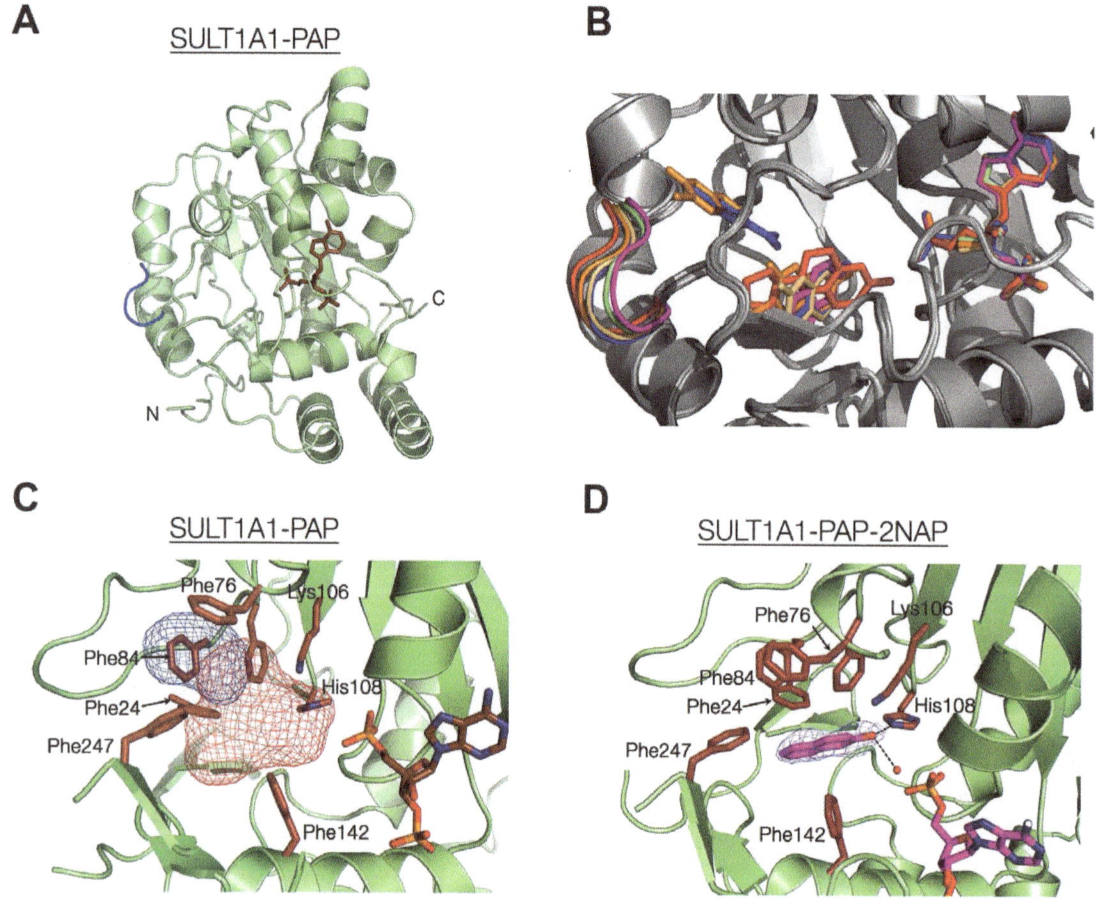


**Figure 1. The overall structure of SULT1A1-PAP and flexibility of the SULT1A1 binding pocket.** (A) View of SULT1A1-PAP with the PAP molecule colored in brown and the gating loop (residues 86–90) colored in blue. (B) Structural flexibility of SULT1A1 is demonstrated by the overlapping of the donor and acceptor binding pockets of all newly and previously determined SULT1A1 structures (see text for details). The colors of the gating loop, the donor and the acceptors are as follows: SULT1A1-PAP-2NAP-pink, SULT1A1-PAP-green, SULT1A1-PAP-3CyC2-orange, SULT1A1-PAP-*p*NP-blue (PDB code 1LS6), SULT1A1-PAP-E2-red (PDB code 2D06) (C) SULT1A1 in complex with PAP shows the formation of a SULT1A1 empty acceptor binding site comprising pocket-1 and pocket-2, depicted as red and blue mesh blobs, respectively. (D) The structure of SULT1A1-PAP-2NAP. The 2NAP acceptor and PAP donor are colored in pink and the 2NAP is outlined with a $F_O$-$F_C$ electron omit map contoured at 2.56. Key residues in panels **C** and **D** are colored red.

PAP-3CyC2 was probably formed due to drop dehydration leading to increased 3CyC concentration. These structures allowed us to examine the changes undergone by SULT1A1 upon binding of one versus two 3CyC molecules. In the SULT1A1-PAP-3CyC2 structure, the first 3CyC molecule (3CyC1) is located in a catalytically competent position with its first phenol ring fitting in between Phe142 and Phe81 in pocket-1, and its hydroxyl group forming hydrogen bonds with the catalytic His108 and Lys106 residues (**Fig. 2A**). The second 3CyC molecule, 3CyC2, is stacked between Phe76 and Phe84 at a 90° degree planar rotation in relation to 3CyC-1 (**Fig. 2A**). The SULT1A1-PAP-3CyC1 structure, containing only one 3CyC acceptor (**Fig. 2B**), demonstrates similar features to the SULT1A1-PAP-3CyC2 structure, although several important differences exist. The most prominent difference was identified at Phe247, which is flipped 100° towards pocket-1 in a similar manner as seen in the SULT1A1-PAP and SULT1A1-PAP-2NAP structures (**Fig. 2C**). The location of Phe247 leads to a vertical shift in the position of the 3CyC phenol ring by 28.7°. Phe247 is thus partially stabilized by a π-π interaction between the cyanide group of 3CyC and Phe247 and by hydrogen bonding to the Phe247 carbonyl (distance 3.76 Å; **Fig. 2B**). The flexibility of 3CyC-1 in pocket-1, affecting Phe247, can lead to the binding of a second 3CyC molecule to pocket-2, thereby affecting the catalytic efficiency of sulfate transfer (**Fig. 2C**). However, it is possible that 3CyC-2 binding is leading to conformational changes in the SULT1A1 active site affecting Phe247.

## Comparing the different SULT1A1 structures

Comparing the four newly determined structures of WT SULT1A1 together with the previously determined structures SULT1A1-PAP-*p*NP and SULT1A1-PAP-E2 [10,11] enabled us to explore conformational changes induced by the binding of diverse acceptors to SULT1A1 (**Fig. 3** and **Table S3**). Identified changes among the different structures were exclusively mapped to those residues shaping pocket-1, representing the catalytically competent binding site (**Fig. 1C**). The most prominent changes were identified at residues 86–90, corresponding to the gating loop (**Fig. 1B**). The gating loop dictates the pocket entrance width and can thus control the entry and evacuation of non-sulfonated and sulfonated molecules, respectively (**Fig. 3**). Comparing the root mean square deviation (RMSD) values of the SULT1A1-PAP structure gating loop with that of SULT1A1-PAP-2NAP, SULT1A1-PAP-3CyC1, SULT1A1-PAP-pNP, SULT1A1-PAP-3CyC2 and SULT1A1-PAP-E2 demonstrated significant differences (1.08, 1.28, 1.4, 1.57 and 2.04 Å, respectively) which correlated with a gradual withdrawal of the loop towards the solvent (**Fig. 1B**). Comparing the dimension properties of the acceptor pockets (i.e., cavity volume, pore area and pore width) among the different structures reveals a general correlation between a withdrawal of the gating loop and an increase in the volume of the cavity (**Fig. 1** and **Table S3**), illustrating the plasticity of the binding site to accommodate diverse acceptors.

Comparing the structures of SULT1A1-PAP-2NAP and SULT1A1-PAP-3CyC1 (**Fig. S1**) reveals that the binding of a

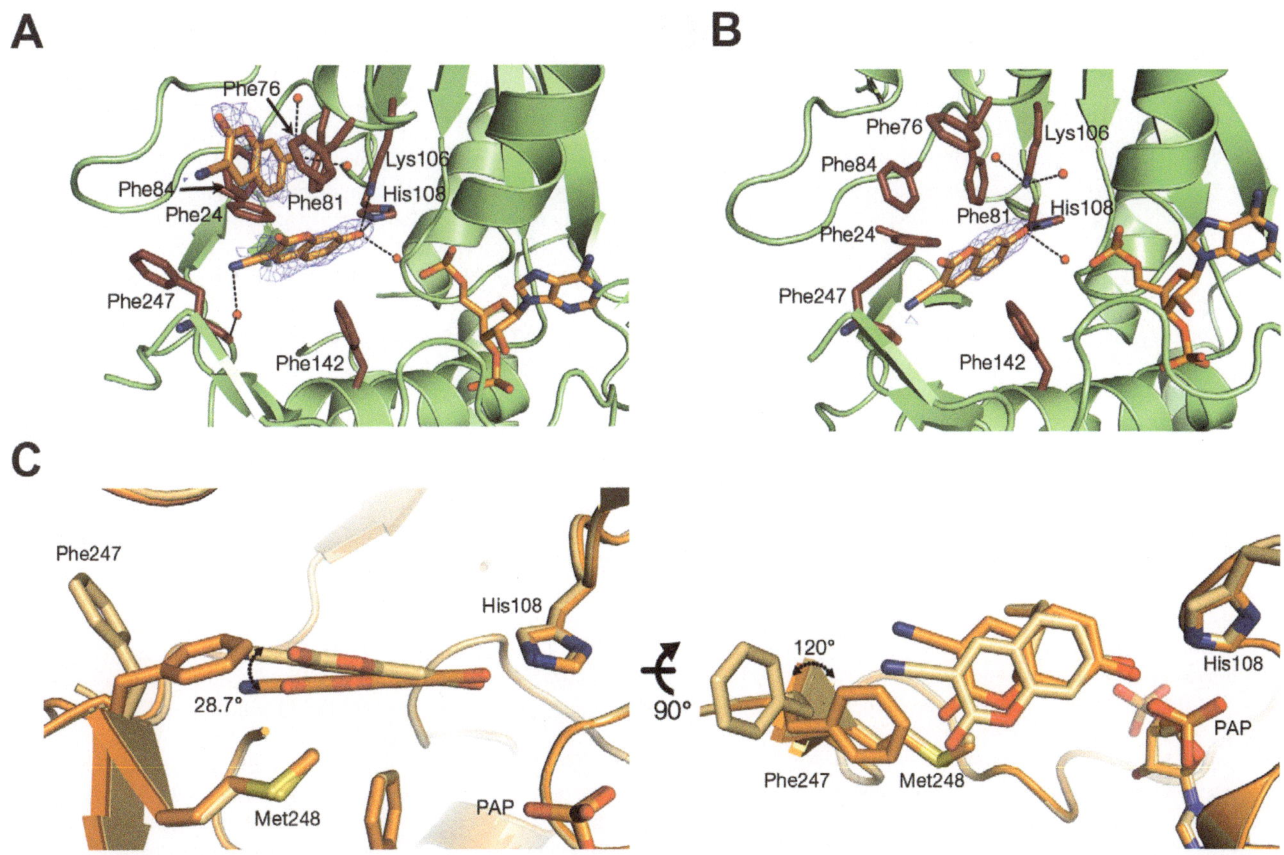


**Figure 2. The molecular basis of SULT1A1 3CyC substrate inhibition.** Structure of SULT1A1 in complex with PAP and two molecules of 3CyC (SULT1A1-PAP-3CyC2, A) or one molecule of 3CyC (SULT1A1-PAP-3CyC1, B) bound at the acceptor-binding site. Key residues are highlighted in red and the 3CyC molecules are outlined with $F_O$-$F_C$ electron omit maps contoured at 2.5σ. (C) Superposition of 3CyC molecules and key residues within the SULT1A1-PAP-3CyC1 and SULT1A1-PAP-3CyC2 structures indicating dramatic movements of the 3CyC cyano group and of Phe247.

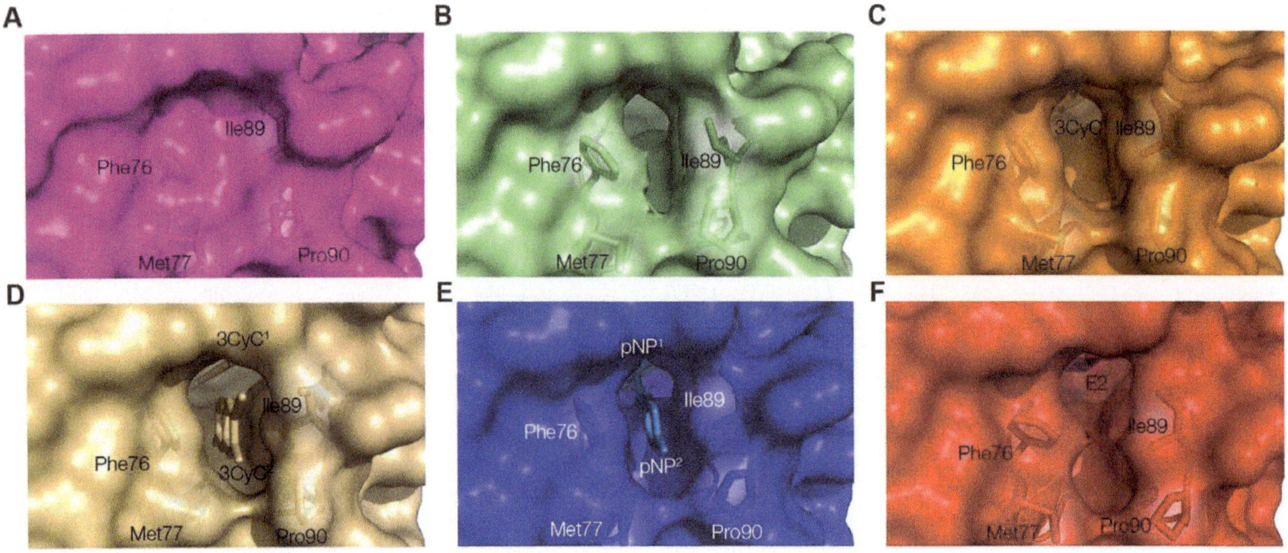


**Figure 3. Surface representation of the binding pocket and pore size of the different SULT1A1 structures.** Ligands and binding pockets are colored according to the displayed SULT1A1 structure: SULT1A1 in complex with PAP and 2NAP (A, pink), PAP (B, green), PAP and 3CyC1 (C, containing one 3CyC molecule, orange), PAP and 3CyC2 (D, containing two 3CyC molecules, grey), PAP and *p*NP (E, blue) or PAP and E2 (F, red). Key residues are highlighted in stick-form. Newly determined structures are A–D.

2NAP molecule, which is more hydrophobic than 3CyC [14], leads to inward movement of the gating loop. This, in turn, dramatically decreases the distance between Ile89 and Phe76 (3.84 versus 6.60 Å in SULT1A1-PAP-2NAP and SULT1A1-PAP-3CyC1, respectively), compacting pocket-2 and the entrance to the pore (see **Fig. S1** and **Table S3**). It thus appears that the hydrophobic nature of 2NAP is sufficient to create additional hydrophobic interactions with Phe84 and Phe247 that drive the closure of the binding pocket in such a way that deflects any additional 2NAP molecules from entering the non-catalytic pocket.

## Directed evolution of SULT1A1

To obtain a deeper understanding of the structure-function relationship of SULT1A1, we employed a directed evolution approach to obtain SULT1A1 mutants showing enhanced thermostability and catalytic activity (**Fig. S2**). The directed evolution approach involves two major steps: (i) the generation of genetic diversity in the gene of interest to obtain large mutant libraries and (ii) the selection of these libraries for the desired catalytic or binding activity [15,16,17]. As a first step in the evolution of SULT1A1 for novel specificity, we sought to improve SULT thermostability. It was previously shown that highly thermostable enzymes that readily express in *E. coli* serve as an excellent starting point for further directed evolution efforts aiming for enhanced catalytic activity and novel specificity [18]. In addition, a trade-off was observed between thermostability and change in enzyme catalytic activity [18]. Thus, mutations that can alter SULT1A1 catalytic properties can lead to reduced thermostability and expression of active protein in *E. coli*.

## Generation of SULT1A1 gene libraries

An effective approach for generating gene libraries that are highly enriched in thermostable mutants involves the targeted mutagenesis of residues that deviate from the consensus sequence of the family back to the consensus sequence [19]. To identify such residues, 34 mammalian SULT1 sequences, corresponding to 8 SULT1A1 homologues, were aligned. Comparison of the SULT sequences revealed that 13 different positions in human SULT1A1 deviate from the consensus sequence (**Table S4**). These residues were thus targeted by mutagenesis. We incorporated the 'back-to-consensus' mutations into SULT1A1 using the ISOR (Incorporation of Synthetic Nucleotide via Gene Reassembly [20]) methodology for partial mutagenesis of the targeted positions. Following library generation, sequencing of 10 random SULT1A1 library variants revealed an average of 6 'back-to-consensus' mutations per gene. Each library variant carried a random, and different, subset of mutated residues, with the entire set being represented in the library (data not shown).

## Screening of the 'back-to-consensus' library for enhanced thermostability

To isolate SULT1A1 mutants with enhanced thermostability from the 'back-to-consensus' library described above, we have developed a simple and rapid high-throughput screening methodology for assessing the transfer of sulfate to 3CyC (**Fig. S2**). This screening assay, performed using *E. coli* crude cell lysates expressing the mutant library clones, follows the quenching of 3CyC fluorescence upon sulfate transfer. We screened ~600 different mutants from the library for sulfate transfer (SULT) to 3CyC activity by following the time-dependent decrease in the 3CyC fluorescent signal. The top 45 of the SULT1A1 variants exhibiting WT or increased SULT activity were further challenged by heat inactivation at 50°C for 15 min and the residual sulfate transfer activity to 3CyC was measured. These conditions significantly reduced the activity of the WT protein and thus, were optimal for identifying thermostable SULT1A1 mutants. The 10 mutants exhibiting the highest thermostability were sequenced. We found combinations of all of the 'back-to-consensus' mutations in the selected SULT1A1 mutants (**Table S5**). This indicates that such mutations are either neutral or beneficial for the catalytic activity and thermostability of the mutants. Selected mutants were over-expressed in *E. coli* cells and purified using Ni-NTA affinity chromatography. Examination of the sulfate transfer activity of these mutants to 3CyC following prolonged incubation at different temperatures revealed an up to 7°C improvement in the heat-inactivation temperature, relative to WT SULT1A1 (**Fig. 4** and **Table 1**). Mapping the 'back-to-consensus' mutations on the SULT1A1-PAP crystal structure revealed that many of the mutations are scattered on the outer surface of the protein (**Fig. S3**). Some of the mutations were characterized by a substitution of hydrophobic residues with charged residues (**Table S5**), resulting in reduced hydrophobicity of the surface and promotion of electrostatic interactions with the solvent. However, due to the large number of mutations, it is difficult to dissect the contribution of each mutation to the overall thermostability. Still, it is clear that enhanced thermostability is attained by diverse contributions from the 'back-to-consensus' mutations.

## Screening and isolation of SULT1A1 mutants with enhanced catalytic activity

To select for SULT1A1 mutants with increased catalytic activity, we employed genes encoding the 10 most highly thermostable versions SULT1A1 as template for the generation of a random mutant library. Next, we screened ~700 mutants for SULT-mediated modification of the pNP and 3CyC acceptors at 100 μM and 10 μM, respectively (see Material and Methods for details). Due to the low $K_M$ of SULT1A1 for pNP [10,21] the screening assay should lead to the isolation of mutants with increase in maximal velocity rather than a decrease in $K_M$ for this substrate (see below). Using the screening assay, we identified several mutants with substantially increased catalytic activity to these acceptors, including the Y240C and D249G, mutations located in a loop region in the vicinity of the SULT1A1 acceptor-binding site (**Table 1** and **Figs. 5** and **6**). To further characterize the effects of these specific mutations, we generated Y240C and the D249G single mutants on the background of WT SULT1A1 through site-directed mutagenesis. We found that the Y240C mutant is prone to aggregation even at room temperature. As such, this mutant was not further analyzed.

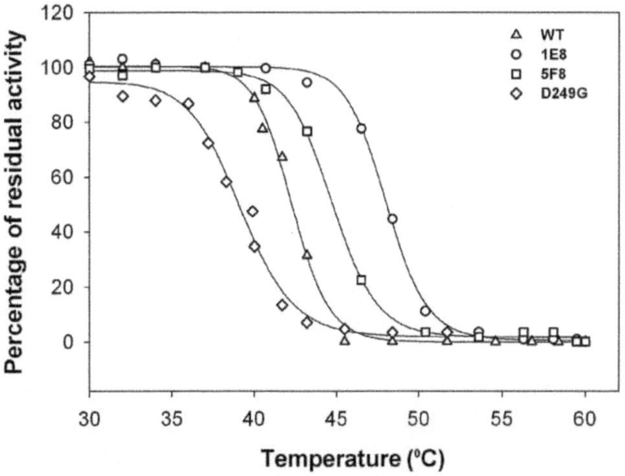


**Figure 4. Heat inactivation curves for the WT SULT1A1 protein and the 5C2, 5F8 and D249G SULT1A1 mutants.** The 1E8 mutant was isolated after the first round of evolution for increased thermostability and the 5F8 was isolated after the second round of evolution for increased catalytic activity. The decrease in stability of the 5F8 mutant, relative to 1E8, and the D249G mutant, relative to the WT protein, highlight the trade-off between thermostability and the acquisition of new catalytic properties (see text for details).

To characterize the thermostability and catalytic activity of the evolved SULT1A1 mutants, we over-expressed the mutants in *E. coli* cells and purified the proteins by affinity chromatography. We found that the mutants displayed lower thermostability, relative to the first generation of highly stable precursor mutants, yet were more stable than the WT protein (**Table 1** and **Fig. 4**). In agreement with these results, the D249G point mutant exhibited lower stability, relative to the WT parent protein (**Fig. 4**). These results highlight the trade-off between the acquisition of increased catalytic activity and thermostability. Next, the catalytic activity of the WT protein and different mutants with the pNP and 3CyC acceptors was measured using steady state kinetics. The kinetic analysis of the evolved SULT1A1 mutants with pNP showed the characteristic increase in SULT1A1 activity at low pNP concentrations and inhibition at high substrate concentrations [10,21] (**Fig. 5, Table 2**). The evolved SULT1A1 5F8 and 5C2 mutants (**Table 2**), as well as the D249G mutant, exhibited similar or lower catalytic efficiency ($k_s$, **Table 2**) but significantly increase in the maximum velocity ($V_p$) relative to the WT enzyme (**Table 2**). Hence, our data together with previous analysis of the WT enzyme using colorimetric and radioactive assays [21,22] indicate that the mutations in these evolved mutants led to increased catalytic activity at high pNP concentrations but reduced affinity to the first and second pNP molecules occupying the SULT1A1 active site. Due to the low $K_M$ of WT SULT1A1 to pNP [21,22] and the relatively low sensitivity of our non-radioactive assay, we could not monitor the activity of WT SULT1A1 at low pNP concentrations.

In contrast, kinetic analysis of the activity towards 3CyC indicated no inhibition at high substrate concentrations for the WT and the mutant proteins. WT SULT1A1 exhibited a much higher $K_M$ of ~20 μM for 3CyC, relative to the pNP, in accordance with the role of SULT1A1 in the sulfonation of small phenolic substrates [8]. The evolved mutants, as well as the D249G point mutant, exhibit higher $K_M$ values, relative to the WT protein, and no saturation at concentrations up to 400 μM 3CyC (**Fig. 5**). Overall, the 5C2, 5F8 and D249G mutants exhibit increased sulfate transfer activity to pNP and 3CyC at high substrate concentrations and reduced affinity for these acceptors.

## Structural analysis of SULT1A1 D249G mutant

To obtain deeper understanding of the structural differences between the D249G mutant and the WT protein, we determined the structure of the D249G mutant in complex with PAP and pNP (D249G-PAP-pNP). Comparing the mutant to the previously determined SULT1A1-PAP-pNP structure [10] indicates high structural similarity, with calculated RMSD values of the total atoms of 0.89Å (**Fig. 6A**). However, we found significant local structural changes in the vicinity of the D249G mutation. The substitution of aspartic acid by glycine leads to a loss of hydrogen bonding with Ser251 and possible destabilization of the loop (**Fig. 6**). This destabilization can further disrupt and break the interaction between His250 and the facing carbonyl of Tyr240 by swiveling the His by 100° toward the solvent, as exemplified by the presence of an additional rotamer (**Fig. 6**). Comparison of many SULTs structures shows that the interaction between His250 and the carbonyl of Tyr 240 is conserved in the majority of SULTs, highlighting the importance of this interaction in controlling loop stability (**Fig. S4**). This loss of connection may increase the flexibility of the SULT1A1 binding site, leading to decreased affinity for pNP and 3CyC molecules (**Fig. 6**). It should be noted that the density surrounding the naturally occurring His250 rotamer is stronger than that of the alternative rotamer. We thus assume that in the current crystal packing, this is the ruling

**Table 1.** Mutations and heat inactivation temperatures[a] of newly evolved SULT1A1 variants.

| SULT1A1 variants | Mutations | Heat inactivation temperatures[a] |
|---|---|---|
| WT | - | 42.3±0.1 |
| 1E8 (R1) | Q56E, A101S, T117S, Q177K, M223I, V243L | 48.1±0.1 |
| 1E9 (R1) | L67V, A101S, Q177K, V211L, F222K, F247I | 48.3±0.4 |
| 5F8 (R2)[b] | Q56E, A101S, T117S, H213R, F222L, V243L, T266N, **L111P, Y240C** | 44.7±0.1 |
| 5C2 (R2)[b] | Q56E, L67V, A101S, T117S, H213R, F222L, V243I **D249G** | 43.3±0.3 |
| D249G[c] | D249G | 39.3±0.2 |

[a]Heat inactivation temperatures were measured by testing SULT1A1 residual activity following incubation at different temperatures (see **Fig. 4** and Material and Methods).
[b]Mutations inserted by random mutagenesis in the second round of evolution are highlighted in bold (see text).
[c]The D249G mutation was generated on the background of the WT protein.

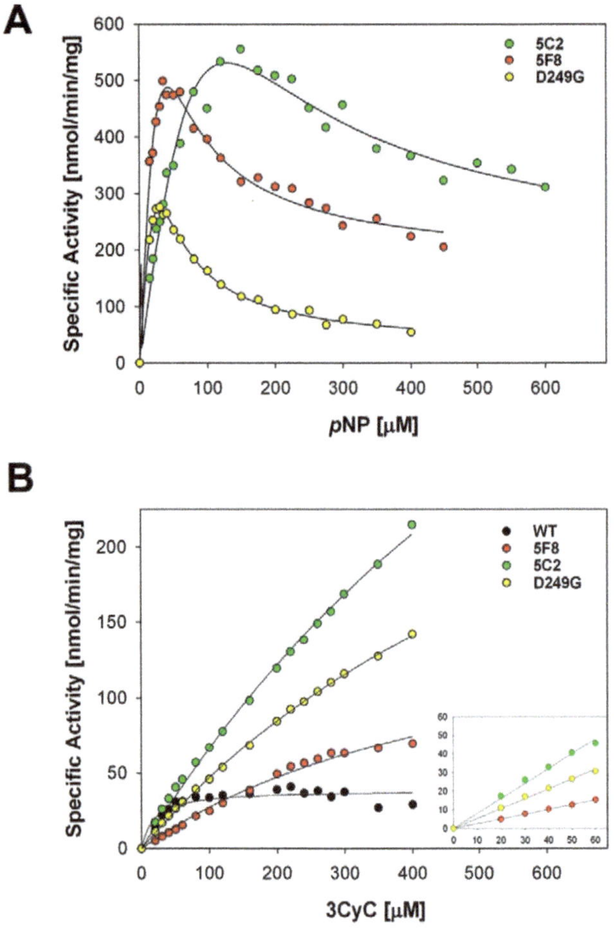


**Figure 5. Kinetic analysis of the sulfate transfer activity of the WT SULT1A1 protein and the 5C2, 5F8 and D249G SULT1A1 mutants to pNP (A) and 3CyC (B).** Each data point represents the mean of three independent experiments. The lines in A represent fit to data obtained with pNP of equation 1 (see Material and Methods) adopted from ref. 21, taking into account the inhibition seen at high pNP concentrations. The lines in B represent fit to Michaelis-Menten equation of data obtained with 3CyC or linear fit to data obtained at low 3CyC concentrations (inset). The $k_{cat}/K_M$ parameters derived from the fits are 1208, 466, 149 and 308 for the WT, 5C2, 5F8 and D249G, respectively, presented as $sec^{-1}$ $M^{-1}$.

conformation that maintains the native position of Tyr240 within the binding pocket (**Fig. 6**). As this phenomenon was not observed in any of the structures of WT SULT1A1 in complex with different acceptors, the importance of Asp249 to the stabilization of the loop and consequently, to the stabilization of Tyr240 within the catalytic pocket, is further supported.

## Discussion

In this work, we employed structural, protein engineering and kinetic approaches to gain detailed understanding of the molecular basis for the broad specificity of SULT1A1. The determination of four new structures of SULT1A1 in complex with the PAP donor and different acceptors enabled us to monitor conformational changes in SULT1A1 upon binding of diverse acceptors. Our results, providing snapshots of SULT1A1 in complex with different acceptors, thus allows us to follow the gradual changes in binding site volume and gating loop movement that occur upon acceptor binding (**Fig. 1** and **Fig. 3**). These results agree with previous structural analysis of different SULT structures determined in complex with different acceptors, demonstrating the high plasticity of the SULT enzyme family [7]. The analysis of the SULT1A1 binary complex with PAP showed that this flexible binding pocket is preformed in the absence of acceptors. Such 'priming' was previously shown for various SULTs and suggests ordered binding in SULT1A1, whereby binding of PAPS precedes that of the acceptor [7]. In contrast, it was shown that in the case of SULT2A1, the donor and acceptors can bind independently and that the binding of the acceptor does not necessarily prime the donor binding site [7].

A common characteristic of SULT activity is inhibition at high substrate concentrations. Elegant structural and kinetic characterization of SULT1A1 with pNP indicated two pNP molecules bound at the active site, albeit in different binding modes [10]. It was shown that one pNP molecule is bound in a catalytic mode while the other is bound at a non-catalytic site, probably leading to substrate inhibition [10]. Our determined SULT1A1 structures in complex with one or two 3CyC acceptor molecules indicate that the formation of the binding site for the second 3CyC molecule (3Cyc-2) is probably induced upon binding of the first molecule to the enzyme active site. Comparing the two structures suggests that subtle movements of the 3CyC-1 in the active site leads to substantial movement of Phe247, leading to the exposure of the second 3CyC-binding site (**Fig. 2**). However, it is possible that the binding of the second 3CyC molecule leads to this movement of Phe247. Our kinetic analysis of WT SULT1A1 with 3CyC does not reveal inhibition at 3CyC concentrations of up to 400 μM, indicating that sulfate transfer at these acceptor concentrations

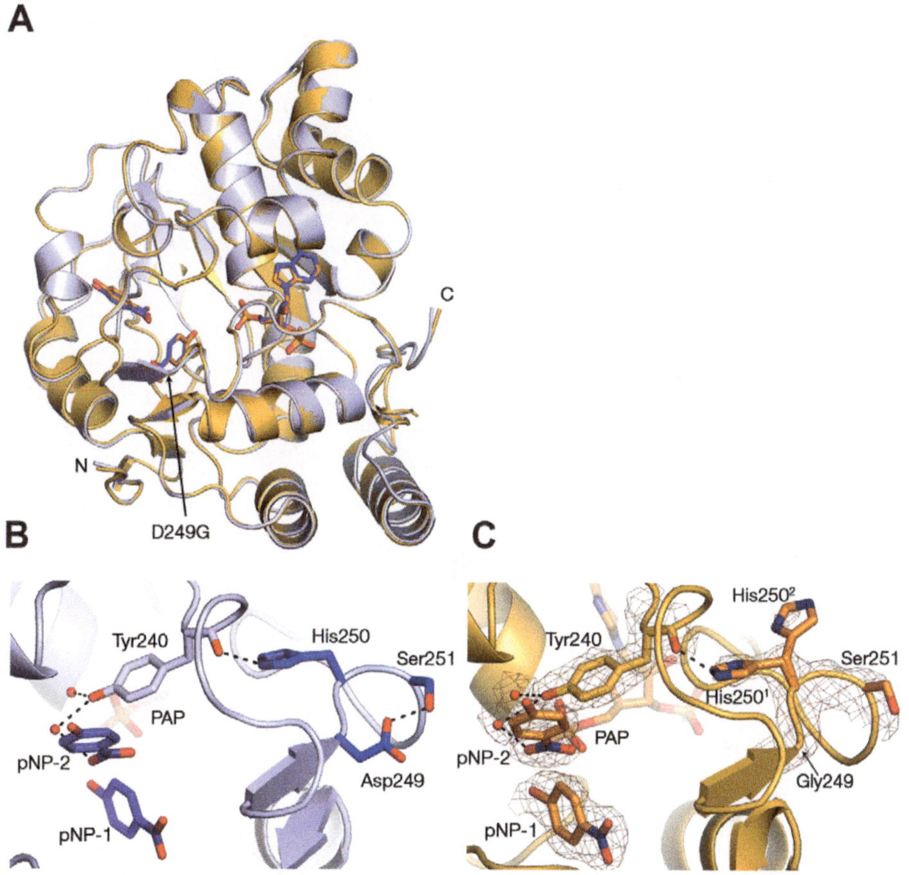


**Figure 6. Comparison of the SULT1A1 and SULT1A1-D249G structures in complex with PAP and *p*NP.** (A) Structural superposition of SULT1A1 (PDB code 1LS6, light blue) and SULT1A1-D249G mutant (gold). (B–C) View of SULT1A1 (B) and the SULT1A1-D249G mutant (C) ligand pocket and the stabilizing loop containing Asp249 or Gly249 (rotamers are indicated with superscript). SULT1A1-D249G key residues and pNP molecule are outlined with a $2F_O$-$F_C$ electron density map contoured at 1.6σ.

takes place in the absence of a second 3CyC molecule bound in a non-productive mode. In contrast, our crystallization efforts performed with SULT1A1 together with 3CyC at high concentration of 1000 μM, yielding crystals of SULT1A1 in complex with one or two 3CyC molecules. These structures, together with the SULT1A1 structure in complex with pNP, indicate a common mechanism for substrate inhibition in SULT1A1 in which a similar binding site is formed at high acceptor concentrations to allow the binding of a second acceptor molecule. Our results are also in correlation with previous studies showing that the SULT1A1 F247L mutation presents substrate inhibition by dopamine, revealing the importance of Phe247 in controlling SULT1A1 substrate inhibition [21]. Interestingly, flipping of the Phe247 was previously identified in SULT1A1 upon binding of estradiol in a non-catalytic orientation [10,11]. In contrast, it was shown that upon binding of two pNP molecules in SULT1A1 the Phe247 is not flipped [10]. These structures together with the structures described in this study shows that the movement of Phe247 is probably dependent on the acceptor size and is not obligatory to allow the binding of a second acceptor molecule in SULT active site.

**Table 2.** Kinetic parameters of SULT1A1 wild-type and newly evolved 5C2, 5F8 and D249G mutants with pNP and 3CyC.

| SULT1A1 | pNP | | | 3CyC |
|---|---|---|---|---|
| | $k_s$[a] *sec⁻¹ M⁻¹* | $V_p$[a] *nmol/min/mg* | $V_\infty$[a] *nmol/min/mg* | $k_{cat}/K_M$[c] *sec⁻¹ M⁻¹* |
| WT[b] | 12857 | 92.5 | - | 1208 |
| 5C2 | 4783±175 | 441±28 | 165±28 | 466 |
| 5F8 | 13591±466 | 393±18 | 174±10 | 149 |
| D249G | 10675±233 | 259±6 | 26±3 | 308 |

[a]Kinetic parameters for sulfate transfer to pNP were determined by fitting the experimental data to the equation described in ref 21 (see also Material and Methods), taking into account the inhibition observed at high pNP concentrations. The parameter $k_s$ is the specificity constant at low pNP concentrations and is equivalent to $k_{cat}/K_M$ in Michaelis-Menten kinetics. $V_p$ is the rate at the peak and $V_\infty$ is the limiting rate when pNP reaches ∞ (Fig. 5).
[b]The WT parameters for pNP are adapted from Alcombri et al.[22] the $k_{cat}/K_M$ is equivalent to $k_s$ and $V_{max}$ is equivalent to $V_p$.
[c]The $k_{cat}/K_M$ values for 3CyC are derived from the fit to Michaelis-Menten equation of data obtained with 3CyC for the WT or linear fit to data obtained at low 3CyC concentrations for the different mutants (Fig. 5).

To provide further insight into the molecular basis of SULT1A1 catalytic activity, we utilized a directed evolution approach. Accordingly, we first generated highly stable SULT1A1 mutants by insertion of 'back-to-consensus' mutations, followed by random mutagenesis to isolate mutants with higher catalytic activity than the WT enzyme. We found that increased SULT1A1 thermostability was probably the result of several stabilizing mutations, rather than being due to one global suppressor, as is the case with the TEM1 β-lactamase M182T mutation [19,23]. The SULTA1 mutations, mainly localized on the protein surface, conferred up to a 7°C increase in heat inactivation temperature, can buffer the deleterious effects of additional mutations, leading to increased activity. This trade-off between stability and the acquisition of new function has been previously reported for several enzymes [18]. It was previously shown that stabilized P450 and TEM1 mutants exhibit higher evolvability upon the introduction of a large variety of mutations without substantial disruption of the native folding of these proteins [19,24]. We found that two mutations, namely D249G and Y240C, both located in the vicinity of the active site, can lead to substantial increases in catalytic activity with 3CyC and pNP at high substrate concentrations but lower affinity for these substrates. Structural analysis of the D249G mutant reveals an increase in the local flexibility of this region, resulting in a dramatic movement of His250 that can affect the SULT1A1 active site. The structural conservation of the His250 in many different SULT structures (**Fig. S4**) may highlight the importance of this region for controlling SULT catalytic activity. It is possible that the higher flexibility allows higher substrate turnover but reduces affinity due to changes in the active site. In accordance, electrostatic surface potential analysis of the D249G mutant relative to the WT enzyme indicates a dramatic change in charge distribution around the active site (**Fig. S5**). These changes can lead to reduced affinity for the different acceptors due to possible differences in acceptor binding or effects on the dynamics and flexibility of SULT1A1.

In summary, our work provides new insight into the molecular basis for the broad specificity of SULT1A1 and highlights the importance of structural flexibility for the recognition of a variety of substrates and for controlling SULT1A1 catalytic properties. The use of structural, engineering and kinetic approaches as employed here can be highly beneficial for understanding the molecular basis for the broad specificity of many other liver enzymes that can detoxify a wide variety of substrates, including amidases, monooxygenases, and other transferases.

## Materials and Methods

### Crystallization conditions, diffraction measurements and structural determination

All crystallization conditions and statistics for the five new SULT1A1 crystal structures are provided in **Tables S1** and **S2**. Grown crystals were mounted in cryoloops at −180°C and collected X-ray diffraction data sets were reduced and scaled using the HKL2000 suite (SULT1A1-PAP and SULT1A1-PAP-2NAP) [25] or iMosflm and Scala (SULT1A1(D249G)-PAP-pNP and SULT1A1-PAP-3CyC) [26,27]. Phases were obtained by molecular replacement with Phaser [28] using the human SULT1A1 structure (PDB code 1LS6) for our diffraction data [10]. Molecular replacement solutions were followed by rigid body refinement, restrained refinement (Refmac5 software package) [29] and manual building (Coot [30]).

### Least-square overlaps and volume calculation

Structural superposition was performed by the SwissPDB viewer alternate domain fitting function [31]. Ligand pocket volume calculations were established by subtracting the molecular surface of protein complexed with donor and acceptor from the molecular surface of protein complexed with donor alone. The cavity volume of the SULT1A1-PAP-2NAP structure was directly calculated by the software. The cavity volume of the SULT1A1-PAP structure was calculated by the inclusion of two pNP molecules according to their location by superimposing 1LS6 on the structure of SULT1A1-PAP. The cavity volume of the total SULT1A1-PAP-3CyC2 structure was calculated by inclusion of molecule 3CyC2 from the SULT1A1-PAP-3CyC1 structure.

### Calculation of electrostatic potential and binding pocket cavity

Electrostatic calculations were performed by PyMOL [32] using the Adaptive Poisson-Boltzmann Solver (APBS) plug-in [33]. The ligand binding pockets of the hSULT1A1-PAP structure were calculated with the Hollow program [34], setting the solvent-accessible surface around the binding pocket to a probe radius of 1.4 Å.

### Plasmids and bacterial strains

The *E. coli* DH5α and Clooni (Lucigen) strains were used for cloning. The *E. coli* BL21 (DE3) strain was used for protein expression and purification. Human cDNA was used as template for amplification of the human SULT1A1 gene. The amplified gene was cloned into bacterial vector pET32tr (a version of pET32 (Novagen) with truncated thioredoxin) using the NdeI and XhoI sites.

### Synthetic shuffling

SULT1A1 amino acid sequences from eight mammalian species were aligned using Muscle software (EMBL-EBI). Oligo-nucleotides used for synthetic shuffling were 31–33 bases long and contained the 'back-to-consensus' single mutation flanked by 15 bases complementary to the SULT1A1 gene at both ends. The *SULT1A1* gene was PCR amplified and 6–10 μg of the PCR products were digested by DNaseI. Fragments (approximate size, 80–120 bp) were extracted and purified. To incorporate the oligonucleotides, the purified fragments were mixed with 5–10 nM of the oligonucleotides and subjected to assembly PCR. The products of the assembly PCR were directly used as template for the amplification of the full length SULT1A1 library containing the spiked oligonucleotides, digested with NdeI and XhoI and cloned into the pET32tr vector as described above.

### Random mutagenesis

Target SULT1A1 sequences were randomly mutagenized by PCR performed with mutagenic dNTP analogs [35]. Mutagenesis stringency was controlled by the number of PCR cycles performed and dNTPs analogs concentration [35].

### Library expression and screening

Single *E. coli* BL21 (DE3) colonies transformed with library or control plasmids were picked to inoculate 600 μl LB media containing 100 μg/ml ampicillin in 96 deep-well plates (Nunc). The plates were incubated with shaking at 37°C overnight. The cells were diluted 1:50 in fresh selective LB media, incubated with shaking at 37°C until an $OD_{600}$ of 0.4–0.6 and induced with 0.1 mM of IPTG (Calbiochem). The plates were incubated with shaking at 30°C for an additional 5 h and centrifuged at 4000 rpm for 15 min. Media was discarded and the cells were resuspended in 150 μl lysis buffer (2.5 mM $MgCl_2$, 0.5 mM $CaCl_2$, 0.2% Triton X-100, 2 U/μl DNaseI (NEB), 1 mg/ml lysozyme, 20 mM HEPES, pH 7.5). The plates were incubated with shaking at 37°C for 30 min, followed by 15 min of centrifugation at 4000 rpm. The supernatant (120 μl) was transferred to 96 V shaped-well plates (Nunc) and stored at 4°C for analysis (see sulfotransferase activity assay, below).

### Protein expression and purification

Single colonies were picked to inoculate 5 ml of LB media containing ampicillin (100 μl/ml) and were grown for 16 h at 37°C, diluted 1:100 with fresh selective LB media and induced with 0.1 mM IPTG (Calbiochem) for an additional 18 h at 20°C. Cells were harvested and centrifuged for 10 min at 6000 g, resuspended in 30 ml 20 mM HEPES, pH 7.5, and recentrifuged for 20 min. Following centrifugation, cells were resuspended in 10 ml/g of binding buffer (1 M NaCl, 40 mM imidazole, 1 mg/ml lysozyme, 20 mM HEPES, pH 7.4). Next, the cells were lysed by sonication, centrifuged and the cleared supernatant was loaded on a pre-equilibrated column containing 2 ml Ni-NTA resin (Qiagen). Following incubation, the resin was washed and SULT1A1 was eluted in 1 ml fractions upon addition of elution buffer. Fractions containing SULT1A1 were analyzed by SDS-PAGE, pooled and dialyzed for 16 h against storage buffer (40 mM HEPES, pH 7.5, 7 mM $MgCl_2$, 1.5 mM DTT and 20% glycerol) at 4°C. Protein concentration was determined with a BCA protein assay kit (Pierce). Wash and elution buffers were based on 500 mM NaCl, 20 mM HEPES, pH 7.4 and supplemented with imidazole, according to the manufacturer's recommendations.

## Sulfotransferase activity assays

Clear lysates or purified proteins were added to a reaction mix consisting of the pNP or 3CyC acceptors at different concentrations, 1 mM PAPS, 20 mM HEPES, pH 7.5, 7 mM $MgCl_2$, and 1.5 mM DTT to a final volume of 200 μl. The decrease in absorbance (405 nm and 408 nm for pNP and 3CyC, respectively) or fluorescence (3CyC, 408 nm excitation and 450 nm emission) was monitored every 15–40 sec, for at least 20 min, using an ELISA plate reader (Infinite-200, Tecan). Screening of the SULT1A1 library for mutants with increased catalytic efficiency or altered specificity was performed using pNP and 3CyC at 10 μM and 100 μM, respectively. Analysis of the kinetic data obtained with pNP was performed according to ref. 21 by fitting the data to the following equation: $v=(k_s[S]+V_\infty[S]^2/K_s^2)/(1+[S]^2/K_s^2)$ (equation 1) using SigmaPlot software. The parameter $k_s$ is the specificity constant at low pNP concentrations and is equivalent to $k_{cat}/K_M$ in Michaelis-Menten kinetics. $V_p$ is the rate at the peak and $V_\infty$ is the limiting rate when pNP reaches ∞. Analysis of the kinetic data obtained with 3CyC was performed by fitting the data to the Michaelis-Menten equation or to a linear equation. For heat inactivation analysis, the clear lysates or the purified protein were incubated for 15 min at the desired temperature using a DNA Engine Peltier Thermal Cycler (Bio-Rad) and then assayed as described above.

## Supporting Information

**Figure S1 Comparison of SULT1A1 structures in complex with PAP and 2NAP or 3CyC.** Superposition of the gating loop and key residues indicate a closure of the gating loop, leading to a smaller cavity volume (see main text and **Table S3** for details).

**Figure S2 Scheme describing the directed evolution process and the acceptors used for the analysis of SULT1A1 specificity.** (**A**) The directed evolution process for the generation of SULT1A1 mutants with increased thermostability (R1) and specificity (R2). (**B–D**) Acceptors used for the directed evolution process of SULT1A1 were pNP (**B**), 3CyC (**C**).

**Figure S3 Surface representation model of the human SULT1A1 structure in complex with PAP.** The surface mutations identified in thermostable SULT1A1 mutants (Table 1 and Table S3) are highlighted in blue. The model was generated using the Swiss PDB viewer program.

**Figure S4 Comparison of SULT structures highlighting the loop region containing Tyr240, Asp249, Ser251 and His250.** The high extent of overlap between the loop location and residues demonstrates the structural conservation of this region. In the SULT1A1 D249G mutant, His250 flips about 100° degrees towards the solvent, leading to the loss of interaction between His250 and the carbonyl of Tyr240. The structures that are overlapped are: WT human SULT1A1 in complex with PAP (1A1 w.t., light green), the SULT1A1 D249G mutant in complex with PAP and pNP (1A1 mutant, orange), human SULT1C2 in complex with PAP (1C2, magenta, PDB code 2GWH), mouse SULT1D1 in complex with PAP (1D1, blue, PDB code 2ZYT), human SULT1B1 in complex with PAP and resveratol (1B1, dark orange, PDB code 3CKL), human SULT1C3 in complex with PAP (1C3, dark green, PDB code 2HK8), and human SULT1A2 in complex with PAP (1A2, yellow, PDB code 1Z29).

**Figure S5 Electrostatic surface representation of SULT1A1 and SULT1A1-D249G showing a dramatic change in the surface electrostatics of the two proteins.** Changes in the electrostatics of the surface can affect the pNP acceptor binding site that is located in the vicinity of Asp249. The approximate locations of Asp249 or G249 and the pNP2 molecule are highlighted by arrows.

**Table S1** Crystallographic statistics of SULT1A1 structures.

**Table S2** Crystallization conditions for SULT1A1.

**Table S3** Comparison of SULT1A1 pore size and cavity volume.

**Table S4** Amino acids in human SULT1A1 that deviate from the family consensus: Comparison to SULT1A1 and SULT1 families.

**Table S5** Mutation distribution in SULT1A1 thermostable mutants.

## Acknowledgments

We thank Melanie Bokstad for her help with the library generation and screening and Dan Tawfik for discussion and advice.

## Author Contributions

Conceived and designed the experiments: IB CG DA RZ AA. Performed the experiments: IB CG DA. Analyzed the data: IB CG DA RZ AA. Contributed reagents/materials/analysis tools: CG RZ. Wrote the paper: AA.

## References

1. Chapman E, Best MD, Hanson SR, Wong CH (2004) Sulfotransferases: structure, mechanism, biological activity, inhibition, and synthetic utility. Angew Chem Int Ed Engl 43: 3526–3548.
2. Gamage N, Barnett A, Hempel N, Duggleby RG, Windmill KF, et al. (2006) Human sulfotransferases and their role in chemical metabolism. Toxicol Sci 90: 5–22.

3. Negishi M, Pedersen LG, Petrotchenko E, Shevtsov S, Gorokhov A, et al. (2001) Structure and function of sulfotransferases. Arch Biochem Biophys 390: 149–157.
4. Wang Y, Spitz MR, Tsou AM, Zhang K, Makan N, et al. (2002) Sulfotransferase (SULT) 1A1 polymorphism as a predisposition factor for lung cancer: a case-control analysis. Lung Cancer 35: 137–142.
5. Blanchard RL, Freimuth RR, Buck J, Weinshilboum RM, Coughtrie MW (2004) A proposed nomenclature system for the cytosolic sulfotransferase (SULT) superfamily. Pharmacogenetics 14: 199–211.
6. Falany CN (1997) Enzymology of human cytosolic sulfotransferases. Faseb J 11: 206–216.
7. Allali-Hassani A, Pan PW, Dombrovski L, Najmanovich R, Tempel W, et al. (2007) Structural and chemical profiling of the human cytosolic sulfotransferases. PLoS Biol 5: e97.
8. Hempel N, Gamage N, Martin JL, McManus ME (2007) Human cytosolic sulfotransferase SULT1A1. Int J Biochem Cell Biol 39: 685–689.
9. Kakuta Y, Pedersen LG, Carter CW, Negishi M, Pedersen LC (1997) Crystal structure of estrogen sulphotransferase. Nat Struct Biol 4: 904–908.
10. Gamage NU, Duggleby RG, Barnett AC, Tresillian M, Latham CF, et al. (2003) Structure of a human carcinogen-converting enzyme, SULT1A1. Structural and kinetic implications of substrate inhibition. J Biol Chem 278: 7655–7662.
11. Gamage NU, Tsvetanov S, Duggleby RG, McManus ME, Martin JL (2005) The structure of human SULT1A1 crystallized with estradiol. An insight into active site plasticity and substrate inhibition with multi-ring substrates. J Biol Chem 280: 41482–41486.
12. Bidwell LM, McManus ME, Gaedigk A, Kakuta Y, Negishi M, et al. (1999) Crystal structure of human catecholamine sulfotransferase. J Mol Biol 293: 521–530.
13. Dajani R, Cleasby A, Neu M, Wonacott AJ, Jhoti H, et al. (1999) X-ray crystal structure of human dopamine sulfotransferase, SULT1A3. Molecular modeling and quantitative structure-activity relationship analysis demonstrate a molecular basis for sulfotransferase substrate specificity. J Biol Chem 274: 37862–37868.
14. Wolfenden R (1985) Affinities of organic compounds for solvent water; hydrophilic and hydrophobic character. J Chem Scie 94: 121–138.
15. Aharoni A, Griffiths AD, Tawfik DS (2005) High-throughput screens and selections of enzyme-encoding genes. Curr Opin Chem Biol 9: 210–216.
16. Arnold FH, Wintrode PL, Miyazaki K, Gershenson A (2001) How enzymes adapt: lessons from directed evolution. Trends Biochem Sci 26: 100–106.
17. Tao H, Cornish VW (2002) Milestones in directed enzyme evolution. Curr Opin Chem Biol 6: 858–864.
18. Tokuriki N, Tawfik DS (2009) Stability effects of mutations and protein evolvability. Curr Opin Struct Biol 19(5): 596–604.
19. Bershtein S, Goldin K, Tawfik DS (2008) Intense neutral drifts yield robust and evolvable consensus proteins. J Mol Biol 379: 1029–1044.
20. Herman A, Tawfik DS (2007) Incorporating Synthetic Oligonucleotides via Gene Reassembly (ISOR): a versatile tool for generating targeted libraries. Protein Eng Des Sel 20: 219–226.
21. Barnett AC, Tsvetanov S, Gamage N, Martin JL, Duggleby RG, et al. (2004) Active site mutations and substrate inhibition in human sulfotransferase 1A1 and 1A3. J Biol Chem 279: 18799–18805.
22. Alcolombri U, Elias M, Tawfik DS (2011) Directed Evolution of Sulfotransferases and Paraoxonases by Ancestral Libraries. J Mol Biol 411(4): 837–53.
23. Wang X, Minasov G, Shoichet BK (2002) Evolution of an antibiotic resistance enzyme constrained by stability and activity trade-offs. J Mol Biol 320: 85–95.
24. Bloom JD, Labthavikul ST, Otey CR, Arnold FH (2006) Protein stability promotes evolvability. Proc Natl Acad Sci U S A 103: 5869–5874.
25. Otwinowski Z, Minor W (1997) Processing of X-ray Diffraction Data Collected in Oscillation Mode. Methods in Enzymology, Volume 276: Macromolecular Crystallography, part A. pp 307–326.
26. (1994) The CCP4 suite: programs for protein crystallography. Acta Crystallogr D Biol Crystallogr 50: 760–763.
27. Evans P (2006) Scaling and assessment of data quality. Acta Crystallogr D Biol Crystallogr 62: 72–82.
28. McCoy AJ, Grosse-Kunstleve RW, Adams PD, Winn MD, Storoni LC, et al. (2007) Phaser crystallographic software. J Appl Crystallogr 40: 658–674.
29. Murshudov GN, Vagin AA, Dodson EJ (1997) Refinement of macromolecular structures by the maximum-likelihood method. Acta Crystallogr D Biol Crystallogr 53: 240–255.
30. Emsley P, Cowtan K (2004) Coot: model-building tools for molecular graphics. Acta Crystallogr D Biol Crystallogr 60: 2126–2132.
31. Guex N, Peitsch MC (1997) SWISS-MODEL and the Swiss-PdbViewer: an environment for comparative protein modeling. Electrophoresis 18: 2714–2723.
32. DeLano WL (2002) The PyMOL Molecular Graphics System. (DeLano Scientific, San Carlos) Release 0.99.
33. Baker NA, Sept D, Joseph S, Holst MJ, McCammon JA (2001) Electrostatics of nanosystems: application to microtubules and the ribosome. Proc Natl Acad Sci U S A 98: 10037–10041.
34. Ho BK, Gruswitz F (2008) HOLLOW: generating accurate representations of channel and interior surfaces in molecular structures. BMC Struct Biol 8: 49.
35. Zaccolo M, Williams DM, Brown DM, Gherardi E (1996) An approach to random mutagenesis of DNA using mixtures of triphosphate derivatives of nucleoside analogues. J Mol Biol 255: 589–603.

# 2

# Conjugation of Proteins by Installing BIO-Orthogonally Reactive Groups at Their N-Termini

**Nagasundarapandian Soundrarajan[1], Sriram Sokalingam[1], Govindan Raghunathan[1], Nediljko Budisa[2], Hyun-Jong Paik[1], Tae Hyeon Yoo[3,4], Sun-Gu Lee[1]***

**1** Department of Polymer Science and Chemical Engineering, Pusan National University, Busan, South Korea, **2** Department of Chemistry, Technische Universität Berlin, Franklinstrasse, Berlin, Germany, **3** Department of Molecular Science and Technology, Ajou University, Suwon, South Korea, **4** Department of Applied Chemistry and Biological Engineering, Ajou University, Suwon, South Korea

## Abstract

N-terminal site-specific modification of a protein has many advantages over methods targeting internal positions, but it is not easy to install reactive groups onto a protein in an N-terminal specific manner. We here report a strategy to incorporate amino acid analogues specifically in the N-terminus of a protein *in vivo* and demonstrate it by preparing green fluorescent protein (GFP) having bio-orthogonally reactive groups at its N-terminus. In the first step, GFP was engineered to be a foldable, internal methionine-free sequence via the semi-rational mutagenesis of five internal methionine residues and the introduction of mutations for GFP folding enhancement. In the second step, the N-terminus of the engineered protein was modified *in vivo* with bio-orthogonally functional groups by reassigning functional methionine surrogates such as L-homopropargylglycine and L-azidohomoalanine into the first methionine codon of the engineered internal methionine-free GFP. The N-terminal specific incorporation of unnatural amino acids was confirmed by ESI-MS analysis and the incorporation did not affect significantly the specific activity, refolding rate and folding robustness of the protein. The two proteins which have alkyne or azide groups at their N-termini were conjugated each other by bio-orthogonal Cu(I)-catalyzed click chemistry. The strategy used in this study is expected to facilitate bio-conjugation applications of proteins such as N-terminal specific glycosylation, labeling of fluorescent dyes, and immobilization on solid surfaces.

**Editor:** Dafydd Jones, Cardiff University, United Kingdom

**Funding:** This research was supported by Basic Science Program (NRF-2011-0021222) and Mid-career Research Program (NRF-2012016803) through the National Research Foundation of Korea funded by the Ministry of Education, Science and Technology. The funders had no role in study design, data collection and analysis, decision to publish, or preparation of the manuscript.

**Competing Interests:** The authors have declared that no competing interests exist.

* E-mail: sungulee@pusan.ac.kr

## Introduction

The N-terminus of a protein is an attractive target site for functionalization to afford protein modifications such as PEGylation, glycosylation and fluorescent labeling; these modified proteins can be valuable sources for development of therapeutics and diagnostics [1–3]. Several chemical methods have been developed for N-terminal functionalization of a protein, but they are generally complicated and involve side reactions which yield heterogeneous products [4–6]. Biological methods for the N-terminal functionalization of a protein have recently been recognized as efficient ways to overcome the problems in chemical N-terminal modification. In particular, an approach based on the methionine (Met) residue substitution method allows the efficient production of proteins with an N-terminal specific functional group *in vivo* [7–9], which would pave the way to generate proteins with novel functions.

The Met residue substitution method introduces unnatural Met analogues into a protein by reassigning the Met codon globally in a protein sequence [5,6]. The simple procedure using the Met auxotroph enables the production of a range of proteins with functional groups on a large scale. Bio-orthogonally reactive groups, such as L-homopropargylglycine (Hpg) and L-azidohomoalanine (Aha), have been incorporated into the Met positions of proteins *in vivo* by adding the Met surrogates instead of Met because the wild-type Met-tRNA synthetase recognizes the unnatural amino acids [5,10]. In addition, engineering of the substrate specificity of Met-tRNA synthetase can expand the scope of this methodology [11,12].

Bacterial proteins are synthesized from Met and the removal process of the start Met can be suppressed by selecting the second residue next to the Met carefully [7,9]. Therefore, Met analogues can be incorporated into the N-termini of proteins using the Met residue substitution method. However, the presence of the internal Met codons in the target sequences limits the successful application of the Met residue substitution method for N-terminal specific functionalization due to the reassignment of unnatural Met surrogates to internal Met codons as well as to the first Met codon [7–9]. This problem can be overcome by engineering the protein sequence to be devoid of internal Met residues. Although this approach sometimes needs time-consuming protein engineering work to find internal Met-free variants having original functions of proteins, to our knowledge, this approach is the only one that makes the N-terminal specific modification of a protein possible.

Our previous report showed that a protein sequence could be engineered to be an internal Met-free using a consensus-based concept [8]. In the study, the internal Met residues of the single chain fragment variable (scFv) antibody sequence were replaced successfully with other conserved amino acids without affecting the activity of the protein. This allowed subsequent N-terminal specific functionalization of the scFv using the Met residue substitution method. The stability of scFv probably contributed to the success of the approach because stability of a protein is known to be related to the resistance to mutations [13,14]. However, it is easily expected that the Met removal based on consensus sequences may not always work, because most proteins are marginally stable and thus cannot withstand multiple changes in their sequences [15]. In particular, hydrophobic residues such as Met are frequently located in the highly packed hydrophobic core, which makes it harder to generate functional Met-free protein sequences.

We here engineered a green fluorescent protein (GFP) to be an internal Met-free protein sequence and demonstrated its N-terminal functionalization using the *in vivo* Met residue substitution method. It was previously reported that mutations of the three Met residues in the core hydrophobic regions of GFP based on consensus approach induced complete misfolding of the protein [16]. In the present study, a GFP devoid of internal Met residues was generated by semi-rational mutagenesis and its folding efficiency was improved by introducing mutations for GFP folding enhancement, which yielded an internal Met-free GFP sequence that can be properly folded. Subsequently, bio-orthogonally reactive amino acid analogues were introduced at the N-terminus of the engineered GFP (Figure 1). Then a protein-protein conjugation was demonstrated using the N-terminally modified GFPs.

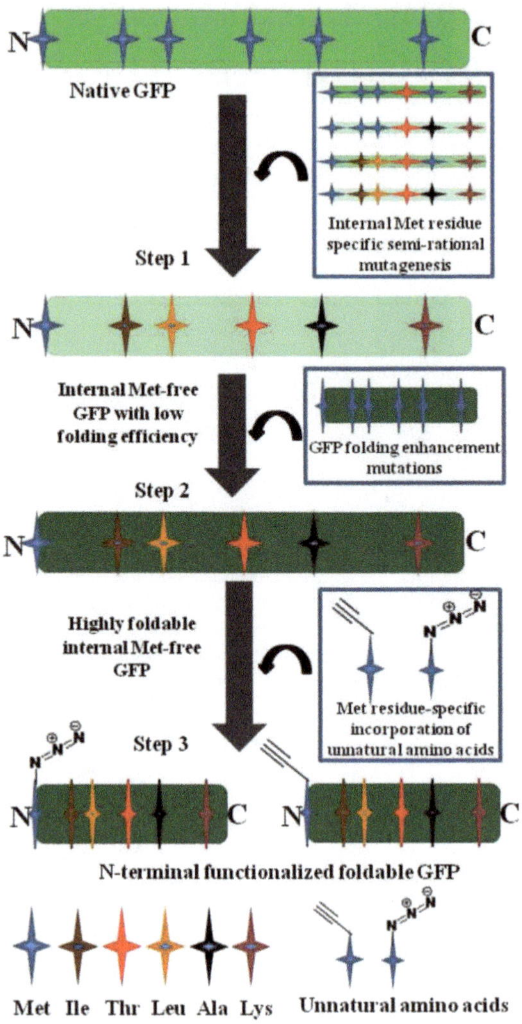


**Figure 1. Schematic illustration of the approach for the generation of N-terminal functionalized green fluorescent protein (GFP).** Step 1: generation of internal methionine (Met) free GFP by semi-rational mutagenesis using similar physicochemical amino acids. Step 2: introduction of folding enhancement mutations to rescue the fluorescence activity of internal Met-free native GFP. Step 3: N-terminal functionalization of active GFP by Met residue specific incorporation methodology.

## Materials and Methods

### Materials

T4 DNA ligase, restriction endonucleases and PCR reagents, were purchased from New England Biolabs (Tokyo, Japan). The isopropyl-D-thiogalactopyranoside (IPTG) and other chemicals were purchased from Sigma chemicals (St. Louis, MO, USA) unless otherwise indicated. Hpg and Aha were purchased from Chiralix (Nijmegen, Netherlands). *Escherichia coli* strain M15A methionine auxotroph was kindly donated by Prof. David A. Tirrell (Caltech, USA) and plasmid pQE80-L was obtained from Qiagen (Valencia, CA, USA).

### Construction of the GFP Variants

The pQE-80-GFP constructed in our previous study [17] was used for the template of $GFP_{nt}$. Mutagenesis of the Met codons in $GFP_{nt}$ were carried out by assembly PCR method [18]. Table S1 shows the primers used in the mutagenesis. The amplification of $GFP_{nt}$ in pQE80-L using two sets of primers ($GFP_{nt}$Out-F and $GFP_{nt}$M153-R, and $GFP_{nt}$M153-F and $GFP_{nt}$OutR) was performed with VentR DNA Polymerase (New England Biolabs) to mutate M153 to Thr. The DNA fragments obtained from these PCR steps were purified on 1.5% agarose gels (Qiagen Gel Extraction Kit). Equimolar quantities of the fragments were used as the template for the assembly PCR using the following pairs of primers: $GFP_{nt}$Out-F and $GFP_{nt}$Out-R. Similarly, M233K mutation was introduced into $GFP_{nt}$ containing M153T using two sets of primers ($GFP_{nt}$Out-F and $GFP_{nt}$M233-R, and $GFP_{nt}$M233-F and $GFP_{nt}$OutR) followed by assembly PCR described above, which generated $GFP_{nt}$-r2M. $GFP_{nt}$-r3M was obtained by the saturation mutagenesis of the M218 position in $GFP_{nt}$-r2M using two sets of primers ($GFP_{nt}$Out-F and $GFP_{nt}$M218-R, and $GFP_{nt}$M218-F and $GFP_{nt}$OutR) and subsequent screening of fluorescent colonies. $GFP_{nt}$-r4M was also obtained by the saturation mutagenesis of the M78 and M88 positions in $GFP_{nt}$-r2M using respective sets of primers and subsequent screening of fluorescent colonies. GFPhs-r5M was commercially synthesized by Genescript Corporation (New Jersey, USA). The $GFP_{nt}$-r2M, $GFP_{nt}$-r3M, $GFP_{nt}$-r4M and $GFP_{nt}$-r5M were cloned into pQE80-L by using *Bam*HI and *Hin*dIII restriction sites, and GFPhs-r5M was cloned into the same plasmid by using *Eco*RI and *Hin*dIII sites. All of the constructs possessed hexa-histidine tags in their C-terminals.

### Expression and Purification of Met Analogues Incorporated GFP Variants

The GFP variants in pQE80-L was transformed into *E. coli* M15A Met auxotroph and expressed in minimal medium according to previously described protocols [14]. Briefly, the limiting concentration of Met (0.035 mM) allowed the cells to attain an $OD_{600}$ 0.6–0.8 and the target proteins were induced with

1 mM IPTG followed by simultaneous addition of either Met or Met analogues (0.5 mM) and allowed the expression for 5 h. The harvested cells were suspended in lysis buffer (50 mM Na-Phosphate buffer pH 7.4 containing protease inhibitor) and disrupted by french press. The suspension was subsequently spun down to collect the soluble and insoluble protein fraction for SDS-PAGE analysis. The remaining soluble protein fractions were purified by Ni-NTA column chromatography (GE Healthcare Bio-Sciences, Sweden) by standard protocol. Elution fractions were analyzed by SDS-PAGE, and those that were enriched in the desired GFP variants were pooled and dialyzed against 1X phosphate buffered saline (PBS).

### Measurement of Fluorescence

Whole cell fluorescence assay was performed on cells with a 0.1 $OD_{600}$, suspended in 1X PBS, by measuring fluorescence intensity by exciting at 485 nm and collecting emission at 515 nm with excitation/emission slits of 5.0 nm using Perkin Elmer/Wallac Victor 2 Multilabel Counter (1420-011). The protein samples were excited at 490 nm and emissions collected at 511 nm with excitation/emission slits of 5.0 nm were recorded on Hitachi FL spectrophotometer (F-4500) equipped with FL solution program for analysis of the spectra.

### Denaturation and Refolding of GFP Variants

Each purified GFP variant (30 μM) was unfolded at 95°C for 5 min in 1X PBS containing 8 M urea and 5 mM DTT. Refolding was carried out by 100-fold dilution of urea denatured samples at room temperature into 1X PBS containing 5 mM DTT. The concentrations of denatured proteins were adjusted to 0.3 μM and recovered fluorescence was measured using Hitachi FL spectrophotometer (490 nm excitation, 511 nm emission, 10 nm excitation/emission slit) for 30 min with an interval of 3 sec. The recovered fluorescence was normalized by dividing final fluorescence after 24 h value. The normalized values were fitted with Sigma Plot (Systat Software Inc., CA) using equations as described by previous report [19].

### Renaturation Equilibrium Measurement

Each purified GFP variant (100 μM) was denatured at 95°C for 5 min in 9 M urea in TNG buffer (25 mM Tris pH 7.5, 0.2 M NaCl, 5% Glycerol, 1 mM DTT). Equilibrium fluorescence values were measured by diluting the urea denatured proteins into refolding buffer (TNG) containing 5 mM DTT to various final concentration of urea (1–6 M), and allowing the refolding to proceed up to 24 h at 15°C. Fluorescence was measured by exciting at 485 nm and emission at 515 nm with excitation/emission slits of 5.0 nm and was recorded on Perkin Elmer/Wallac Victor 2 Multilabel Counter (1420-011). $C_{0.5}$ was calculated by measuring the concentration of urea at which the 50% of initial fluorescence was recovered after 24 h of incubation, and the values were determined by sigmoidal fit using sigma plot (Systat Software Inc., CA) [19].

### Protein-protein Conjugation

Copper (I)-catalyzed cycloaddition reaction between Hpg incorporated GFPhs-r5M and Aha incorporated GFPhs-r5M was performed in a reaction mixture composed of 0.5 mg/ml of each protein (250 μl), 50 mM Tris.HCl pH 8 (100 μl), 50 mM $CuSO_4$ (50 μl), 50 mM L-ascorbic acid (50 μl), and $H_2O$ (300 μl). Control was prepared in similar way without the addition of $CuSO_4$ and L-ascorbic acid. The reaction mixture was shaken for 24 h at 4°C, and dialyzed against 1X PBS thrice at 4°C. Finally, the reaction mixtures were analyzed by SDS-PAGE.

## Results

### Structural Role of Methionine Residues in GFP

The wild-type GFP is a 27 kDa protein with a single chain polypeptide containing 238 amino acids. The protein folds into an 11 stranded β-barrel with a single α-helix running through the barrel to form a chromophore. The chromophore cyclization reaction begins once the GFP achieves its near-native like structure [20]. The fluorescent chromophore is directly correlated with the folding status of the protein, which makes GFP an excellent model protein for protein engineering study [21]. In the present study, $GFP_{mut3.1b}$ with a fast folding property [22] bearing one N-terminal Met and five internal Met (M78, M88, M153, M218, and M233) was used as a native GFP variant, designated as $GFP_{nt}$. Among the five internal Met residues, three Met residues (M78, M88, and M218) are located in the hydrophobic core of the protein and the remaining Met residues (M153 and M233) are exposed to solvent. The buried internal Met218 plays a major role in folding process by interacting with single Trp57 through sulfur-aromatic interactions, which is essential for the GFP folding process [23].

### Engineering of Internal Met-free GFP Sequences

To generate the internal Met-free GFP sequence, mutations of internal Met residues were attempted by considering the Met locations in the GFP structure. First, mutation of the surface exposed two Met residues (M153 & M233) in GFP was performed. Since the amino acid residues on the protein surface are relatively insensitive to mutations [24,25], the two Met residues were mutated simply based on previous results instead of rigorous mutation studies. It was reported that the M153T mutation of GFP suppressed the aggregation of GFP [26]. For the M233 residue, our previous study informed that it was tolerable to the mutation of Lys [16]. Thus, the M153T and M233K variant of $GFP_{nt}$, designated as $GFP_{nt}$-r2M, was generated and the effect of mutation on the GFP activity and productivity were examined. $GFP_{nt}$-r2M showed similar whole cell fluorescence (Figure 2) and soluble expression level (Figure S1A) with $GFP_{nt}$, indicating that the mutations did not affect the GFP folding and activity significantly as expected.

As next step, we attempted to replace the remaining three internal Met residues (M78, M88, and M218) of $GFP_{nt}$-r2M in the core hydrophobic region with other amino acids. Actually it was reported that the three Met residues can be mutated into other residues without affecting significantly the fluorescence of cells expressing the GFP variant (GFPrm_AM), in which M78, M88, and M218 were changed into Leu, Phe, and Ala respectively [12]. Therefore, we introduced the three mutations into $GFP_{nt}$-r2M, but the fluorescence of cells expressing the GFP variant (M78L, M88F, M153T, M218A, and M233K) was much lower than that of cells expressing $GFP_{nt}$-r2M. Since different GFP sequences were used as template to generate GFPrm_AM and the Met-free GFP in this study, we decided to change the three Met residues of $GFP_{nt}$-r2M step by step. The M218 residue that plays an important role in folding [23] was randomized into hydrophobic amino acids (Leu, Ile, Val, Phe, and Ala) using oligonucleotides having degenerate codons. The clones showing fluorescence were selected manually based on their fluorescence, and a $GFP_{nt}$-r2M variant having the M218A mutation, designated as $GFP_{nt}$-r3M, showed the highest fluorescent among the mutants obtained. The whole cell fluorescence of the $GFP_{nt}$-r3M was approximately 5

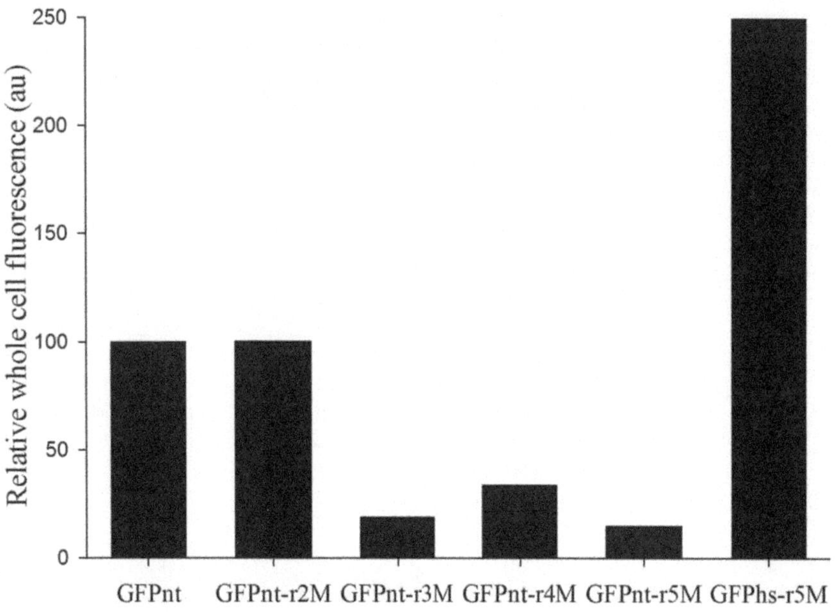


**Figure 2. Comparison of functional productivity of $GFP_{nt}$ and its variants.** The data show the whole cell fluorescences of $GFP_{nt}$ (containing all the 5 internal Met), $GFP_{nt}$-r2M (containing M153T and M233K mutations), $GFP_{nt}$-r3M (containing M153T, M233K and M218A mutations), $GFP_{nt}$-r4M (M78I, M88L, M153T and M233K mutations), $GFP_{nt}$-r5M (M78I, M88L, M153T, M218A and M233K mutation) and GFPhs-r5M ($GFP_{nt}$-r5M containing mutations for folding enhancement). The relative fluorescence (in arbitrary units) is the fluorescence of whole cells compared with the fluorescence of cells expressing $GFP_{nt}$.

times lower than that of $GFP_{nt}$-r2M (Figure 2). SDS-PAGE analysis confirmed that the reduced fluorescence of $GFP_{nt}$-r3M was caused by a misfolding of the protein (Figure S1B), which highlights the importance of the M218 residue in the folding of GFP. Similarly, the other two internal Met positions (M78 and M88) in $GFP_{nt}$-r2M were randomized at the same time with hydrophobic amino acids (Leu, Ile, Phe, Val, and Ala). A $GFP_{nt}$-r2M variant having the M78I and M88L mutations, designated as $GFP_{nt}$-r4M, showed the highest fluorescence; cells expressing $GFP_{nt}$-r4M exhibited around 3-fold lower fluorescence than those expressing $GFP_{nt}$-r2M (Figure 2). This result suggests that the M78 and M88 residues in the hydrophobic core are also important in GFP folding. All the three mutations, M78I, M88L, and M218A, were introduced into $GFP_{nt}$-r2M, which resulted in a complete internal Met-free GFP sequence, $GFP_{nt}$-r5M. However, the whole cell fluorescence of $GFP_{nt}$-r5M was approximately 7 times lower than that of $GFP_{nt}$-r2M (Figure 2), and $GFP_{nt}$-r5M was mostly expressed as an insoluble form (Figure S1C). This confirms that the three Met residues in the hydrophobic core are very important in the formation of active GFP structure. Although it was not successful to generate an internal Met-free protein with preserved initial activity, these results suggest that the semi-rational approach based on similar physicochemical amino acids can be a handy tool for engineering a protein devoid of internal Met.

Both the three mutations M78L, M88F, and M218A in GFPrm_AM, and the mutations found in this study (M78I, M88L, and M218A) did not result in an active internal Met-free GFP variant. One thing that needs to be noted is that the starting GFP sequence to generate GFPrm_AM is a GFP variant (L024_3-3) that exhibited higher expression, better refolding behavior and higher stability than normal GFP [27], and thus we suspected that the properties of template GFP sequence could be an important factor for succeeding in generating an internal Met-free GFP variant. Since L024_3-3 was engineered to make GFP fluorescent with 5,5,5-trifluoroleucine, we turned to another GFP variant, superfolder GFP [19], which also showed improved folding properties and much more resistance to mutations than a wild type GFP. We introduced the mutations of superfolder GFP (S30R, Y39N, F64L, F99S, N105T, Y145F, M153T, V163A, I171V, and A206V) into $GFP_{nt}$-r5M. It was also reported that N149K [28] and S208L [29] affected the folding efficiency of GFP positively, although their effects were not significant. The two mutations (N149K and S208L) were additionally introduced, and the resulting variant was named GFPhs-r5M. As shown in the Figure 2, the whole cell fluorescence of GFPhs-r5M was much higher than that of $GFP_{nt}$-r5M, and approximately 2.5 times higher than $GFP_{nt}$. SDS-PAGE analysis of the expressed protein confirmed that the soluble expression level of the GFPhs-r5M protein was improved significantly compared to that of $GFP_{nt}$-r5M and higher than that of $GFP_{nt}$ (Figure S1D), suggesting that the introduced mutations improved the folding efficiency of $GFP_{nt}$-r5M remarkably. Table S2 shows the protein sequence of the soluble and active internal Met-free variant, i.e. GFPhs-r5M.

## N-terminal Functionalization of the Internal Met-free GFP

The GFPhs-r5M variant obtained from the above study is expressed as a functional form, and contains a Met residue only in its N-terminus, which suggests that the expression of the gene for GFPhs-r5M using the Met residue substitution method may enable the production of N-terminal functionalized GFP *in vivo*. To demonstrate this, the gene for GFPhs-r5M was expressed in the Met auxotrophic *E. coli* with the addition of Met surrogates, Hpg or Aha, according to the previously reported procedures [14]. Hpg and Aha are unnatural amino acids containing alkyne and azide groups respectively, which are illustrated in Figure S2. The soluble expression of GFPhs-r5M with Hpg or Aha was confirmed by SDS-PAGE (Figure 3A) and the corresponding active fluorescent proteins were produced despite an approximately 20% decrease in whole cell fluorescence compared to GFPhs-r5M with Met (Figure 3B). The proteins produced were purified and analyzed by ESI-MS to identify the incorporation of Hpg or Aha. The Hpg or Aha incorporated proteins showed an exact mass shift of −22 and −5 Da corresponding to one Met residue substitution of the respective unnatural amino acids (Figure S3). The ESI-MS data in the Figure S3 also showed an incorporation efficiency of >90%. These results clearly show that active GFP with N-terminal specific functional groups with high yield could be produced using the engineered GFPhs-r5M and Met residue substitution method.

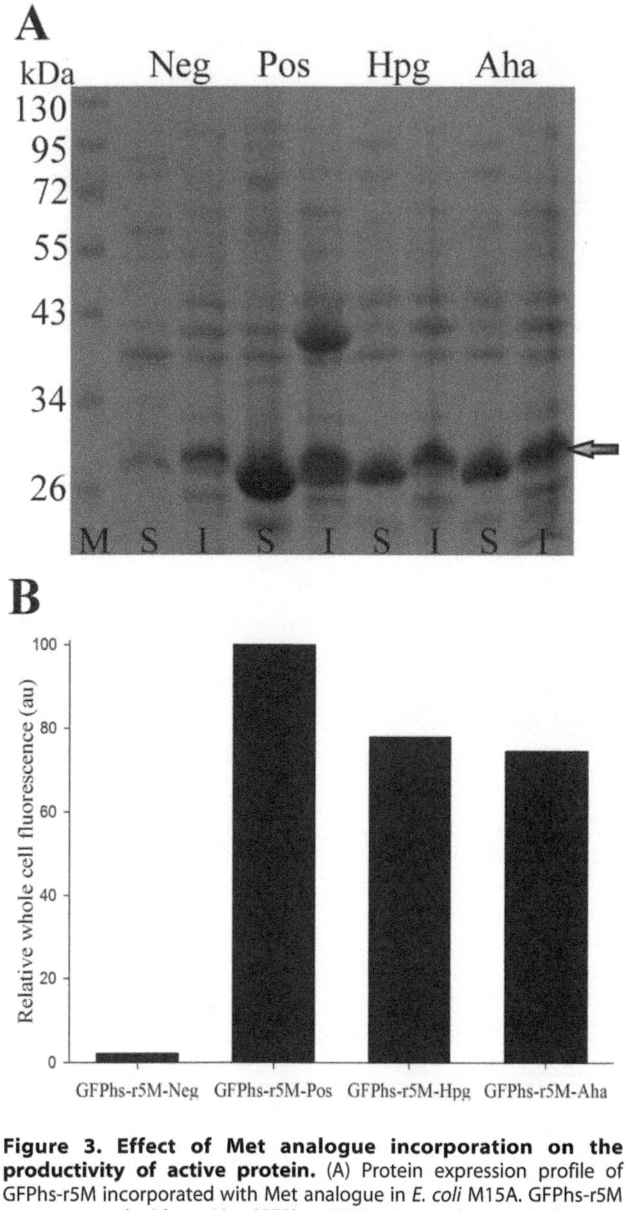


**Figure 3. Effect of Met analogue incorporation on the productivity of active protein.** (A) Protein expression profile of GFPhs-r5M incorporated with Met analogue in *E. coli* M15A. GFPhs-r5M was expressed without Met (GFPhs-r5M-Neg) or with Met (GFPhs-r5M-Pos) or with Met analogues such as Hpg (GFPhs-r5M-Hpg) and Aha (GFPhs-r5M-Aha) in minimal media containing 19 amino acids. The expected size is indicated by an arrow (S, soluble fraction; I, insoluble fraction; M, molecular weight marker). (B) Whole cell fluorescence of GFPhs-r5M expressed in minimal medium without Met (GFPhs-r5M-Neg) or with Met (GFPhs-r5M-Pos) or with Met analogues such as Hpg (GFPhs-r5M-Hpg) and Aha (GFPhs-r5M-Aha). The relative fluorescence (in arbitrary units) is the fluorescence of whole cells compared with the fluorescence of cells expressing GFPhs-r5M-Pos.

## Characterization of the Functionalized GFP Variants

The specific fluorescence, refolding rate and folding robustness of the N-terminal functionalized GFPhs-r5M with Hpg or Aha (designated as GFPhs-r5M-Hpg and GFPhs-r5M-Aha respectively) were compared with those of GFPhs-r5M to examine the biophysical effects of N-terminal functionalization on the protein. The biophysical properties of $GFP_{nt}$ were also examined and compared as a control.

As shown in Figure 4, GFPhs-r5M, GFPhs-r5M-Hpg and GFPhs-r5M-Aha exhibited similar specific fluorescence activities, which suggest that the addition of alkyne or azide on the N-terminus of the protein did not affect the protein activity negatively. On the other hand, the specific fluorescent activities of GFPhs-r5M and its derivatives were approximately 1.5–2 fold higher than that of $GFP_{nt}$. This indicates that the mutations introduced into $GFP_{nt}$-r5M for folding enhancement had influence on the spectral properties of protein in addition to the folding efficiency. This also suggests that the higher whole cell fluorescence of GFPhs-r5M than that of $GFP_{nt}$ in Figure 2 was caused by an enhancement of the specific fluorescent activity as well as by an increase in the soluble expression level.

Figure 5 shows the refolding kinetics of the $GFP_{nt}$, GFPhs-r5M, and GFPhs-r5M with Hpg or Aha. Both GFPhs-r5M-Hpg and GFPhs-r5M-Aha showed similar folding rates in both the fast and slow phases compared to GFPhs-r5M, which were 4–5 fold higher folding rate compared to $GFP_{nt}$. These results are correlated with the soluble expressions level of $GFP_{nt}$ and GFPhs-r5M (Figure S1).

The study on folding robustness was carried out by estimating the refolding tolerance of the four GFP variants to protein denaturant. The fractions of recovered fluorescence under different urea concentrations were determined after 24 hours and their $C_{0.5}$ were estimated from the refolding equilibrium plot (Figure 6). The estimated $C_{0.5}$ values of the GFP variants suggest that the incorporation of the unnatural amino acids has little effect on the folding robustness.

Overall, the GFPhs-r5M and its variants with N-terminal specific functional groups showed comparable biophysical properties, and their specific activity, refolding rate and folding robustness were higher than $GFP_{nt}$. This suggests that the functionalized unnatural proteins produced were active and stable enough for further study such as bio-conjugation through the introduced functional groups. The results also support the possibility that the introduced mutations for GFP folding enhancement provided the GFP sequence with sufficient folding robustness to withstand the cumulative effects of internal Met-free mutations.

## Protein-protein Conjugation Using the Functionalized Proteins

Azides and alkynes are highly energetic functional groups with a particularly narrow distribution of reactivity. In addition, the copper (I) catalyzed cycloaddition reaction of azide and alkyne yields 1,4-disubstituted 1,2,3-triazole linked conjugates under very mild conditions such as room temperature and in an aqueous buffer. Moreover, the reaction is highly regiospecific, chemoselective and tolerant to a wide range of functional groups [1,30,31]. These outstanding features of click chemistry have been extended to various bio-conjugation applications.

Protein-protein bio-conjugation reaction based on the click chemistry described above was performed to evaluate the possibility of bio-conjugation using the N-terminal specific functionalized GFPs produced *in vivo*. The purified GFPhs-r5M-Hpg and GFPhs-r5M-Aha containing an alkyne and azide

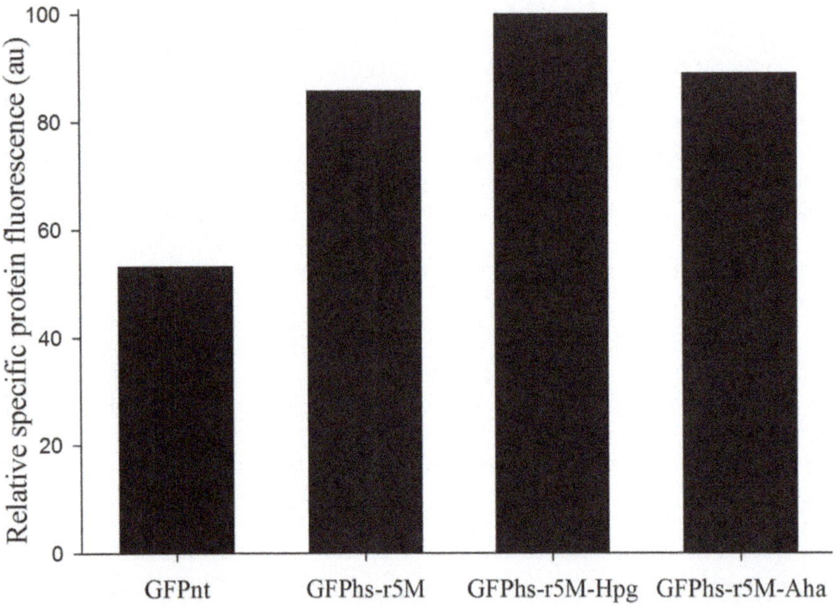


**Figure 4. Relative specific fluorescene of $GFP_{nt}$, GFPhs-r5M, GFPhs-r5M-Hpg and GFPhs-r5M-Aha of purified proteins.** Relative specific activity (in arbitrary units) is the fluorescence of purified protein compared with the fluorescence of purified GFPhs-r5M-Hpg.

groups, respectively, on their N-termini were incubated in the presence of $CuSO_4$ and L-ascorbic acid to carry out the cycloaddition (Figure 7A). The bio-conjugation was analyzed by SDS-PAGE. Formation of the GFPhs-r5M dimer, 55 kDa in size, was observed with a yield of approximately 50% (Figure 7B), whereas the control reaction performed without $CuSO_4$ and L-ascorbic acid did not produce such dimer band. This suggests that the protein-protein conjugation between the N-terminal specific functionalized proteins was achieved in a site-specific manner. In the protein-protein conjugation reaction, we could observe some protein aggregation and this hampered further characterization of the conjugated protein. This problem should be solved to use the protein-protein conjugation method more efficiently.

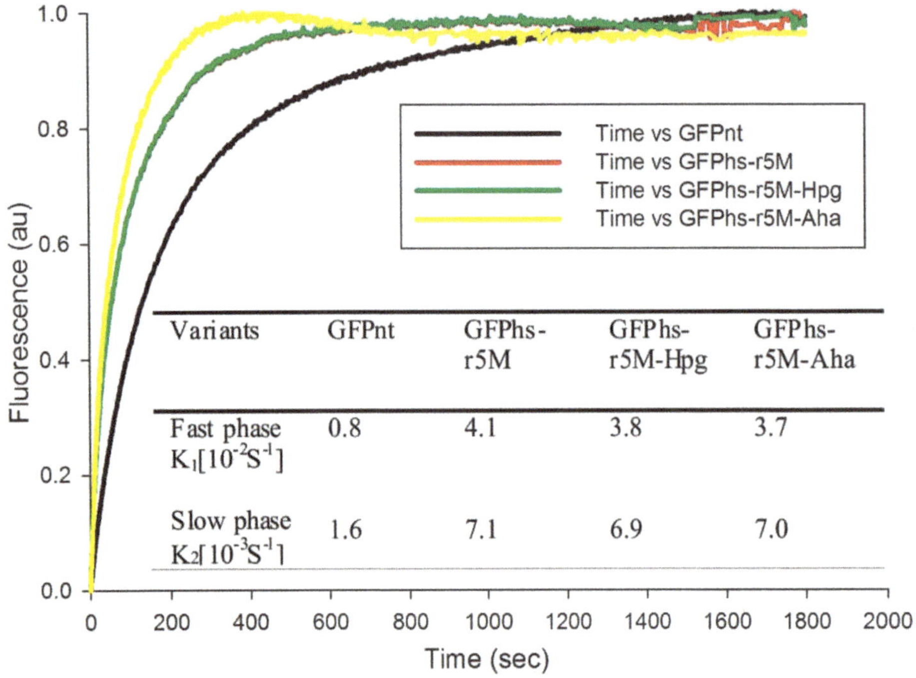


| Variants | GFPnt | GFPhs-r5M | GFPhs-r5M-Hpg | GFPhs-r5M-Aha |
|---|---|---|---|---|
| Fast phase $K_1[10^{-2}S^{-1}]$ | 0.8 | 4.1 | 3.8 | 3.7 |
| Slow phase $K_2[10^{-3}S^{-1}]$ | 1.6 | 7.1 | 6.9 | 7.0 |

**Figure 5. Refolding kinetics of the variants $GFP_{nt}$, GFPhs-r5M, GFPhs-r5M-Hpg and GFPhs-r5M-Aha.** Refolding kinetics was measured after denaturation in 8M urea followed by renaturation by dilution. Inlet table shows the refolding rates of $GFP_{nt}$ and GFPhs-r5M variants for fast and slow phase of refolding progress. Normalized fluorescence in arbitrary units was plotted against time.

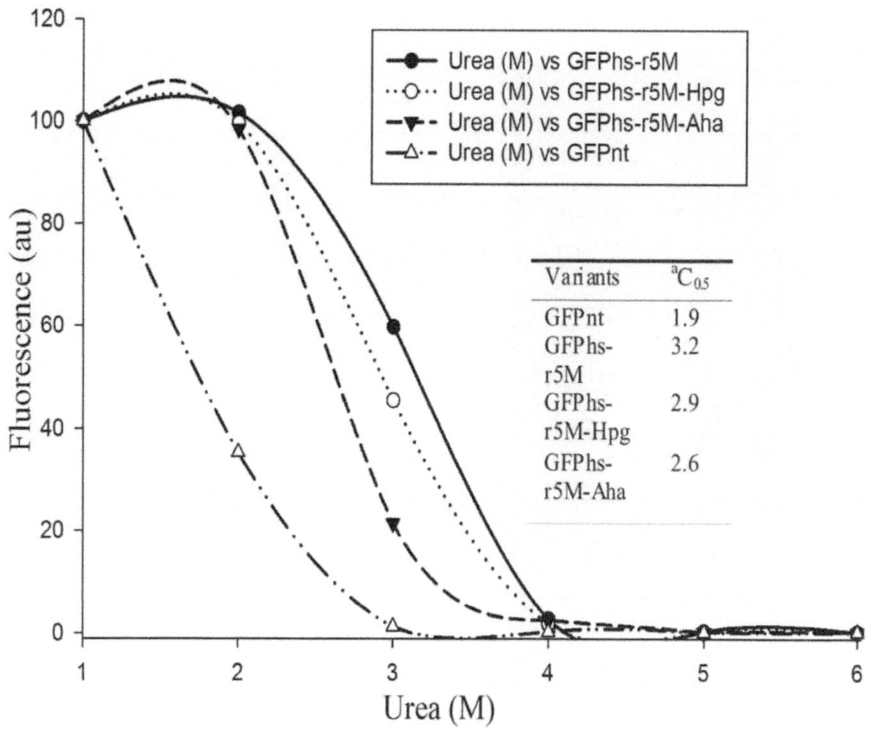


**Figure 6. Equilibrium renaturation plots of GFP$_{nt}$ and GFPhs-r5M variants (fraction of recovered fluorescence).** Urea-denatured protein samples were diluted in different concentrations of urea in refolding buffer and recovered fluorescence was normalized by dividing it by fluorescence of corresponding non-denatured samples diluted in similar fashion. Inlet table shows the [a]concentration of urea at which the 50% of fluorescence is recovered during renaturation process under urea-unfolded conditions.

## Discussion

This study has demonstrated the biological N-terminal site-specific modification of GFP with bio-orthogonally reactive groups and its application to conjugation of the protein. The preparation of GFPs with N-terminal specific functional groups could be achieved by generating the internal Met-free GFP sequence that can be properly folded and subsequent use of the Met residue-specific substitution method. The produced unnatural GFPs were sufficiently homogeneous and showed almost comparable activity and folding efficiency to the natural GFP. Finally, the protein-protein conjugation using the functionalized GFPs was successfully demonstrated. Even though it was not done in this study, the N-terminally functionalized GFPs, GFPhs-r5M-Hpg or GFPhs-r5M-Aha, can be conjugated to the N-terminally functionalized scFvs, scFv-Aha or scFc-Hpg, which were prepared in our previous study [8], and the resulting GFP-scFv conjugates are fluorobodies which can be used for developing biosensors and diagnostics [32]. It is expected that the strategy employed in our study may not only enable the preparation of artificial N-terminal functionalized proteins on a large scale but also facilitate the various protein related bio-conjugation studies such as N-terminal specific glycosylation, PEGylation and labeling.

Although *in vivo* site-specific incorporations of unnatural amino acids into target proteins using stop codons have been successfully achieved by suppressor based methodology [33], this approach has not been demonstrated for the N-terminus of proteins. The Met residue-specific methodology used in this study is probably the only method to install amino acid analogues in the N-terminus of a protein, and this approach is a handy choice for N-terminal specific functionalization of target proteins when the internal Met residues of the proteins can be changed into other ones. Sometimes the internal Met residues of proteins could be replaced with other hydrophobic amino acids such as Leu, Ile, and Ala [7,8]. However, it is easily expected that Met residues could play an important role in folding and stability of proteins, and these Met residues cannot be mutated to other ones without hurting the function or structure of the proteins. One of the key points in our study is that even the protein sequences such as GFP, of which the Met residues are very sensitive to mutations, could be engineered to be an internal Met-free sequence by introducing mutations known to enhance the folding efficiency and robustness of the protein. Although the mutations previously identified were used for the enhancement of GFP folding and stability in this study, current protein engineering approaches such as directed evolution and computational protein engineering can be efficiently employed in the identification of such folding enhancement mutations for other proteins [24]. This implies that the generation of the internal Met-free sequences which can be properly folded may not be a serious problem anymore in the preparation of the N-terminal functionalized proteins through the *in vivo* Met-residue specific substitution method. This also indicates that it is possible to artificially manipulate the incorporation sites of target proteins by genetically reassigning the Met codons to any sites of the internal Met-free protein sequence, which would allow the selective site-specific functionalization of a protein. In the case that the unnatural amino acids incorporated into the first Met codon is not required, it can be removed by engineering the penultimate residue with non-bulky amino acids such as Gly, Ala, Cys [7,9,34].

There are some general or specific limitations in the proposed method, which should be considered before applying the method to bio-conjugations. For example, the method may be very

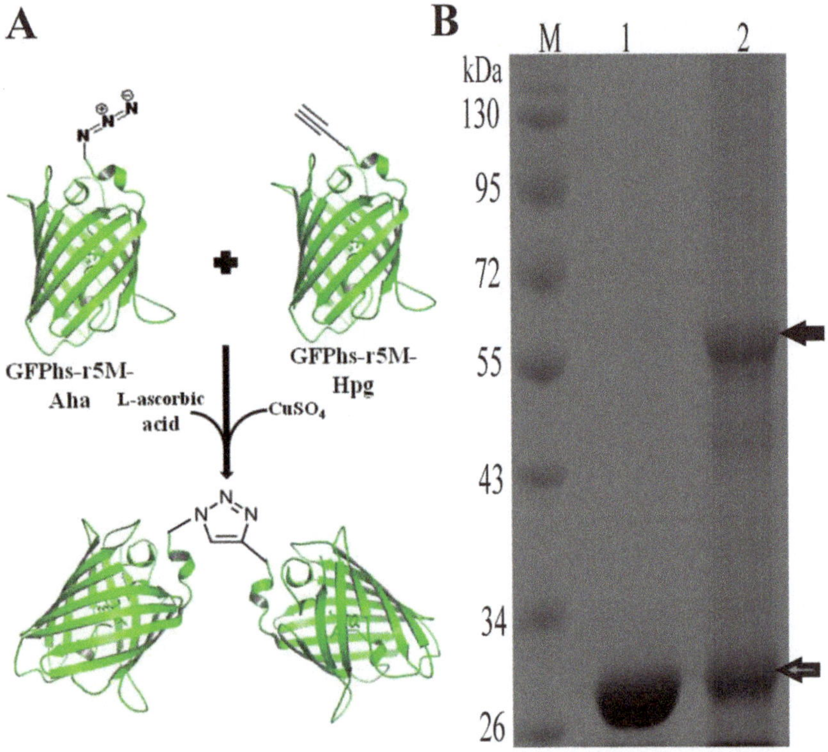


**Figure 7. Protein-protein bio-conjugation of GFPhs-r5M-Hpg and GFPhs-r5M-Aha.** (A) Copper (I)-catalyzed cycloaddition (CCCA) reaction between azide and alkyne incorporated to GFPhs-r5M resulted in the formation of triazole-linked protein-protein dimer bio-conjugation. (B) SDS-PAGE analysis of CCCA reaction between GFPhs-r5M proteins incorporated with Hpg (alkyne) and Aha (azide group). Lane 1: CCCA reaction without catalysis agents, $CuSO_4$ and L-ascorbic acid; lane 2: CCCA reaction with catalysis agents, $CuSO_4$ and L-ascorbic acid. This result shows the formation of triazole-linked protein-protein bio-conjugation of GFPhs-r5M dimer. M is molecular weight marker, thick arrow indicates the protein-protein conjugated GFPhs-r5M dimer of 55.2 kDa and grey arrow indicates the 27.6 kDa monomer of GFPhs-r5M containing Hpg and Aha respectively.

inefficient for the proteins with N-terminal signal sequences which can be cleaved *in vivo* or with hidden N-termini where the incorporated non-natural amino acids cannot be accessed once incorporated. In addition, the target proteins need to be purified to execute highly specific bio-conjugation reactions because the unnatural amino acids can also be slightly incorporated into endogenous proteins.

In our study, the mutations of the Met residues in the buried hydrophobic core regions of GFP significantly lowered the folding efficiency of GFP, which was rescued by introducing the mutations for GFP folding enhancement, the majority of which were from the superfolder GFP [19]. According to the structural analysis of the superfolder GFP, the mutations resulted in the higher folding rate and folding robustness by inducing new noncovalent interactions involving ionized residues [19]. For instance, the S30R mutation contributed the formation of double salt bridges with E17 and E32 and intramolecular ionic network through four residues (E17, E32, R122 and E115) located in four different adjacent β-sheets in the structure. It is presumed that this kind of superfolder mutation effect compensated the destabilization effect caused by the mutations of the three Met residues in the hydrophobic-core [19]. The higher folding efficiency and folding robustness of GFPhs-r5M than those of $GFP_{nt}$ indicates that the superfolder mutations might presumably provide $GFP_{nt}$-r5M with more stabilization energy than such compensating energy. On the other hand, we presume that the higher specific fluorescence of GFPhs-r5M than $GFP_{nt}$ might be caused by the mutations such as F64L, F99S and N149K mutations which can change the spectral properties of GFP by enhancing the hydrogen bonding networks around the chromophore [22,26,28,29]. Further mutagenesis and structural studies need to be performed to understand the improved folding and spectral properties of the variants more exactly.

## Supporting Information

**Figure S1** A. SDS-PAGE analysis of the soluble and insoluble protein fractions of GFPnt and GFPnt-r2M. (M: molecular weight marker, lane 1: insoluble fraction of GFPnt, lane 2: soluble fraction of GFPnt, lane 3: soluble fraction of GFPhs-r2M, lane 4: insoluble fraction of GFPhs-r2M) B. SDS-PAGE analysis of the soluble and insoluble protein fractions of GFPnt-r3M. (S, soluble fraction; I, insoluble fraction). C. SDS-PAGE analysis of the soluble and insoluble protein fractions of GFPnt-r5M. (S, soluble fraction; I, insoluble fraction). D. SDS-PAGE analysis of the soluble and insoluble protein fractions of GFPhs-r5M and GFPnt. (S, soluble fraction; I, insoluble fraction; M, molecular weight marker). The expected size of 27.6 kDa is indicated by arrow in the figures.

**Figure S2 Chemical structure of natural L-methionine (Met) and their unnatural surrogates L-homopropargyl-glycine (Hpg) and L-azidohomoalanine (Aha) (Mol. Wt: molecular weight).**

**Figure S3 ESI-MS analysis of GFPhs-r5M incorporated with Hpg and Aha. GFPhs-r5M (A), GFPhs-r5M-Hpg (B)**

**and GFPhs-r5M-Aha (C) incorporated with Met, Hpg and Aha, respectively.** *Inset* table of each spectra shows calculated and found masses. The peaks corresponding to found masses with Met, Hpg and Aha incorporated proteins might be due to cleavage of 8 residues. We generally could observe these peaks with almost all of the samples of GFPhs-r5M variants. The GFPhs-r5M containing Hpg and Aha showed the mass shift of −22 and −5 Da respectively, compared to GFPhs-r5M with Met.

**Table S1 Oligonucleotides used for saturation mutagenesis of internal Met-free GFP construction.**

**Table S2 Amino acid sequence of the GFPhs-r5M.** Red indicates Met replacement mutations, and green indicates the mutations for folding enhancement. The variant expressed as recombinant protein contains a hexahistidine tag sequence in the C-terminus of the protein for Ni-NTA purification.

## Author Contributions

Conceived and designed the experiments: NS SGL. Performed the experiments: NS SS GR. Analyzed the data: NS SGL HJP THY. Contributed reagents/materials/analysis tools: SGL NB. Wrote the paper: NS THY SGL.

## References


1. Merkel L, Beckmann HSG, Wittmann V, Budisa N (2008) Efficient N-terminal glycoconjugation of proteins by the N-end rule. ChemBioChem 9: 1220–1224.
2. Saerens D, Huang L, Bonroy K, Muyldermans S (2008) Antibody fragments as probe in biosensor development. Sensors 8: 4669–4686.
3. Tomizaki KY, Usui K, Mihara H (2005) Protein-detecting microarrays: Current accomplishments and requirements. ChemBioChem 6: 783–799.
4. Baker DP, Lin EY, Lin K, Pellegrini M, Petter RC, et al. (2005) N-terminally PEGylated human interferon-β-1a with improved pharmacokinetic properties and in vivo efficacy in a melanoma angiogenesis model. Bioconj Chem 17: 179–188.
5. Budisa N (2004) Prolegomena to future experimental efforts on genetic code engineering by expanding its amino acid repertoire. Angew Chem 43: 6426–6463.
6. Link AJ, Tirrell DA (2005) Reassignment of sense codons in vivo. Methods 36: 291–298.
7. Merkel L, Cheburkin Y, Wiltschi B, Budisa N (2007) In vivo chemoenzymatic control of N-terminal processing in recombinant human epidermal growth factor. ChemBioChem 8: 2227–2232.
8. Selvakumar E, Rameshkumar N, Lee S-G, Lee S-J, Park H-S (2010) In vivo production of functional single-chain Fv fragment with an N-terminal-specific bio-orthogonal reactive group. ChemBioChem 11: 498–501.
9. Wang A, Winblade Nairn N, Johnson RS, Tirrell DA, Grabstein K (2008) Processing of N-terminal unnatural amino acids in recombinant human interferon-β in *Escherichia coli*. ChemBioChem 9: 324–330.
10. Kiick KL (2001) Incorporation of azides into recombinant proteins for chemoselective modification by the Staudinger ligation. Proc Natl Acad Sci USA 99: 19–24.
11. Link AJ (2006) Discovery of aminoacyl-tRNA synthetase activity through cell-surface display of noncanonical amino acids. Proc Natl Acad Sci USA 103: 10180–10185.
12. Yoo TH, Tirrell DA (2007) High-throughput screening for methionyl-tRNA synthetases that enable residue-specific incorporation of noncanonical amino acids into recombinant proteins in bacterial Cells. Angew Chem 46: 5340–5343.
13. Bloom JD, Labthavikul ST, Otey CR, Arnold FH (2006) Protein stability promotes evolvability. Proc Natl Acad Sci USA 103: 5869–5874.
14. Nagasundarapandian S, Merkel L, Budisa N, Govindan R, Ayyadurai N, et al. (2010) Engineering protein sequence composition for folding robustness renders efficient noncanonical amino acid incorporations. ChemBioChem 11: 2521–2524.
15. Taverna DM, Goldstein RA (2002) Why are proteins marginally stable? Proteins: Struct Funct Genet 46: 105–109.
16. Soundrarajan N, Edwardraja S, Lee S-G, Yun H, Ayyadurai N (2012) Enhancing the productivity of soluble green fluorescent protein through methionine-residue specific consensus approach. Afr J Biotechnol 11: 1059–1064.
17. Ayyadurai N, Neelamegam R, Nagasundarapandian S, Edwardraja S, Park HS, et al. (2009) Importance of expression system in the production of unnatural recombinant proteins in Escherichia coli. Biotechnol Bioprocess Eng 14: 257–265.
18. Bessette PH, Mena MA, Nguyen AW, Daugherty PS (2003) Construction of designed protein libraries using gene assembly mutagenesis. 29–37.
19. Pédelacq J-D, Cabantous S, Tran T, Terwilliger TC, Waldo GS (2006) Engineering and characterization of a superfolder green fluorescent protein. Nat Biotechnol 24: 79–88.
20. Ormö M, Cubitt AB, Kallio K, Gross LA, Tsien RY, et al. (1996) Crystal structure of the Aequorea victoria green fluorescent protein. Science 273: 1392–1395.
21. Reid BG, Flynn GC (1997) Chromophore formation in green fluorescent protein. Biochemistry 36: 6786–6791.
22. Cormack BP, Valdivia RH, Falkow S (1996) FACS-optimized mutants of the green fluorescent protein (GFP). Gene 173: 33–38.
23. Budisa N, Pal PP, Alefelder S, Birle P, Krywcun T, et al. (2004) Probing the role of tryptophans in Aequorea victoria green fluorescent proteins with an expanded genetic code. Biol Chem 385: 191–202.
24. Roodveldt C, Aharoni A, Tawfik DS (2005) Directed evolution of proteins for heterologous expression and stability. Curr Opin Struct Biol 15: 50–56.
25. Tokuriki N, Stricher F, Schymkowitz J, Serrano L, Tawfik DS (2007) The stability effects of protein mutations appear to be universally distributed. J Mol Biol 369: 1318–1332.
26. Andreas Crameri, Whitehorn EA, Tate E, Stemmer WP (1996) Improved green fluorescent protein by molecular evolution using DNA shuffling. Nat Biotechnol 14: 315–318.
27. Yoo TH, Link AJ, Tirrell DA (2007) Evolution of a fluorinated green fluorescent protein. Proc Natl Acad Sci USA 104: 13887–13890.
28. Teerawanichpan P, Hoffman T, Ashe P, Datla R, Selvaraj G (2007) Investigations of combinations of mutations in the jellyfish green fluorescent protein (GFP) that afford brighter fluorescence, and use of a version (VisGreen) in plant, bacterial, and animal cells. Biochim Biophys Acta 1770: 1360–1368.
29. Ito Y, Suzuki M, Husimi Y (1999) A novel mutant of green fluorescent protein with enhanced sensitivity for microanalysis at 488 nm excitation. Biochem Biophys Res Commun 264: 556–560.
30. Huisgen R (1963) 1.3-Dipolare Cycloadditionen Rückschau und Ausblick. Angew Chem 75: 604–637.
31. Rostovtsev VV, Green LG, Fokin VV, Sharpless KB (2002) A Stepwise Huisgen cycloaddition process: Copper(I)-catalyzed regioselective "ligation" of azides and terminal alkynes. Angew Chem 41: 2596–2599.
32. Pavoor TV, Cho YK, Shusta EV (2009) Development of GFP-based biosensors possessing the binding properties of antibodies. Proc Natl Acad Sci USA 106: 11895–11900.
33. Wang L, Brock A, Herberich B, Schultz PG (2001) Expanding the genetic code of Escherichia coli. Science 292: 498–500.
34. Jean-Paul Boissel, Thomas J Kasper, Bunn HF (1988) Cotranslational amino-terminal processing of cytosolic poteins. J Biol Chem 263: 8443–8449.

# 3

# Directed Evolution of Aminoglycoside Phosphotransferase (3′) Type IIIa Variants That Inactivate Amikacin but Impose Significant Fitness Costs

**Joseph R. Kramer, Ichiro Matsumura***

Emory University School of Medicine, Department of Biochemistry, Rollins Research Center, Atlanta, Georgia, United States of America

## Abstract

The rules that govern adaptive protein evolution remain incompletely understood. Aminoglycoside aminotransferase (3′) type IIIa (hereafter abbreviated APH(3′)-IIIa) is a good model enzyme because it inactivates kanamycin efficiently; it recognizes other aminoglycoside antibiotics, including amikacin, but not nearly as well. Here we direct the evolution of APH(3′)-IIIa variants with increased activity against amikacin. After four rounds of random mutation and selection in *Escherichia coli*, the minimum inhibitory concentration of amikacin rose from 18 micrograms/mL (wild-type enzyme) to over 1200 micrograms/mL (clone 4.1). The artificially evolved 4.1 APH(3′)-IIIa variant exhibited 19-fold greater catalytic efficiency ($k_{cat}/K_M$) than did the wild-type enzyme in reactions with amikacin. *E. coli* expressing the evolved 4.1 APH(3′)-IIIa also exhibited a four-fold decrease in fitness (as measured by counting colony forming units in liquid cultures with the same optical density) compared with isogenic cells expressing the wild-type protein under non-selective conditions. We speculate that these fitness costs, in combination with the prevalence of other amikacin-modifying enzymes, hinder the evolution of APH(3′)-IIIa in clinical settings.

**Editor:** Floyd Romesberg, The Scripps Research Institute, United States of America

**Funding:** The authors were supported by a grant from the National Instutite of Health (1 R01 GM086824). The funders had no role in study design, data collection and analysis, decision to publish, or preparation of the manuscript.

**Competing Interests:** The authors have declared that no competing interests exist.

* E-mail: imatsum@emory.edu

## Introduction

We strive to understand how enzymes originate and evolve. A better understanding of the rules that govern these adaptive processes would fill a knowledge gap in the Darwinian Paradigm, and guide protein engineers toward more efficient design algorithms. We have learned much from previous laboratory evolution experiments. Evolutionary theory suggests that new biological systems are products of contingency, when "a feature evolved long ago for a different use has fortuitously permitted survival during a sudden and unpredictable change in rules" [1]. At the molecular level, an enzyme that originally evolved to catalyze the conversion of a particular substrate into a particular product might also react similarly with other substrates (substrate ambiguity) or accelerate a different chemical transformation (catalytic promiscuity). These weak secondary functions could serve as seeds for subsequent evolutionary innovation.

Studies of individual enzymes *in vitro* showed that wild-type enzymes could be multi-functional [2,3]. Pioneering directed evolution experiments showed that weak secondary activities can be physiologically relevant and evolvable [4,5,6]. In general, though with notable exceptions [7,8,9,10,11] artificially evolved enzymes are not as specific or efficient as their respective wild-type ancestors. Why does directed evolution usually fall short of the natural process? This question is not easily addressed [12], even after 20 years of directed protein evolution experiments [4,13]. Protein engineers value speed, so they may be inclined to impose exceedingly stringent selections upon small populations for a relatively small number of generations. We also suspect that laboratory selection conditions are more uni-dimensional than those in the wild.

The TEM-1 beta-lactamase catalyzes the hydrolysis of beta-lactam antibiotics. It readily evolves, *in vitro* and *in situ*, to recognize cephalosporin antibiotics, so it serves as a model system for studies of protein evolution [4,14,15,16,17]. The aminoglycoside phosphotransferases (APH) also present attractive but under-utilized systems for evolutionary studies, as they diverged in nature to recognize and inactivate a wide range of commercially available, clinically relevant antibiotics [18]. APHs are expressed in the cytoplasm, unlike beta-lactamase, which is secreted into the periplasm where its activity can benefit other cells. Previous workers have shown that double mutations in one family member, aminoglycoside phosphotransferase (2″)-IIa, could increase the resistance of host cells to amikacin and isepamecin eight-fold relative to the wild-type (as measured by differences minimum inhibitory concentrations) [19]. Changes of that magnitude are clinically relevant, but we believe that quantitatively larger adaptations are possible in the laboratory and that such experiments would be informative.

We chose aminoglycoside phosphotransferase (3′)-IIIa, or APH(3′)-IIIa, for our studies. It is broader in specificity than is APH(2″)-IIa [18], and is therefore potentially more evolvable. APH(3′)-IIIa is well characterized [20,21,22], and was the first APH to be crystallized [23], so the biochemical mechanisms of mutations that accumulate during directed evolution can be inferred. We show here that the enzyme readily evolves to recognize amikacin, which is used to treat multidrug resistant gram negative pathogens, and that it is relatively easy to purify, store and assay *in vitro*. We also show that some mutations that are beneficial under selections are associated with significant fitness costs under non-selective conditions.

## Materials and Methods

### Chemicals and Reagents

Restriction enzymes (EcoRI, SalI, NdeI, HindIII), molecular biology enzymes (T4 DNA ligase), and reagents (1 kb DNA ladder) were from New England Biolabs (Ipswitch, MA). Taq DNA polymerase was over-expressed in *E. coli* and partially purified via heat treatment and dialysis [24]. The dNTPs were from Roche Applied Science (Indianapolis, IN). DNA purification kits (for plasmid, PCR and gel extraction) were from Qiagen (Valencia, CA). IPTG was from Gold Biotechnology (St. Louis, MO). Antibiotics (amikacin, ampicillin, chloramphenicol, kanamycin), buffer components (Tris-HCl), salts (KCl, $MgCl_2$, $MnCl_2$, imidazole), assay enzymes (pyruvate kinase, lactate dehydrogenase), and substrates (adenosine triphosphate, nicotinamide adenine dinucleotide, phosphoenolpyruvic acid) were from Sigma-Aldrich (St. Louis, MO). DNA oligonucleotides were synthesized by Integrated DNA Technologies (Coralville, IA). Lonza Seakem LE agarose and Difco granulated agar were from Thermo Fisher Scientific (Waltham, MA). EMD Millipore Luria Broth was from VWR (Radnor, PA). Acrylamide for SDS gels was from Bio-Rad (Hercules, CA).

### Mutagenesis and Cloning

The *aph(3')-IIIa* gene in the pBAV1K plasmid [25] was converted into a "BioBrick" (DNA sequence flanked by unique standard restriction sites [26]) by PCR amplification using primers 1 and 2 (Table 1). The product was subsequently recombined into the multiple cloning site of the pQBAV3c vector via overlap extension PCR cloning [25,27]. Outside primers 3 and 4 were subsequently used to introduce random mutations in error prone PCRs (standard conditions with an extra 0.2 mM dCTP, 0.2 mM dTTP, 1.2 mM $MgCl_2$ and 0.125 mM $MnCl_2$, as previously described [28]), or to randomly recombine alleles in staggered extension process PCR (80 cycles with ultra-short 15 second extensions at sub-maximal 60°C temperatures) [29]. Outside primers 3 and 4 were used in combination with internal primers 5 and 6 in two separate PCRs (one with primers 3 and 6, the other with 4 and 5), followed by an overlap extension PCR (using primers 3 and 4), to introduce the D190A mutation into APH(3')-IIIa. After each of these PCR reactions, restriction enzymes EcoRI and SpeI were used to clone the amplification products back into pQBAV3c.

We wanted to over-express APH(3')-IIIa with an N-terminal $his_6$-tag, but did not want to use an expression system that produced its own aminoglycoside phosphotransferase. The *aph(3')-Ia* gene of pET28a+ (Novagen) was thus replaced with the *bla* gene encoding TEM-1 beta-lactamase gene from pET20b+ using restriction enzymes AlwNI and XhoI. The *aph(3')-IIIa* gene was PCR amplified with primers 7 and 8. The product was cloned into the modified pET28 vector using restriction enzymes NdeI and HindIII restriction sites, thereby fusing the *aph(3')-IIIa* gene to sequence encoding a $his_6$-tag on the N-terminus of the protein.

### Genetic Selection of Aph(3')-IIIa Variants that Confer Decreased Amikacin Susceptibility

*E. coli* InvαF' carrying the *aph(3')-IIIa*-pQBAV3c plasmid were spread on LB agar plates with varying concentrations of amikacin after each round of selection. In the first round, 10,000 colonies were spread on LB agar supplemented with 18 micrograms/mL amikacin ("wild-type MIC"). Colonies that formed were then grown to saturation in liquid LB supplemented with chloramphenicol, diluted $10^{-5}$-fold; 50 microliters were spread on 10×LB plates containing 22, 26, 30, 35, 40, 45, or 50 micrograms/mL amikacin. In the second round, 10,000 colonies were spread on LB supplemented with 80 micrograms/mL amikacin. Colonies that formed were subsequently restruck as above on LB plates containing 160, 170, 180, 200 micrograms/mL amikacin. In the third round, colonies were selected on LB supplemented with 220 micrograms/mL amikacin, and those that formed were then diluted $5\times10^{-4}$-fold and restruck on LB plates containing 250, 280, 310, 340, 370, 400, 425, 450, 500, 550, 600, 650, 700, 750, 800, 825, 850, 875, 900, 950, 1000, 1050, 1100, 1150, 1200, 2000 micrograms/mL amikacin. After the fourth round, colonies were spread (at approximately 1000 CFU/plate on 10 plates) on LB with 1200 micrograms/mL amikacin. The reported susceptibility of each mutant (Table 2) is the minimum concentration at which it was unable to form any colonies.

### Purification of APH(3')-IIIa Enzyme and Mutants

*E. coli* BL21(DE3) carrying the modified, beta-lactamase producing pET28 plasmid encoding *his*$_6$*-aph(3')-IIIa* were propagated to saturation overnight in a 4 mL culture of LB broth containing ampicillin. The starter culture was then added to a 400 mL culture of LB-ampicillin broth and grown for 4.5 hours at 37°C. At mid-log phase, protein expression was induced overnight by the addition 1 mM IPTG. The preparation and purification of was performed at 4°C. The 400 mL of the fully grown cultures were collected by centrifugation at 1540×*g* for 10 minutes. The cells were resuspended in 30 mL of 50 mM Tris, pH 7.5 and sonicated on ice for 5 minutes (5 cycles of 30 seconds on/off sonication at 6 watts by a Misonix Sonicator 3000). The cell lysate was eliminated by centrifugation at 20,100×*g* for 40 minutes, and the supernatant was saved. A his-trap column connected to an AKTA purifier (GE Healthcare) was washed with buffer B (50 mM Tris pH 7.5, 400 mM imidazole) and then equilibrated with buffer A (50 mM Tris pH 7.5) before the supernatant was run through the column. The column was washed with buffer A before applying a step gradient of buffer B to elute the $his_6$-APH(3')-IIIa enzyme. Fractions containing the enzyme were dialyzed (Spectra #132720– MWCO: 3,500) in 50 mM Tris pH 7.5 and analyzed by SDS-polyacrylamide gel electrophoresis to verify purification.

### Enzyme Assays

A coupled assay with pyruvate kinase/lactate dehydrogenase was used to measure the formation of ADP [30,31]. Assay buffer (975 microliters of 50 mM Tris pH 7.6, 40 mM KCl, 10 mM $MgCl_2$, 0.25 mg/mL NADH, 2.5 mM PEP and 1 mM ATP in a quartz cuvette, Fisher #14-385-914A) was mixed with 5 microliters of pyruvate kinase/lactate dehydrogenase (600–1000 u/mL PK, 900–1400 u/mL LDH) and 10 microliters of the appropriate concentration of aminoglycoside, then incubated at 37°C for 15 minutes. The reaction was initiated with the addition of 10 microliters of stock solution (1 micromolar) of $his_6$-APH(3')-IIIa. Absorbance at 340 nm, indicating oxidation of NADH, was monitored with a Shimadzu UV-1601 spectropho-

**Table 1.** Primers used in this study.

| Primer | Name | Sequence |
|---|---|---|
| 1 | KAN_STDBB_F | ggaattcgcggccgcttctagagaaattctatcataattgtggtttcaa |
| 2 | KAN_STDBB_R | ctgcagcggccgctactagtattattaaaacaattcatccagtaaaatataa |
| 3 | pQBAV3_MCS_−48_ecori_F | gcaaacaaaccaccgctggtagcg |
| 4 | pQBAV3_MCS_+278_psti_R | gccgtaatatccagctgaacggtctggttatag |
| 5 | aph_D190A_for | cttgtcttttcccacggcgccctgggagac |
| 6 | aph_D190A_rev | gtctcccagggcgccgtgggaaaagacaag |
| 7 | aph_pet28_F_ndel_nhis | ggggtatctttaaatactgtagaaaagaggaaggaaataacatatggct |
| 8 | aph_pet28_R_hindIII_nhis | gtctgcagcggccgctactaaagcttttaaaacaattcatc |

tometer for 5 minutes. The rate of oxidation of NADH was determined using its extinction coefficient (6220 $mol^{-1}$ $cm^{-1}$) [32]. This rate was converted into the initial velocity of the APH(3′)-IIIa-catalyzed phosphotransfer reaction. The data were fit to standard kinetic models by the methods of least squares using Kaleidagraph v3.5 (Synergy Software).

$$v = \frac{V_{\max} * [S]}{K_M + [S]}$$

$$v = \frac{V_{\max} * [S]}{K_M + [S] + \frac{[S]^2}{K_i}}$$

### Fitness Assays

*E. coli* InvαF' carrying the wild-type or variant *aph(3′)-IIIa*-pQBAV3c plasmids were grown overnight to saturation in 2 mL LB broth with 34 micrograms/mL chloramphenicol or 50 micrograms/mL kanamycin. A 200 microliter sample of each culture was transferred into the well of a clear, flat bottom 96 well plate (VWR # 62409-068), and the optical density at 600 nm of each culture was measured in a plate reader (Biotek Synergy 2). Each culture was serially diluted three times, each time with 20 microliters of the sample being added to 980 microliters of LB broth; 50 microliters of the final serial dilution were spread on LB plates containing 34 micrograms/mL chloramphenicol or 50 micrograms/mL kanamycin. Plates were incubated at 37°C overnight, and the visible colonies were counted.

*E. coli* InvαF' containing the *aph(3′)-IIIa*-pQBAV3c plasmid with was grown overnight to saturation in 2 mL LB broth with 34 micrograms/mL chloramphenicol or 50 micrograms/mL kanamycin. Each culture was serially diluted twice, each time with 20 microliters of the sample being added to 980 microliters of LB broth. A 200 microliter sample of each culture was transferred into a clear, flat bottom 96 well plate. The plate was agitated at a medium speed at 37°C for 24 hrs in a Biotek Synergy2 microtiter plate reader; the optical density at 600 nm was measured every 30 minutes.

## Results

### APH(3′)-IIIa Readily Adapts to Amikacin

APH(3′)-IIIa is well-suited for our study of adaptive enzyme evolution because, as we explained above, its weak activities against a range of antibiotics enables facile selections in *Escherichia coli* [25]. We chose amikacin as the "novel" substrate because it is nearly identical in structure to its "native" (most reactive) substrate, kanamycin, from which it is chemically synthesized [33]. The *aph(3′)-IIIa* gene was amplified and randomly mutated in an error-prone PCR; the resulting library cloned into pQBAV3c [27], which also encodes chloramphenicol acetyltransferase. *E. coli* strain InvαF' was transformed with the plasmid-borne library; 10,000 colony-forming units were spread on 10×(100×15 mm) Petri dishes containing LB agar supplemented with chloramphenicol and 18 micrograms/mL amikacin. Isogenic control cells transformed with the ancestral *aph(3′)-IIIa*-pQBAV3c plasmid do not grow under these conditions, but 43 colonies formed among the approximately 10,000 that expressed mutant APH(3′)-IIIa proteins. The 43 selected *aph(3′)-IIIa* alleles were sequenced (Table 2); most contained just 1 or 2 nucleotide mutations in the open reading frame (ORF) (27/43 mutants). Nearly all alleles also contained mutations in the upstream 5′ region, including those without ORF mutations. Each mutant was restruck on fresh plates containing higher concentrations of amikacin (22–50 micrograms/mL) in order to measure the minimum inhibitory concentration (MIC, Table 2).

The 43 variant *aph(3′)-IIIa*-pQBAV3c plasmids were pooled, along with the ancestral plasmid, and used as templates for staggered extension process (StEP) recombination [29]. The resulting recombinant library was ligated back into the pQBAV3c plasmid. *E. coli* InvαF' was transformed with the library and spread on LB agar plates supplemented with 80 micrograms/mL amikacin. Seven colonies out of approximately 10,000 formed under these more stringent conditions (Table 3). We sequenced these *aph(3′)-IIIa* alleles and found that this small population was dominated by two new mutations (I40T and D193N) and two others (S194R and K255R) that were selected in the first round. Most of the selected mutants unexpectedly contained only single amino acid changes (4/7 mutants). It is possible that the increase in amikacin resistance of these mutants in this round was due to mutations outside of the ORF; mutants 2.4 and 2.5 had the same ORF mutation as 1.18 but additional non-ORF mutations (most notably −58(T-C) and −35(T-C)). It is also possible that the amikacin resistance can increase from the elimination of slightly deleterious mutations; mutant 2.3 shares a mutation with mutants 1.27 and 1.33, but is missing some mutations unique to them.

The seven selected *aph(3′)-IIIa* alleles were pooled, amplified, and mutated in an error-prone PCR. The mutated genes were cloned back into the pBAV3c plasmid. InvαF' cells were transformed with the cloned library, and spread on LB agar plates supplemented with 220 micrograms/mL amikacin. Four

**Table 2.** Aph(3′)-IIIa sequences and minimum inhibitory concentrations (round 1).

| Round of Mutagenesis | Mutant | Missense Mutations | Silent Mutations | Non-Coding Mutations | Amikacin MIC (μg/mL) |
|---|---|---|---|---|---|
| 1 | 1 | **H78**Y, **V96A** | | n.d. | 30 |
| 1 | 2 | **N38**D | | | 26 |
| 1 | 3 | I254M | **R211** | −192(A-G), −154(A-G), −135(T-C) | 26 |
| 1 | 4 | **Q236R** | | −51(A-G) | 30 |
| 1 | 5 | **E9K, K12R, L136I** | | −137(T-A), −20(A-G) | 30 |
| 1 | 6 | **K12E** | D261 | −134(A-G), −53(T-G) | 35 |
| 1 | 7 | **D193N** | | −134(A-G) | 35 |
| 1 | 8 | **V96I, K248E** | | n.d. | 30 |
| 1 | 9 | F79Y, **Q236R** | **D167** | −174(T-C) | 30 |
| 1 | 10 | | | −113(T-C), −7(A-G) | 30 |
| 1 | 11 | M1V | G69 | −177(T-C), −162(T-G), −157(T-C), −75(A-T) | 26 |
| 1 | 12 | I258T | | | 30 |
| 1 | 13 | Y102C | E235 | −31(A-T) | 30 |
| 1 | 14 | D167G | | −31(A-G) | 30 |
| 1 | 15 | E16G, **K255R** | | −182(T-A) | 26 |
| 1 | 16 | K21E, **V198**A | L244 | −161(C-T) | 26 |
| 1 | 17 | | | −35(T-C) | 26 |
| 1 | 18 | **S194R** | | −86(A-G) | 35 |
| 1 | 19 | **L136I** | E80 | | 26 |
| 1 | 20 | T177A, D231G, **K248E** | **S27**, L259 | −133(A-T), −112(T-G), −35(T-C) | 26 |
| 1 | 21 | **I40T** | **L10** | | 26 |
| 1 | 22 | | **L41** | −46(T-C) | 30 |
| 1 | 23 | E15K | L140 | n.d. | 26 |
| 1 | 24 | **E160D** | | −188(T-C), −29(T-C) | 30 |
| 1 | 25 | | E234 | −58(T-C) | 26 |
| 1 | 26 | E103G | L175, V185 | −35(T-A) | 35 |
| 1 | 27 | **I112V**, **V198**M | | −199(A-G), −58(T-C) | 45 |
| 1 | 28 | | | −77(G-A), −65 (G-C) | 30 |
| 1 | 29 | E181G | K21 | | 30 |
| 1 | 30 | **E9G**, K179R, **K255R** | P28 | −23(A-G) | 30 |
| 1 | 31 | | **E9**, D153 | −142(A-T), −67(T-A) | 30 |
| 1 | 32 | D104V | | −59(T-C), −49(C-T), −26(T-C) | 26 |
| 1 | 33 | **I112V**, E161G | **E234** | −21(A-G) | 26 |
| 1 | 34 | **H78**T, A152V | **E24**, **L41** | −6(T-A) | 50 |
| 1 | 35 | **N38**S | T55, **D144** | −132(A-G) | 26 |
| 1 | 36 | G36R, **E234G**, **K255E** | D190 | −150(T-C), −98(T-C), −58(T-C) | 26 |
| 1 | 37 | | **R211** | −35(T-C) | 26 |
| 1 | 38 | | D94 | n.d. | 30 |
| 1 | 39 | | | −105(T-C) | 30 |
| 1 | 40 | | D137 | −150(T-C), −52(T-C) | 26 |
| 1 | 41 | A2T | K43 | −185(T-C), −181(G-A), −123(A-T), −104(T-C) | 26 |
| 1 | 42 | | K253, E262 | −74(T-C) | 26 |
| 1 | 43 | | | −31(A-G) | 30 |

Recurring mutations are written in **bold** text.

colonies (out of approximately 10,000 transformants) formed under these selection conditions. DNA sequencing of the four associated alleles showed that each contained the S194R and three of the four mutants (3.1, 3.3 and 3.4) also had a mutation in the 40th residue, although there were two different mutations (I40V and I40T) at this position. The D193N mutation, carried by two

**Table 3.** Aph(3′)-IIIa sequences and minimum inhibitory concentrations (rounds 2–4).

| Round of Mutagenesis | Mutant | Missense Mutations | Silent Mutations | Non-Coding Mutations | Amikacin MIC (μg/mL) |
|---|---|---|---|---|---|
| 2 | 1 | **I40T, K255R** | | −35(T-A) | 160 |
| 2 | 2 | **I40T, S194R, K255R** | **L10, V73** | −31(A-G) | 200 |
| 2 | 3 | **I112V** | | −58(T-C), −31(A-G) | 160 |
| 2 | 4 | **S194R** | | −86(A-G), −58(T-C) | 160 |
| 2 | 5 | **S194R** | | −35(T-C), −24(T-C) | 160 |
| 2 | 6 | **D193N** | | −92(A-G), −77(G-A), −65(G-C) | 170 |
| 2 | 7 | **D193N** | **E24** | −31(A-G), −6(T-A) | 160 |
| 3 | 1 | K3R, I6M, **I40T, D144G, E160G, K176R, S194R, I196F, K255R** | **E9, L10, S27, V73**, V96 | −185(T-C), −172(A-G), −31(A-G) | 280 |
| 3 | 2 | **I112V, S194R** | | −58(T-C), −32(A-G) | 700 |
| 3 | 3 | **E24V, I40V, S194R** | E68, L140 | −90(T-C), −87(A-T), −86(A-G0, −58(T-C) | 1200 |
| 3 | 4 | K11N, **I40T, R120K, C156R, S194R, K255R** | **L10, V73** | −102(A-T), −31(A-G) | 500 |
| 4 | 1 | **E24V, I40T, R120K, C156R, K176R, S194R, I196F**, Y219H, **K255R** | **V73**, V76 | −102(A-T), −31(A-G) | - |

mutants from the second round of evolution went extinct in the third.

The four selected plasmids, and their ancestor, which encodes the wild-type *aph(3′)-IIIa* gene, were pooled and used as templates for StEP recombination. The recombinant library was cloned and used to transform InvαF'. The transformants were challenged with LB agar supplemented with 1200 micrograms/mL amikacin. One colony (out of approximately 10,000 transformants) formed. The DNA sequence of mutant 4.1 (E24V, I40T, R120K, C156R, K176R, S194R, I196F, Y219H, K255R; Figure 1) showed the persistence of three mutations that appeared in the first two rounds, I40T, S194R, and K255R (Table 2). We speculate upon the biochemical mechanisms of these adaptations in the Discussion section (*vide infra*).

## The 4.1 Aph(3′)-IIIa Variant is as Catalytically Efficient as its Wild-type Progenitor

The wild-type and 4.1 *aph(3′)-IIIa* alleles were subcloned into a modified version of pET28a+ that encodes the TEM-1 beta-lactamase in place of the usual *aph(3′)-Ia* [18]. The two recombinant plasmids were separately used to transform the *E. coli* production strain BL21(DE3). The proteins were over-expressed and purified by virtue of N-terminal hexahistidine tags encoded by the pET28a+ vector. The kinetic parameters of the two enzyme variants in reactions with the "native" substrate (kanamycin) and "novel" substrate (amikacin) were measured with a coupled pyruvate kinase/lactate dehydrogenase assay (Table 4, Figure 2). The $K_M$ and $k_{cat}$ of the his-tagged enzyme, $his_6$-APH(3′)-IIIa in reactions with kanamycin and amikacin were similar to the published values of the native untagged form [21]. The evolved $his_6$-4.1 APH(3′)-IIIa exhibited substantially higher $k_{cat}$ and lower $K_M$ in reactions with amikacin (and detectable substrate inhibition, $Ki > 2$ mM), when compared to the $his_6$-wildtype APH(3′)-IIIa. The second order rate constant ($k_{cat}/K_M$) was similar to that of the wild-type enzyme in reactions with kanamycin. The evolved enzyme retains some catalytic activity in reactions with kanamycin, but its steady state enzyme kinetics could not be fit to the Michaelis-Menten equation or to any substrate inhibition model [34]. The interaction between $his_6$-4.1 APH(3′)-IIIa and kanamycin is likely complex; we speculate that multiple non-productive binding modes compete.

## Fitness Costs of Beneficial Mutations

We noticed during our directed evolution experiments that some of the selected clones formed fewer colonies than did the isogenic ancestral strain. Fresh InvαF' cells were transformed with *aph(3′)-IIIa*-pQBAV3c plasmids encoding the wild-type, evolved 2.3, 3.1 or 4.1 alleles (or pBC or pACYC Duet as controls). The transformants were propagated in parallel under non-selective conditions (liquid LB supplemented with 34 micrograms/mL chloramphenicol or 50 micrograms/mL kanamycin). The optical density (600 nm) of each culture was measured; the cultures were serially diluted, then spread on LB agar plates supplemented with either chloramphenicol or kanamycin. We observed significant and reproducible differences among isogenic transformants in growth rates (during log phase in liquid culture) and colony forming ability (Table 5).

Two to eight-fold decreases in fecundity would almost certainly be decisive in nature, so we consider them worthy of further study. We first wondered whether the proliferation of untransformed cells in liquid culture, those that absorb light at $OD_{600}$ but fail to form colonies on agar plates containing chloramphenicol, could explain the observed differences in fitness. If mutations in the *aph(3′)-IIIa* gene could affect plasmid stability, a disparity in colony forming ability would be revealed by growing *E. coli* InvαF' containing pQBAV3c on plates both with and without selection for plasmid retention. We observed little or no plasmid loss in cells expressing the wild-type or 4.1 variants of APH(3′)-IIIa (Table 6). This result suggests that the fitness differences we observed (Table 5) are consequences of sequence differences in the *aph(3′)-IIIa* alleles themselves, rather than of indirect effects upon plasmid stability.

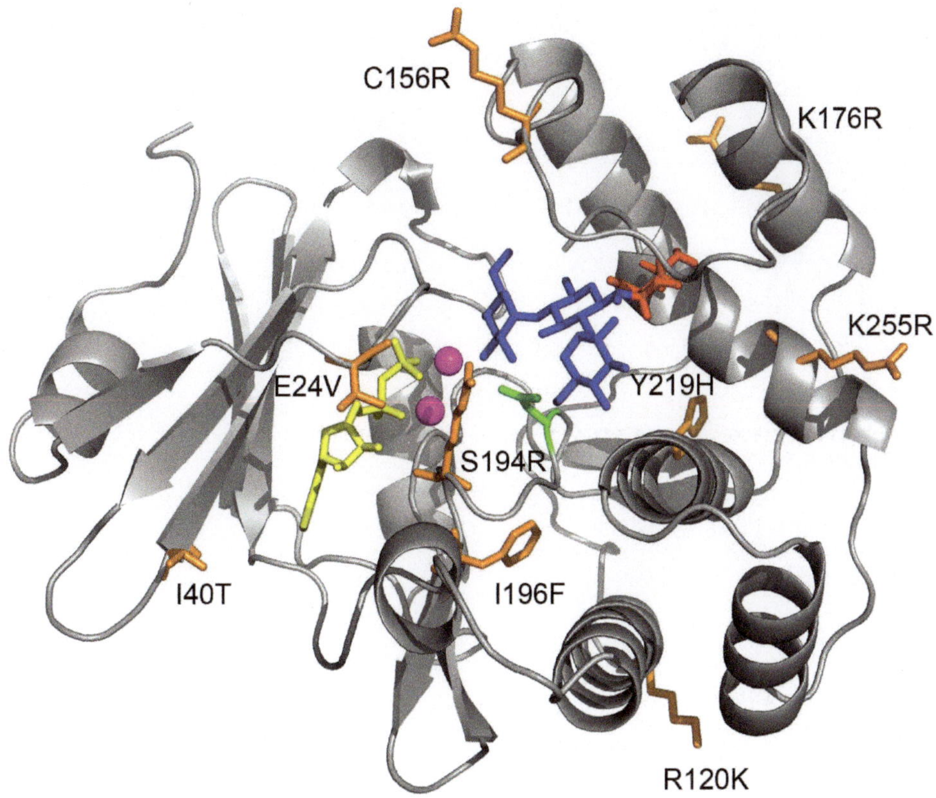


**Figure 1. Model of artificially evolved aminoglycoside phosphotransferase (3′)-IIIa, based upon crystal structure (1L8T) of the wild-type enzyme [23] rendered in PyMOL.** The putative beneficial mutations are colored orange, the catalytic D190 residue is green), adenosine diphosphate is yellow and magnesium is pink. Kanamycin is blue, while the "extra" moiety that differentiates amikacin from kanamycin is red.

We wondered whether the fitness costs correlate with improvements in activity against amikacin. We already have circumstantial evidence against this hypothesis. The intermediate 2.3 *aph(3′)-IIIa*-pQBAV3c imparts decreased fitness (relative to isogenic cells carrying its wild-type *aph(3′)-IIIa*-pQBAV3c ancestor), while mutant 3.1 did not (Table 5). Furthermore, the fitness associated with that APH(3′)-IIIa variants that we tested was unaffected by kanamycin (Table 5), suggesting that active-site occupancy apparently does not affect the fitness. To investigate our hypothesis more decisively, a single point mutation was made in the 190th residue of the wild-type and 4.1 APH(3′)-IIIa variants, changing the catalytic aspartic acid into alanine. This well characterized mutation has no significant effect on the structure or stability of APH(3′)-IIIa, but it abrogates detectable catalytic activity [20]. As expected, *E. coli* InvαF' expressing the D190A or 4.1+ D190A APH(3′)-IIIa failed to grow in LB medium supplemented with kanamycin (Table 5) or amikacin (data not shown).

To our surprise, however, the fitness effects of the D190A mutation were context-dependent. Cells expressing the D190A APH(3′)-IIIa protein were significantly less fit than those expressing the wild-type protein under non-selective conditions. In contrast, cells expressing the 4.1+ D190A APH(3′)-IIIa were much fitter than isogenic cells expressing the 4.1 variant (Table 5). We don't know whether the D190A and 4.1 proteins debilitate fitness through different biochemical mechanisms, or whether these protein variants act through a common non-catalytic mechanism that is sensitive to epistatic interactions among mutations. Many have observed that chromosomal mutations that impart resistance to other antibiotics come with a fitness cost. The population biology of these phenomena may be broadly similar

**Table 4.** Kinetic Parameters of the wild-type and evolved 4.1 APH(3′)-IIIa.

| | wild-type | | | | 4.1 | | | |
|---|---|---|---|---|---|---|---|---|
| **Substrate** | $k_{cat}$ (s$^{-1}$) | $K_m$ (μM) | $K_i$ (mM) | $k_{cat}/K_m$ (M$^{-1}$ s$^{-1}$) | $k_{cat}$ (s$^{-1}$) | $K_m$ (μM) | $K_i$ (mM) | $k_{cat}/K_m$ (M$^{-1}$ s$^{-1}$) |
| kanamycin | 2.02±0.05 | 12.1±1.8 | n.d.[a] | $1.67\times10^5$ | n.f.[b] | n.f.[b] | n.f.[b] | n.f.[b] |
| amikacin | 1.95±0.14 | 232±41 | n.d.[a] | $8.42\times10^3$ | 6.86±0.8 | 42.9±13 | 2.50±1.47 | $1.60\times10^5$ |

[a]n.d. = not detected.
[b]n.f. = not fit.

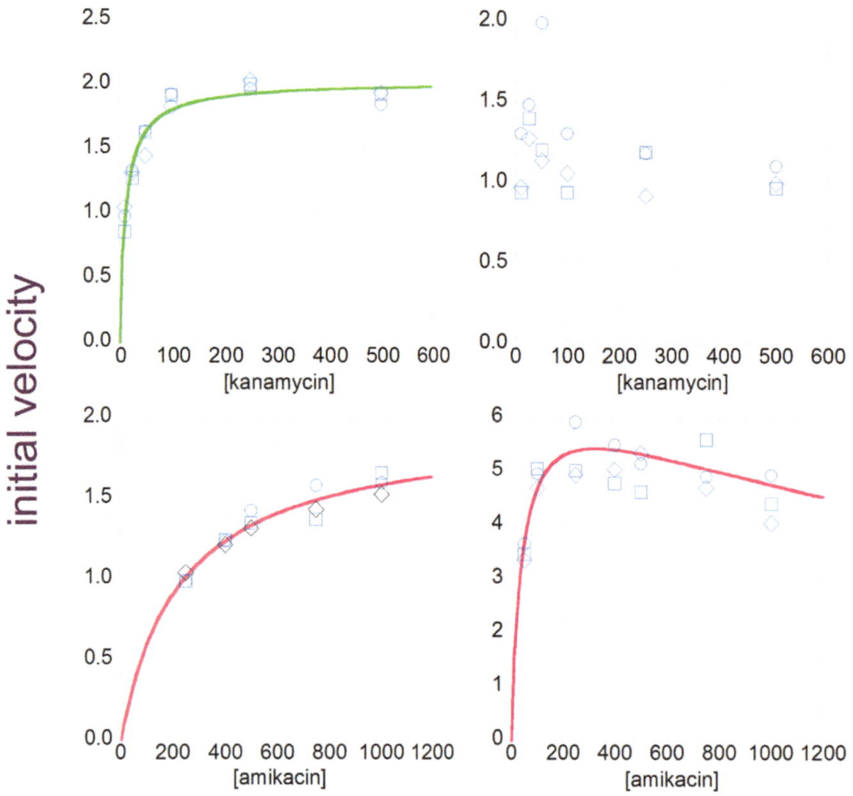


**Figure 2. Michaelis-Menten plots of the wild-type and 4.1 APH(3′)-IIIa variants.** The plots show the dependence of initial velocity upon substrate concentration for the following reactions: a) wild-type APH(3′)-IIIa with kanamycin, b) artificially evolved 4.1 variant with kanamycin, c) wild-type APH(3′)-IIIa with amikacin, and d) 4.1 variant with amikacin. Substrate concentration is in units of micromolar and reaction velocity is in moles of substrate/moles of enzyme/second. Each series of reactions was conducted in triplicate. The average initial velocity values were fit to the Michaelis-Menten equation (a, c) or a simple substrate inhibition model (d) as described in the Methods; the derived kinetic parameters are presented in Table 4.

[35,36,37,38,39], but the biochemical mechanisms are almost certainly idiosyncratic.

## Discussion

### Structural Hypotheses

The evolved 4.1 APH(3′)-IIIa contained nine amino acid replacements relative to its wild-type ancestor: E24V, I40T, R120K, C156R, K176R, S194R, I196F, Y219H and K255R. All but one, Y219H, first appeared in previous rounds of selection, suggesting that they are beneficial with respect to amikacin recognition. Two, namely I40T and S194R, are apparently beneficial in isolation. Amikacin is structurally identical to kanamycin, except that it contains an extra bulky modification (represented as red sticks in Figure 1) that creates a steric clash with the dynamic aminoglycoside binding loop (residues 147–170), at least in its kanamycin-binding conformation. The E157, N158 and E160 residues in that loop form hydrogen bonds with the amine groups in middle saccharide ring of kanamycin (including the one modified in amikacin) [23]. We therefore hypothesize that

**Table 5.** Fitness of transformed isogenic *Escherichia coli.*

| Plasmid | colony formation chloramphenicol (cfu/mL/$OD_{600}$) | colony formation kanamycin (cfu/mL/$OD_{600}$) | growth rate chloramphenicol ($OD_{600}$/min) |
|---|---|---|---|
| pBC (*lacZα*) | $2.7\times10^{8}\pm1.3\times10^{8}$ | 0 | $0.0039\pm0.00066$ |
| pACYC Duet (lower copy #, *lacI*) | $7.3\times10^{8}\pm1.8\times10^{8}$ | 0 | $0.0051\pm0.00059$ |
| WT aph(3′)-IIIa-pQBAV3c | $4.9\times10^{8}\pm1.7\times10^{8}$ | $5.2\times10^{8}\pm5.8\times10^{7}$ | $0.0026\pm0.00053$ |
| D190A aph(3′)-IIIa-pQBAV3c | $6.2\times10^{7}\pm2.4\times10^{7}$ | 0 | $0.0016\pm0.0011$ |
| 2.3 aph(3′)-IIIa-pQBAV3c | $8.0\times10^{7}\pm1.0\times10^{7}$ | $6.3\times10^{7}\pm2.1\times10^{7}$ | $0.0024\pm0.0012$ |
| 3.1 aph(3′)-IIIa-pQBAV3c | $4.6\times10^{8}\pm4.4\times10^{7}$ | $3.8\times10^{8}\pm1.7\times10^{7}$ | $0.0019\pm0.0017$ |
| 4.1 aph(3′)-IIIa-pQBAV3c | $1.2\times10^{8}\pm2.5\times10^{7}$ | $9.1\times10^{7}\pm1.5\times10^{7}$ | $0.0017\pm0.00020$ |
| 4.1+ D190A aph(3′)-IIIa-pQBAV3c | $7.3\times10^{8}\pm2.2\times10^{7}$ | 0 | $0.015\pm0.000099$ |

**Table 6.** Stability of ancestral and evolved *aph(3')-IIIa*-pQBAV3c plasmids.

| | No Plasmid | wild-type APH(3')-IIIa | evolved 4.1 APH(3')-IIIa |
|---|---|---|---|
| CFU/OD/mL (LB only) | $1.41 \pm 0.19 \times 10^8$ | $1.82 \pm 0.31 \times 10^8$ | $9.71 \pm 2.41 \times 10^7$ |
| CFU/OD/mL (Chloramphenicol) | 0 | $1.75 \pm 0.24 \times 10^8$ | $8.93 \pm 2.53 \times 10^7$ |
| Plasmid Retention | – | 97±9.0% | 92±8.4% |

the C156R mutation increases the conformational flexibility of the loop, enabling it to accommodate amikacin. Two other mutations, Y219H and K255R, occur in alpha-helices that interact with the binding loop, and could therefore influence its conformation.

The D190 residue, which we mutated, is in different active site loop (residues 188–195), and forms a hydrogen bond with the hydroxyl group that the enzyme later phosphorylates [23]. The S194R and I196F mutations in that loop could increase its conformational flexibility, so that amikacin can bind in an orientation different than that of kanamycin. The E24 residue, located in yet another active-site loop (residues 22–29), forms a hydrogen bond with neomycin B, but not with kanamycin. We hypothesize that the E24V mutation destabilizes an unproductive binding mode. Other mutations in the evolved 4.1 APH(3')-IIIa, namely I40T, I120K and K176R, occurred in residues more distant from the active-site, so it is more difficult to speculate about their effects upon amikacin recognition.

Most wild-type proteins are only marginally stable. Most amino acid changes are destabilizing, so the evolvability of most proteins is limited by conformational stability [40]. Mutations that alter the molecular recognition properties of an enzyme are particularly likely to be destabilizing [41]. Two active-site mutations that we observed, S194R and I196F, probably destabilize the active conformation by introducing new steric clashes with adjacent residues. The other seven amino acid changes occurred in surface residues, so their effects upon thermostability, if any, are less obvious. Global suppressor mutations, such as M182T in the TEM-1 beta-lactamase [16], can offset the destabilizing effects of beneficial mutations, but we made no deliberate effort to select for such mutations [42] nor did we see evidence for any. These hypotheses could be tested by calorimetric measurement of the thermodynamic parameters of the wild-type and mutant proteins.

## Clinical Implications

The *aph(3')-IIIa* gene has appeared in clinical samples tested for resistance to kanamycin and other aminoglycosides [43,44], but not amikacin [45,46,47]. We showed here, however, that four rounds of mutation and selection were sufficient to direct the evolution of an APH(3')-IIIa variant (4.1) that conferred resistance to 60 times higher concentrations of amikacin than did the wild-type. The MIC exceeded 1200 micrograms/mL, substantially higher than the highest serum level (35 micrograms/mL) recommended in humans [48]. The aminoglycoside modifying enzymes diverged to adapt to different aminoglycosides in nature [49], but no mutants of these enzymes have been identified in clinical isolates [19]. It is tempting to speculate that the fitness costs observed in our study, in combination with the plethora of existing aminoglycoside modifying enzymes that efficiently confer resistance to amikacin, including AAC(6')-Iad, ANT(4')-II, and APH(3')-VI [45,46,47] collude to prevent the adaptive evolution of APH(3')-III in the clinic.


## Acknowledgments

The work described here was inspired in part by an earlier effort by Kristen Marshall Thompson. We thank Anton Bryksin for the parental *aph(3')-IIIa*-pQBAV3c plasmid, and for his advice on protein purification. We are grateful to Sylvia Yip for her advice pertaining to the enzyme assays.


## Author Contributions

Conceived and designed the experiments: JRK IM. Performed the experiments: JRK. Analyzed the data: JRK IM. Contributed reagents/materials/analysis tools: JRK IM. Wrote the paper: JRK IM.

## References


1. Gould SJ (1989) Wonderful Life: The Burgess Shale and the Nature of History New York, NY: W. W. Norton & Co. 347 p.
2. O'Brien PJ, Herschlag D (1999) Catalytic promiscuity and the evolution of new enzymatic activities. Chem Biol 6: R91–R105.
3. Copley SD (2003) Enzymes with extra talents: moonlighting functions and catalytic promiscuity. Curr Opin Chem Biol 7: 265–272.
4. Stemmer WP (1994) Rapid evolution of a protein in vitro by DNA shuffling. Nature 370: 389–391.
5. Yano T, Oue S, Kagamiyama H (1998) Directed evolution of an aspartate aminotransferase with new substrate specificities. Proc Natl Acad Sci U S A 95: 5511–5515.
6. Gerlt JA, Babbitt PC (1998) Mechanistically diverse enzyme superfamilies: the importance of chemistry in the evolution of catalysis. Curr Opin Chem Biol 2: 607–612.
7. Fasan R, Chen MM, Crook NC, Arnold FH (2007) Engineered alkane-hydroxylating cytochrome P450(BM3) exhibiting nativelike catalytic properties. Angew Chem Int Ed Engl 46: 8414–8418.
8. Fasan R, Meharenna YT, Snow CD, Poulos TL, Arnold FH (2008) Evolutionary history of a specialized p450 propane monooxygenase. J Mol Biol 383: 1069–1080.
9. Varadarajan N, Gam J, Olsen MJ, Georgiou G, Iverson BL (2005) Engineering of protease variants exhibiting high catalytic activity and exquisite substrate selectivity. Proc Natl Acad Sci U S A 102: 6855–6860.
10. O'Loughlin TL, Greene DN, Matsumura I (2006) Diversification and specialization of HIV protease function during in vitro evolution. Mol Biol Evol 23: 764–772.
11. Sarkar I, Hauber I, Hauber J, Buchholz F (2007) HIV-1 proviral DNA excision using an evolved recombinase. Science 316: 1912–1915.
12. Goldsmith M, Tawfik DS (2012) Directed enzyme evolution: beyond the low-hanging fruit. Curr Opin Struct Biol 22: 406–412.
13. Moore JC, Arnold FH (1996) Directed evolution of a para-nitrobenzyl esterase for aqueous-organic solvents. Nat Biotechnol 14: 458–467.
14. Barlow M, Hall BG (2002) Predicting evolutionary potential: in vitro evolution accurately reproduces natural evolution of the tem beta-lactamase. Genetics 160: 823–832.
15. Bershtein S, Segal M, Bekerman R, Tokuriki N, Tawfik DS (2006) Robustness-epistasis link shapes the fitness landscape of a randomly drifting protein. Nature 444: 929–932.
16. Petrosino J, Cantu C, (1998) beta-Lactamases: protein evolution in real time. Trends Microbiol 6: 323–327.
17. Weinreich DM, Delaney NF, Depristo MA, Hartl DL (2006) Darwinian evolution can follow only very few mutational paths to fitter proteins. Science 312: 111–114.
18. Wright GD, Thompson PR (1999) Aminoglycoside phosphotransferases: proteins, structure, and mechanism. Front Biosci 4: D9–21.

19. Toth M, Frase H, Chow JW, Smith C, Vakulenko SB (2010) Mutant APH(2'')-IIa enzymes with increased activity against amikacin and isepamicin. Antimicrob Agents Chemother 54: 1590–1595.
20. Kaul M, Barbieri CM, Srinivasan AR, Pilch DS (2007) Molecular determinants of antibiotic recognition and resistance by aminoglycoside phosphotransferase (3')-IIIa: a calorimetric and mutational analysis. J Mol Biol 369: 142–156.
21. McKay GA, Thompson PR, Wright GD (1994) Broad spectrum aminoglycoside phosphotransferase type III from Enterococcus: overexpression, purification, and substrate specificity. Biochemistry 33: 6936–6944.
22. McKay GA, Wright GD (1996) Catalytic mechanism of enterococcal kanamycin kinase (APH(3')-IIIa): viscosity, thio, and solvent isotope effects support a Theorell-Chance mechanism. Biochemistry 35: 8680–8685.
23. Fong DH, Berghuis AM (2002) Substrate promiscuity of an aminoglycoside antibiotic resistance enzyme via target mimicry. EMBO J 21: 2323–2331.
24. Grimm E, Arbuthnot P (1995) Rapid purification of recombinant Taq DNA polymerase by freezing and high temperature thawing of bacterial expression cultures. Nucleic acids research 23: 4518–4519.
25. Bryksin AV, Matsumura I (2010) Rational design of a plasmid origin that replicates efficiently in both gram-positive and gram-negative bacteria. PLoS One 5: e13244.
26. Shetty RP, Endy D, Knight TF, Jr. (2008) Engineering BioBrick vectors from BioBrick parts. J Biol Eng 2: 5.
27. Bryksin AV, Matsumura I (2010) Overlap extension PCR cloning: a simple and reliable way to create recombinant plasmids. Biotechniques 48: 463–465.
28. Matsumura I, Ellington AD (2002) Mutagenic polymerase chain reaction of protein-coding genes for in vitro evolution. Methods Mol Biol 182: 259–267.
29. Zhao H, Giver L, Shao Z, Affholter JA, Arnold FH (1998) Molecular evolution by staggered extension process (StEP) in vitro recombination. Nat Biotechnol 16: 258–261.
30. Easterby JS (1973) Coupled enzyme assays: a general expression for the transient. Biochim Biophys Acta 293: 552–558.
31. Lindsley JE (2001) Use of a real-time, coupled assay to measure the ATPase activity of DNA topoisomerase II. Methods Mol Biol 95: 57–64.
32. Atwal KS, Wang P, Rogers WL, Sleph P, Monshizadegan H, et al. (2004) Small molecule mitochondrial F1F0 ATPase hydrolase inhibitors as cardioprotective agents. Identification of 4-(N-arylimidazole)-substituted benzopyran derivatives as selective hydrolase inhibitors. J Med Chem 47: 1081–1084.
33. Kawaguchi H (1976) Discovery, chemistry, and activity of amikacin. J Infect Dis 134 SUPPL: S242–248.
34. Verhamme IM, Van Dedem GW, Lauwers AR (1988) Ionic-strength-dependent substrate inhibition of the lysis of Micrococcus luteus by hen egg-white lysozyme. Eur J Biochem 172: 615–620.
35. Andersson DI, Levin BR (1999) The biological cost of antibiotic resistance. Curr Opin Microbiol 2: 489–493.
36. Levin BR, Perrot V, Walker N (2000) Compensatory mutations, antibiotic resistance and the population genetics of adaptive evolution in bacteria. Genetics 154: 985–997.
37. Maisnier-Patin S, Andersson DI (2004) Adaptation to the deleterious effects of antimicrobial drug resistance mutations by compensatory evolution. Res Microbiol 155: 360–369.
38. Rozen DE, McGee L, Levin BR, Klugman KP (2007) Fitness costs of fluoroquinolone resistance in Streptococcus pneumoniae. Antimicrob Agents Chemother 51: 412–416.
39. Schrag SJ, Perrot V, Levin BR (1997) Adaptation to the fitness costs of antibiotic resistance in Escherichia coli. Proc Biol Sci 264: 1287–1291.
40. Bloom JD, Labthavikul ST, Otey CR, Arnold FH (2006) Protein stability promotes evolvability. Proc Natl Acad Sci U S A 103: 5869–5874.
41. Tokuriki N, Stricher F, Serrano L, Tawfik DS (2008) How protein stability and new functions trade off. PLoS Comput Biol 4: e1000002.
42. Bershtein S, Goldin K, Tawfik DS (2008) Intense neutral drifts yield robust and evolvable consensus proteins. J Mol Biol 379: 1029–1044.
43. Kobayashi N, Alam M, Nishimoto Y, Urasawa S, Uehara N, et al. (2001) Distribution of aminoglycoside resistance genes in recent clinical isolates of Enterococcus faecalis, Enterococcus faecium and Enterococcus avium. Epidemiol Infect 126: 197–204.
44. Schmitz FJ, Fluit AC, Gondolf M, Beyrau R, Lindenlauf E, et al. (1999) The prevalence of aminoglycoside resistance and corresponding resistance genes in clinical isolates of staphylococci from 19 European hospitals. J Antimicrob Chemother 43: 253–259.
45. Vila J, Ruiz J, Navia M, Becerril B, Garcia I, et al. (1999) Spread of amikacin resistance in Acinetobacter baumannii strains isolated in Spain due to an epidemic strain. J Clin Microbiol 37: 758–761.
46. Doi Y, Wachino J, Yamane K, Shibata N, Yagi T, et al. (2004) Spread of novel aminoglycoside resistance gene aac(6')-Iad among Acinetobacter clinical isolates in Japan. Antimicrob Agents Chemother 48: 2075–2080.
47. Jacoby GA, Blaser MJ, Santanam P, Hachler H, Kayser FH, et al. (1990) Appearance of amikacin and tobramycin resistance due to 4'-aminoglycoside nucleotidyltransferase [ANT(4')-II] in gram-negative pathogens. Antimicrob Agents Chemother 34: 2381–2386.
48. Sarubbi FA, Jr., Hull JH (1978) Amikacin serum concentrations: prediction of levels and dosage guidelines. Ann Intern Med 89: 612–618.
49. Shaw KJ, Rather PN, Hare RS, Miller GH (1993) Molecular genetics of aminoglycoside resistance genes and familial relationships of the aminoglycoside-modifying enzymes. Microbiol Rev 57: 138–163.

# 4

# Tracking Molecular Recognition at the Atomic Level with a New Protein Scaffold Based on the OB-Fold

**John D. Steemson[1], Matthias Baake[1], Jasna Rakonjac[2], Vickery L. Arcus[1], Mark T. Liddament[3]***

**1** Department of Biological Sciences, University of Waikato, Hamilton, New Zealand, **2** Institute of Molecular BioSciences, Massey University, Palmerston North, New Zealand, **3** OBodies Limited, Hamilton, New Zealand

## Abstract

The OB-fold is a small, versatile single-domain protein binding module that occurs in all forms of life, where it binds protein, carbohydrate, nucleic acid and small-molecule ligands. We have exploited this natural plasticity to engineer a new class of non-immunoglobulin alternatives to antibodies with unique structural and biophysical characteristics. We present here the engineering of the OB-fold anticodon recognition domain from aspartyl tRNA synthetase taken from the thermophile *Pyrobaculum aerophilum*. For this single-domain scaffold we have coined the term OBody. Starting from a naïve combinatorial library, we engineered an OBody with 3 nM affinity for hen egg-white lysozyme, by optimising the affinity of a naïve OBody 11,700-fold over several affinity maturation steps, using phage display. At each maturation step a crystal structure of the engineered OBody in complex with hen egg-white lysozyme was determined, showing binding elements in atomic detail. These structures have given us an unprecedented insight into the directed evolution of affinity for a single antigen on the molecular scale. The engineered OBodies retain the high thermal stability of the parental OB-fold despite mutation of up to 22% of their residues. They can be expressed in soluble form and also purified from bacteria at high yields. They also lack disulfide bonds. These data demonstrate the potential of OBodies as a new scaffold for the engineering of specific binding reagents and provide a platform for further development of future OBody-based applications.

**Editor:** Stefan Dübel, Technical University of Braunschweig, Germany

**Funding:** This project was funded by a grant from the Marsden Fund of New Zealand with additional funds from the Maurice Wilkins Centre for Molecular Biodiscovery, New Enterprise Research Fund, AgResearch Limited, The University of Waikato and OBodies Ltd. The funders had no role in study design, data collection and analysis, decision to publish, or preparation of the manuscript.

* E-mail: mliddament@obodies.co.nz

## Introduction

Molecular recognition is a crucial aspect of a successful biological system. High-affinity interactions govern immune function, while relatively transient events are seen in signal transduction. Each type of interaction is dependent on the adaptability of protein folds during evolution. We have explored this phenomenon using a widely-distributed protein domain, the OB-fold, originally named for observed functions of oligosaccharide and oligonucleotide binding [1].

The OB-fold is a 5-stranded β-barrel domain that presents a concave binding face. In virtually all cases where the domain is present, this same face is used for binding ligand. A survey of the SCOP database reveals many OB-folds which are heavily modified with additional loops or entirely new domains inserted [2]. Examples of OB-folds can be found in diverse organisms, including archaea, bacteria, yeast and mammals, with no detectable sequence conservation across the superfamilies and representing a diverse range of natural ligands, including proteins, oligonucleotides, oligosaccharides and small molecules. These combined factors led Murzin to suggest that the OB-fold is ancient and tolerant to mutation, with an easily-adaptable binding face [1]. Ultimately the unifying feature of the OB-fold is not sequence, but structure and topology, with architecture capable of supporting a very wide range of sequences and modifications [3]. These observations suggested that the plasticity of the OB-fold might be exploited *in vitro* through protein engineering for specific molecular recognition.

Antibodies and antibody fragments are currently the dominant class of engineered proteins for molecular recognition; very large combinatorial libraries (in excess of $10^{11}$ individual members) have been made and selected *in vitro* to obtain binding reagents with very high affinity and specificity (reviewed in [4–6]). However, some general limitations of the antibody format exist, such as their large size, requirement for glycosylation (in the case of full-length antibodies) and a critical dependence on intrachain disulfide bonds, a limitation shared by even the smallest fragments of antibodies (individual V domains, termed domain antibodies [7]). Driven by a desire to circumvent these limitations, a need to expand the useful target range that can be addressed pharmacologically and also to avoid potentially limiting issues of intellectual property, the field of non-antibody protein scaffolds has developed rapidly over the last decade (recently reviewed in [8–10]). Affibodies (based on a three-helix-bundle Z domain from staphylococcal protein A), have achieved high affinities against a range of protein targets [11–13]. Anticalins are high-affinity small-

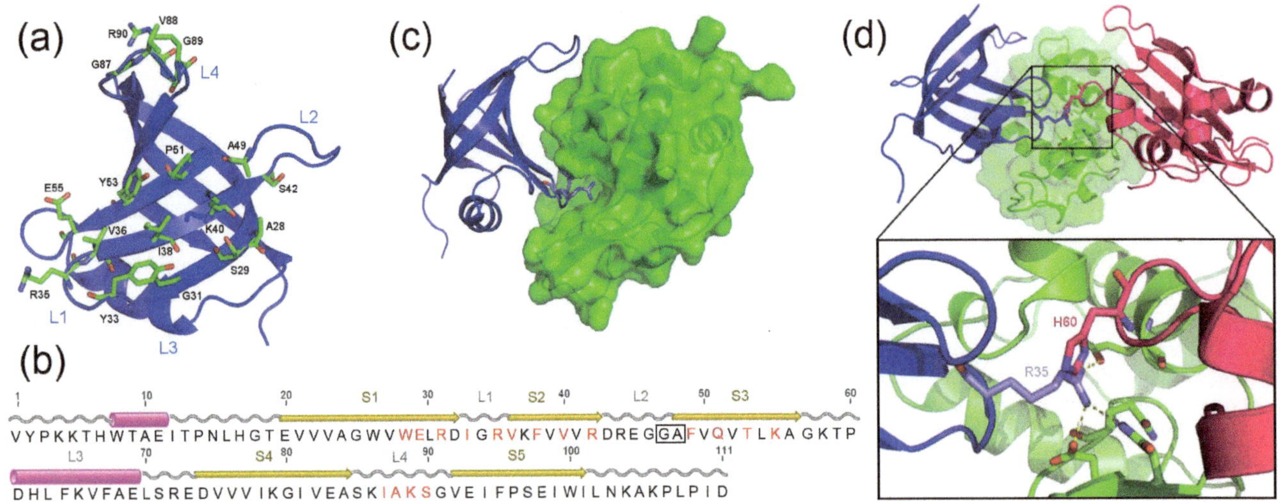


**Figure 1. Naïve library design and NL8 structure.** (a) Model of *Pyrobaculum aerophilum* AspRS, derived from the crystal structure of NL8 and showing the naïve library design. The structure is displayed in blue cartoon representation, with the 17 residues randomised shown as green stick models. (b) Schematic of naïve library design. Residues randomised in the library by NNK codons are shown in red. The annotated secondary structural elements, derived from the crystal structure of NL8, are shown above the sequence as yellow arrows (β-strands; S1–5), pink cylinders (α-helices) and grey wavy lines (loops; L1–4). The GA residues (positions 46 and 47; boxed) are the result of a 6 bp insertion in the wild-type gene immediately following wild-type residue 46, done to facilitate library construction. (c) The NL8 OBody (blue) is shown in complex with HEL (green surface model). Highlighted as stick models are R35 and Y33, interacting with the substrate-binding groove of HEL. (d) The R35 of the NL8 Obody (blue) binds into the HEL active site, in a similar manner to H60 from lysozyme inhibitor protein YkfE from *E. coli* (pink).

molecule binders that were selected from libraries of a mutant lipocalin fold [14]. Designed Ankyrin repeat proteins (DARPins) have been engineered with sub-nanomolar affinities for protein targets [15]. The Sac7d DNA-binding protein, an SH3-like five-stranded incomplete β-barrel (mistakenly identified as an OB-fold [16]) from the hyperthermophile *Sulfolobus acidocaldarius* has successfully been used as a scaffold to produce high-affinity binders to a range of targets – these were initially called Affitins, but later renamed Nanofitins [17]. Engineered derivatives of variable lymphocyte receptors, termed Repebodies, have also recently been described with affinities in the nanomolar range [18]. Given the large number of proteins investigated for engineering purposes, there are relatively few crystal structures of engineered protein-protein complexes. Hogbom et al [19] claimed the first example with an Affibody, and other examples have been published, most notably those of an Anticalin [20], a DARPin [21] and two Repebodies [18].

In principle, OB-folds possess features granting potential benefits to an engineered OB-fold domain over and above antibodies based on the immunoglobulin fold; they are small, single-domain binding modules and generally lack disulphide bonds. Due to their ubiquitous nature, potential OB-fold-based scaffolds may be sourced from widely diverse organisms, including human proteins for therapeutic and *in vivo* diagnostic use, or thermophilic microorganisms for more general applications. Given that OB-folds have a unique antigen binding site (formed from a combination of concave β-sheet and loops), it may be possible to favour different binding solutions than those seen in scaffolds where loops are the principal mediators of specificity and affinity, such as immunoglobulins.

In this study we have designed and tested the properties of a new protein scaffold based on the OB-fold of the aspartyl tRNA synthetase (AspRS) from the thermophile *Pyrobaculum aerophilum*. We term these small, single-domain engineered binding proteins OBodies. To demonstrate the amenability of this new scaffold to engineering for new binding specificity, we developed a series of OBodies with increasing affinity to a model target protein, hen egg-white lysozyme (HEL). Parallel crystal structures of OBody:-HEL complexes have given us a molecular-level view of the affinity maturation process, as we follow binders from micromolar to nanomolar affinity. Sharing a common range of epitopes centred on the HEL active site, these OBodies were also found to be efficient inhibitors of HEL activity. These data demonstrate the potential of OBodies as a new scaffold and suggest that a broad new class of binding reagents based on other OB-folds may be feasible.

## Results

### Design, Construction, Selection and Validation of a Naïve Library Based on the OB-fold from the Pyrobaculum Aerophilum Aspartyl tRNA Synthetase

To generate a naïve library, 17 specific sites on the surface of the domain were selected based on surface accessibility and their interaction with the native tRNA substrate, using a homology model built by the Swiss Model automated server [22]. These residues were randomised by PCR using multiple overlapping oligonucleotides containing degenerate NNK codons (Figure 1a and b) in phagemid pRpsp2 [23], generating a library comprising $10^8$ independent clones. To facilitate library construction using this strategy, 6 bp (encoding the residues GA) were inserted into loop 2 of the wild-type AspRS OB-fold gene immediately following residue 46 to provide adequate annealing overlap for the mutagenic oligonucleotides used in that region (Figure 1b, Figure S4). It was reasoned that an insertion here would be well-tolerated and unlikely to destabilise the AspRS OB-fold domain, an assumption that was proven correct by the subsequent performance of the library in selections and also crystal structures described below.

Initial solid-phase phage display selections were done over five rounds, using immobilised HEL. Monoclonal phage isolates were screened for binding by phage ELISA, and positive clones were expressed in *E. coli*. Surface plasmon resonance (SPR) analysis with immobilised HEL and purified OBodies identified binders of moderate affinity (best NL8, $K_D = 35\ \mu M$; Figure S1). To investigate both the nature of the OBody-HEL complex and provide a structural platform for further development of the domain as a scaffold, the crystal structure of OBody NL8 in complex with HEL was solved to 2.75 Å (Figure 1c and d, Table 1). The structure showed, in atomic detail, a classical protein-protein interaction surface; burial of ~1600 $Å^2$ total solvent-accessible area (823 $Å^2$ from the OBody) with a central hydrophobic patch made up of OBody residues Y33, V36 and I38, surrounded by residues making polar contacts (11 hydrogen bonds) and three inter-domain complementary electrostatic interactions. The concave β-sheet binding surface of the OBody wrapped around the HEL target and inserted β-strand one (β1) into the HEL substrate binding groove. The active site of the enzyme was occupied by R35 from the OBody, in a manner similar to H60 from the known HEL inhibitor YkfE [24] (Figure 1d). Interdomain contacts made around the edges of the binding face were made by wild-type residues. In particular, acidic residues D32 and E93 of the OBody mediated electrostatic bonds with complementary basic residues on the bound HEL. The loop 4 (L4) residues of the OBody (positions 87–91) were ordered poorly and did not appear to make any substantive contacts with HEL. However, immediately adjacent to the randomised region, K86 made hydrophobic contacts via the aliphatic carbons on the lysine side chain, and a hydrogen bond via the backbone carbonyl. Computational analysis by Robetta alanine scanning [25] and KFC2 [26] servers predicted the position of hotspots, indicating that binding was focused around Y33, R35 and E93.

## Affinity Maturation of HEL-binding OBodies: First Round

Examination of the NL8-HEL structure showed a poorly-ordered L4, with electron density visible only for the backbone atoms, suggesting it made little contribution to binding. This loop was therefore targeted for randomisation and also the region randomised was extended by two residues to improve its chances of making contact with HEL.

At the β-sheet face, non-contacting residues S29 and A56 were identified as being small and close to the interface, suggesting that selection of larger residues at these positions may enable additional interactions with HEL. Residues K37 and P51 were identified as making possible negative contributions to binding; P51 because it is a β-breaking residue located on a β-strand, and K37 because of proximity to the guanidinyl group of R61 from HEL, leading to a like-charge clash.

A total of 10 residues in NL8 were therefore randomised (Figure S5) by PCR using multiple overlapping oligonucleotides containing degenerate NNK codons, with the exception of residue 29, which was randomised using an NRK codon. The subset of amino acids encoded by NRK is biased towards bulkier residues, which it was thought would be beneficial at this position. The resulting library in pRpsp2 comprised $10^7$ independent clones. Solid-phase selections using immobilised HEL resulted in fixation of five unique sequences after round six (Figure 2a). All five variants showed improvements in affinity for HEL, ranging from 1–8 μM (determined by SPR; Figure S2). A consensus was clear in the β-sheet face residues, but absent in the L4 residues. The crystal structure of one variant, AM1L10 ($K_D = 5\ \mu M$; Figure 2a), was solved in complex with HEL by molecular replacement to 1.95 Å (Figure 2b and c).

Analysis of the AM1L10 structure gave a clear rationale for the observed improvement in affinity: the substitution A56Y introduced a new hydrogen bond and increased buried surface area. However, analysis by the Robetta [25] virtual alanine scanning server indicated that the predicted energetic contribution to binding by Y56 ($\Delta\Delta G = 0.86$ kcal/mol) was not sufficient to completely account for the observed change in affinity. Comparison with the NL8-HEL structure showed a moderate rigid-body shift in the arrangement of the OBody-HEL interaction relative to the parent OBody NL8-HEL interface, constituting a rotation centred on R35 of approximately 10° (Figure 2b and c). Residue K37 was re-selected to parental type, but the like-charge clash appeared to be ameliorated by a conformational shift in R61 away from the interface. The L4 residues remained predominantly disordered in the AM1L10 structure (residues D89, M90, H91 and N92 could not be resolved), suggesting that these residues were probably not involved in significant interactions with HEL, with the possible exception of pre-binding long-range electrostatic associations. Surprisingly, substitution P51S did not result in an appreciable structural change, which suggested that any advantage this substitution granted over parent-type may have been a manifestation of different dynamics at that position as well as surrounding residues.

For the next round of affinity maturation, AM1L200 was chosen as the parental clone over AM1L10. This was done for two main reasons: firstly, AM1L10 appeared only once in sequencing of individual phage clones from outputs, compared to multiple instances for all other unique AM1 sequences, suggesting possible problems with AM1L10 display or expression. Secondly, AM1L200 appeared to satisfy better the rational basis for selection of the randomised residues; S29H and A56R introduced large, polar residues and K37M removed the like-charge clash.

## Affinity Maturation of HEL-binding OBodies: Second Round

It was hypothesised that sparse random sampling of a wider sequence space throughout the whole AM1L200 OBody might yield improvements in affinity by introducing beneficial substitutions at sites other than the natural binding face or L4. Accordingly, full-length AM1L200 was subjected to error-prone (EP)-PCR. The resulting library was cloned into phagemid pRpsp2 and comprised $10^8$ independent members; sequencing showed that there were an average of 4 nucleotide substitutions per 1000 bp. Solid-phase selections using immobilised HEL resulted in the isolation of clone AM2EP06, with an improved $K_D$ of 250 nM, as determined by SPR (Figure 3b and c).

The crystal structure of AM2EP06 was solved in complex with HEL to a resolution of 1.86 Å. With three amino acid substitutions compared to the parent AM1L200 (T19S, M37K and K86E), it showed a binding face that was very similar to AM1L10 and NL8, with differences at the periphery. In a manner very similar to the AM1L10 structure, a small conformational change in R61 from HEL moved its guanidinyl group away from the interface.

In contrast to the previous two HEL-complex structures, L4, with an entirely new sequence, was moderately well-defined as electron density in the structure. The loop showed some contact with HEL, increasing buried surface area by ~100 $Å^2$, although no additional hydrogen bonds or other polar contacts were noted. The contribution of L4 to ΔG of binding to HEL appear to be improving, which was suggested by the crystallographic B-factors. In the previous structures, atoms in the interface had higher B-factors than the average B-factor across the structure, whereas in AM2EP06, while L4 still showed B-factors consistent with a flexible surface loop (well above the average atomic B-factor), B-factors were slightly reduced compared to that seen for L4 in NL8 and AM1L10(Figure 4). The L4 residues from AM2EP06 also made extensive contacts with the β-sheet binding face via S85 and

**Table 1.** Data collection and refinement statistics for all OBody-HEL complexes reported here.

| | NL8 | AM1L10 | AM2EP06 | AM3L15 | AM3L09 |
|---|---|---|---|---|---|
| **Data collection** | | | | | |
| Space Group | $p4_12_12$ | $p2_12_12_1$ | $p2_12_12_1$ | $p2_1$ | p1 |
| Cell dimensions | | | | | |
| a, b, c (Å) | 76.759,76.759, 166.344 | 60.54, 186.26, 245.70 | 50.43, 58.33, 81.82 | 51.99, 58.83, 95.18 | 59.00, 69.38, 76.36 |
| α, β, γ (°) | 90 | 90 | 90 | 90, 95.43, 90 | 72.17, 69.46, 77.55 |
| Wavelength (Å) | 0.95666 | 0.95666 | 1.54179 | 1.54179 | 0.95369 |
| Resolution (Å) | 34.9 - 2.69 (2.76 - 2.69)[a] | 29.76 - 1.95 (2.05 - 1.95 | 34.58 - 1.86 (1.96 - 1.86) | 35.09 - 1.86 (1.96 - 1.86) | 51.50 - 2.57 (2.71 - 2.57) |
| Measured Reflections | 144772 | 2883973 | 143392 | 130001 | 558386 |
| Unique Reflections | 14295 (908) | 201772 (27775) | 20797 (2819) | 46187 (6490) | 32670 (4634) |
| Multiplicity | 9 | 14.3 (12.4) | 6.9 (6.5) | 2.8 (2.7) | 7.8 (7.7) |
| Completeness (%) | 99.15 (92.8) | 99 (95.0) | 99.4 (94.1) | 95.7 (92.3) | 96.2 (93.6) |
| *R*merge (%)† | 7.3 (54.3) | 6.4 (48.4) | 3.9 (22.6) | 8.2 (53.3) | 11.3 (42.2) |
| $<I/\sigma(I)>$ | 32.5 (3.6) | 24.5 (5.0) | 29.2 (7.5) | 8.1 (1.8) | 14 (4.9) |
| Wilson B (Å2) | 65 | 28.19 | 26.4 | 22.9 | 29.7 |
| Mosaicity (°) | 0.6 | 0.2 | 0.5 | 0.7 | 0.7 |
| **Refinement** | | | | | |
| Resolution (Å) | 27.5–2.75 (2.82 - 2.75) | 29.76- 1.95 (2.01 - 1.95) | 34.58 - 1.86 (1.91 - 1.86) | 34.58 - 1.86 (1.91 - 1.86) | 44.83 - 2.57 (2.67 - 2.57) |
| *R* factor (%)‡ | 22.9 (26.5) | 18.8 (21.34) | 17.18 (20.8) | 17.32 (28.10) | 20.14 (23.63) |
| Reflections | 12,789 (908) | 191,052 (12,500) | 19654 (1229) | 43800 (3090) | 31848 (2736) |
| $R_{free}$ (%) | 29.64 (37.6) | 22.97 (27.03) | 21.5 (29.4) | 20.24 (30.8) | 24.48 (31.43) |
| Free reflections | 667 (55) | 10136 (674) | 1061 (57) | 2336 (165) | 1656 (116) |
| Refined Atoms | 3379 | 18,959 | 2101 | 4381 | 8155 |
| Protein Chains | 4 | 18 | 2 | 4 | 8 |
| r.m.s.d bond lengths (Å) | 0.013 | 0.007 | 0.021 | 0.023 | 0.004 |
| r.m.s.d bond angles (°) | 1.452 | 1.052 | 1.854 | 1.863 | 0.762 |
| B factor (mean) | 51.15 | 30.20 | 37.77 | 26.34 | 26.8 |

[a]Values in parentheses are for highest-resolution shell.
†$Rmerge = \Sigma|I_{obs} - <I>|/\Sigma I_{obs}$.
‡$R = \Sigma||F_{obs}| - |F_{calc}||/\Sigma|F_{obs}|$.

(a)

| | β-sheet face | | | | Loop 4 | | | | | | |
|---|---|---|---|---|---|---|---|---|---|---|---|
| | 29 | 37 | 51 | 56 | 87 | 88 | 89 | 90 | 91 | 92 | $K_D$ (μM) |
| NL8 | S | K | P | A | G | V | - | - | G | R | 35.3 |
| AM1L05 | N | K | S | Y | E | P | H | G | Q | D | 6.9 |
| AM1L06 | S | Q | P | L | W | T | D | E | L | M | 8.2 |
| AM1L10 | S | K | S | Y | A | A | D | M | H | N | 4.6 |
| AM1L18 | S | K | D | A | S | V | A | A | N | A | 7.5 |
| AM1L200 | H | M | P | R | Q | Y | G | S | D | T | 1.2 |

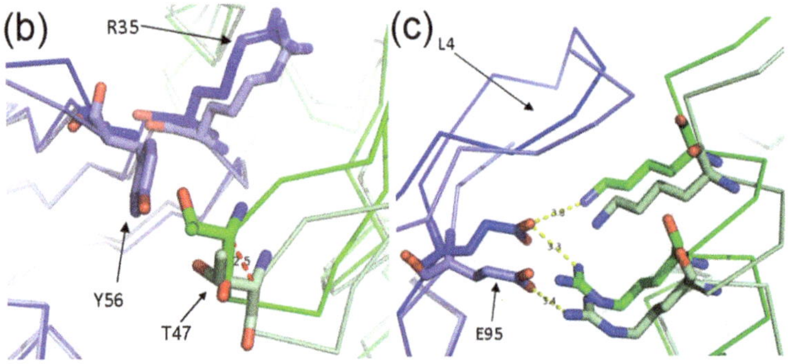


**Figure 2. Sequences and affinities of OBodies from first round of affinity maturation and AM1L10 structure.** (a) OBody sequences from the first round of affinity maturation, showing only the mutations at the targeted residues and labelled according to their sequence position in the library. For comparison, the equivalent positions from NL8 are also shown. Residues are coloured by polarity or charge (yellows = non-polar, greens = polar, red = acidic, blue = basic). (b) AM1L10 (pale blue Cα trace) in complex with HEL (pale green), superposed on to NL8 (dark blue, r.m.s.d. 0.45 Å) with the NL8-bound HEL also visible (dark green). Substitution A56Y introduces a hydrogen bond with HEL T47 and the subsequent relative shift in HEL position is evident, with a 2.5 Å shift in Cα position at that residue (indicated by red dashed line). (c) AM1L10 (pale blue Cα trace) in complex with HEL (pale green), superposed on to NL8 (dark blue, r.m.s.d. 0.45 Å) with the NL8-bound HEL also visible (dark green). At the top of the binding face, E95 accommodates the shift with a conformational change, maintaining close contact with a lysine and arginine from HEL. Dashed yellow lines are potential hydrogen or electrostatic bonds, labelled with distances in angstroms.

(a)

| Clone | ELISA $A_{450}$ | Fold increase over AM2EP06 | Loop 4: 85 | 86 | 87 | 88 | 89 | 90 |
|---|---|---|---|---|---|---|---|---|
| AM2EP06 | 0.17 | - | S | E | Q | Y | G | S |
| AM3L09 | 1.76 | 10.2 | T | T | V | T | R | W |
| AM3L05 | 1.31 | 7.6 | S | L | P | S | P | D |
| AM3L15 | 0.90 | 5.2 | G | W | P | V | A | L |
| AM3L10 | 0.70 | 4.1 | T | L | T | T | T | Y |
| AM3L04 | 0.69 | 4.0 | P | L | P | I | P | P |
| AM3L13 | 0.66 | 3.8 | S | I | M | S | L | Y |
| AM3L02 | 0.57 | 3.3 | S | G | W | D | Y | W |
| AM3L08 | 0.52 | 3.0 | T | V | Y | D | S | W |
| AM3L14 | 0.51 | 3.0 | P | P | P | A | P | P |
| AM3L01 | 0.48 | 2.8 | T | Y | F | T | T | F |
| AM3L07 | 0.48 | 2.8 | V | L | P | S | P | T |
| AM3L16 | 0.46 | 2.7 | S | L | M | S | P | H |
| AM3L12 | 0.42 | 2.4 | T | I | N | N | Q | I |
| AM3L03 | 0.40 | 2.3 | D | I | D | W | L | R |
| AM3L11 | 0.39 | 2.2 | T | A | - | D | P | L |
| AM3L06 | 0.38 | 2.2 | T | L | T | D | H | I |

non-polar polar basic acidic

(b)

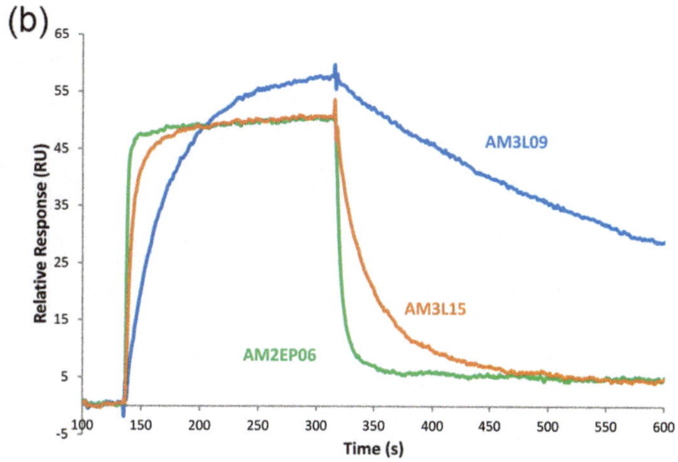


(c)

| OBody | $k_{on}$ ($M^{-1}s^{-1}$) | $k_{off}$ ($s^{-1}$) | $K_D$ (M) |
|---|---|---|---|
| **AM3L09** | $8.53 \times 10^5$ | $2.49 \times 10^{-3}$ | $2.92 \times 10^{-9}$ |
| **AM3L15** | $8.23 \times 10^5$ | 0.0256 | $3.11 \times 10^{-8}$ |
| **AM2EP06** | - | - | $2.53 \times 10^{-7}$ |

**Figure 3. Phage ELISA, SPR and kinetic data for top hits from third round ofaffinity maturation.** (a) Top hits from the third round of affinity maturation ranked by phage ELISA signal and showing the corresponding L4 sequence. (b) SPR sensorgrams for OBodies AM3L09 (blue line), AM3L15 (orange line) and AM2EP06 (green line) binding to HEL. For visual comparison, sensorgrams are shown only for the highest single concentration of each OBody analysed (AM3L09 at 32 nM, AM3L15 at 128 nM and AM2EP06 at 800 nM). These data were produced on a single chip and are representative of multiple independent analyses, performed with separate chips and protein samples. (c) Kinetic data for each OBody. The AM3L09 and AM3L15 kinetic data were calculated using Graphpad Prism association/dissociation modelling, whereas the affinity of AM2EP06 was calculated using an equilibrium model of maximum response. While the $k_{on}$ for both AM3L15 and AM3L09 are essentially the same, the difference in dissociation constant can be attributed to a substantial decrease (10-fold) in $k_{off}$. Separate on and off rates could not be determined from the AM2EP06 data.

Q87, potentially introducing conformational restraints to HEL-binding residues (Figure S3). Similarly, at the C-terminal end of L4 an intrachain hydrogen bond between S90 and T92 also restrained the loop.

## Affinity Maturation of HEL-binding OBodies: Third Round

Examination of the AM2EP06-HEL complex structure showed that few strong contacts with HEL were made by L4 residues. The third affinity maturation library therefore randomised six residues in L4 of AM2EP06 using a modified version of this sequence as the

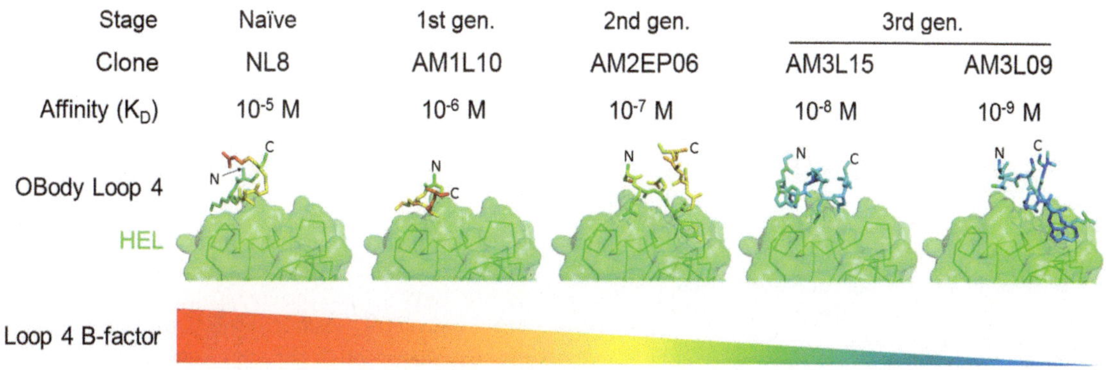


**Figure 4. Contribution of L4 to binding in five OBody-HEL complexes.** HEL is shown in green as a surface representation in identical orientations in each image and the L4 residues from each OBody variant, coloured according to B-factor relative to the average in each individual case; blue is low B-factor, graduating through green, yellow then red as it increases. The lower-affinity variants NL8 and AM1L10 both show poorly-ordered L4 residues. Note that residues D89, M90, H91 and N92 are missing from the model of AM1L10 L4, as they could not be resolved. Relative stabilisation is evident in L4 of the higher-affinity variants AM3L15 and AM3L09, compared to the parent clone AM2EP06, implying increased involvement in binding. Although the L4 structures of AM3L09 and AM2EP06 have superficially similar configurations, with Y88 and W90 binding in similar positions, W90 L4 from AM3L09 makes a greater number of contacts and is packed more closely. The alpha carbons of N- and C-terminal L4 residues have been labelled with 'N' and 'C' respectively, to denote the anchor points for the loop in each case.

template, in which a stop codon was introduced by site-directed mutagenesis at position 90 in L4 (AM2EP06-stop). This construct ensured that only clones with a randomised L4 sequence would be functional in the resulting library. The pRpsp2 phagemid was also modified to remove a 2-residue duplication in the pelB leader and revert it to wild-type sequence, creating phagemid pAS1. Residues S85, E86, Q87, Y88, G89 and S90 of AM2EP06 were then randomised using oligonucleotides containing NNS codons by Kunkel mutagenesis [27] in phagemid pAS1, to generate a library comprising $10^9$ independent members. It was estimated by sequencing that L4 was randomised in 75% of clones; the remainder of the library was AM2EP06-stop.

Three rounds of solution-phase selection were done using decreasing concentrations of biotinylated HEL (1 μM, 10 nM and 1 nM across rounds 1–3, respectively). A total of 192 randomly-picked variants were screened by phage ELISA from rounds two and three of selection. Sixteen unique clones were identified with an ELISA signal at least 2-fold higher than the parent clone AM2EP06 (AM3L01-16; Figure 3a). Inspection of the L4 sequences showed a general paucity of charged species, high incidence of proline and convergence towards hydrophobic (L4 residues 86, 87 and 89) or small hydrophilic residues (L4 residue 85). These data suggested a common selective pressure at the randomised positions, but no clear consensus emerged. Based on the ELISA response, two of the top four hits (AM3L09 and AM3L15) were analysed in detail by SPR alongside the parent, AM2EP06 (Figure 3b). The highest affinity attained was 3 nM for variant AM3L09. To examine the structural basis for the observed affinity improvement, this variant along with AM3L15 ($K_D = 31$ nM), was crystallised in complex with HEL and the structures solved to 2.57 Å and 1.86 Å, respectively.

## Anatomy of a nM HEL-binding Obody

The interface between AM3L09 ($K_D = 3$ nM) and HEL buries a total 1800 $Å^2$ solvent-accessible surface area (SASA; 923 $Å^2$ from the OBody) and maintains the intimate association with the HEL active site and substrate-binding groove as first established in the NL8-HEL complex. In a general sense, the primary interface residues are the same: a hydrophobic patch at the centre of the interface composed of Y33, V36 and I38, surrounded by an inter-molecular hydrogen bond network and electrostatic interactions (Figure 5a). In contrast to the NL8-HEL interface, the AM3L09-HEL interface shows an increased number of polar interactions, (going from 10 to 15 inter-molecular hydrogen bonds) and the addition of two salt bridges. Residues W90 and D91, located in L4 of AM3L09, together account for three hydrogen bonds, a salt bridge and burial of 195 $Å^2$ at the interface. A very clear rigid-body shift relative to the NL8 binding position is also maintained in AM3L09 (Figure 5d).

In AM3L15 ($K_D = 35$ nM) L4 has adopted a helical character, with no similarity to L4 in AM2EP06. The N-terminus of the helix in L4 is capped by a tryptophan cis-prolyl peptide bond (W86, P87). Intriguingly, this type of bond has been highly conserved across the proteins in which it has appeared during evolution (across all families of protein fold), underscoring its vital functional role [28]. In this case P87 appears to make hydrophobic contacts with HEL, while W86 mediates an inter-molecular hydrogen bond. Two L4 residues (D91 and T92), although not mutated in this library or previously directly involved in binding HEL were recruited for close involvement at the binding interface. In a position similar to Q87 in AM2EP06, the sidechain of T92 makes polar contacts with OBody interface residues of the β-sheet face, while neighbouring residue D91 introduces an inter-molecular salt bridge with HEL. These new contacts are the product of L4 rearrangement brought about by this affinity maturation step.

In both AM3L09 and AM3L15, complementary electrostatic charges are concentrated in two major patches. One consists solely of R35, which associates directly with the negatively charged active site of HEL. The other involves D91, which despiteL4 adopting completely different conformations in these two variants, occupies the same position in both structures, becoming one of three acidic residues (E83, E95 and D91) arranged in a line across the top of the interface, creating a negatively charged patch which binds to a lysine/arginine pair from HEL (Figure 5b and c).

In addition to the protein atoms which interact directly, the AM3L09-HEL interface has a complement of highly-ordered water molecules which mediate both intra- and inter-domain interactions. In particular, two clusters of ordered solvent are closely involved with interface residues, with one anchoring the C-terminal end of L4, and another larger cluster arranged between critical binding residues Y33 and R35 (Figure 5e). Comparison with the AM3L15 and AM2EP06 structures show analogous interactions across the interface in very similar positions at L4, but mediated by protein atoms in both cases. Overall, the improvement in affinity, from 250 nM in AM2EP06, to 35 nM in AM3L15 and finally 3 nM in AM3L09 is correlated with the stabilisation and recruitment of L4 into progressively more substantive contact with HEL. This observation is supported by changes in relative crystallographic B-factors: L4 residues from AM2EP06, AM1L10 and NL8 are generally above the average B-factor or disordered, consistent with a relatively unrestrained surface loop. In contrast, L4 residues from AM3L15 and L09 are at or below the average B-factor, which is more consistent with core or interface residues (Figure 4).

## Thermal Stability of HEL-binding Obodies

Differential scanning fluorimetry [29] was used to estimate the melting temperature ($T_m$) of a range of HEL-binding OBodies and a control non-HEL-binding OBody (Figure 6a). These data show that although stability has reduced slightly as the number of mutations increase, the thermophilic character has been maintained despite mutation of up to 22% of the protein.

## Inhibition of HEL Enzymatic Activity

HEL was originally chosen simply as a model protein for selection of OBody binders. However, the similarity of the binding to known HEL inhibitors with respect to interface statistics (Table 2) as well as mode of binding suggested that, with sufficient affinity, this lineage of OBodies had the potential to inhibit HEL enzyme activity. The capacity of AM2EP06, AM3L15 and AM3L09 to inhibit HEL activity was therefore measured by titration against a constant concentration of HEL in the presence of *Micrococcus lysodeikticus* cells to determine the $IC_{50}$. The ranking observed in $IC_{50}$ mirrors that seen for affinity: in the presence of 162 nM HEL, AM3L09 had an $IC_{50}$ of 261 nM, AM3L15 362 nM and AM2EP06 1.1 μM (Figure 6b). The negative control was an OBody with no HEL-binding activity (U81), which showed no effect on HEL activity over the same concentration ranges. Thus, as expected based on the X-ray structural data, the OBodies were able to inhibit specifically HEL enzymatic activity.

# Discussion

The development of OBodies has followed a similar path to other engineered non-antibody scaffolds, with the initial library design and selection producing binders of moderate affinity, in the micromolar range [11,30]. The subsequent improvements in

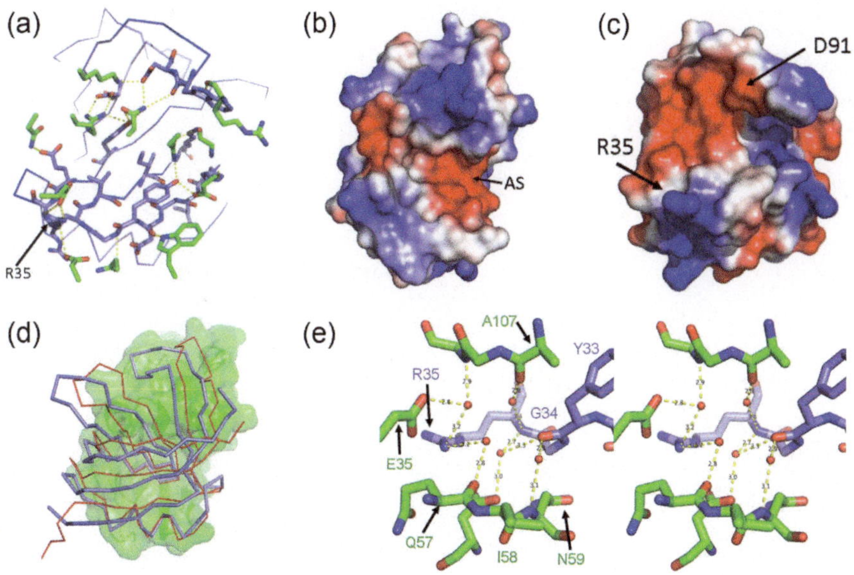


**Figure 5. Anatomy of the AM3L09-HEL interface.** (a) AM3L09 is coloured blue with interface residues shown as stick models. HEL residues calculated to make a hydrogen bond with AM3L09 are shown as green stick models. Potential hydrogen bonds are indicated by a dashed yellow line. (b) HEL electrostatic surface. The highly electronegative HEL active site (AS) is filled by R35 from AM3L09. (c) AM3L09 electrostatic surface, shown in the same orientation as panel a. The negatively charged patch containing D91 associates with a complementary positively charged patch on the HEL interface. (d) Comparative binding positions of AM3L09 (blue, thick Cα trace) and NL8 (red, narrow Cα trace) to HEL (green surface). (e) The AM3L09-HEL interface, shown in wall-eye stereo. Bridging water molecules between AM3L09 (blue) and HEL (green) are shown as red spheres. Potential hydrogen bonds are indicated by a dashed yellow line labelled with the length in angstroms.

affinity achieved through phage selection of combinatorial libraries have shown that the OBody scaffold is capable of monomeric affinities in the low nanomolar range, and therefore worthy of further development for specific applications. But, uniquely amongst other engineered non-antibody scaffolds, the progressive affinity maturation process undergone by the HEL-binding OBodies has been thoroughly documented by atomic-resolution X-ray crystallographic structures of representative OBodies in complex with HEL.

Through examination of these structures, we were able to propose a rationale for the effect of the mutations selected at each step, from fine- to broad-scale changes. Selections from the naïve library isolated OBody NL8 ($K_D = 35$ μM); the crystal structure of the NL8-HEL complex at 2.75 Å showed an OBody bound to the active site and substrate-binding groove of HEL, with a central hydrophobic patch made up of Y33, V36 and I38, surrounding polar contacts (11 hydrogen bonds) and three inter-domain complementary electrostatic charges. This structure yielded two critical pieces of information. Firstly, binding to HEL was mediated by residues targeted in the library, showing that the library design was successful, even though it was based on a scaffold for which there was no available structural data. This is unusual for a new scaffold, as crystallographic structural data is usually considered vital as a starting point. The NL8-HEL interface made no apparent use of the four randomised positions in L4 of NL8 (out of a total of 17 distributed throughout the domain), which immediately suggested that the recruitment of L4 into binding interactions might be a viable strategy for affinity maturation. Secondly, the structure of the complex showed that the beta-sheet binding face had statistics which were not excessively different from other HEL-binding proteins, most notably antibodies, strongly suggesting that with appropriate maturation steps the affinity could be improved.

Selections from the first affinity maturation library yielded the NL8 variants AM1L10 ($K_D = 5$ μM) and AM1L200 ($K_D = 1$ μM), representing 6- and 35-fold improvements in affinity over NL8, respectively. The crystal structure of the AM1L10-HEL complex at 1.95 Å revealed two major factors in the improvement in binding: the introduction of additional inter-domain contacts at Y56, and a consequent moderate shift in binding orientation compared to the NL8-HEL complex. To some extent the pattern-recognition function of hydrogen bond networks is similar in both intra- and inter-molecular interactions [31]. Where they differ is in the degree to which they can be optimised. Norel and co-workers [32,33] showed that geometric complementarity can be used successfully as the sole docking criteria for finding binding sites, even when using structures of monomers not solved in the presence of their ligand. This strongly implies that, at least in the cases examined, the binding surface shape is largely determined before binding, essentially restricting unsatisfied hydrogen bonds on the interface of a folded structure to a rigid-body search. Consequently, bonds formed between proteins present a much broader range of angles and distances than found in other contexts [34,35], impacting negatively on bond stability as they diverge from the theoretical ideal. There are cases where large conformational shifts can be seen on binding, most obviously in domain-swapping dimers or natively unfolded proteins (e.g. βγ-crystallin on calcium binding [36]) but it has been noted that antibody-antigen complexes exhibit comparatively little conformational changes as a result of binding [37], and that amongst the available degrees of freedom for a particular residue, a strongly-binding but infrequently-sampled conformational isomer contributes poorly to the free energy of binding [38,39]. Thus, the shift in binding orientation seen in AM1L10, and later in AM2EP06, can be viewed as an overall optimisation of bonding across the entire binding face, with the OBody "settling" on to the HEL ligand.

(a)

| OBody | $T_m$ (°C) | No. residues mutated vs. WT | % mutant residues vs. WT |
|---|---|---|---|
| NL8 | 79.6 | 17 | 15% |
| U81 | 65.8 | 19 | 17% |
| AM1L10 | 80.8 | 20 | 18% |
| AM2EP06 | 79.2 | 23 | 21% |
| AM3L15 | 66.4 | 24 | 22% |
| AM3L09 | 71.8 | 24 | 22% |

(b)

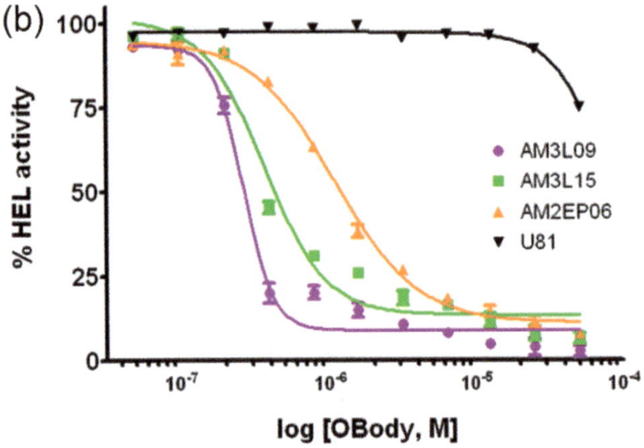


**Figure 6. Differential scanning calorimetry thermal denaturation and HEL inhibition assay for HEL-binding and control OBodies.** (a) Thermal denaturation by differential scanning fluorimetry of HEL-binding OBodies and a control non-HEL-binding OBody U81. Calculated $T_m$ values are shown alongside the number of amino acid mutations as compared to the wild-type AspRS OB-fold domain from *P. aerophilum*. (b) HEL activity assay showing inhibition by OBodies AM3L09, AM3L15 and AM2EP06 and negative control U81. Lines show the nonlinear fit of a variable-slope dose-response model. Error bars show the 95% confidence interval derived from triplicate data points. The data shown are representative of that obtained from multiple independent repeats of the same assay.

For the second phase of affinity maturation, AM1L200 was used as the template for an EP-PCR library, selections from which produced variant AM2EP06 ($K_D = 250$ nM), representing a 4-fold improvement in affinity over AM1L200. The crystal structure of the AM2EP06-HEL complex was solved at 1.86 Å. In this case the exact molecular determinants of the improvement in affinity were less clear, but the most likely explanation seemed to be a change in L4 dynamics through the K86E substitution. Residue E86 from AM2EP06 made limited contacts with HEL, and its sidechain was poorly-ordered. Although some role in long-range electrostatic attraction may be hypothesised, we think it most likely that this mutation increased the occupation of binding-favourable conformations of L4 by removing competing interactions with HEL. Indeed, the NL8 and AM1L10 structures both showed K86 in a HEL-contacting position which could not be accommodated at the same time as the L4 arrangement following the K86E substitution in AM2EP06.

In the third and final phase of affinity maturation, six residues in L4 of AM2EP06 were randomised and selections from this library resulted in the isolation of AM3L15 ($K_D = 30$ nM) and AM3L09 ($K_D = 3$ nM), constituting 8- and 83-fold improvements in affinity over AM2EP06, respectively. In this case, selections were performed in solution using biotinylated target and trypsin-sensitive helper phage KM13 [40], which we expect greatly increased their efficiency over selections performed with previous libraries (given that the latter were done with solid-phase selections and VCSM13 helper phage). The crystal structures of the AM3L15-HEL and AM3L09-HEL complexes were solved at 1.86 Å and 2.57 Å, respectively. These two final structures showed unequivocally that the improvements in affinity were due to recruitment of conformationally-restrained L4 residues into direct contact with HEL (Figure 4). They also demonstrated that the core OBody fold is capable of accommodating very different loop sequences, including a cis-prolyl bond capping a short α-helix in the case of AM3L15, as well as the extended conformations seen in AM2EP06 and AM3L09. With HEL-binding OBodies in the low nM-range, inhibition of HEL activity could be shown, both demonstrating a model application for engineered OBodies and providing further confirmation of binding independent of SPR-based measurements.

While the naïve library selections were clearly successful, indicating that the library design was sufficient, the structures solved subsequently showed an important property of the designed interface; interactions with HEL relied heavily on, and were consequently limited by, the involvement of wild-type residues. This is, at least partially, a consequence of the way NL8 bound its first β-strand into the HEL substrate binding groove, which forced non-randomised residues on the underside of the β-strand to make contact with HEL residues. In addition, the two regions targeted, namely L4 and the β-sheet binding face, did not form a continuous surface, which resulted in non-randomised residues from β-strands 4 and 5 also being forced into contact. Both of these non-randomised regions contain charged residues; especially important is E95 on β-strand 5, positioned directly between L4 and the β-sheet interface. While this may or may not be an issue for binding

**Table 2.** Interface statistics of OBody HEL-binding compared to other known HEL-binding proteins, calculated by PDBePISA.

| Molecule | Buried solvent-accessible area ($A^2$) | H-bonds | Salt bridges | $K_D$ |
|---|---|---|---|---|
| NL8 | 821 | 10 | 5 | 35 μM |
| AM1L10 | 834 | 10 | 4 | 5 μM |
| AM2EP06 | 945 | 10 | 3 | 250 nM |
| AM3L15 | 974 | 14 | 6 | 30 nM |
| AM3L09 | 923 | 16 | 6 | 3 nM |
| Fab' (1FDL) | 680 | 14 | 0 | 22 nM |
| Camelid VHH (1JTP) | 772 | 8 | 0 | 50 nM |
| YkfE (1GPQ) | 768 | 15 | 3 | 1 nM |

epitopes on a given target, we may be able use this most recent structural information to design an improved binding surface, by repositioning the scaffold binding region and removing charged groups which might otherwise interfere.

In addition, these five structures provide an empirical basis for design of OBody libraries in the future, enabling precise selection of target residues for mutations and assessment of resulting hypothetical binding surfaces without the need for further low-throughput structural studies of the kind employed here.

OBodies can also be produced easily by expression in *E. coli* and without performing any optimisation we have been able to achieve purified yields of 100–200 mg per litre of culture.

We have shown that an OBody scaffold is capable of binding in monomeric form with nanomolar affinity to HEL, a model target ligand. Detailed structural information at each step, from naïve clone and through the affinity maturation process showed an iterative improvement in interface statistics, and demonstrated the scaffold's ability to maintain highly diverse sequences (mutating up to 22% of the protein) with no appreciable change in core residue arrangement (Cα r.m.s.d. across all of the structures of 0.2 Å, excluding surface loops ) and maintenance of the thermophilic character of the ancestral OB-fold domain ($T_m$ 72°C for AM3L09). Due to their small size (13-fold smaller than IgGs), lack of disulfides and superior biophysical characteristics, OBodies derived from thermophilic bacterial OB-fold domains may be useful for extended-shelf-life *in vitro* diagnostics or as tools for detection or bioseparation. Further applications in therapeutics may yet be possible, either with OBodies derived from human OB-folds or engineered de-immunised versions of the bacterial OBody scaffold presented here.

## Materials and Methods

### Naïve and First Affinity Maturation Library Construction

These libraries were constructed by PCR dissection of the ancestor gene into overlapping fragments followed by overlap-extension assembly incorporating degenerate oligonucleotides. The naïve OBody library based on the OB-fold from *Pyrobaculum aerophilum* aspartyl tRNA synthetase (GenBank ID NP_558783.1; residues 1–109) was constructed in the pRpsp2 phagemid [23] using PCR and overlapping oligonucleotides containing NNK (N = G, A, T or C; K = G or T) codons, randomising the amino acids at 17 positions across the domain and introducing a 6 bp insertion immediately following wild-type residue 46 which encoded the residues GA, to facilitate library construction (Figure 1b and Figure S5). For the naïve library, mutagenic oligos 051, 052 and 053 were used (see Table S1 for oligo sequences and Figure S4 for detailed construction plan). The first affinity maturation library was made using PCR and overlapping oligonucleotides containing NNK codons and a single NRK (R = G or A) codon in pRpsp2. For this library, mutagenic oligos 183, 184 and 163 were used (see Table S1 for oligo sequences and Figure S5 for detailed construction plan). The intact gene libraries were digested with *Nco* I and *Not* I and ligated into phagemid pRpsp2 using a T4 ligase reaction with an insert:vector ratio of 3:1 (9 pmol vector, 3 pmol library insert). DNA was purified from the ligation by ethanol precipitation to reduce the volume, then on a column using a Roche HiPure PCR Purification kit. The ligated DNA was transformed into freshly-prepared electrocompetent TG1 *E. coli* in 50 µL aliquots. Library size was estimated by serial dilution and colony count.

### Second Affinity Maturation Library Construction

Error-prone (EP)-PCR [41] was used to generate a randomly-mutated library. EP-PCR was performed using *Taq* DNA polymerase and primer pair 005 and 006 (Table S1). Polymerase errors were induced by the introduction of 0.5 mM $MnCl_2$ to reduce polymerase specificity. Also, deoxynucleotide-triphosphates thymidine and cytidine were disproportionately increased, from 0.2 mM to 1 mM, to increase the likelihood of misincorporation. $MgCl_2$ was increased to 7 mM to reduce primer specificity. To reduce the chances of non-specific products, the template gene was introduced as a freshly-prepared PCR product, at 10 fmol per 100 µL reaction. Taq polymerase concentration was also increased, to 5 U per reaction. The mixture was split into 10 µL aliquots for cycling to prevent dominance of an early mutation in the library as a whole. The EP-PCR product was gel-purified and the whole library amplified in a normal PCR reaction with phagemid cloning primers 192 and 040 (Table S1) to produce sufficient quantities for cloning. The final PCR product library was gel-purified and digested with *Nco* I and *Not* I for cloning into the pRpsp2 phagemid.

### Third Affinity Maturation Library Construction

The third affinity maturation library was made using Kunkel mutagenesis with NNS randomisation in the pAS1 phagemid, essentially as described elsewhere [42]. Single-stranded template DNA containing uracil (dU-ssDNA) was generated from phagemid pAS1-AM2EP06, following growth in and isolation from, *E. coli* CJ236. A single colony was used to inoculate 200 mL LB, supplemented with 0.25 µg/mL uridine and 100 µg/mL carbenicillin, and grown at 250 rpm, 37°C until $OD_{600}$ = 0.1–0.2. Helper phage KM13 [40] was added to the culture at a multiplicity of infection of 20 and the culture incubated at 250 rpm, 37°C for 1 h. Kanamycin was added to 50 ug/mL and the culture incubated overnight at 250 rpm, 37°C. The cells were removed by centrifugation and discarded. Phage were precipitated from the supernatant by the addition of PEG 8,000 to 4% (w/v) and NaCl to 3% (w/v) and incubation on ice for 1 h. The precipitate was harvested by centrifugation at 15,000 × *g* for 15 min, the supernatant discarded and the pellet dissolved in 4 mL PBS pH 7.4. The final solution was centrifuged at 15,000 × *g* for 15 min, then filtered through a 0.45 µm membrane. dU-ssDNA was isolated from phage using a Qiagen M13 preparation kit.

Generation of heteroduplex DNA was performed using degenerate oligonucleotide 247 (Table S1). The oligonucleotide was phosphorylated by treatment with T4 Polynucleotide kinase, annealed to the template dU-ssDNA, then extended and ligated using Klenow fragment and T4 Ligase. The heteroduplex DNA was purified using a Roche High Pure PCR Purification kit and transformed into freshly prepared electrocompetent TG1 *E. coli* in 400 µL aliquots. Library size was estimated by serial dilution and colony counting.

### Solid-phase Phage Selection

Purified phage were generated using VCSM13 helper phage (Agilent Technologies, USA), packaging the pRpsp2 phagemid [23]. Nunc immunotubes (ThermoFisher Scientific, Waltham MA, U.S.A.) were used as the initial, solid-phase method of ligand immobilisation for panning experiments. Ligand in 2 mL PBS was incubated with constant shaking overnight at room temperature in parafilm-sealed immunotubes. Each tube containing ligand was matched to a control tube with an equal concentration of BSA. After the overnight incubation, each tube was blocked for 3 h in 4 mL PBS with 1% BSA (w/v) and rinsed five times with 5 mL PBS. Input phage samples were pre-adsorbed with PBS with 0.1%

BSA for at least 1 h, then panned in 2 mL aliquots per immunotube for 1 h with gentle stirring. The supernatant was discarded and each tube washed 10 times with 5 mL PBS with 0.1% Tween-20 (PBS-T), then five times with 5 mL PBS. Bound phage were eluted with 1 mg/mL trypsin in PBS with 1 mM $CaCl_2$ and orbital mixing.

Eluted phage from a single selection were added to 10 mL log-phase TG1 *E. coli* (grown in 2TY) and incubated for 1 h at 37°C, 150 rpm. Phage titre was determined by plating 100-fold serial dilutions of the infected culture on 2TY agar with 100 μg/mL carbenicillin and counting the resultant colonies. The remainder of each phage output was reduced in volume to ~1–2 ml by centrifugation at 4000 g for 10 min before plating on a bioassay dish (Nunc) containing the same medium and incubation at 30°C overnight. For the second and subsequent rounds of selection, 100 μL of filtered (0.45 μm) culture supernatant was used as the input phage sample. To determine the input titre, 10 μL of the supernatant was diluted 10,000-fold in PBS and used to infect 990 μL log-phase TG1 for 1 h at 37°C, 150 rpm, then plated and incubated as above.

## Solution-phase Phage Selection

Purified phage were generated using KM13 helper phage [40], packaging the pRpsp2 phagemid [23]. For a single selection experiment, 100 μL of Dynal M280 streptavidin magnetic beads (Life Technologies) were washed three times with 1 mL PBS and blocked for 1 h with 1 mL PBS-T with 3% (w/v) skim milk powder (PBS-TM) on a rotator at room temperature. The input phage were deselected in PBS-TM with 50 μL blocked streptavidin beads by incubation for 1 h on a rotator at room temperature. Streptavidin beads were removed from the input phage and biotinylated ligand was added to a final concentration of 1 μM for the first-round selections, 10 nM for the second-round selections and 1 nM for the third-round of selections. Solution-phase binding was allowed to proceed for 2 h on a rotator at room temperature. To recover ligand-bound phage complexes, a 50 μL aliquot of blocked streptavidin beads at 10 mg/ml was added to the selection for 10 min on a rotator at room temperature. The beads were then collected and washed eight times with 1 mL PBS-T. Bound phage were eluted by the addition of 1 mL PBS with 1 mM $CaCl_2$ and 1 mg/mL Trypsin and incubation for 1 h at 37°C with orbital mixing. Eluted phage were infected and titrated as indicated above for solid-phase selections.

## Phage ELISA

To generate monoclonal phage samples and perform phage ELISAs for screening selections from the third round of affinity maturation, individual clones were grown in 100 μL 2YT with 100 μg/mL carbenicillin in a 96-well tissue culture plate overnight at 37°C, 150 rpm. The resulting saturated cultures were seeded into 500 μL 2YT with 100 μg/mL carbenicillin in 96-deepwell (2 mL) blocks and incubated for approximately 5 h at 37°C, 250 rpm, or until turbidity was evident. The cultures were infected with KM13 helper phage at ~$10^9$ pfu and incubated at 37°C, 150 rpm, for 1 h. Media was exchanged by centrifugation for 10 min at 3,000 *g*, carefully discarding the supernatant and resuspension in fresh 2YT with 100 μg/mL carbenicillin and 50 μg/mL kanamycin. Plates were grown overnight at 25°C, 250 rpm. As a negative control, phage displaying the wild-type AspRS OB-fold domain were produced in a separate plate by the same method. To prepare the phage in each well for ELISA, 500 μL of PBS with 6% w/v skim milk powder was added to each well, incubated for 1 h, then the cells pelleted. Nunc Maxisorp 96-well plates were coated with HEL or BSA by incubation with 50 μL at 50 μg/mL overnight at 4°C. Each plate was rinsed 3 times with 300 μL PBS, blocked with 300 μL PBS containing 3% (w/v) skim milk powder (PBS-3M) for 1 h and washed again as above.

From the deepwell plate, 50 μL of the prepared phage solution was transferred in duplicate to the ligand and control plates then incubated at room temperature for 1 h. Anti-HEL human antibody (HuCAL Clone 11397, AbD Serotec, Oxford, UK) 50 μL in PBS-3M at 1 μg/mL was used as a positive control. A blank negative control well with 50 μL PBS-3M only was also used. The wells were washed three times with 300 μL PBS-T, then probed for 1 h with 50 μL monoclonal HRP-conjugated anti-M13 antibody (dilution 1:5000, GE Healthcare), or HRP-conjugated polyclonal goat anti-human Fc (dilution 1:1000, AbD Serotec) in the case of the positive control. The plates were washed 3 times as above. For visualisation, 50 μL 3,3′,5,5′-tetramethylbenzidine (TMB) peroxidase substrate (Sigma) per well was added and incubated for 30 min at room temperature, then the reaction stopped with 50 μL 2 M $H_2SO_4$ per well and read in in a spectrophotometer at 450 nM.

To generate monoclonal phage samples and perform phage ELISAs for screening all other selections, 96-well ELISA plates were coated with 5 μg/ml hen egg white lysozyme or 1% BSA, in PBS at 4°C overnight. After two washes with PBS, plates were blocked with PBS containing 5% (w/v) skim milk powder (PBS-5M) for 1 h at RT before helper phage ($10^9$ per well; ΔgIII helper phage VCSM13d3 [43]) were added in PBS-T containing 2.5% (w/v) skim milk powder. Plates were incubated for 2 hours at RT with agitation. After 10 washes with PBS, monoclonal HRP-conjugated anti-M13 antibody (GE Healthcare) diluted in PBS-5M was added and incubated for 1 h at RT. Plates were washed 4 times with water and HRP-conjugated rabbit-anti-mouse antibody (Pierce) in PBS-5M was added to the wells and incubated for 1 h at RT. Plates were washed 4 times with water and 50 μl substrate solution (1 mg/ml o-phenylene-diamine in PBS and 0.03% $H_2O_2$) added per well. The reaction was stopped after ~15 min by addition of 25 μl 2.5 M $H_2SO_4$ and the absorbance was recorded at 492 nm.

## Differential Scanning Fluorimetry

Measurement of changes in protein stability was performed using the general method for fluorescent detection with SYPRO Orange and an RT-PCR instrument, outlined elsewhere [29]. Briefly, protein samples were melted using a Corbett Rotogene 6000 RT-PCR machine, in duplicate at both 150 and 300 μg/mL, with SYPRO Orange at 100x concentration in a total volume of 25 μL per sample. The temperature was ramped at 1°C per minute, from 25°C to 99°C, with excitation at 470 nM and emission at 555 nM. Curve fitting to obtain the $T_m$ was performed with GraphPad Prism, using the following equation, where yMin and yMax are the minimum and maximum fluorescence values, h is Hill's slope and $T_m$ is the point of inflection:

$$y = yMin + \frac{yMax - yMin}{1 + 10^{h(Tm-x)}}$$

## Affinity Measurements

Dissociation constants of the various Obodies for HEL were measured using a Biacore 3000 surface plasmon resonance instrument (GE Healthcare). HEL was immobilised on to a CM5 sensor chip in small steps to give approximately 100 response units (RU). Obodies were measured in 2-fold dilution series, beginning with at least 10-times the $K_D$, where known. Analysis runs were performed at a standard 50 µL/min using HBS-EP running buffer (GE Healthcare). Chip surfaces were regenerated after each analyte injection with two 10 µL injections of 1 M NaCl. Where possible, kinetic data was modelled to obtain the $k_{on}/k_{off}$ using global fitting to a 1:1 Langmuir binding model in BIAevaluation software version 3.2, otherwise steady-state dissociation constants were modelled by plotting the maximum response at each concentration vs. OBody concentration in mol/L and fitting a Langmuir saturation binding curve to the data, using GraphPad Prism with the following equation, where Rmax is the maximum response, $K_D$ is the dissociation constant and for the linear component, m is the gradient and c is the y-intercept of the linear portion:

$$y=\frac{(R\max\cdot x)}{(K_D+x)+mx+c}$$

## Inhibition of HEL Enzymatic Activity

Assays were carried out in triplicate, using flat-bottomed 96-well untreated ELISA plates (Greiner). OBodies suspended in PBS pH 7.4 were first diluted 1:2 serially over 11 steps, starting with a final assay concentration of 50 µM. A 25 µL aliquot of each OBody dilution was either added to 25 µL hen egg-white lysozyme (Sigma) at 648 nM (162 nM final concentration in assay) or 25 µL PBS pH 7.4 and allowed to equilibrate at room temperature for 15 min. To each OBody dilution, a 50 µL aliquot of substrate solution containing inactivated *Micrococcus lysodeikticus* cells (Sigma, M3770) suspended in PBS pH 7.4 was added to a final concentration of 0.4 mg/mL and plates were incubated at room temperature for 45 min before reading the absorbance at 450 nm in a BMG Fluostar Optima plate reader. The data were normalised and processed using Prism v5.04 (Graphpad) software and non-linear regression performed using the following model [Y = Bottom+(Top-Bottom)/(1+10^((LogIC$_{50}$−X)*HillSlope))] to obtain $IC_{50}$ values.

## X-ray Crystallography

All crystal were grown using the sitting-drop vapour-diffusion method. Conditions were as follows: NL8, 0.2 M HEPES pH 7.3, 7% MPEG 5,000; AM1L10, 0.2 M HEPES pH 7.4, 9% MPEG 5,000; AM2EP06, 0.2 M HEPES pH 7.0, 13% MPEG 5,000; AM3L15, 0.2 M HEPES pH 7.4, 5% MPEG 5,000; AM3L09, 0.2 M HEPES pH 7.8, 9% MPEG 5,000. Purified, concentrated OBody in PBS pH 7.4 was combined with equimolar HEL in 10 mM sodium acetate pH 5.0 to yield concentrations of 40 mg/mL (OBody) and 45 mg/mL (HEL). Diffraction data was collected using the home source at the Maurice Wilkins Centre, University of Auckland, the Stanford Synchrotron Radiation Lightsource, or the Australian Synchrotron MX2 beamline. Data were integrated using MOSFLM [44], reduced using SCALA [45]/TRUNCATE [46], phased by molecular replacement with PHASER [47], refined using CCP4 [48], PHENIX [49] and COOT [50] and visualised with PyMOL [51]. TLS parameters for refinement were determined using the TLSMD server [52,53].

## Biotinylation of Hen Egg-white Lysozyme

Biotinylated HEL was produced using an amine-reactive biotinylation reagent from Qanta Biodesign, which attaches a biotin with a polyethylene linker via an N-hydroxysuccinimide group. The reagent was dissolved in dry dimethyl sulfoxide (DMSO) to make a 100 mM stock solution and stored at −20°C. The HEL was obtained from Roche and purified on a Superose 75 10/300 size exclusion column and incubated with a 5-fold molar excess of the biotinylation reagent for 1 h. The reaction was halted with the addition of ethanolamine to 1 mM. The sample was dialysed into ultrapure water and analysed for biotinylation levels by MALDI-TOF mass spectrometry.

## Protein Purification

Selected OBodies were subcloned by restriction/ligation into expression vector pRoEx Htb (Life Technologies) using generic OBody primers 005 and 006 (Table S1). Cultures for expression were inoculated from seeder cultures into baffled flasks of Luria Broth (LB) with 100 µg/mL carbenicllin and grown at 37°C. Induction of expression in host *E. coli* strain DH5α was at an $OD_{600}$ of 0.5–0.8 with the addition of IPTG to 1 mM, overnight at 37°C. OBodies were purified as follows: cell lysis by sonication in PBS, solution fraction separated by centrifugation at 15,000 *g* for 20 min, followed by nickel IMAC with a 5 mL HisTrap FF column (GE Healthcare), removal of the 6xHis tag by digestion with recombinant tobacco etch virus protease and size exclusion purification with a Superose 75 10/300 analytical-grade column (GE Healthcare).

## Accession Numbers

Coordinates and structure factors for the reported crystal structures have been deposited at the Protein Data Bank under accession codes 4GLA (NL8-HEL), 4GN3 (AM1L10-HEL), 4GN4 (AM2EP06-HEL), 4GLV (AM3L09-HEL) and 4GN5 (AM3L15-HEL).

## Supporting Information

**Figure S1 SPR equilibrium analysis of naïve HEL-binding OBodies.** Affinities were calculated using Graphpad Prism software and an equilibrium model of maximum response, as described in Materials and Methods.

**Figure S2 SPR equilibrium analysis of HEL-binding OBodies from AM1 selections.** Affinities were calculated using Graphpad Prism software and an equilibrium model of maximum response, as described in Materials and Methods.

**Figure S3 AM2EP06-HEL complex structure detail, showing AM2EP06 L4-HEL contacts.** Viewed from within the bound HEL (green Cα trace), L4 contacts with E95 at the top of the AM2EP06 (pale blue Cα trace) β-sheet interface are shown. Dashed yellow lines are potential hydrogen or electrostatic bonds, labelled with distances in angstroms.

**Figure S4 Sequence plan showing detailed construction method for the naïve library.** Blue arrows represent mutagenic oligonucleotides. Green arrows represent non-mutagenic oligonucleotides used for amplifying individual fragments and performing final gene reconstruction. Refer to Supporting Information Table S1 for oligonucleotide sequences.

**Figure S5 Sequence plan showing detailed construction method for the first affinity-maturation library.** Blue arrows represent mutagenic oligonucleotides. Green arrows represent non-mutagenic oligonucleotides used for amplifying individual fragments and performing final gene reconstruction. Refer to Table S1 for oligonucleotide sequences.

**Table S1 Oligonucleotide sequences.**

## Acknowledgments

The authors would like to thank Abby Sharrock for help with construction of pAS1. We thank David Goldstone for assistance with X-ray data collection and Meredith Ross for critical reading of the manuscript. We also thank Ted Baker and the Maurice Wilkins Centre for Molecular Biodiscovery for the use of their X-ray facilities.

## Author Contributions

Conceived and designed the experiments: MTL JDS VA MB. Performed the experiments: MTL JDS MB. Analyzed the data: MTL JDS MB VA JR. Contributed reagents/materials/analysis tools: JR. Wrote the paper: MTL JDS VA.

## References

1. Murzin AG (1993) OB(Oligonucleotide Oligosaccharide Binding)-fold - common structural and functional solution for nonhomologous sequences. Embo Journal 12: 861–867.
2. Murzin AG, Brenner SE, Hubbard T, Chothia C (1995) SCOP - a structural classification of proteins database for the investigation of sequences and structures. Journal of Molecular Biology 247: 536–540.
3. Arcus V (2002) OB-fold domains: a snapshot of the evolution of sequence, structure and function. Current Opinion in Structural Biology 12: 794–801.
4. Hoogenboom HR (2005) Selecting and screening recombinant antibody libraries. Nature Biotechnology 23: 1105–1116.
5. Dufner P, Jermutus L, Minter RR (2006) Harnessing phage and ribosome display for antibody optimisation. Trends in Biotechnology 24: 523–529.
6. Cobaugh CW, Almagro JC, Pogson M, Iverson B, Georgiou G (2008) Synthetic antibody libraries focused towards peptide ligands. Journal of Molecular Biology 378: 622–633.
7. Jespers L, Schon O, James LC, Veprintsev D, Winter G (2004) Crystal structure of HEL4, a soluble, refoldable Human V-H single domain with a germ-line scaffold. Journal of Molecular Biology 337: 893–903.
8. Binz HK, Pluckthun A (2005) Engineered proteins as specific binding reagents. Current Opinion in Biotechnology 16: 459–469.
9. Nygren PA, Skerra A (2004) Binding proteins from alternative scaffolds. Journal of Immunological Methods 290: 3–28.
10. Gebauer M, Skerra A (2009) Engineered protein scaffolds as next-generation antibody therapeutics. Current Opinion in Chemical Biology 13: 245–255.
11. Nord K, Gunneriusson E, Ringdahl J, Stahl S, Uhlen M, et al. (1997) Binding proteins selected from combinatorial libraries of an alpha-helical bacterial receptor domain. Nature Biotechnology 15: 772–777.
12. Orlova A, Magnusson M, Eriksson TLJ, Nilsson M, Larsson B, et al. (2006) Tumor Imaging using a picomolar affinity HER2 binding affibody molecule. Cancer Research 66: 4339–4348.
13. Malm M, Kronqvist N, Lindberg H, Gudmundsdotter L, Bass T, et al. (2013) Inhibiting HER3-mediated tumor cell growth with affibody molecules engineered to low picomolar affinity by position-directed error-prone PCR-like diversification. Plos One 8: e62791.
14. Beste G, Schmidt FS, Stibora T, Skerra A (1999) Small antibody-like proteins with prescribed ligand specificities derived from the lipocalin fold. Proceedings of the National Academy of Sciences of the United States of America 96: 1898–1903.
15. Steiner D, Forrer P, Pluckthun A (2008) Efficient Selection of DARPins with Sub-nanomolar Affinities using SRP Phage Display. Journal of Molecular Biology 382: 1211–1227.
16. Mouratou B, Schaeffer F, Guilvout I, Tello-Manigne D, Pugsley AP, et al. (2007) Remodeling a DNA-binding protein as a specific in vivo inhibitor of bacterial secretin PulD. Proceedings of the National Academy of Sciences of the United States of America 104: 17983–17988.
17. Cinier M, Petit M, Pecorari F, Talham DR, Bujoli B, et al. (2012) Engineering of a phosphorylatable tag for specific protein binding on zirconium phosphonate based microarrays. Journal of Biological Inorganic Chemistry 17: 399–407.
18. Lee SC, Park K, Han J, Lee JJ, Kim HJ, et al. (2012) Design of a binding scaffold based on variable lymphocyte receptors of jawless vertebrates by module engineering. Proc Natl Acad Sci U S A 109: 3299–3304.
19. Hogbom M, Eklund M, Nygren PA, Nordlund P (2003) Structural basis for recognition by an in vitro evolved affibody. Proceedings of the National Academy of Sciences of the United States of America 100: 3191–3196.
20. Schonfeld D, Matschiner G, Chatwell L, Trentmann S, Gille H, et al. (2009) An engineered lipocalin specific for CTLA-4 reveals a combining site with structural and conformational features similar to antibodies. Proceedings of the National Academy of Sciences of the United States of America 106: 8198–8203.
21. Binz HK, Amstutz P, Kohl A, Stumpp MT, Briand C, et al. (2004) High-affinity binders selected from designed ankyrin repeat protein libraries. Nature Biotechnology 22: 575–582.
22. Kiefer F, Arnold K, Kunzli M, Bordoli L, Schwede T (2009) The SWISS-MODEL Repository and associated resources. Nucleic Acids Research 37: D387–D392.
23. Beekwilder J, Rakonjac J, Jongsma M, Bosch D (1999) A phagemid vector using the E-coli phage shock promoter facilitates phage display of toxic proteins. Gene 228: 23–31.
24. Monchois V, Abergel C, Sturgis J, Jeudy S, Claverie JM (2001) Escherichia coli ykfE ORFan gene encodes a potent inhibitor of C-type lysozyme. Journal of Biological Chemistry 276: 18437–18441.
25. Kim DE, Chivian D, Baker D (2004) Protein structure prediction and analysis using the Robetta server. Nucleic Acids Research 32: W526–W531.
26. Zhu XL, Mitchell JC (2011) KFC2: A knowledge-based hot spot prediction method based on interface solvation, atomic density, and plasticity features. Proteins-Structure Function and Bioinformatics 79: 2671–2683.
27. Kunkel TA (1985) Rapid and efficient site-specfic mutagenesis without phenotypic selection. Proceedings of the National Academy of Sciences of the United States of America 82: 488–492.
28. Lorenzen S, Peters B, Goede A, Preissner R, Frommel C (2005) Conservation of cis prolyl bonds in proteins during evolution. Proteins-Structure Function and Bioinformatics 58: 589–595.
29. Niesen FH, Berglund H, Vedadi M (2007) The use of differential scanning fluorimetry to detect ligand interactions that promote protein stability. Nature Protocols 2: 2212–2221.
30. Koide A, Bailey CW, Huang XL, Koide S (1998) The fibronectin type III domain as a scaffold for novel binding proteins. Journal of Molecular Biology 284: 1141–1151.
31. Xu D, Lin SL, Nussinov R (1997) Protein binding versus protein folding: The role of hydrophilic bridges in protein associations. Journal of Molecular Biology 265: 68–84.
32. Norel R, Lin SL, Wolfson HJ, Nussinov R (1995) Molecular-surface complementarity at protein-protein interfaces - the critical role played by surface normals at well placed, sparse, points in docking. Journal of Molecular Biology 252: 263–273.
33. Norel R, Petrey D, Wolfson HJ, Nussinov R (1999) Examination of shape complementarity in docking of unbound proteins. Proteins-Structure Function and Genetics 36: 307–317.
34. Lin SL, Tsai CJ, Nussinov R (1995) A study of 4-helix bundles - investigating protein-folding via similar architectural motifs in protein cores and in subunit interfaces. Journal of Molecular Biology 248: 151–161.
35. Xu D, Tsai CJ, Nussinov R (1997) Hydrogen bonds and salt bridges across protein-protein interfaces. Protein Engineering 10: 999–1012.
36. Srivastava AK, Sharma Y, Chary KVR (2010) A Natively Unfolded beta gamma-Crystallin Domain from Hahella chejuensis. Biochemistry 49: 9746–9755.
37. Lo Conte L, Chothia C, Janin J (1999) The atomic structure of protein-protein recognition sites. Journal of Molecular Biology 285: 2177–2198.
38. Gallicchio E, Lapelosa M, Levy RM (2010) Binding Energy Distribution Analysis Method (BEDAM) for Estimation of Protein Ligand Binding Affinities. Journal of Chemical Theory and Computation 6: 2961–2977.
39. Lowman HB, Bass SH, Simpson N, Wells JA (1991) Selecting high-affinity binding-proteins by monovalent phage display. Biochemistry 30: 10832–10838.
40. Kristensen P, Winter G (1998) Proteolytic selection for protein folding using filamentous bacteriophages. Folding & Design 3: 321–328.
41. Cadwell RC, Joyce GF (1992) Randomization of genes by PCR mutagenesis. PCR Methods Appl 2: 28–33.
42. Sidhu SS, Lowman HB, Cunningham BC, Wells JA (2000) Phage display for selection of novel binding peptides. Applications of Chimeric Genes and Hybrid Proteins, Pt C. pp. 333–363.
43. Rakonjac J, Jovanovic G, Model P (1997) Filamentous phage infection-mediated gene expression: construction and propagation of the gIII deletion mutant helper phage R408d3. Gene 198: 99–103.
44. Leslie AGW (1992) Recent changes to the MOSFLM package for processing film and image plate data. Joint CCP4+ ESF-EAMCB Newsletter on Protein Crystallography 26.
45. Evans P (2006) Scaling and assessment of data quality. Acta Crystallographica Section D-Biological Crystallography 62: 72–82.
46. French S, Wilson KS (1978) On the treatment of negative intensity observations. Acta Crystallographica A34: 517–525.
47. McCoy AJ, Grosse-Kunstleve RW, Adams PD, Winn MD, Storoni LC, et al. (2007) Phaser crystallographic software. Journal of Applied Crystallography 40: 658–674.
48. Bailey S (1994) The CCP4 suite - programs for protein crystallography. Acta Crystallographica Section D-Biological Crystallography 50: 760–763.

49. Adams PD, Afonine PV, Bunkoczi G, Chen VB, Davis IW, et al. (2010) PHENIX: a comprehensive Python-based system for macromolecular structure solution. Acta Crystallographica Section D-Biological Crystallography 66: 213–221.
50. Emsley P, Cowtan K (2004) Coot: model-building tools for molecular graphics. Acta Crystallographica Section D-Biological Crystallography 60: 2126–2132.
51. Schrodinger LLC (2010) The PyMOL Molecular Graphics System, Version 1.3r1.
52. Painter J, Merritt EA (2006) Optimal description of a protein structure in terms of multiple groups undergoing TLS motion. Acta Crystallographica D62: 439–450.
53. Painter J, Merritt EA (2006) TLSMD web serve for the generation of multi-group TLS models. Journal of Applied Crystallography 39: 109–111.

5

# Industrial Systems Biology of *Saccharomyces cerevisiae* Enables Novel Succinic Acid Cell Factory

**José Manuel Otero[1,2¤a], Donatella Cimini[3], Kiran R. Patil[2¤b], Simon G. Poulsen[2], Lisbeth Olsson[1,2], Jens Nielsen[1,2]***

**1** Department of Chemical and Biological Engineering, Chalmers University of Technology, Gothenburg, Sweden, **2** Center for Microbial Biotechnology, Department of Systems Biology, Technical University of Denmark, Lyngby, Denmark, **3** Department of Experimental Medicine, Second University of Naples, Naples, Italy

## Abstract

*Saccharomyces cerevisiae* is the most well characterized eukaryote, the preferred microbial cell factory for the largest industrial biotechnology product (bioethanol), and a robust commerically compatible scaffold to be exploitted for diverse chemical production. Succinic acid is a highly sought after added-value chemical for which there is no native pre-disposition for production and accmulation in *S. cerevisiae*. The genome-scale metabolic network reconstruction of *S. cerevisiae* enabled *in silico* gene deletion predictions using an evolutionary programming method to couple biomass and succinate production. Glycine and serine, both essential amino acids required for biomass formation, are formed from both glycolytic and TCA cycle intermediates. Succinate formation results from the isocitrate lyase catalyzed conversion of isocitrate, and from the α-keto-glutarate dehydrogenase catalyzed conversion of α-keto-glutarate. Succinate is subsequently depleted by the succinate dehydrogenase complex. The metabolic engineering strategy identified included deletion of the primary succinate consuming reaction, Sdh3p, and interruption of glycolysis derived serine by deletion of 3-phosphoglycerate dehydrogenase, Ser3p/Ser33p. Pursuing these targets, a multi-gene deletion strain was constructed, and directed evolution with selection used to identify a succinate producing mutant. Physiological characterization coupled with integrated data analysis of transcriptome data in the metabolically engineered strain were used to identify 2$^{nd}$-round metabolic engineering targets. The resulting strain represents a 30-fold improvement in succinate titer, and a 43-fold improvement in succinate yield on biomass, with only a 2.8-fold decrease in the specific growth rate compared to the reference strain. Intuitive genetic targets for either over-expression or interruption of succinate producing or consuming pathways, respectively, do not lead to increased succinate. Rather, we demonstrate how systems biology tools coupled with directed evolution and selection allows non-intuitive, rapid and substantial re-direction of carbon fluxes in *S. cerevisiae,* and hence show proof of concept that this is a potentially attractive cell factory for over-producing different platform chemicals.

**Editor:** Scott E. Baker, Pacific Northwest National Laboratory, United States of America

**Funding:** This work was in part supported by a Merck Doctoral Fellowship to J.M.O. from Merck Research Labs, Merck & Co., Inc. USA. No additional external funding was received for this study. The funders had no role in study design, data collection and analysis, decision to publish, or preparation of the manuscript.

**Competing Interests:** The authors have the following interests: Jose Manuel Otero is employed by Merck, the funder of this study. During his PhD study in Nielsen lab, he was financed by a PhD scholarship by Merck. The research work of his PhD is of no commercial interest to Merck. There are no patents, products in development or marketed products to declare.

* E-mail: nielsenj@chalmers.se

¤a Current address: Vaccine & Biologics Process Development, Bioprocess Research & Development, Merck Research Labs, West Point, Pennsylvania, United States of America
¤b Current address: Structural and Computational Biology Unit, European Molecular Biology Laboratory, Heidelberg, Germany

## Introduction

Industrial biotechnology is a promising alternative to traditional petrochemical production of chemicals focused on developing commercially sustainable and environmentally favorable processes [1]. Metabolic engineering, the directed genetic modification of cellular reactions, aims to change the metabolic architecture of microorganisms to efficiently produce target chemicals [2]. Although examples of metabolic engineering successes exist, there has yet to be developed a pipeline where preferred industrial hosts are rapidly engineered to produce a non-native accumulating target metabolite. Recent advances in systems biology enabled *in silico* genome-scale metabolic network reconstructions to guide metabolic engineering strategies [1], [3], [4]. Here we describe a pipeline where a microbial strain was constructed, physiologically characterized, and genomic tools were used to verify the predictions. An essential part of the pipeline is the use of genome-scale metabolic models for initial guiding of the metabolic engineering, which has been shown to be useful also in earlier studies [3], [5], [6]. This approach was repeated and complemented with traditional directed evolution and selection until a proof of concept microbial cell factory was reached. This pipeline resulted in the construction of a non-intuitive *Saccharomyces cerevisiae* cell factory over-producing succinic acid, a building block chemical.

*S. cerevisiae* is the most well characterized eukaryote and is unique in its broad application as an industrial production platform for a large portfolio of products including foods and beverages, bioethanol, vaccines, and therapeutic proteins [1].

Many systems biology tools, including high-throughput genome sequencing, transcriptional profiling, metabolomics, carbon flux estimations, proteomics, *in silico* genome-scale modeling, and bioinformatics driven data integration were first applied to *S. cerevisiae* [4]. Metabolic engineering has benefited from each of these tools; however, relatively few examples exist where cumulative integration has resulted in a generalized pipeline, in particular for the production of a target compound that the organism does not accumulate significantly naturally.

Succinic acid, systematically identified as butanedioic acid ($pKa_1$ 4.21, $pKa_2$ 5.72), is a value-added chemical building block, with an estimated 15,000 t/year world-wide demand predicted to expand to commodity chemical status with 270,000 t/year [7], [8] representing a potential >2 billion USD annual market. There are several elegant examples of bio-based production of succinate in *Anaerobiospirillium succiniciproducens*, *Actinobacillus succinogenes*, *Succinivibrio dextrinosolvens*, *Corynebacterium glutanicum*, *Prevotella ruminocola*, a recently isolated bacterium from bovine rumen, *Mannheimia succiniciproducens*, and a metabolically engineered succinic acid over-producing *E. coli* [7], [8], [9], [10], [11], [12]. All of the hosts described are prokayotic that grow at neutral pH, and consequently secrete the salt, succinate, requiring a cost-intensive acidification and precipitation to reach the desired succinic acid. This concern is not specific to succinic acid production, but rather universal when considering organic acid producing microbial cell factories [13]. *S. cerevisiae* represents a well-established, generally regarded as safe, robust, scalable (1L to 100,000L) industrial production host capable of growth on diverse carbon sources, chemically defined medium, both aerobic and anaerobic, and a wide pH operating range (3.0–6.0). Unlike the hosts described above, succinate is not naturally produced by *S. cerevisiae*; but as there are many factors of importance for the choice of a microbial cell factory it is not uncommon that the chosen cell factory lacks predisposition to produce the target chemical of choice [14]. As industrial biotechnology progresses forward, and the concept of biorefineries are gaining increased importance, platform technologies including microbial cell factories that can plug-and-play into existing infrastructures must be developed [15]. *S. cerevisiae* is uniquely positioned as a platform technology as it is already used widely for bioethanol production, but also because of the extensive library of genetic engineering tools, a very well annotated genome, many omics tools available, and well established complimentary approaches for directed evolution and selection. We therefore addressed the question whether it is possible to metabolically engineer *S. cerevisiae* such that the carbon fluxes are redirected towards succinic acid, and hereby establish proof-of-concept of using this yeast as a general cell factory platform for chemical production. The final strain emerging from this study requires significant further metabolic engineering and process development prior to consideration for commercialization, but the approach and integration of methods demonstrated supports the hypothesis that highly regulated central carbon metabolism in ideal production hosts can be reconfigured to produce target chemicals, relatively quickly and with minimal resources.

## Results

In the present study, our objective was to evaluate the use of genome-scale metabolic models for *in silico* identification of gene deletion targets. We therefore used results from a previous study where an evolutionary programming method, termed OptGene, was developed for identification of deletion targets to couple biomass formation (or other biological objective function that is applicable to the organism under question) with the design objective function, such as yield (or other linear/non-linear objective) [16]. These results guided the metabolic engineering strategy described in Figure 1. Glycine, serine, and threonine, all representing essential amino acids required for biomass formation, may be formed from both glycolytic and tricarboxylic acid cycle intermediates. Succinate formation results from the isocitrate lyase, Icl1p, catalyzed conversion of isocitrate to equimolar glyoxylate and succinate, and from the α-keto-glutarate dehydrogenase complex, Kgd1p/Kgd2p/Lpd1p, catalyzed conversion of α-keto-glutarate to equimolar succinate, with a net production of $CO_2$, NADH, and ATP. Succinate is subsequently depleted by the succinate dehydrogenase complex, Sdh1p/Sdh2p/Sdh3p/Sdh4p to equimolar fumarate with the net production of protonated ubiquinone. The metabolic engineering strategy identified by OptGene included deletion of the primary succinate consuming reaction encoded by *Sdh3* (cytochrome b subunit of the succinate dehydrogenase complex, essential for function), and interruption of glycolysis derived serine by deletion of 3-phosphoglycerate dehydrogenase, Ser3p/Ser33p (isoenzymes). The remaining pathway for serine synthesis must originate from glycine, and glycine synthesis is largely derived from the alanine:pyruvate aminotransferase, Agx1p, converting glyoxylate and alanine to glycine and pyruvate. With this strategy, glycine and serine biomass requirements are directly coupled to succinate formation via the glyoxylate cycle. Substantial succinate accumulation (defined as >0.1 g $L^{-1}$) in the culture broth is not observed in wild-type *S. cerevisiae*, and deletion of *sdh3* has not resulted in appreciable succinate accumulation [17]; a conclusion also found by chemical inhibition of the succinate dehydrogenase complex with titration of malonate (Figure S1), a chemical inhibtor of this complex [18].

The mutant resulting from the *in silico* strategy, referred to as 8D (*Δsdh3 Δser3 Δser33*), required supplementation with 500 mg $L^{-1}$ glycine to be able to grow. When evaluated in well controlled, aerobic, batch stirred tank reactors supplemented with 20 g $L^{-1}$ glucose in chemically defined medium, it exhibited a 13-fold improvement in succinate secreted titer (0.03 v 0.40 g $L^{-1}$), 14-fold improvement in succinate biomass yield (0.01 v 0.14 g-succinate g-biomass$^{-1}$), and a modest 33% reduction in the specific growth rate. Thus, the *in silico* guided metabolic engineering strategy was shown to work, representing a proof-of-concept of the use of model guided metabolic engineering. However, in order to obtain a protothropic strain directed evolution was employed to screen and select for 8D mutants that did not require glycine supplementation. Specifically, six consecutive shake flask cultivations in media supplemented with decreasing concentrations of glycine, from an initial 500 mg $L^{-1}$ to 0 mg $L^{-1}$ (see Figure 2) were performed. The resulting strain demonstrated a 7.7-fold improvement in succinate yield on biomass (0.09 v 0.69 g-succinate g-biomass$^{-1}$), strongly suggesting the direct coupling of glycine formation from glyoxylate and succinate formation. The resulting strain had a relatively low specific growth rate, 0.03 $h^{-1}$, and was therefore subsequently cultivated in shake flasks and transferred across six shake flasks (only first three shake flasks shown in Figure 2) to improve the specific growth rate. Finally, a specific growth rate of 0.14 $h^{-1}$ was reached, however, resulting in a decreased succinate yield on biomass (0.69 v 0.27 g-succinate g-biomass$^{-1}$). The final strain, referred to as 8D Evolved, was shown to exhibit a 60-fold improvement in biomass coupled succinate production (0.01 v 0.30 g-succinate g-biomass$^{-1}$), and 20-fold improvement in succinate titer (0.03 v 0.60 g $L^{-1}$) relative to the reference strain when grown in aerobic batch cultivations.

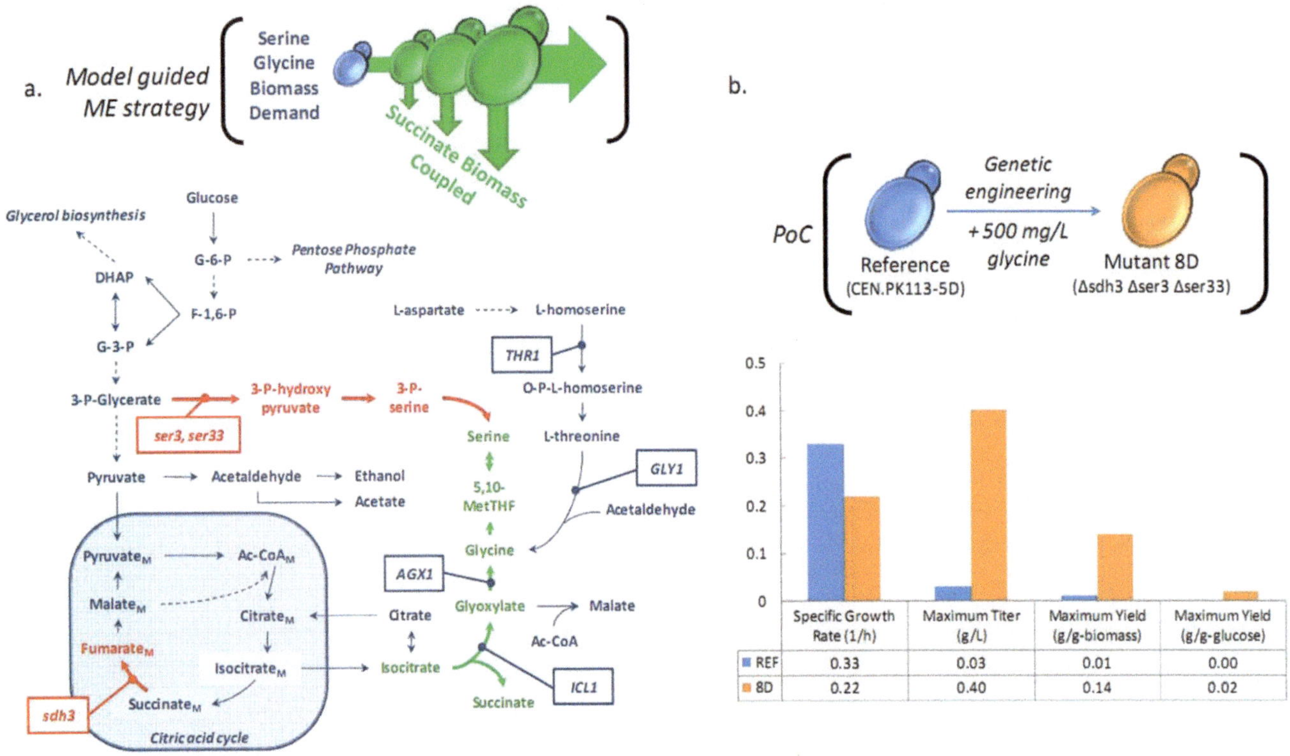


**Figure 1. Proof-of-concept: Successful metabolic engineering strategy guided by genome-scale metabolic modeling.** Panel a shows the central carbon metabolism of *S. cerevisiae*, and the model-guided metabolic engineering strategy for succinate over-production. Succinate production is directly coupled to biomass formation based on three gene deletions: *sdh3* (cytochrombe b subunit of succinate dehydrogenase complex), and *ser3/ser33* (3-phosphoglycerate dehydrogenase isoenzymes). The remodeling of central carbon flux towards succinate requires minimizing the conversion of succinate to fumarate, and forcing the biomass-required amino acids L-glycine and L-serine to be produced from glyoxylate pools. Production of glyoxylate results from isocitrate conversion by Icl1p, producing equimolar succinate. As the biomass yield increases, the demand for L-glycine and L-serine increase proportionally, driving biomass-coupled succinate production. Legend: native reactions (blue solid line), lumped native reactions (blue dashed line), interrupted reactions (red solid line), up-regulated reactions (green solid line). Panel b demonstrates the proof of concept. The reference strain and genetically engineered mutant strain, 8D, supplemented with 500 mg $L^{-1}$ glycine were physiologically characterized in 2L well-controlled stirred-tank fermentations. There was a 13.3X improvement in succinate titer.

To investigate the apparent decoupling of succinate coupled biomass formation, and potentially identify second-round metabolic engineering strategies, the transcriptome was measured in aerobic, glucose-limited, mid-exponential phase grown batch cultivations of 8D Evolved and the reference strain. Continous cultivations, both carbon-limited and nitrogen-limited chemostats were attempted with the 8D Evolved mutant; however, in both cases steady-state was not reached and wash-out occured, even at relatively low dilution rates ($D = 0.03\ h^{-1}$ compared to $\mu_{max} = 0.14\ h^{-1}$). It was expected that 8D Evolved would not support cultivation in carbon-limited continuous culture due to the down-regulation of the TCA cycle (*Δsdh3*), and consequently, reduced capacity for respiratory metabolism and oxidative phosphorylation. Therefore, batch cultivations were employed acknowledging the significant differences in specific growth rate (0.33 v 0.13 $h^{-1}$), and glucose uptake rate (90 v 26 C-mmol g-$DCW^{-1}\ h^{-1}$), while maintaining relatively similar biomass yields (0.18 v 0.19 C-mol biomass C-mol $glucose^{-1}$).

Several studies have shown that significant differences in specific growth rate directly impact transcriptome interpretation, with anywhere between 268 and 2400 genes classified as potentially growth-related [19], [20], [21]. Previously generated continuous cultivation transcriptome data for both carbon-limited (glucose, respiratroy growth) and nitrogen-limited (ammonium sulfate, respiro-fermentative growth) conditions at dilution rates of 0.03, 0.1, and 0.2 $h^{-1}$ were therefore used to identify statistically differentially expressed growth-related genes [21]. A total of 6 and 7 differentially expressed genes were identified within the carbon-limited and nitrogen-limited data sets as being growth-related ($p\text{-}value_{B\text{-}H} < 0.1$, $n = 3$ at each dilution rate), respectively, and a total of 66 differentially expressed genes were identified when comparing carbon-limited and nitrogen-limited data sets, paired at each dilution rate ($p\text{-}value_{B\text{-}H} < 0.1$, $n = 3$ at each dilution rate). Of the total 2406 differentially expressed genes between the 8D Evolved and reference strain ($p\text{-}value_{B\text{-}H} < 0.01$, $|log\text{-}fold\ change| > 0.5$, $n = 3$ biological replicates, $n = 2$ DNA microarray duplicates), 36 unique growth-related genes were identified suggesting that few of the genes with a significant change in transcription in 8D Evolved are due to changes in the specific growth rate (see Figure 3). However, a total of 8 of the top 20 $p\text{-}value_{B\text{-}H}$ ranked differentially expressed genes identified from pair-wise comparison of 8D Evolved and the reference strain, are growth-related genes (*ARO9, SER3, JLP1, HMALPHA1, ARO10, MFALPHA2,* and two uncharacterized genes, *YPL033c* and *YLR267w*).

The top 2000 (there were no metabolic genes in the remaining 406 genes nor were there any biological process annotations available as determined by gene ontology, and therefore they were not included in further analysis) differentially expressed genes were selected for further analysis, and after removal of the 36 growth-related genes, a list of 1964 genes was submitted for metabolic

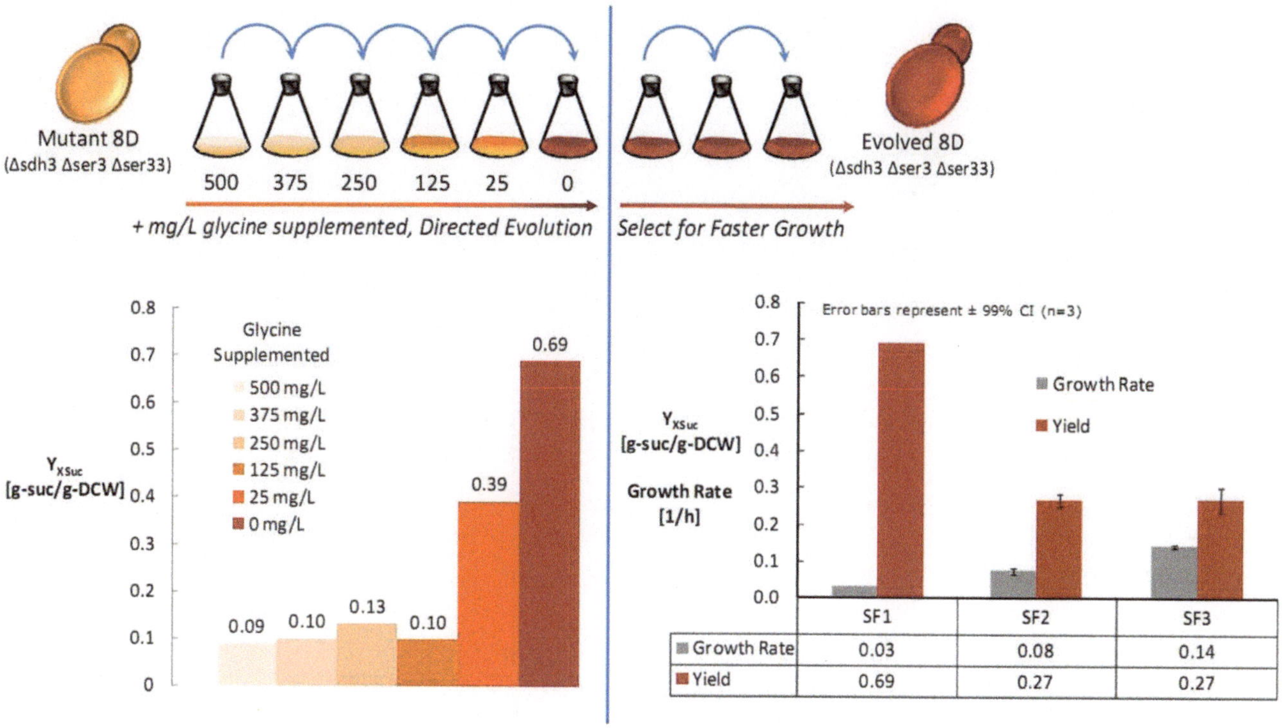


**Figure 2. Metabolic engineering enhanced by directed evolutions.** Cell populations were transferred across six shake flask cultures until a glycine prototroph was isolated. Subsequently, successive cultures were used to select for faster growth. From the final shake flask (SF3) the strain Evolved 8D was isolated. The succinate yield on biomass is plotted for each shake flask culture, demonstrating a 7.8X increase. The right plot shows the profile of specific growth rate and succinate yield on biomass for the final selection of faster growing cells.

pathway visualization and characterization to the Expression Viewer [22] available at the Yeast Genome Database [23] (see Figure 3). The log-fold change of the 8D Evolved:Reference expression ratio was mapped onto the metabolic map of *S. cerevisiae* strain S288c, version 12.0, composed of 140 pathways, 925 enzymatic reactions, and a total of 675 compounds (see Figure S2). A total of 315 genes mapped to a specific metabolic pathway on the expression viewer, with a mean log-fold expression ratio value of 0.3±1.3 (*n* = *315*, ± SD).

Three biologial insights were immediately apparent (see Figure S2). First, *SDH3*, *SER3*, and *SER33* had negative log-fold expression ratios (log-fold change <-8.0) confirming the gene deletions targeted in the 8D strain and the maintained low expression through the directed evolution. Second, when examining the glycine, serine, and threonine metabolism, *AGX1* was 4.3 log-fold change upregulated in the 8D Evolved strain, confirming significant upregulation of glycine synthesis from glyoxylate pools, as predicted by the original metabolic engineering strategy. However, there was no upregulation of *SHM2*, *SHM1*, the genes encoding pathways for L-serine formation from L-glycine pools. Most surprisingly *GLY1*, encoding threonine aldolase, was significantly up-regulated (log-fold change 1.6). In the genome-scale metabolic network reconstructions of *S. cerevisiae* iFF708, iND750, and iIN800, upon which the 8D metabolic engineering strategy is based, Gly1p encodes the irreversible conversion of glycine and acetaldehyde to threonine [24], [25], [26], leading to the prediction that threonine biosynthesis from glycolytic intermediates could be down-regulated, and provided for from glycine pools. This consequently leads to a greater biomass-coupled drive for glyoxylate synthesis from isocitrate, yielding equimolar succinate. Levaraging this over-all strategy, another *S. cerevisiae* mutant was constructed, referred to as 20G (*Δsdh3, Δser3, Δthr1*), where Thr1p, encoding homoserine kinase that is required for threonine biosynthesis, was deleted. However, this strain required threonine supplementation and after several extensive attempts at adaptive evolution, the threonine auxotrophy persisted, suggesting the irreversibility of *Gly1* was incorrect and the aldolase strongly favors glycine formation (Figure S3). The significant up-regulation of *Gly1* therefore provides a strong hypothesis for why 8D Evolved had an attenuation of succinate production, even under increasing specific growth rate, suggesting a decoupling of biomass coupled succinate production. It should be noted that in the most recent update of the genome-scale metabolic reconstruction of *S. cerevisiae*, iMM904, the directionality of *Gly1* was corrected to now indicate threonine irreversible conversion to glycine and acetaldehyde [27].

The transcriptome not only provides for a global, rapid, and quantitative assessment of the predicted *in silico* metabolic engineering strategy and insight into the genetic engineering modifications that result from directed evolution and selection, but also provides a source for identification of second round metabolic engineering targets not previously predicted. Several targets were identified, but of particlar interest was *ICL1*, encoding isocitrate lyase, converting isocitrate to glyoxylate and succinate in equimolar concentrations. All tricarboxylic acid cycle genes are up-regulated, with the exception of *SDH3* (target gene deletion), and *ICL1*, providing a clear metaboic engineering target for up-regulation in the 8D Evolved strain. Therefore, native *ICL1* was PCR amplified and cloned into the 2 μm ori plasmid pRS426CT containing the strong constitutive *TEF1* promoter and *CYC1* terminator [28], and then transformed into the reference, 8D, and 8D Evolved strain (strains transformed with the constructed plasmid pRS426T-ICL1-C are referred to as "with pICL1"). All

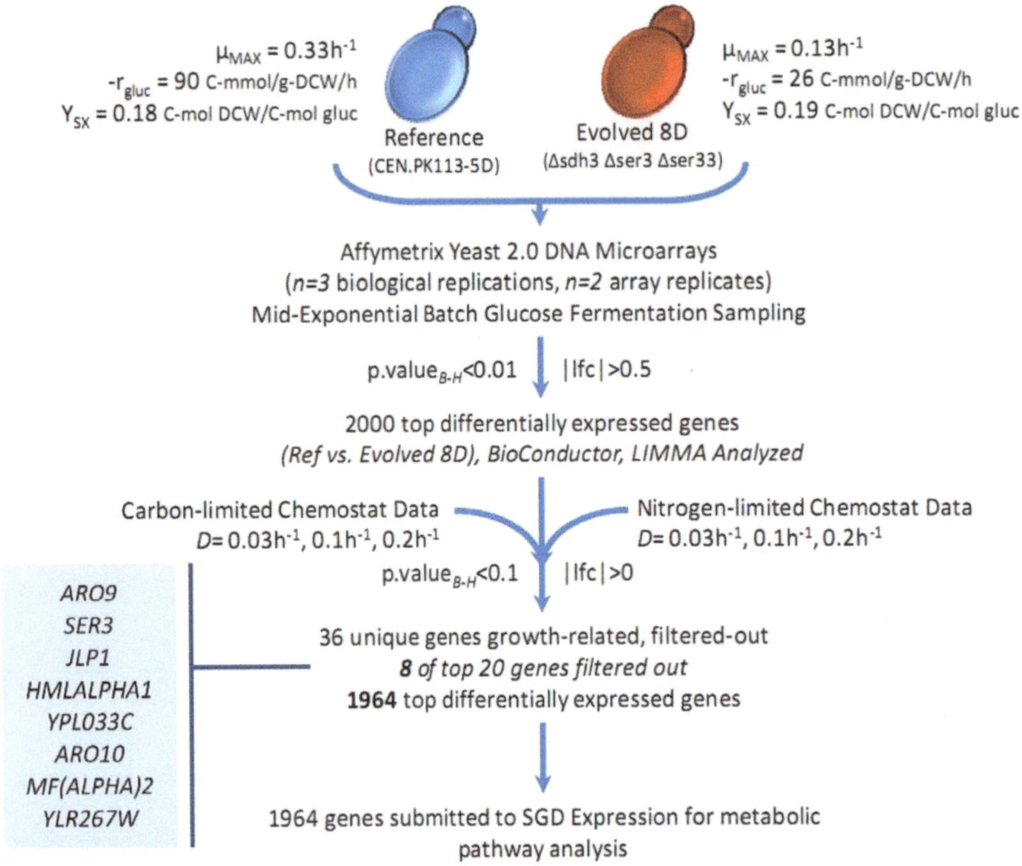


**Figure 3. Transcriptome guided metabolic engineering – Analysis.** Affymetrix Yeast 2.0 DNA microarrays were used for transcriptome analysis of each strain cultured in well-controlled glucose batch fermentations. The top 2000 differentially expressed genes had an adjusted *p.value<0.01* and *log-fold change (lfc)>0.5*. A carbon-limited and nitrogen-limited chemostat transcriptome data set using the reference strain, surveyed at dilution rates (*D*) of 0.03, 0.1, and 0.2 $h^{-1}$ was used to determine which genes are growth-related under each condition. A total of 36 unique growth-related genes were identified from statistical analysis of each data set and with a total of 8 growth-related genes being among the top 20 differentially expressed genes between the reference and evolved 8D strain. After removal of the 36 genes, a total of 1964 genes were carried further for pathway analysis.

strains were evaluated in aerobic, glucose-supplemented batch fermentations, and only 8D Evolved with pICL1 exhibited a change in succinate production (see Figure 4). Specifically, the succinate titer, biomass yield, and glucose yield were 0.90 g $L^{-1}$, 0.43 g-succinate g-biomass$^{-1}$, and 0.05 g-succinate g-glucose$^{-1}$, respectively, representing a 1.5-fold, 1.4-fold, 1.7-fold improvement over 8D, respectively (see Figure 4).

## Discussion

A *S. cerevisiae* strain capable of succinate production, requiring redirection of carbon flux from typically produced $C_2$ (ethanol, acetate) and $C_3$ (glycerol, pyruvate) over-flow metabolites to the target $C_4$ succinic acid was achieved through metabolic engineering, requiring integration of systems biology methods and directed evolution. Clearly, the resulting strain (8D Evolved with pICL1), while being a successful demonstration of a multi-round metabolic engineering strategy, still requires significant process development and further enhancement to compete commercially with existing bacterial platforms.

The resulting strain, 8D Evolved with pICL1, represents a 30-fold improvement in succinate titer, and a 43-fold improvement in succinate yield on biomass, with only a 2.8-fold decrease in the specific growth rate compared to the reference strain. Despite success of using simple stoichiometric-based calculations for driving metabolic engineering, it is interesting to note that regulatory mechanisms not captured in these models are likely playing a significant role in the succinate production observed. The biomass requirements for glycine and serine are 0.290 and 0.185 mmol g-DCW$^{-1}$ [24]. Assuming that all glycine, and all glycine and serine combined demands are supplied from the glyoxylate pool, then the theoretical production of succinate would amount to 0.034 and 0.056 g-succinate g-DCW$^{-1}$, respectively. The 8D and 8D Evolved strains are producing 0.30 and 0.43 g-succinate g-biomass$^{-1}$, respectively, suggesting a nearly 8-fold higher succinate production than required to meet biomass amino acid demands. A potentially 3$^{rd}$ metabolic engineering target would be deletion of *GLY1* to further minimize alternative biosynthetic routes of glycine production, thereby isolating all glycine production to be dependent on glyoxylate formation, and consequently succinate formation. Yet, it's clear that any increase in succinate formation would not be due to biomass requirements, but rather regulatory (e.g., non-stoichiometric driven) mechanisms. Therefore, while the strategy presented and demonstrated here is likely to be a major component of an over-all succinate production cell factory, complimentary strategies focussing on the other major succinate production pathway, TCA cycle, will be required. Examples of malic acid production, that included

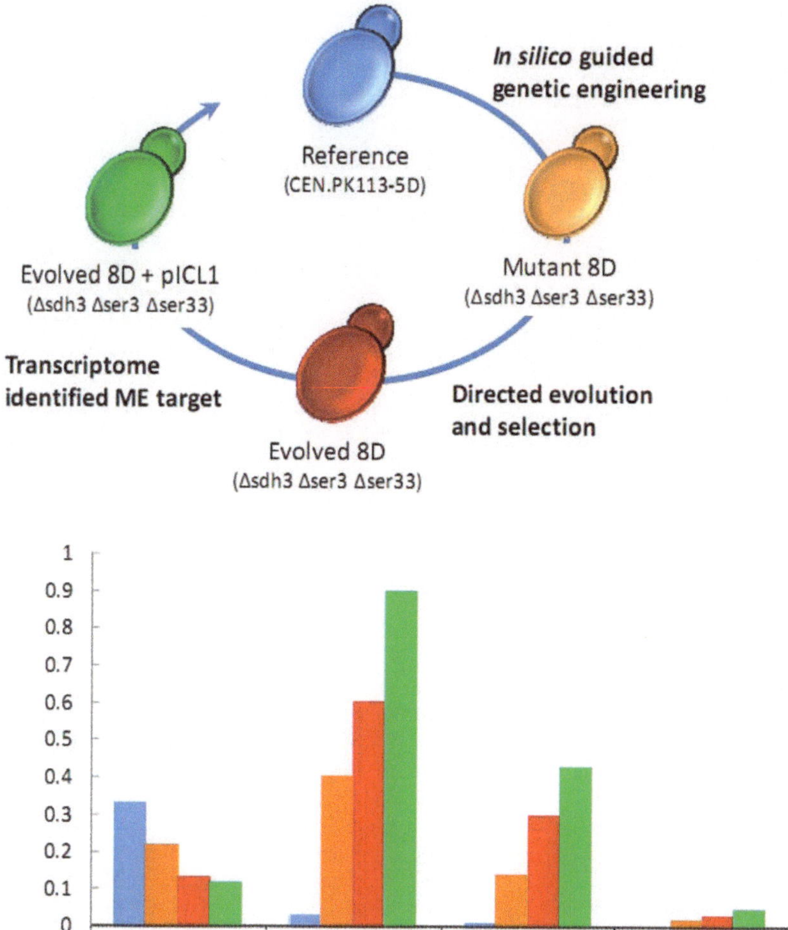


| | Specific Growth Rate (1/h) | Maximum Titer (g/L) | Maximum Yield (g/g-biomass) | Maximum Yield (g/g-glucose) |
|---|---|---|---|---|
| REF | 0.33 | 0.03 | 0.01 | 0.00 |
| 8D | 0.22 | 0.40 | 0.14 | 0.02 |
| 8D Evolved | 0.13 | 0.60 | 0.30 | 0.03 |
| 8D Evolved + pICL1 | 0.12 | 0.90 | 0.43 | 0.05 |

**Figure 4. Summary of succinate microbial cell factory construction.** The specific growth rate (1/h), maximum succinate titer (g/L), maximum succinate yield on biomass (g/g-biomass), and maximum yield on glucose (g/g-glucose) are reported for the reference strain, 8D, 8D evolved, and 8D evolved with pICL1. A 43-fold improvement in succinate yield on biomass was observed across the full cycle of metabolic engienering that included *in silico* guided approaches, directed evolution, and transcriptome based identification of a 2nd round of metabolic engineering targets.

engineering of pyruvate carboxylation (overexpression of *PYC2*), oxaloacetate reduction (overexpression of *MDH3*), and malate export (functional expression of the non-native Sp*MAE1*), resulted in a *S. cerevisiae* strain capable of producing 59 g-malate $L^{-1}$ and 0.42 mol malate mol-glucose$^{-1}$ [29]. A similar approach, requiring yet further engineering and understanding of the reductive TCA cycle to convert malate to succinate is likely required, but a major hurdle with this strategy is the conversion of fumarate to succinate by fumarate reductase which is thermodynamically favoured in the direction of fumarate.

The transcriptome analysis performed, specifically consideration of continous culture data sets at different dilution rates to filter growth-related genes was an integral part of identifying the 2nd round of metabolic engineering targets. Although a relatively small number of growth related genes were identified, they were of high-value. For example, *ARO9* and *ARO10*, encoding key enzymatic conversion steps in aromatic amino acid metabolism may have incorrectly pointed towards tryptophan, tyrosine, and phenylalanine catabolism or phosphenolpyruvate decarboxylase activity as metabolic areas of interest for understanding physiological differences between 8D Evolved and the reference strains. This approach may be extended to future efforts and other organisms, where continuous cultivation of engineered strains may not be possible, as in this case, or applied industrially where the dominant and preferred processing mode is batch.

Furthermore, this work clearly demonstrated that obvious genetic targets did not result in increased succinate formation. Specifically, deletion of the primary succinate consuming pathway (*Δsdh3*) [17], or constitutive over-expression of one of two of the primary succinate formation pathways (*ICL1*) did not result in any increased succinate production (See Figure S1). It is further interesting to note that the 8D with pICL1 strain also did not result in any increased succinate production, but rather only in the 8D Evolved with pICL1 strain. The ability to measure transcriptome on a strain that underwent targetted genetic engineering and directed evolution was critical to identifying pICL1 as a 2nd metabolic engineering target, which would have been discarded if selected based on intuition.

The approach employed represents an integration of diverse methods for rapid metabolic engineering proof-of-concept. The strain selection process thus need not be limited to considering organisms showing a predisposition to the production of the metabolite of interest, but rather, should include hosts most suitable for large-scale, robust, and biorefinery processing. With such hosts, carbon and redox flux redistribution requiring multi-gene approaches can be predicted, tested, and supplemented with directed evolution, screening, and selection. These strains are then transcriptionally characterized and optimized until commercially viable titers, productivities, and yields are reached. It is only through whole-process optimization and elimination of severe constraints such as forced use of non-industrially favorable strains, that the promise of a bio-based economy may be fully realized.

## Materials and Methods

### Strain Construction

The reference strain *Saccharomyces cerevisiae* CEN.PK113-5D (*Mat a MAL2-8C SUC2 URA3-52*) [30] was used for the construction of the *Δsdh3 Δser3 Δser33* knockout strain, referred to as the 8D mutant, and for the construction of the *Δsdh3 Δser3 Δthr1* knockout strain, referred to as the 20G mutant, through the cloning-free PCR-based allele replacement method previously described [31]. The upstream *SDH3* fragment was amplified by PCR from genomic DNA using the primers SDH3_Up_Fw (sequence 5′-CGAAATATGGTAAGAGAAAATG-3′) and SDH3_Up_Rv (sequence 5′- CAGGGATGCGGCCGCTGACGACATCG TTTATTATTCTTAGAGC-3′). Similarly, the downstream *SDH3* fragment was amplified using the primers SDH3 _Dw_Fw (sequence 5′- CCGCTGCTAGGCGCGCCGTGCTTTAT-GATTCTTTAAGGCGACGC-3′) and SDH3_Dw_Rv (sequence 5′- GTAATCTGTTATCGATAATCTGCC -3′). The upstream *THR1* fragment was amplified by PCR from genomic DNA using the primers THR1_Up_Fw (sequence 5′-GCAGTTC TTGCTCAGTAATCTTAG-3′) and THR1_Up_Rv (sequence 5′-GCAGGGATGCGGCCGCTGACCCATA TCTTTCGA-GATGATGACTC-3′). Similarly, the downstream *THR1* fragment was amplified using the primers THR1 _Dw_Fw (sequence 5′-CCGCTGCTAGGCGCGCCGTGCATACTGTAATT-GACCGTTAACGG-3′) and THR1_Dw_Rv (sequence 5′-CCAATCATGGATGAACCAGTAATG-3′). The upstream *SER3* fragment was amplified by PCR from genomic DNA using the primers SER3_Up_Fw (sequence 5′- CTCACAATCGAG-TAA TGCCTTTG-3′) and SER3_Up_Rv (sequence 5′-GCAGGGATGCGGCCGCTGACCATTGCTGTCGA TTTTTCTGTGG-3′). Similarly, the downstream *SER3* fragment was amplified using the primers SER3 _Dw_Fw (sequence 5′-CCGCTGCTAGGCGCGCCGTGGATAGAAGAATGCTT-GAGGC-3′) and SER3_Dw_Rv (sequence 5′- CGAATTT-GATTGTACCTGGTGC-3′). The upstream *SER33* fragment was amplified by PCR from genomic DNA using the primers SER33_Up_Fw (sequence 5′- GTACTCTTTATGG-GAGTCTTTAGC -3′) and SER33_Up_Rv (sequence 5′-GCAGGGATGCGGCCGCTGACGCAGCTGAATAAGA-CATGTTAGG- 3′). Similarly, the downstream *SER33* fragment was amplified using the primers SER33 _Dw_Fw (sequence 5′-GCAGGGATGCGGCCGCTGACGCAGCTGAATAAGA-CATGTTAGG- 3′) and SER33_Dw_Rv (sequence 5′-CTATT CTGGGTGGTCTTTTACTGG- 3′). The lithium acetate tranformation method was used [32]. As described previously, *URA3* from *Kluyvermyces lactis* was used as the selection marker in the transformation process [31]. With this approach tranformants are easily selected on uracil depleted media supplemented with 5-fluoroorotic acid. The knockout was confirmed by restriction analysis followed by sequencing (MWG Biotech AG, Ebersberg, Germany).

The plasmid pRS426T-ICL1-C was constructed and transformed into 8D Evolved, described earlier and used for constitutive *S. cerevisiae ICL1* overexpression. The parent plasmid, pRS426CT (6347 bp), was previously constructed in our laboratory by inserting the strong constitutive *TEF1* promoter (gene encoding *S. cerevisiae* translation-elongation factor 1α) and the *CYC1* transcription terminator into pRS426 [28]. This original backbone plasmid is a 5726 bp yeast episomal plasmid (YEp)-type shuttle vector with a high copy number of about 20 per cell [33]. The plasmid contains the 2 μm ori and pUC ori for independent episomal replication in *S. cerevisiae* and *E. coli*, respectively, and *URA3* and *ampR* (*bla*, beta-lactamase) genes. The final plasmid size was 8074 bp, with 2484 containing the *TEF1* promoter, the *ICL1* insert, and the *CYC1* transcription terminator sequence, verified by sequencing (MWG Biotech AG, Ebersberg, Germany).

A total of eight primers were required for amplification of the native *ICL1* gene from the reference strain, sequencing of the constructed plasmid pRS426-ICL1-C, and PCR to verify plasmid presence in the transformed reference and 8D Evolved strains (referred to as 8D Evolved with pICL1). The PCR amplification of *ICL1* was carried out using the Phusion™ High-Fidelity DNA Polymerase (Finnzymes Oy, Espoo, Finland) according to the manufacturer's protocol. The native *ICL1* was amplified from genomic DNA using the up- and downstream primers ICL1_Sp1 (sequence 5′-GCCTGCCA|CTAGTCAACGAAAAATGCC-TATCCCCG-3′), and ICL1_Asp1 (sequence 5′-GCCTCGACCCGGGCTAGAGAAAGGCATTCTTG-CACGG-3′ ), respectively. The amplicon length was 1915 bp. The fragment was cut with restriction endonucleases (REN) *Spe*I, the restriction site of which was *de novo* introduced on primer ICL1_Sp1, and *Ngo*MIV, and then ligated with pRS426CT cut with *Spe*I and *Xma*I. By using the non-compatible RENs in either end of the insertion, the direction of the insert is secured and furthermore the sole parent plasmid *Xma* site is lost. This allowed for an *in vitro* pre-selection for the correct pRS426-ICL1-C construct prior to transformation.

The four sequencing primers for construct verification included M13_rev_-29 (sequence 5′-CAGGAAACAGCTATGACC-3′), ICL1_In_1f (sequence 5′-CTGGTTGGCAGTGTTCATCA-3′), ICL1_In_2f (sequence 5′-CATCCCACAGAGAAGCCAAG-3′), and M13_uni_-21 (sequence 5′-TGTAAAACGACGGCCAGT-3′). The two primers used for plasmid verification via PCR (Taq DNA Polymerase of *Thermus aquaticus* from Sigma, St. Louis, MO, were ICL1_part_Sense (sequence 5′-TCCTGTTCAGATTTCT-CAAATGGC-3′) and ICL1_CYC_Antisense (sequence 5′-AAATTAAAGCCTTCGAGCGTCCC-3′) and these were used for analytical PCRs according to the instruction manual's recommendations). Plasmid transformation of electrocompetent *E. coli* DH5α were completed as described previously, as was plasmid transformation of the *S. cerevisiae* reference strain and 8D Evolved using the lithium acetate method [28], [31], [32].

### Medium Formulation

A chemically defined minimal medium of composition 5.0 g $L^{-1}$ $(NH_4)_2SO_4$, 3.0 g $L^{-1}$ $KH_2PO_4$, 0.5 g $L^{-1}$ $MgSO_4{\cdot}7H_2O$,

1.0 mL $L^{-1}$ trace metal solution, 300 mg $L^{-1}$ uracil, 0.05 g $L^{-1}$ antifoam 204 (Sigma-Aldrich A-8311), and 1.0 mL $L^{-1}$ vitamin solution was used for all shake flask and 2L well-controlled fermentations [34]. The trace elment solution included 15 g $L^{-1}$ EDTA, 0.45 g $L^{-1}$ $CaCl_2{\cdot}2H_2O$, 0.45 g $L^{-1}$ $ZnSO_4{\cdot}7H_2O$, 0.3 g $L^{-1}$ $FeSO_4{\cdot}7H_2O$, 100 mg $L^{-1}$ $H_3BO_4$, 1 g $L^{-1}$ $MnCl_2{\cdot}2H_2O$, 0.3 g $L^{-1}$ $CoCl_2{\cdot}6H_2O$, 0.3 g $L^{-1}$ $CuSO_4{\cdot}5H_2O$, 0.4 g $L^{-1}$ $NaMoO_4{\cdot}2H_2O$. The pH of the trace metal solution was adjusted to 4.00 with 2 M NaOH and heat sterilized. The vitamin solution included 50 mg $L^{-1}$ d-biotin, 200 mg $L^{-1}$ *para*-amino benzoic acid, 1 g $L^{-1}$ nicotinic acid, 1 g $L^{-1}$ Ca•pantothenate, 1 g $L^{-1}$ pyridoxine HCl, 1 g $L^{-1}$ thiamine HCl, and 25 mg $L^{-1}$ m•inositol. The pH of the vitamin solution was adjusted to 6.5 with 2 M NaOH, sterile-filtered and the solution was stored at 4°C. The final formulated medium, excluding glucose and vitamin solution supplementation, is adjusted to pH 5.0 with 2 M NaOH and heat sterilized. For carbon-limited cultivations the sterilized medium is supplemented with 20 g $L^{-1}$ glucose, heat sterilized separately, and 1.0 mL $L^{-1}$ vitamin solution is added by sterile filtration (0.20 µm pore size Ministart®-Plus Sartorius AG, Goettingen, Germany). For cultures where glycine or threonine auxotrophic strains were cultivated the final culture medium was supplemented with glycine 500 mg $L^{-1}$ or 100 mg $L^{-1}$ threonine added by sterile filtration.

## Shake Flask Cultivations and Stirred Tank Fermentations

Shake flask cultivations were completed in 500 mL Erlenmeyer flasks with two diametrically opposed baffles and two side-necks with septums for sampling by syringe. Flasks were heat sterilized with 100 mL of medium, inoculated with a single colony, and incubated at 30°C with orbital shaking at 150 RPM. Stirred tank fermentations were completed in well-controlled, aerobic, 2.2L Braun Biotech Biostat B fermentation systems with a working volume of 2L (Sartorius AG, Goettingen, Germany). The temperature was controlled at 30°C. The fermenters were outfitted with two disk-turbine impellers rotating at 600 RPM. Dissolved oxygen was monitored with an autoclavable polarographic oxygen electrode (Mettler-Toledo, Columbus, OH). During aerobic cultivation the air sparging flow rate was 2 vvm. The pH was kept constant at 5.0 by automatic addition of 2 M KOH. Off-gas passed through a condenser to minimize the evaporation from the fermenter. The fermenters were inoculated from shake flask precultures to an initial $OD_{600}$ 0.01.

## Fermentation Analysis

**Off-gas Analysis.** The effluent fermentation gas was measured every 30 seconds for determination of $O_2$(g) and $CO_2$(g) concentrations by the off-gas analyzer Brüel and Kjær 1308 (Brüel & Kjær, Nærum, Denmark).

**Biomass Determination.** The optical density (OD) was determined at 600 nm using a Shimadzu UV mini 1240 spectrophotometer (Shidmazu Europe GmbH, Duisberg, Germany). Duplicate samples were diluted with deionized water to obtain $OD_{600}$ measurements in the linear range of 0–0.4 $OD_{600}$ Samples were always maintained at 4°C post-sampling until $OD_{600}$ and dry cell weight (DCW) measurements were performed. DCW measurements were determined through the exponential phase, until stationary phase was confirmed according to $OD_{600}$ and off-gas analysis. Nitrocellulose filters (0.45 µm Sartorius AG, Goettingen, Germany) were used. The filters were pre-dried in a microwave oven at 150W for 10 min., and cooled in a dessicator for 10 min. 5.0 mL of fermentation broth were filtered, followed by 10 mL DI water. Filters were then dried in a microwave oven for 20 min. at 150W, cooled for 15 min. in a desiccator, and the mass was determined.

**Metabolite Concentration Determination.** All fermentation samples were immediately filtered using a 0.45 µm syringe-filter (Sartorius AG, Goettingen, Germany) and stored at −20°C until further analysis. Glucose, ethanol, glycerol, acetate, succinate, pyruvate, fumarate, citrate, oxalate, and malate were determined by HPLC analysis using an Aminex HPX-87H ion-exclusion column (Bio-Rad Laboratories, Hercules, CA). The column was maintained at 65°C and elution performed using 5 mM $H_2SO_4$ as the mobile phase at a flow rate of 0.6 mL min.$^{-1}$. Glucose, ethanol, glycerol, acetate, succinate, citrate, fumarate, malate, oxalate were detected on a Waters 410 differential refractometer detecter (Shodex, Kawasaki, Japan), and acetate and pyruvate were detected on a Waters 468 absorbance detector set at 210 nm.

## Transcriptomics

**RNA Sampling and Isolation.** Samples for RNA isolation from the late-exponential phase of glucose-limited batch cultivations were taken by rapidly sampling 25 mL of culture into a 50 mL sterile Falcon tube with 40 mL of crushed ice in order to decrease the sample temperature to below 2°C in less than 10 seconds. Cells were immediately centrifuged (4000 RPM at 0°C for 2.5 min.), the supernatant discarded, and the pellet frozen in liquid nitrogen and it was stored at −80°C until total RNA extraction. Total RNA was extracted using the FastRNA Pro RED kit (QBiogene, Carlsbad, USA) according to manufacturer's instructions after partially thawing the samples on ice. RNA sample integrity and quality was determined prior to hybridization with an Agilent 2100 Bioanalyzer and RNA 6000 Nano LabChip kit according to the manufacturer's instruction (Agilent, Santa Clara, CA).

**Probe Preparation and Hybridization to DNA Microarrays.** Messenger RNA (mRNA) extraction, cDNA synthesis, labeling, and array hybridization to Affymetrix Yeast Genome Y2.0 arrays were performed according to the manufacturer's recommendations (Affymetrix GeneChip® Expression Analysis Technical Manual, 2005–2006 Rev. 2.0). Washing and staining of arrays were performed using the GeneChip Fluidics Station 450 and scanning with the Affymetrix GeneArray Scanner (Affymetrix, Santa Clara, CA).

**Microarray Gene Transcription Analysis.** Affymetrix Microarray Suite v5.0 was used to generate CEL files of the scanned DNA microarrays. These CEL files were then processed using the statistical language and environment R v5.3 (R Development Core Team, 2007, www.r-project.org), supplemented with Bioconductor v2.3 (Biconductor Development Core Team, 2008, www.bioconductor.org) packages Biobase, affy, gcrma, and limma [35], [36]. The probe intensities were normalized for background using the robust multiarray average (RMA) method only using perfect match (PM) probes after the raw image file of the DNA microarray was visually inspected for acceptable quality. Normalization was performed using the qspline method and gene expression values were calculated from PM probes with the median polish summary. Statistical analysis was applied to determine differentially expressed genes using the limma statistical package. Moderated *t*-tests between the sets of experiments were used for pair-wise comparisons. Emperical Bayesian statistics were used to moderate the standard errors within each gene and Benjamini-Hochberg's method was used to adjust for multi-testing. A cut-off value of adjusted $p<0.05$ was used for statistical significance. Furthermore, principal component analysis (PCA) was performed in order to elucidate the relative importance of substrate limitation (carbon vs. nitrogen) and growth rate (0.03 $h^{-1}$, 0.1 $h^{-1}$, 0.2 $h^{-1}$), previously described [21], when compared with the gene expression of the reference and 8D Evolved strain. To select genes whose expression levels were related to these factors, the moderated *t*-statistics were followed up with *F*-distributions to yield a statistic referred to as $F_g$, which is simply the usual *F*-statistic from linear model theory but with the posterior variance substituted for the sample variance in the denominator, as described else where [35]. The cut-off value of adjusted $p<0.1$ was used for statistical significance.

All microarray data is MIAME compliant and the raw data has been deposited in ArrayExpress (http://www.ebi.ac.uk/microarray-as/ae/).

## Supporting Information

**Figure S1 Inhibition of the succinate dehydrogenase complex with malonate supplementation in shake flask cultures was evaluated.** The reference and *Δsdh3* strain, previously described [17], were cultured in minimal media supplemented with 10 g $L^{-1}$ glucose and no succinate accumulation was detected (Panel a). The reference strain was cultured with 0.1, 1.0, 5.0, 10.0, and 50.0 mM malonate supplementation. Under no supplementation conditions succinate accumulation was observed (Panel b). In order to confirm that the concentration of malonate in the culture was effectively inhibiting succinate dehydrogenase activity, residual ethanol in the culture broth was monitored. Succinate dehydrogenase activity, as previously described, catalyzes the conversion of succinate to fumarate with net production of protonated ubquinone. Ethanol is a carbon source readily catabolized by *S. cerevisiae* using respiro-fermentative pathways and requiring succinate dehydrogenase activity. Panel c shows the residual glucose concentration in the culture broth at 0, 17, 22, and 37h post-inoculation for no supplementation of malonate (reference) and then 0.1, 1.0, 5.0, 10.0, and 50.0 mM malonate supplementation. These growth profiles were generated using the reference strain (CEN.PK113-7D). As expected, full catabolism of glucose was observed at all malonate concentrations with the exception of 50.0 mM, thereby considered an upper limit. Similarly, in panel d, is the ethanol concentration in the culture broth for the same malonate concentrations and sample times. At 37 h, as expected, the reference strain had consumed nearly all ethanol produced during the glucose consumption phase. Malonate concentrations of 1.0, 5.0, and 10.0 mM malonate resulted in significant ethanol respiration inhibition compared to no supplementation and 0.1 mM malonate, confirming that respiro-fermentative catabolism was inhibited. Under no circumstances was succinate accumulation observed. Furthermore, the *Δsdh3* strain was supplemented with 50.0 mM malonate to ensure no unexpected interaction between the genetic modification and malonate supplementation (panel e).

**Figure S2 A total of 1964 genes were submitted to the Saccharomyces Genome Database tool, Pathway Expression Viewer.** The resulting Pathway Expression map shows the relative log-fold change of all *S. cerevisiae* metabolic reactions (Evolved 8D vs. Reference). Three key results are high-lighted from the transcriptome. First, isocitrate lyase (*ICL1*) was amongst the few genes not up-regulated in the Evolved 8D strain, thereby becoming a 2nd round metabolic engineering target. Second, alanine:glyoxylate aminotransferase (*AGX1*) was 4.3 log-fold higher in the Evolved 8D strain, confirming the predicted model-guided strategy of up-regulated glycine formation from glyoxylate pools. Third, threonine aldolase (*GLY1*) was 1.6 log-fold higher in the Evolved 8D strain. The genome-scale model reconstruction used for predictions annotated Gly1p as catalyzing the reversible conversion of threonine to glycine. This reaction has since been shown to be irreversible, converting threonine to glycine, consuming equimolar acetaldehyde. The transcriptome data suggests that the Evolved 8D strain demonstrated de-coupling of succinate and biomass production because alternative reactions (e.g., Gly1p) were supplying glycine pools.

**Figure S3 Panel a briefly describes the mutant construction of 20G, *Δsdh3 Δser3 Δthr1*, from the reference strain and initially supplemented with 100 mg $L^{-1}$ threonine and 500 mg $L^{-1}$ glycine to satisfy the resulting auxotrophies.** All growth challenges were evaluated in shake flasks supplemented with minimal medium, 300 mg $L^{-1}$ uracil, 10 g $L^{-1}$ glucose, and either threonine and/or glycine added, as indicated. The mutant 20G was not capable of sustaining growth in the absence of threonine, and therefore a working cell bank was prepared. Panel b describes the shake flask experiments and progression followed to evaluate the strain's ability to be evolved from threonine supplementation to glycine supplementation. When 20G culture was inoculated from threonine supplemented medium to glycine only supplemented medium, no growth was observed up to 14d post-inoculation (2 samples per day measuring $OD_{600}$). On day 14, a shake flask culture of 20G only supplemented with glycine, was then supplemented with 100 mg $L^{-1}$ threonine, and growth was immediately restored. It was therefore concluded that the mutant 20G was incapable of catalyzing glycine to threonine to satisfy threonine cellular demands, given that threonine synthesis was interrupted with the deletion of *thr1*. This experimental conclusion further supports that *Gly1* encoding threonine aldolase, originally believed to catalyze the conversion of glycine to threonine, catalyzes the reverse reaction and thus cannot meet threonine cellular demands from glycine pools.

## Acknowledgments

We are grateful to Gaëlle Lettier, Uffe Mortensen, Songsak Wattanachaisaereekul, Mohammad Asadollahi, and Jérôme Maury of the Center for Microbial Biotechnology, Technical University of Denmark, for their experimental assistance with yeast strain construction, plasmid construction, and yeast genetics methods.

## Author Contributions

Conceived and designed the experiments: JMO DC KRP LO JN. Performed the experiments: JMO DC SGP. Analyzed the data: JMO DC JN. Wrote the paper: JMO JN.

## References

1. Otero JM, Panagiotou G, Olsson L (2007) Fueling Industrial Biotechnology Through Bioethanol. Adv Biochem Eng Biotechnol 108: 1–40.
2. Nielsen J (2001) Metabolic engineering. Appl Microbiol Biotechnol 55(3): 263–283.
3. Covert MW, Schilling CH, Famili I, Edwards JS, Goryanin II, et al. (2001) Metabolic modeling of microbial strains in silico. Trends Biochem Sci 26(3): 179–186.
4. Nielsen J, Jewett MC (2008) Impact of systems biology on metabolic engineering of *Saccharomyces cerevisiae*. FEMS Yeast Res 8(1): 122–131.

5. Bro C, Regenberg B, Förster J, Nielsen J (2005) *In silico* aided metabolic engineering of *Saccharomyces cerevisiae* for improved bioethanol production. Metabolic Engineering 8(2): 102–111.
6. Edwards JS, Ibarra RU, Palsson BØ (2001) *In silico* predictions of *Eschericia coli* metabolic capabilities are consistent with experimental data. Nature Biotechnology 19(2): 125–130.
7. McKinlay JB, Vieille C, Zeikus JG (2007) Prospects for a bio-based succinate industry. Appl Microbiol Biotechnol 76(4): 727–740.
8. Wilke T, Vorlop KD (2004) Industrial bioconversion of renewable resources as an alternative to conventional chemistry. Appl Microbiol Biotechnol 66(2): 131–142.
9. Zeikus JG, Jain MK, Elankovan P (1999) Biotechnology of succinic acid production and markets for dervied industrial products. Appl Microbiol Biotechnol 51: 545–552.
10. Song H, Lee SY (2006) Production of succinic acid by bacterial fermentation. Enzy Microb Techn 39(3): 352–361.
11. Jantama K, Haupt MJ, Svoronos SA, Zhang X, Moore JC, et al. (2008) Combining metabolic engineering and metabolic evolution to develop nonrecombinant strains of *Escherichia coli* C that produce succinate and malate. Biotechnol Bioeng 99(5): 1140–1153.
12. Lee PC, Lee SY, Hong SH, Chang HN (2002) Isolation and characterization of a new succinic acid-producing bacterium, *Mannheimia succiniciproducens* MBEL55E, from bovine rumen. Appl Microbiol Biotechnol 58(5): 663–668.
13. Sauer M, Porro D, Mattanovich D, Braunduardi P (2008) Microbial production of organic acids: expanding the markets. Trends Biotechnol 26(2): 100–108.
14. Adrio JL, Deman AL (2006) Genetic improvement of processes yielding microbial products. FEMS Microbiol Rev 30(2): 187–214.
15. Lynd LR, Wyman CE, Gerngross TU (1999) Biocommodity Engineering. Biotecnol Prog 15(5): 777–793.
16. Patil KR, Rocha I, Förster J, Nielsen J (2005) Evolutionary programming as a platform for *in silico* metabolic engineering. BMC Bioinformatics 6: 308–319.
17. Cimini D, Patil KR, Schiraldi C, Nielsen J (2009) Global transcritional response of *Saccharomyces cerevisiae* to the deletion of SDH3. BMC Systems Biology 3(1): 17–28.
18. Aliverdieva DA, Mamaev DV, Lagutina LS, Scholtz KF (2006) Specific features of changes in levels of endogenous respiration substrates in *Saccharomyces cerevisiae* cells at low temperature. Biochemistry (Moscow) 71(1): 39–45.
19. Regenberg B, Grotkjær T, Winther O, Fausbøll A, Akesson M, et al. (2006) Growth-rate regulated genes have profound impact on interpretation of transcriptome profiling in *Saccharomyces cerevisiae*. Genome Biol 7(11): R107–119.
20. Castrillo JI, Zeef LA, Hoyle DC, Zhang N, Hayes A, et al. (2007) Growth control of the eukaryote cell: a systems biology study in yeast. J Biol 6(2): 4–28.
21. Fazio A, Jewett MC, Daran-Lapujade P, Mustacchi R, Usaite R, et al. (2008) Transcription factor control of growth rate dependent genes in *Saccharomyces cerevisiae*: A three factor design. BMC Genomics 9: 341–354.
22. Paley SM, Karp PD (2006) The Pathway Tools cellular overview diagram and Omics Viewer. Nucleic Acids Research 34(13): 3771–3778.
23. Fisk DG, Ball CA, Dolinski K, Engel SR, Hong EL, et al. (2006) *Saccharomyces cerevisiae* S288C genome annotation: a working hypothesis. Yeast 23(12): 857–865.
24. Förster J, Famili I, Fu P, Palsson BØ, Nielsen J (2003) Genome-scale reconstruction of the *Saccharomyces cerevisiae* metabolic network. Genome Res 13(2): 244–253.
25. Duarte NC, Harrgård MJ, Palsson BØ (2004) Reconstruction and validation of *Saccaromyces cerevisiae* iND750, a fully compartmentalized genome-scale metabolic model. Genome Res 14(7): 1298–1309.
26. Nookaew I, Jewett MC, Meechai A, Thammarongtham C, Laoteng K, et al. (2008) The genome-scale metabolic model iIN800 of *Saccharomyces cerevisiae* and its validation: a scaffold to query lipid metabolism. BMC Systems Biology 7(2): 71–85.
27. Herrgård MJ, Swainston N, Dobson P, Dunn WB, Arga KY, et al. (2008) A consensus yeast metabolic network reconstruction obtained from a community approach to systems biology. Nature Biotechnology 26 (10): 1155–1160.
28. Wattanachaisaereekul S, Lantz AE, Nielsen ML, Nielsen J (2008) Production of the polyketide 6-MSA in yeast engineered for increased malonyl-CoA supply. Metab Eng 10(5): 246–254.
29. Zelle RM, de Hulster E, van Winden WA, de Waard P, Dijkema C, et al. (2008) Malic acid production by *Saccharomyces cerevisiae*: Engineering of pyruvate carboxylation, oxaloacetate reduction, and malate export. Appl Environ Microbiol 74(9): 2766–2777.
30. van Dijken JP, Bauer J, Brambilla L, Duboc P, Francois JM, et al. (2000) An interlaboratory comparison of physiological and genetic properties of four *Saccharomyces cerevisiae* strains. Enzyme and Microbial Technology 26(9–10): 706–714.
31. Eredeniz N, Mortensen UH, Rothstein R (1997) Cloning-free PCR-based allele replacement methods. Genome Res 7: 1174–1183.
32. Gietz RD, Woods RA (2002) Transformation of yeast by lithium acetate/single-straded carrier DNA/polyethylene glycol method. Methods Enzymol 350: 87–96.
33. Christianson TW, Sikorski RS, Dante M, Shero JH, Hieter P (1992) Multifunctional yeast high-copy-number shuttle vectors. Gen 110(1): 119–122.
34. Verudyn C, Postma E, Scheffers WA, van Dijken JP (1992) Effect of benzoic acid on metabolic fluxes in yeasts: a continuous culture stody on the regulation of respiration and alcoholic fermentation. Yeast 8: 501–517.
35. Smyth GK (2004) Linear models and empirical Bayes methods for assessing differential expression in microarray experiments. Statistical Applications in Genetics and Molecular Biology 3, 1(3).
36. Smyth GK (2005) Limma: linear models for microarray data. In *Bioinformatics and Computational Biology Solutions using R and Bioconductor*, Gentleman R, Carey V, Dudoit S, Irizarry R, Huber W (eds.), 397–420, New York: Springer.

6

# Improving Kinetic or Thermodynamic Stability of an Azoreductase by Directed Evolution

**Vânia Brissos[1], Nádia Gonçalves[1], Eduardo P. Melo[2], Lígia O. Martins[1]***

1 Instituto de Tecnologia Química e Biológica, Universidade Nova de Lisboa, Oeiras, Portugal, 2 Institute for Biotechnology and Bioengineering, Center for Molecular and Structural Biomedicine, Universidade do Algarve, Faro, Portugal

## Abstract

Protein stability arises from a combination of factors which are often difficult to rationalise. Therefore its improvement is better addressed through directed evolution than by rational design approaches. In this study, five rounds of mutagenesis/recombination followed by high-throughput screening (≈10,000 clones) yielded the hit 1B6 showing a 300-fold higher half life at 50°C than that exhibited by the homodimeric wild type PpAzoR azoreductase from *Pseudomonas putida* MET94. The characterization using fluorescence, calorimetry and light scattering shows that 1B6 has a folded state slightly less stable than the wild type (with lower melting and optimal temperatures) but in contrast is more resistant to irreversible denaturation. The superior kinetic stability of 1B6 variant was therefore related to an increased resistance of the unfolded monomers to aggregation through the introduction of mutations that disturbed hydrophobic patches and increased the surface net charge of the protein. Variants 2A1 and 2A1-Y179H with increased thermodynamic stability (10 to 20°C higher melting temperature than wild type) were also examined showing the distinctive nature of mutations that lead to improved structural robustness: these occur in residues that are mostly involved in strengthening the solvent-exposed loops or the inter-dimer interactions of the folded state.

**Editor:** Danilo Roccatano, Jacobs University Bremen, Germany

**Funding:** This work was funded by European Union grant BIORENEW, FP6-2004-NMP-NI-4/026456; Fundação para a Ciência e Tecnologia, Portugal: PEst-OE/EQB/LA0004/2011 and PTDC/QUI-BIQ/119677/2010. V.B. acknowledges a Post-doc fellowship (SFRH/BPD/46808/2008) from FCT, Portugal. The funders had no role in study design, data collection and analysis, decision to publish, or preparation of the manuscript.

* E-mail: lmartins@itqb.unl.pt

## Introduction

Thermal stability is relevant for biological function and molecular evolution of proteins. The thermal denaturation process of proteins is usually complex, but often for monomeric proteins can be simplified to the classical two step process: N↔U→D where N, U and D are the native, the reversible unfolded and the irreversible denatured enzyme. The first step involves unfolding of the polypeptide's native structure. The unfolded protein may refold to the native conformation or in a second step undergo irreversible denaturation to permanent inactivation. This may result from protein aggregation, misfolding and covalent changes such as the deamidation of asparagine or glutamine residues and oxidation of cysteine or methionine residues [1]. Enzyme thermostability encompasses thermodynamic and kinetic stabilities [2,3]. Thermodynamic stability is defined by the enzymes's free energy of stabilization ($\Delta G_{stab}$ reflecting the difference between the free energies of the folded and the unfolded states of the protein) and by its melting temperature ($T_m$, the temperature at which 50% of the protein is unfolded). Kinetic or long-term stability depends on the energy barrier to irreversible inactivation and is generally expressed as the enzyme's half-life ($t_{1/2}$) at a defined temperature. Most frequently the two stabilities correlate since increasing the enzyme resistance to unfolding (higher $T_m$) also increases its resistance to inactivation (higher $t_{1/2}$); firstly, an increase in the stability of the native state leads to slower accumulation of the unfolded state and secondly, the unfolded state is usually the ground state leading to irreversible denaturation/inactivation.

Directed evolution is considered to be the most powerful approach for improving the thermostability of proteins. In fact, comparative studies performed with hyperthermostable enzymes and their mesophilic counterparts have shown nearly superimposable three-dimensional structures suggesting that in nature extreme thermostability seems to be achieved by distributing different types of additional intramolecular interactions throughout the protein [4]. Moreover our understanding of these interactions is incomplete and often does not allow to reliably predicting how they combine to yield a more stable protein. Therefore, rational approaches such as site-directed mutagenesis shows often a limited efficiency and the random introduction of a small number of amino acid changes by error prone PCR or DNA shuffling emerges as the most appropriate methodology to improve protein stability. In recent years these methods have been fine-tuned and a different number of properties in various target enzymes have been successfully improved using directed evolution approaches [5–7]. Thermal stability is a critical property for many biotechnological applications of proteins as it implies longer life-times and frequently higher tolerance to the presence of organic co-solvents, extreme pH values and high salt concentration or pressures. Several examples of successfully evolved lipases, β-glucuronidases, ligninolytic oxidoreductases, xylanases, cytochrome P450 peroxygenases, phytases and glucose dehydrogenase have been reported [8–15].

Flavin-dependent azoredutases have been identified in a wide range of synthetic dye decolourising bacteria including *Escherichia coli*, *Bacillus* sp. SF, *Bacillus* sp. OY1–2, *Pseudomonas aeruginosa*, *Enterococcus feacalis*, *Salmonella typhimurium* [16–20]. Enzymatic bioreduction of synthetic azo dyes or nitroaromatics, prevalent anthropogenic pollutants, are environmental friendly strategies for bioremediation of e.g. dye-containing effluents from textile industries [21–23]. These enzymes share some similarities with regard to sequence, structure, and reaction mechanism with the larger family of flavin-dependent quinone reductases such as, Lot6p from *Saccharomyces cerevisiae* or the mammalian NQO1 [24]. Azoreductases are proposed to take part in the organism's enzymatic general detoxification systems; e.g. in the cellular response to thiol-specific stress [25,26] or in the response to oxidative stress [27,28]. These enzymes require two cycles of NADPH-dependent reduction of FMN to $FMNH_2$ for reducing the azo substrate to two amines and the quinone substrate to a hydroquinone. *Pseudomonas putida* MET94 is a bacteria that degrades a wide range of structurally distinct azo dyes with high efficiency and the azoreductase PpAzoR was shown to play a key role in this process [29]. Its broad substrate specificity makes it attractive for bioremediation processes [30] but its low kinetic stability impairs exploitation of its full potential for environmentally related applications.

In this study, we generated thermostable PpAzoR variants by directed evolution. We investigated the molecular determinants associated with the thermostability of 1B6 variant which reveals a significantly increased kinetic stability when compared to the wild type. Thermoresistant variants (2A1 and 2A1-Y179H) that denature irreversibly similarly to the wild type but show an increased stability of the native state were also characterized. Based in the recently solved crystal structure of the homodimeric wild type enzyme [31] we rationalize the molecular basis behind the increased kinetic or thermodynamic stability of PpAzoR enzyme.

## Materials and Methods

### Bacterial strains, plasmids and media

*Escherichia coli* strain DH5α (Novagen) was used for routine propagation and amplification of plasmid constructs. *E. coli* Tuner (DE3, Novagen) and KRX (Promega) strains were used to express the *ppAzoR* and variant genes cloned in pET-21a (+) plasmid (Novagen). In the Tuner strain the target genes are under the control of T7 promotor, induced by isopropyl β-D-1-thiogalactopyranoside (IPTG) and in the KRX strain the genes are under the control of the $rhaP_{BAD}$ promoter, induced by rhamnose. Luria-Bertani medium (LB) and Terrific Broth medium (TB) were used for the maintenance and growth of *E. coli* strains, supplemented with appropriate antibiotics when needed.

### Growth of recombinant strains and *ppAzor* overexpression

The plasmid pLP-1 [29], containing the *ppAzoR* gene, was transformed into *E. coli* Tuner (DE3, Novagen) or *E. coli* KRX (Promega), producing LOM528 and LOM531 strains, respectively. Single colonies were used to inoculate 20 mL of LB medium supplemented with 100 μg/mL ampicillin, grown overnight at 37°C, 160 rpm. Fresh cultures were transferred to 100 mL of TB medium supplemented with 100 μg/mL ampicillin, in order to start the growth with an $OD_{600\ nm} = 0.05$. Cultures were incubated at 30°C, 160 rpm and when $OD_{600\ nm} \approx 0.6$, 100 μM IPTG was added to LOM528, and 0.1% rhamnose to LOM531. After 24 h of cultivation, cells were collected by centrifugation (10,000 rpm, 15 min at 4°C). The cell pellets were suspended in 1.5 mL of 20 mM Tris-HCl buffer, pH 7.6, containing 5 mM $MgCl_2$, 1 U/mL of DNAse I, and 2 μL/mL of a mixture of protease inhibitors: antipain and leupeptin. Cells were disrupted by French Press (Thermo EFC) and then centrifuged at 18,000 rpm for 2 h at 4°C. The supernatants (cell crude extracts) were collected and used to perform enzymatic assays. The protein concentration was determined using the Bradford assay with bovine serum albumin (BSA) as standard. SDS-PAGE electrophoresis was performed to visualize protein overproduction in crude extracts.

### Enzymatic assays

Enzymatic activities were measured using five quinone substrates, anthraquinone-2-sulfonic acid (AQS), 1,4-benzoquinone (BZ), 1,2-dihydroxybenzene (catechol), 2-hydroxy-1,4-naphthoquinone (Lawsone), 1,2-naphthoquinone-4-sulfonate (NSA), at 50–100 μM, in the presence of 250 μM NADPH. Enzymatic assays were performed at 30°C, in 100 mM sodium phosphate buffer, pH 7. The reactions were initiated by the addition of crude extracts and followed by monitoring the decrease in absorbance of NADPH at 340 nm ($\varepsilon = 6{,}220\ M^{-1}\ cm^{-1}$) on a Nicolet Evolution 300 spectrophotometer (Thermo Industries).

### Random mutagenesis by error-prone PCR (ep-PCR) and mutant library construction

Variation in the ppAzoR gene was generated by using ep-PCR. Primers 5′-GGAGAGTCATATGAAACTGTTGC-3′ (PpaF) and 5′-CAACCAAAGGATCCCTTGATCAGG-3′ (PpaR) were used for amplification. Nucleotides for restriction sites of NdeI and BamHI are underlined. ep-PCR was carried out in 50 μL reaction volumes containing 3 ng of DNA template (plasmid pLP-1, where ppAzoR gene [29] and variants are cloned), 0.5 μM of primers, 200 μM of dNTPs, 7 mM $MgCl_2$, Taq polymerase buffer, and 5 U of Taq polymerase (Fermentas). The effect of $MnCl_2$ was tested at 0.1–0.25 mM concentrations. After an initial denaturation period of 10 min at 94°C, the following steps were repeated for 30 cycles in a thermal cycler (MyCycler™ thermocycler, Biorad): 1 min at 94°C, 1 min at 55°C and 45 s at 72°C followed by a final 10 min period at 72°C. The amplified products were purified using GFX PCR DNA and Gel Band Purification kit (GE Healthcare). The final PCR products were digested with NdeI/BamHI (Fermentas) and cloned into pET-21a (+) (Novagen). Ligations were performed with T4 DNA ligase (Fermentas) using a 1:8 vector to insert ratio. Reaction mixtures were incubated overnight at room temperature, incubated at 65°C for 10 min, and then used to transform electrocompetent E. coli KRX cells.

### Recombination by DNA Shuffling and mutant library construction

DNA shuffling was performed as described previously with some modifications [32,33]. Selected genes coding for the ppAzoR variants were amplified by PCR using primers pET21D (5′-CTTCCCCATCGGTGATGTCGGCGATATAG-3′) and pET21R (5′-CCAAGGGGTTATGCTAGTTATTGCTCAG-3′). A mixture containing 200 ng of each parental gene was digested with 2.5 U/mL of DNase I in a 200 mM Tris-HCl buffer, pH 7 with 80 mM $MnCl_2$ for 15 min at 15°C in a thermocycler (MyCycler™ Thermal Cycler, Biorad). Digestion was stopped by adding 6 μL of 0.5 M EDTA. The PCR reassembly was carried out in a 20 μL reaction volume containing 3 μL of DNA fragments, 200 μM of dNTPs, NZYProof polymerase buffer and 2.5 U of NZYProof polymerase (NZYTech). After an initial denaturation period of 3 min at 96°C, the following steps were repeated for 45 cycles in a thermal cycler (MyCycler$^{TM}$ thermocycler, Biorad): 1 min at 94°C, 90 s at 59°C, 90 s at 56°C, 90 s at 53°C, 90 s at 50°C, 90 s at 47°C, 90 s at 44°C, 90 s at 41°C, and 1 min + 5 s/cycle at 72°C followed by a final 10 min period at 72°C. The PCR reassembly products were amplified by PCR using primers PpaF (5′-GGAGAGTCATATGAAACT-GTTGC-3′) and PpaR (5′-CAACCAAAGGATCCCTTGAT-CAGG-3′). PCR was carried out in a 50 μL reaction volume containing 1 μL of PCR reassembly products, 1 μM of primers, 200 μM of dNTPs, NZYProof polymerase buffer, and 2.5 U of NZYProof polymerase (NZYTech). After an initial denaturation period of 3 min at 94°C, the following steps were repeated for 20 cycles in a thermal cycler (MyCycler$^{TM}$ thermocycler, Biorad): 30 s at 94°C, 1 min at 55°C, 90 s at 72°C followed by a final 10 min period at 72°C. The amplified products were purified using GFX PCR DNA and Gel Band Purification kit (GE Healthcare). The final PCR products were digested with NdeI/BamHI (Fermentas) and cloned into pET-21a (+) (Novagen) as described above. Ligation reaction mixtures were used to transform electrocompetent E. coli KRX cells.

## Site directed mutagenesis

Single amino acid substitutions were created using the QCM protocol developed by Strategene. Plasmid containing the 2A1 variant gene was used as template and the forward 5′-CCCACGGCCTGGCCCATGGCCCGGAGCAG-3′ and reverse 5′-CCCACGGCCTGGCCCATGGCCCGGAGCAG-3′ primers were used to generate the 2A1+Y179H variant. Plasmid containing the 1B6 variant gene was used as a template and the forward 5′-GGCTGCCGATCCCATCCCCCACTTCTCCG-3′ and reverse 5′-CGGAGAAGTGGGGGATGGGATCGGCA-GCC-3′ primers were used to generate the mutant 1B6+A48P. The presence of the desired mutations in the resulting plasmids and the absence of unwanted mutations in other regions of the insert were confirmed by DNA sequence analysis. PCR was carried out in a 50 μL reaction volume containing 3 ng of DNA template, 2 μM of primers, 200 μM of dNTPs, NZYProof polymerase buffer, and 1.25 U of NZYProof polymerase (NZY-Tech). After an initial denaturation period of 2 min at 95°C, the following steps were repeated for 30 cycles in a thermal cycler (MyCycler$^{TM}$ thermocycler, Biorad): 1 min at 95°C, 1 min at 68°C, 7 min at 72°C followed by a final 10 min period at 72°C. The amplified product was purified using GFX PCR DNA and Gel Band Purification kit (GE Healthcare). The final PCR product was used to transform electrocompetent *E. coli* KRX cells.

## Overexpression of *ppAzoR* variants in *E. coli*

From a fresh agar plate, individual colonies were randomly picked and transferred to a 96 well-plate containing 200 μL of LB medium supplemented with ampicillin (100 μg $L^{-1}$). In order to avoid evaporation, only the interior wells were used for cell growth while perimeter wells were filled with water and the plates were sealed with a foil and a plastic cover, closed with parafilm. Four wells in each plate were used to inoculate the parent strain of each generation. Cultures were incubated at 30°C for 24 h, at 750 rpm in a Titramax 1000 shaker (Heidolph). Twenty microliters of these cultures were used to inoculate new 96 well-plates containing 180 μL TB medium supplemented with ampicillin (100 μg $L^{-1}$) and incubated at 30°C for 3 h at 750 rpm. After this period 0.1% rhamnose was added to induce gene expression and cells were harvested by centrifugation (4,000 rpm for 20 min at 4°C) after 24 h of growth.

## Cell disruption in 96-well plates

The 96 well-plates were submerged in liquid nitrogen and then thawed at room temperature for 5 min. After 3 cycles of freeze and thaw, cell pellets were resuspended in 100 μL of 20 mM Tris-HCl buffer, pH 7.6. After cell disruption, plates were centrifuged at 4,000 rpm for 30 min at 4°C and supernatants (cell crude extracts) used for enzymatic activity measurements.

## High-throughput screening for thermostability

Cell crude extracts (20 μL) were transferred into two replica 96 well-plates. One plate was assayed for initial activity ($A_i$) by adding 180 μL of 100 mM sodium phosphate buffer, pH 7 containing 100 μM AQS and 250 μM NADPH. The decrease of NADPH absorption was followed during 5 min at 340 nm ($\varepsilon = 6{,}220\ M^{-1}\ cm^{-1}$) on a Synergy 2 (Biotek) micro plate reader. The crude extracts in the second plate were incubated at different temperatures for a defined time period, cooled on ice for 5 min, incubated at room temperature for 5 min and afterwards assayed for residual activity ($A_r$) using the same conditions indicated for assaying the initial activity. Thermostability was assessed using the ratio of the residual activity to the initial activity of the variant (v), normalized to the parent type (p) – $(A_r/A_i)_v/(A_r/A_i)_p$. The variants activity relative to the parent was calculated using the ratio of the initial activity of the variant to the parent type ($Ai_v/Ai_p$). Variants exhibiting either the highest initial activity or the highest thermostability were re-screened to rule out false positives. Mutations were verified by DNA sequencing analysis using T7 terminator universal primers. In each generation, the variant with the highest stability yet unchanged activity was chosen to be the parent for the next generation and the initial and residual activity was measured in tree different 96-well plates, ensuring the reproducibility.

## Production and purification of selected variants

Plasmids of *ppAzoR* variants were introduced into the host *E. coli* Tuner (DE3, Novagen) and recombinant enzymes were produced and purified as previously described [29]. Enzyme concentration was estimated using the $Abs_{455\ nm}$ value [29].

## Kinetic analysis

Enzymatic activities of purified variants were measured at 30°C, in 100 mM sodium phosphate buffer, pH 7, with 100 μM AQS and 250 μM NADPH as substrates. Reactions were initiated by the addition of enzyme and followed by monitoring the decrease in absorbance of NADPH at 340 nm ($\varepsilon = 6{,}220\ M^{-1}\ cm^{-1}$) on a Nicolet Evolution 300 spectrophotometer (Thermo Industries). Optimal temperatures were determined between 23 and 65°C.

## Enzyme stability assays

Residual activities were determined at 30°C after incubating purified enzyme preparations at a range of temperatures (40–85°C) for 1 h. Kinetic stability studies were performed as described by Martins et al. [34]. In brief, the enzymes were incubated at 50°C in 20 mM Tris-HCl buffer, pH 7.6 and, at fixed time intervals, sample aliquots were withdrawn and tested for activity at 30°C. Thermodynamic stability was assessed by steady-state fluorescence measured with a Carry Eclipse spectrofluorimeter using excitation wavelengths of 280 nm and 296 nm and an emission wavelength of 340 nm [35,36]. Samples containing PpAzoR and its variants ($Abs_{280} = 0.1$) in 20 mM Tris–HCl buffer,

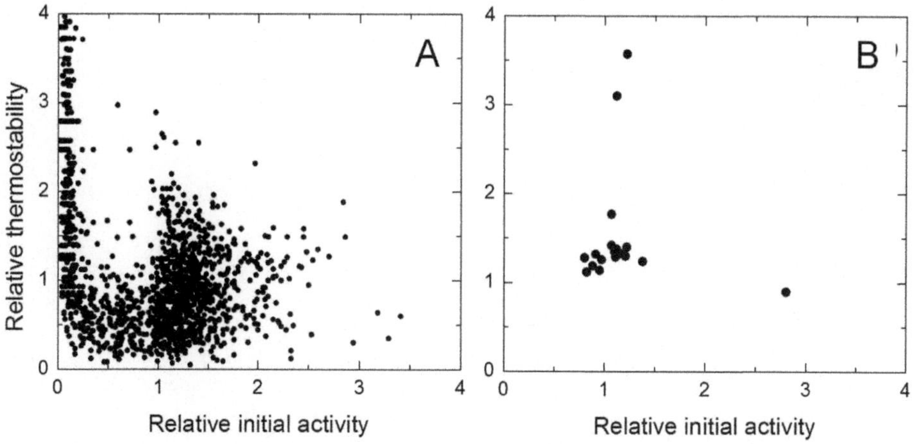


**Figure 1. Directed evolution landscape for the first generation mutant library.** (A) Initial activity *vs* thermostability of 2214 clones screened relative to the wild type. (B) Re-screening of the best mutants identified. Stability was measured by the ratio of residual activity following incubation at 55°C for 60 min to initial activity.

pH 7.6 were placed onto a thermostatically controlled thermal block and then heated at a rate of 1°C/min up to 100°C. For the chemical stability studies, increased guanidinium hydrochloride (GdnHCl) concentrations were used to induce protein unfolding, monitored through a combination of fluorescence intensity and emission maximum as previously described. The thermodynamic stability of both enzymes was analyzed based on a two-state process using the equations described in Durão et al [35] and Fernandes et al [36]. Protein aggregation occurring upon thermal unfolding was monitored by measuring static light scattering at 500 nm as excitation and emission wavelengths. Differential scanning calorimetry (DSC) measurements were carried out by

**Table 1.** Summary of library screening conditions, amino acid substitutions accumulated in PpAzoR variants and initial activity and thermostability relative to their parents.

| Generation | Temperature and incubation period | Variants | Mutations | Initial activity relative to parent[1)] | Thermostability relative to parent[2)] |
|---|---|---|---|---|---|
| 1st | 55°C, 60 min | **B1G6** | Q192R | 1.2±0.1 | **3.7±0.5** |
| | | K7E3 | Y179H | **2.8±0.2** | 0.8±0.3 |
| 2nd | 55°C, 90 min | **16B7** | Q192R, A46P, V159A | 0.7±0.1 | **3.2±0.9** |
| | | 12B8 | Q192R, Y179H | **5.4±0.5** | 0.6±0.1 |
| 3rd | 60°C, 45 min | **23C10** | Q192R, A46P, V159A, C129S | 1.4±0.1 | **2.5±0.1** |
| | | 2A1 | Q192R, A46P, V159A, A48P | 0.5±0.2 | **2.6±0.4** |
| | | 19E4 | Q192R, A46P, V159A, Y179H | **3.2±0.2** | 1.4±0.1 |
| 4th | 80°C, 60 min | **13G10** | Q192R, A46P, V159A, C129S, D7H, A178D | 0.8±0.1 | **3.9±0.2** |
| | | **6F10** | Q192R, A46P, V159A, C129S, N14D, L143Q | 1.0±0.1 | **3.8±0.2** |
| | | **27E4** | Q192R, A46P, V159A, C129S, L161M, L169P | 0.9±0.1 | **3.1±0.2** |
| | | **23C5** | Q192R, A46P, V159A, C129S, K74E, A88G | 0.9±0.1 | **2.9±0.4** |
| | | **1C11** | Q192R, A46P, V159A, C129S, E36D, L143Q | 1.6±0.4 | **2.8±0.3** |
| | | **14D4** | Q192R, A46P, V159A, C129S, A77T, N131D | 1.0±0.1 | **2.5±0.2** |
| | | 32F5 | Q192R, A46P, V159A, C129S, I6V, T79R, Y179H | **2.5±0.2** | 1.4±0.2 |
| | | 23E4 | Q192R, A46P, V159A, C129S, Y179H | **2.1±0.2** | 0.9±0.2 |
| 5th | 85°C, 150 min | 1B6 | Q192R, A46P, V159A, C129S, A178D, A88G, N131D, L143Q | 1.8±0.1 | **2.2±0.2** |
| | | 2E4 | Q192R, A46P, V159A, C129S, A178D, K74E, L143Q | 1.7± 0.01 | **2.2±0.2** |
| | | 2F11 | Q192R, A46P, V159A, C129S, A178D, A31S, K74E, A88G, L143Q | 1.9±0.1 | **2.2±0.4** |
| | | 6F11 | Q192R, A46P, V159A, C129S, A178D, N131D, L143Q | 1.7±0.1 | **2.3±0.2** |
| | | 3B9 | Q192R, A46P, V159A, C129S, A178D, A77T, F98L, N131D | 1.4±0.4 | **2.3±0.3** |

[1)]Ratio of the initial activity of the variant to the parent type ($Ai_v/Ai_p$),
[2)]Ratio of the residual activity to the initial activity of the variant (v), normalized to the parent type (p) – $(A_r/A_i)_v/(A_r/A_i)_p$
The parents for the next generations are in bold.

VP-DSC from MicroCal at a scan rate of 1°C/min. The experimental calorimetric trace was obtained with 0.2 mg/mL of protein in 20 mM Tris-HCl buffer, pH 7.6. The data were processed and fitted using Origin® software supplied by the DSC manufacturer. The progress baseline-subtracted and concentration-normalized DSC curve was fitted to non-two-state transitions.

## Results and Discussion

### Validation of high-throughput screenings

The measurement of enzymatic activity in crude extracts of *E. coli* strains overexpressing *ppAzoR* was tested using AQS, BZ, catechol, Lawsone and NSA as substrates. AQS was selected based on the highest activity differences (40 times) in induced *vs.* non induced strains. LOM531 (*E. coli* KRX) strain was selected for further studies since 1) the differences in activity are 8 times higher as compared with LOM528 (*E. coli* Tuner) and 2) KRX is a cloning as well as an expressing strain, allowing avoiding additional steps of plasmid transfer during the laboratory evolution process. Coefficients of variance (CV = standard deviation/mean x 100%) of ≈15% were achieved for final $OD_{600\ nm}$ of cell cultures as well as for total protein content and maximal rates of enzymatic activity in cell crude extracts using six different 96-well plates, ensuring the aimed reproducibility of the *ppAzoR* expression system. For the set-up of the thermostability assays the temperature and the time of incubation (55°C/60 min for the screening of the 1st generation) were selected based on conditions that resulted in a residual activity of about one-third of the initial activity [37]; the thermostability assays showed a CV of ≈17%.

### Directed evolution of PpAzoR

Around 10,000 clones were examined in five rounds of laboratory evolution for improved thermostability. The libraries were constructed to attain 1–3 amino acid changes which correspond roughly to 30 to 45% of the total number of clones with less than 10% activity of wild type [38]. For this purpose $MnCl_2$ concentrations were varied and 0.2 mM $MnCl_2$ was selected. In the 1st generation a total of 2214 clones were screened (Figure 1A) and 18 variants were identified with increased thermostability or activity and re-screened to rule out false positives (Figure 1B). The B1G6 variant with one amino acid substitution (Q192R) showed around 3.5-fold higher thermostability while maintaining a similar enzymatic activity when compared to the wild type and was selected as parent for the 2nd generation (Table 1 and Figure 2). The heat treatment was adjusted to 55°C for 90 min and 2052 clones were screened and hit 16B7 was identified showing a 3.2-fold higher thermostability as compared with the parent. In the 3rd generation the heat treatment was further adjusted to 60°C for 45 min and the screening of 2160 clones resulted in the identification of two hits, 2A1 and 23C10, exhibiting similar (2.5-fold) improvements in thermostability. The selection of the 23C10 as parent for the 4th generation was based on its higher enzymatic activity in comparison to variant 2A1. Both variants shared mutations Q192R, A46P, V159A from the previous generations and while 23C10 acquired the additional mutation C129S, 2A1 acquired mutation A48P. In the 4th generation the heat treatment was further adjusted to 80°C for 60 min and the screening of 2160 clones resulted in 6 variants with additional 1–3 different new mutations but showing similar thermostability (Table 1). Therefore a recombination approach through DNA Shuffling using the 6

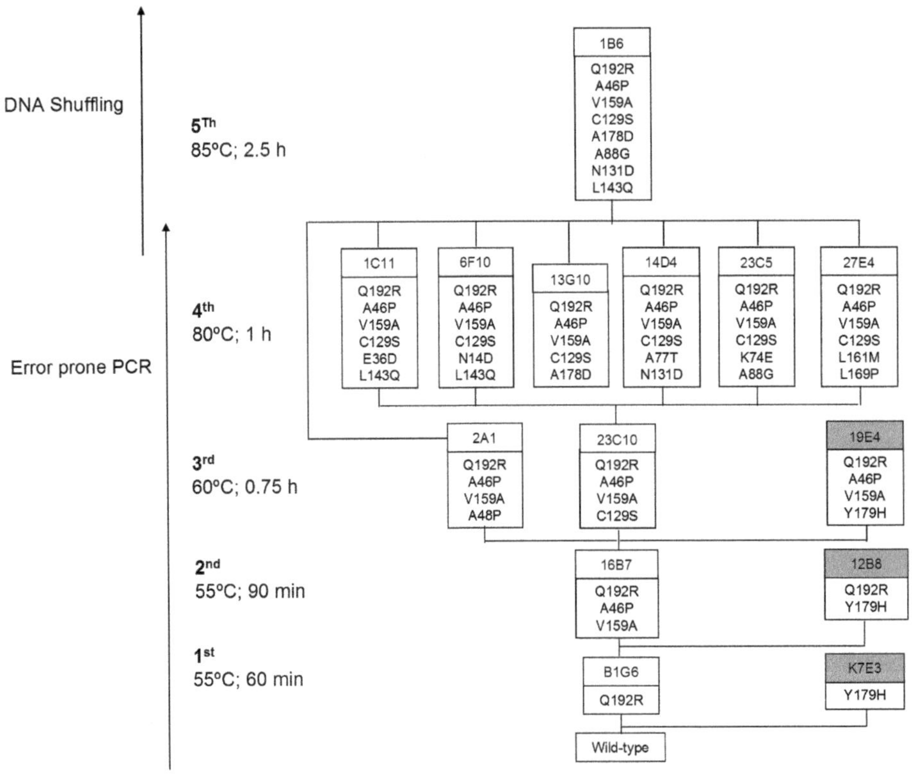


**Figure 2. Lineage of PpAzoR variants generated in this study.** Only non-synonymous mutations are shown. The mutants with higher stability are in white and the mutants with higher activity are in grey.

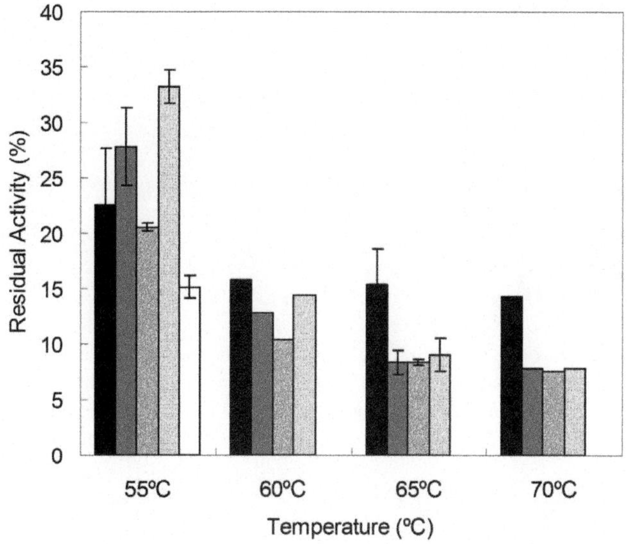


**Figure 3. Residual activity of variants from the 5th generation.** Activity was measured after incubation at different temperatures (55–70°C) for 30 min: 1B6 (black), 2E4 (grey), 2F11 (diagonal), 6F11 (dot), 3B9 (white). Reactions were performed in 100 mM sodium phosphate buffer pH 7, 100 μM AQS and 250 μM NADPH.

variants and 2A1 from the previous generation was followed (Figure 2). A total of 702 clones were screened with a heat treatment adjusted to 85°C for 150 min and 1B6 was selected after re-screening as the most thermostable variant (Table 1, Figure 3) and was purified.

## Characterization of 1B6, a variant resistant to aggregation

Purified 1B6 variant maintains its full activity after incubation for 1 h at temperatures between 40 and 85°C in clear contrast with the wild type that loses activity after one hour at 50–55°C (Figure 4A). First, kinetic stability, that quantifies the amount of enzyme that loses activity irreversibly during incubation at a certain temperature, was measured (Figure 4B). 1B6 is remarkably more stable than the wild type with a 300-fold longer half-life ($t_{1/2}$), i.e. retaining 50% of activity after 68 h at 50°C, while the wild type enzyme half-life is around 13 min. The optimum temperature of 1B6 (around 30°C) is however slightly lower than that of the wild type (around 35°C) (Figure 4C).

The thermodynamic stability of wild type enzyme and 1B6 variant was first addressed by differential scanning calorimetry (DSC) (Figure 5A). Besides the measurement of protein unfolding temperatures DSC gives unique information on thermal stability of proteins based on heat changes. The DSC thermogram of wild type shows the typical protein aggregation fingerprint characterized by an abrupt drop of the baseline after the endothermic peak leading to 100% irreversibility, a $T_m$~47°C could be roughly estimated. This feature remains at least up to the 3rd generation of evolution as clearly shown by the profile DSC trace of 2A1 variant (Fig. 5 dashed line). The variant 1B6 does not aggregate after thermal unfolding (no drop in the base line) and displays two distinct endothermic transitions (Figure 5B). For a homodimeric protein as PpAzoR two distinct endothermic transitions were assigned to the dimer dissociation ($T_m$(1B6) = 39°C, around 8°C lower than that of wild type) and monomer unfolding ($T_m$(1B6) = 55°C). Interestingly, thermal unfolding of 1B6 variant followed by fluorescence, apparently displayed a single transition

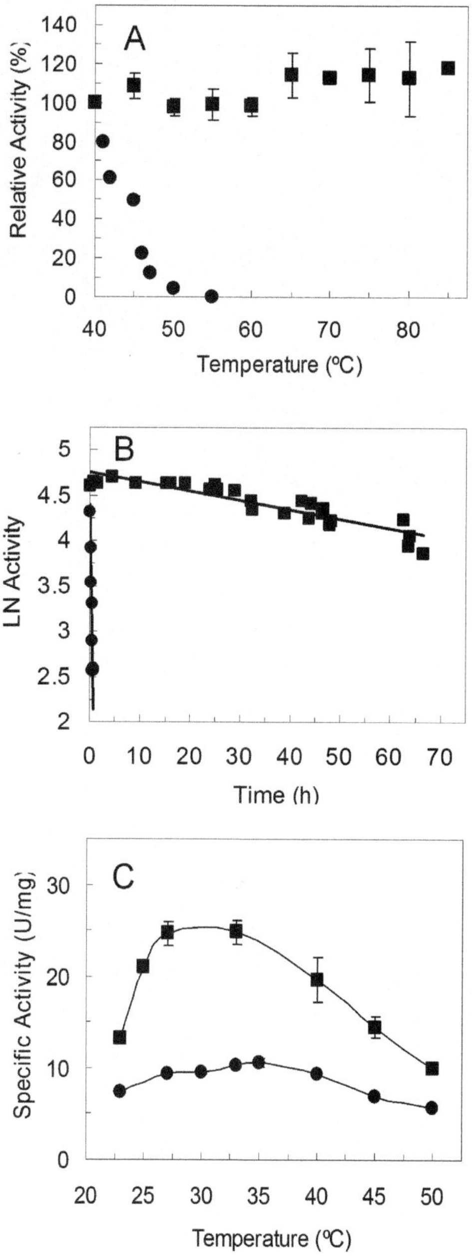


**Figure 4. 1B6 variant characterization.** (A) Activity remaining at 30°C following heating to the indicated temperature for 1 h of wild type PpAzoR (circles) and 1B6 variant (squares). Reactions were performed in 100 mM sodium phosphate buffer pH 7, 100 μM AQS and 250 μM NADPH. (B) Thermal inactivation of wild type PpAzoR (circles) and 1B6 variant (squares). Enzyme samples were incubated at 50°C and activity measured at 30°C in aliquots taken at different time intervals. (C) Temperature dependence of wild type PpAzoR (circles) and 1B6 variant (squares).

with a melting temperature of 44±1°C (Figure 5B inset). This $T_m$ value is in between dimer dissociation and monomer unfolding temperatures probed by DSC, indicating that tryptophan fluorescence was affected upon dimer dissociation. Moreover, this variant showed nearly complete reversibility of thermal unfolding. The wild type could not be followed by fluorescence due to protein aggregation upon temperature increase. However, two endothermic transitions are still perceptible in the wild type thermogram

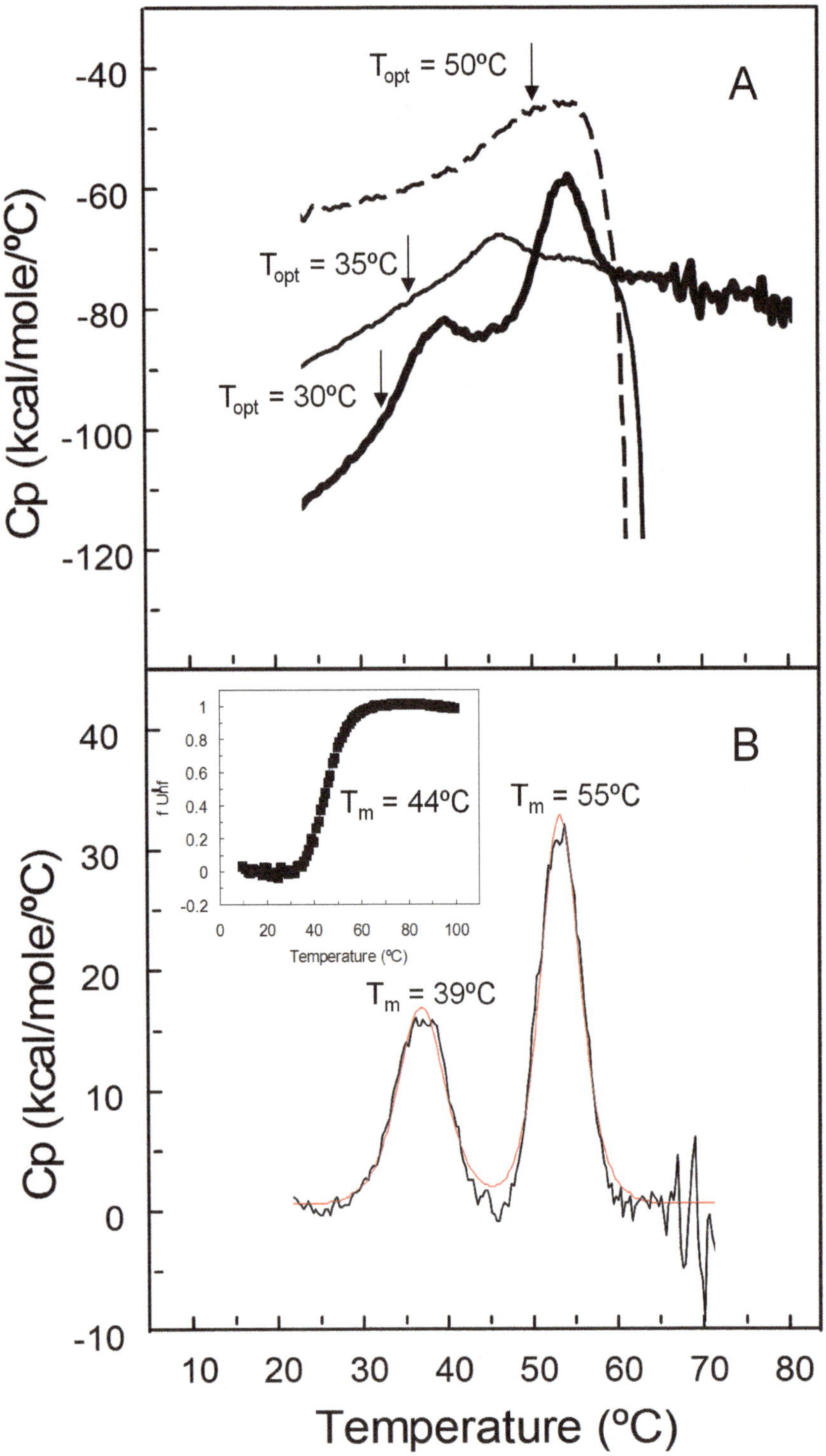


**Figure 5. Differential scanning thermograms.** (A) Wild type PpAzoR (thin line), 2A1 variant (dash line) and 1B6 variant (thick line). Arrows indicate the optimal temperature for activity which occurs in the initial part of the endothermic peak. (B) 1B6 variant fitted by non-two-state transitions (red line); Inset - Fraction of 1B6 variant unfolded ($f_{Unf}$) by temperature at pH 7.6 as measured by fluorescence emission. The solid line is the fit according to the equation $f_U = e^{(-\Delta G^\circ/RT)/(1+e^{(-\Delta G^\circ/RT)})}$.

(Figure 5A). Nevertheless, protein aggregation occurring concomitantly with monomer unfolding abolished almost completely the second transition. The inactivity of the monomer was confirmed by enzymatic assays after its generation by size-exclusion chromatography in the presence of 0.35 M GdnHCl (data not shown).

Based on the data discussed above, thermal deactivation of PpAzoR can be described by the following pathway $N_2 \leftrightarrow 2N \leftrightarrow 2U \rightarrow D$ (where $N_2$ is the homodimer, N the folded

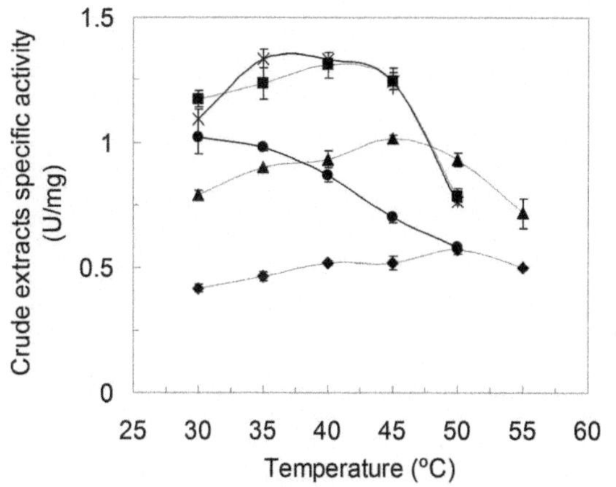


**Figure 6. Temperature dependence of PpAzoR wild type and variants measured in crude cell extracts.** Wild type PpAzoR (circles), B1G6 from 1st generation (squares), 16B7 from 2nd generation (triangles), 2A1 from 3rd generation (diamonds) and 23C10 from 3rd generation (stars). Reactions were performed using crude extracts in 100 mM sodium phosphate buffer, pH 7, in the presence of 100 μM AQS and 250 μM NADPH.

monomer, U the unfolded monomer) that lead to the irreversible aggregate state D in the wild type enzyme. The stabilization of 1B6 variant, leading to the large increase in the kinetic stability, is achieved by preventing the aggregation of the unfolded monomers (i.e. $N_2 \leftrightarrow 2N \leftrightarrow 2U$ can be proposed for the studied T interval)). A slight simultaneous decrease of the stability of the dimeric folded state of the enzyme (lower optimal temperature ($T_{opt}$) and lower $T_m$ of the first endothermic peak) as compared with the wild type was also observed.

## Identification and characterization of thermodynamically resistant variants

In order to find hit variants with increased thermodynamic stability and explore further the stability pathways of PpAzoR, we have measured the optimal temperature of several variants found in the 1st, 2nd and 3rd generation (Figure 6). An upward shift in the optimum temperature by approx. 10°C in the hit of the 1st generation ($T_{opt}$ B1G6 = 40°C), 15°C in the hit of the 2nd generation ($T_{opt}$ 16B7 = 45°C) and 20°C in one of the hits of the 3rd generation ($T_{opt}$ 2A1 = 50°C) was observed. It is now clear that the similar thermal stabilities measured for 23C10 and 2A1 (Table 2) arise apparently from different phenomena: in 23C10 ($T_{opt}$ of 35–40°C) from the impeded aggregation of the unfolded state and in 2A1 from the stabilization of the folded state.

The successive improvements in "thermostability" as indicated by the increased optimal temperature of hits in the first generations arises with a simultaneous decrease in activity (Figure 6), revealing a trade-off between high activity and high stability [39]. This is particularly evident in 2A1 variant which, in spite, of the significant up-shift of 20°C of its $T_{opt}$ shows a 2-fold decrease in the enzymatic activity when compared to the wild type enzyme. In order to improve its activity the mutation Y179H was introduced by site-directed mutagenesis since all variants showing increased activity share this mutation (Figure 2 and Table 1). Tyr179 is located in the surroundings of the substrate binding site close to the FMN cofactor and it is likely that the smaller volume of the histidine residue as compared to tyrosine may account for a slight enlargement of the cavity, contributing to an increased activity. Remarkably, the resulting variant 2A1-Y179H indeed

**Table 2.** Thermodynamic stability of the tertiary structure of PpAzoR wild type, 2A1, 2A1-Y179H and 1B6 variants as assessed by fluorescence spectroscopy.

| | Wt | 2A1 | 2A1-Y179H | 1B6 |
|---|---|---|---|---|
| **$\Delta G^{water}$ (kcal/mol)** | 4.2±0.4 | 7.5±0.7 | 3.9±0.3 | 2.3±0.2 |
| ***m* (kcal/mol M)** | 3.3±0.2 | 6.9±0.6 | 3.9±0.3 | 4.2±0.4 |
| **$[GdnHCl]_{50\%}$ (M)** | 1.1±0.2 | 1.3±0.2 | 1.0±0.3 | 0.6±0.1 |

shows an 8-fold higher activity than 2A1 corresponding to a 4-fold higher specific activity in relation to wild type (Figure 7A). A compromise between high activity and high stability is also present in 2A1-Y179H since it has a lower optimal temperature in relation to variant 2A1; however it still shows a 10°C up-shift in the $T_{opt}$ in relation to the wild type and 1B6 variant (Figure 7A). Aggregation impeded probing the unfolding process by fluorescence in 2A1 and 2A1-Y179H variants but static light scattering confirmed that unlike 1B6 variant, 2A1 variant as wild type aggregate at the highest temperatures tested (Figure 7B). Furthermore, we observed only one endothermic peak in the DSC thermogram, suggesting the simultaneous occurrence of dimer dissociation and monomer unfolding with a increased melting temperature of around 8°C over that of the wild type ($T_m$~55°C, Figure 5A). Overall, our data reveal that although variant 2A1-Y179H unfolds irreversibly similarly to the wild type, it shows a higher stabilization of the dimeric folded state with a higher melting temperature (and thus up-shifted $T_{opt}$) and 100-fold increased half-life at 50°C ($t_{1/2}$ = 23 h; Figure 7C).

## Characterization of chemical stability of variants

The thermodynamic stability of the tertiary structure of PpAzoR wild type and hit variants upon addition of GdnHCl was assessed by fluorescence (Figure 8). The wavelengths at the emission maxima reflect clearly the exposure of tryptophan residues to the high polarity of water molecules at the surface of the protein upon unfolding [40]. The wild type PpAzoR displays a GdnHCl concentration of 1.1 M at the mid-point (where 50% of the molecules are unfolded) and the native state is more stable than the unfolded state by 4.2 kcal/mol at 25°C (Table 2). The unfolding process can be accurately described according to a two-state process with folded and unfolded states of the monomers being the only ones that accumulate significantly. Hit variants follow approximately the same trend upon increased GdnHCl concentrations, however noteworthy 50% of the 1B6 molecules unfolded at half of the GdnHCl concentrations required to unfold the wild type. The native state of 1B6 is more stable than the unfolded state by only 2.3 kcal/mol. This result shows that mutations leading to the drastic reduction of the irreversible aggregation result in an overall folded structure which is less stable than the wild type enzyme, as also observed previously for thermal inactivation by DSC (Figure 5). Variant 2A1 is clearly more stable against chemically induced unfolding displaying the largest free-energy difference between the folded and the unfolded state (7.5 kcal/mol). This large free-energy gap may result from a more restricted conformational freedom (lower entropy) of the unfolded state due to the introduction of two proline residues (A46P and particularly A48P which is not present in variant 1B6). Interestingly the cooperativeness of the unfolding transition (*m* parameter) also increases very significantly in the variant 2A1. This parameter is a measure of the degree of solvent exposure for a given

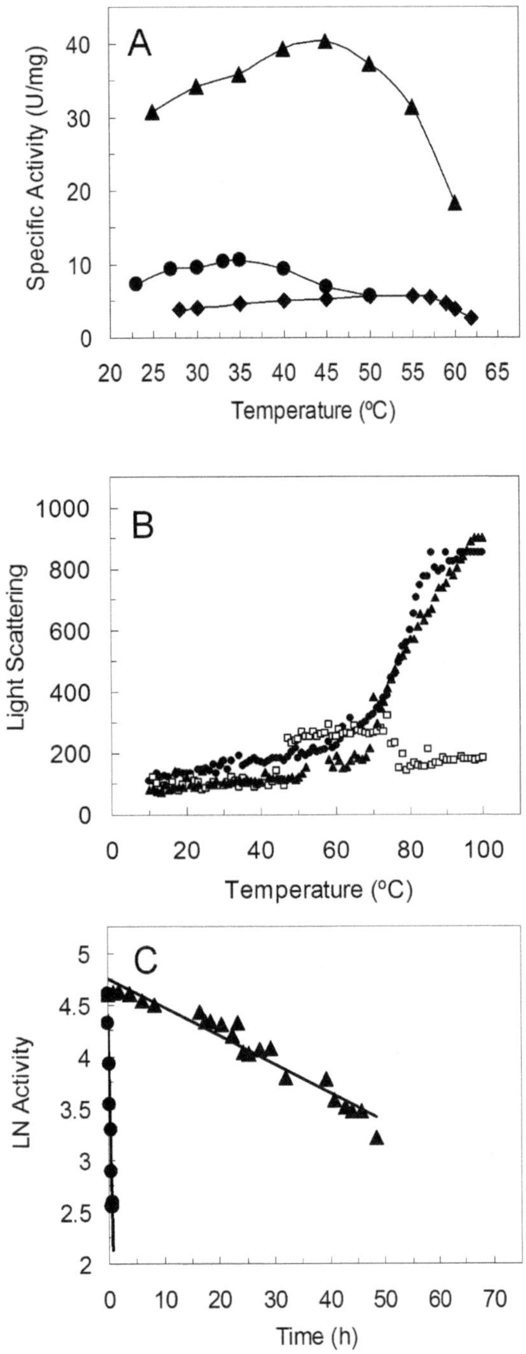


**Figure 7. 2A1 and 2A1-Y179H variants characterization.** (A) Temperature dependence of purified wild type PpAzoR (circles), 2A1 (diamonds) and 2A1-Y179H (triangles) variants. Reactions were performed in 100 mM sodium phosphate buffer, pH 7, in the presence of 100 μM AQS and 250 μM NADPH. (B) Light scattering of wild type PpAzoR (circles), 1B6 (squares) and 2A1-Y179H (triangles) variants. (C) Thermal inactivation of wild type PpAzoR (circles) and 2A1-Y179H (triangles) variant; enzyme samples were incubated at 50°C and activity measured at 30°C in aliquots taken at different time intervals.

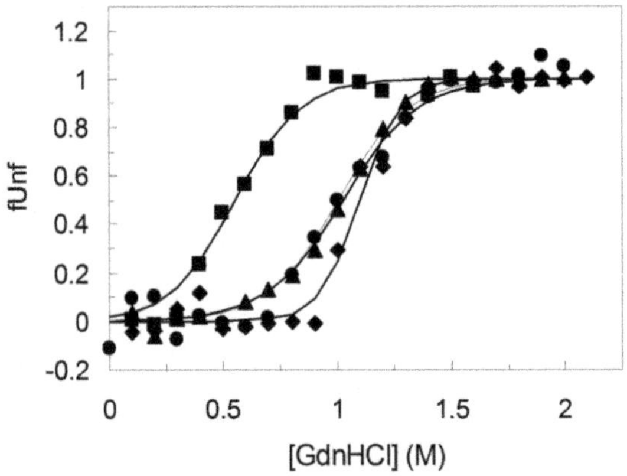


**Figure 8. Chemical stability.** Unfolded fraction ($f_{Unf}$) of wild type PpAzoR (circles) and 1B6 (squares), 2A1 (diamond) and 2A1-Y179H (triangles) variants by GdnHCl at pH 7.6 as measured by fluorescence emission. The solid line is the fit according to the equation $f_U = e^{(-\Delta G°/RT)}/(1+e^{(-\Delta G°/RT)})$

transition; the higher value in 2A1 shows that the level of the solvent exposure is larger than in the wild type and in the other variants analysed. Since the conformational freedom of the unfolded state may be restricted by the introduction of the two prolines, the larger *m* value may be rationalized by the higher degree of compactness of variant 2A1 in the folded state. These two residues are located in the interface of the two monomers [31] and the introduction of a proline in position 48 most probably enhances the homodimer stabilization. The higher exposition to the solvent (*m* value) observed in the variant 2A1 may then relate to an almost simultaneous dimer dissociation and chemical unfolding while in the other proteins the dissociation of the homodimer occurs first followed by monomer unfolding in accordance with DSC data (Figure 5A). The replacement of Y179 by a histidine residue in this variant, despite the improvement in activity, results in decreased free energy differences and cooperativeness of transition (*m*) to values similar to wild type.

## Molecular details of mutations that improve stability

PpAzoR (PDB code 4C0W) is a homodimer and its tertiary structure adopts a flavodoxin-like fold characterized by a central twisted five parallel β-sheet connected by α-helices, which flank the sheet from the front and the back [31,41]. The arrangement of the α-helices and β-stands is identical to structures of azoreductases from *E. coli* (PDB code 2Z98) [16] *Pseudomonas aeruginosa* (PDB code 2V9C) [20], *Enterococcus feacalis* (PDB code 2HPV) [17] and *Salmonella typhimurium* (PDB code 1T5B). Moreover, it contains the conserved motif patterns of flavin-dependent azoreductases, i.e. the sequence involved in the binding of FAD/FMN cofactors, the sequence involved in the dimerisation of the two monomers of the enzyme and the possible putative NAD(P)H binding motif [29,31].

All the mutations identified in both the aggregation or thermodynamically-resistant hit variants are spread on or near the surface of the protein (Figure 9), underscoring the importance of protein surface for stability as found in most of the enzymes showing improved thermostability [42,43]. In addition the majority of mutations occur in loops in accordance with previous studies that show that reduction of flexibility of surface loops generates more stable proteins [44,45]. It is expected that the less ordered regions, loops and turns, would be more accommodative to the substitutions while disruption of packing in the core may interfere with the folding of the protein.

Aggregation-resistant 1B6 hit shares mutations with thermodynamically-resistant 2A1 variant introduced in the early generations of directed evolution (Figure 9). The substitution Q192R from the

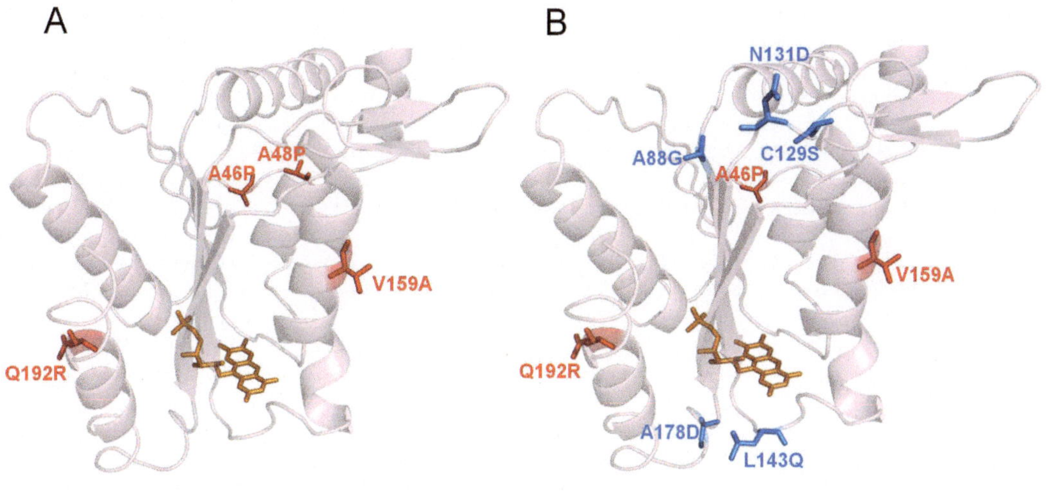


**Figure 9. Mapping of key amino acids for protein thermostabilization in PpAzoR crystal structure (PDB code 4C0W).** (A) 2A1 and (B) 1B6 variants. Mutations responsible for the thermodynamic stability are in red; mutations responsible for aggregation-resistance are in blue; the FMN is in orange.

1st generation (Figure 2 and Table 1) most probably relates to the well documented deamidation of glutamine and asparagine residues triggered by high temperatures that contributes to inactivation of irreversibly denaturated proteins [4]. This substitution in an α-helix has an additional advantage as arginine is known to enhance polar interactions, formation of hydrogen bonds and/or salt bridges, resulting in an increased overall stability. A46P and V159A mutations were picked up in the 2nd generation. V159A is located in a α-helix and the replacement of a valine by a smaller alanine, considered to be the best helix-forming residue, may enhance stability by increasing the helical propensity [4]. The A46P mutation occurs in the longest stretch of amino acid residues (20) of the enzyme without any regular secondary structure [31]. In protein structures, loops are flexible regions, and protein unfolding supposedly starts from loops. Proline can adopt only a few conformations showing the lowest conformational entropy among amino acids, restricting therefore the backbone configurations which result in the stabilization of the native state [46–48]. The hits from the 1st and 2nd generations have increased optimal temperatures in respect to that of the wild type (Figure 6) therefore it is expectable that they also presented higher melting temperatures. Therefore we can conclude that the mutations Q192R, A46P and V159A stabilised mainly the dimeric native state of the enzyme.

Variant 2A1 from the 3rd generation showing the highest melting temperature and stability of the dimeric native state carries a second Pro (A48P) in the same loop as A46P. A fraction of this loop (Ile47, His49, Phe50) is involved in a complex net of hydrogen bonding interactions with the same stretch of residues of the second monomer of the PpAzoR dimer [31]. Thus, it is probable that Pro48 apart from imposing constraints in backbone conformations of the native state, by decreasing flexibility, contribute critically to the stabilization of the homodimer in solution in accordance with the DSC (Figure 5A) and chemical stability (Figure 8) results.

The aggregation-resistant 1B6, the variant with the longest half-life and dramatically less aggregation-prone, shows not A48P but C129S mutation (Figure 2 and Table 1), located in the putative NAD(P)H binding motif [31]. Cysteines are the most reactive amino acids in proteins which is the reason behind their involvement in specific interactions within proteins (e.g., disulfide bridges or metal ligand binding) and/or their localization in positions inaccessible to the solvent [4]. Importantly, serine with a hydroxyl group allows for the potential formation of additional hydrogen bonds within the protein structure or with the solvent.

**Table 3.** Mechanisms of protein thermostabilization present in PpAzoR variants created by directed evolution [4,46–48,54–57].

| Protein thermostabilization | Mutation |
|---|---|
| **Increasing structure robustness** | |
| replacing temperature-sensitive residues by polar or charged residues | C129S; Q192R; N131D |
| strengthening solvent-exposed loops by restricting the number of available main-chain conformations | A46P; A48P |
| dimer stabilization | A48P |
| increasing helical propensity | V159A |
| conformational strain release | A88G |
| **Resistance to aggregation** | |
| residual repulsive interactions between partially unfolded structures | N131D; L143Q; A178D |
| disruption of surface hydrophobic patches | L143Q; A178D |

Interestingly, three out of four mutations introduced in the 4th generation are substitutions to surface charged (N131D, A178D) or polar (L143Q) residues. Notably, five out of the six hit variants from the 4th generation have at least one mutation introducing a charged residue and all variants from the 5th generation have 3 to 4 charged residues in a total of 7 to 8 mutations (Table 1). It has been proposed that the presence of charged residues on the protein surface may lead to residual repulsive interactions increasing the solubility and preventing native state aggregation [49]. The introduction of the positively charged residue R28 was identified as the key determinant of aggregation resistance of the unfolded state of human antibody variable domains created by directed evolution [50] while A281E and V221D mutations inhibited protein aggregation and facilitated refolding of the catalytic active conformation of a *Candida antartica* lipase [51]. In the case of a lipase from *Bacillus subtilis* the simultaneous improvement in thermodynamic stability and non-aggregating properties was achieved by the introduction of two proline residues that led to a better anchoring of loops, disturbing the exposed hydrophobic patches of 6B lipase and one charged residue, E134 [52]. Moreover, we observe here that mutations L143Q and A178D in 1B6 variant are close to a hydrophobic patch present on the protein surface in the surroundings of the FMN molecule, potentially decreasing the association of exposed hydrophobic patches and the formation of aggregate compatible structures. Therefore the combination of the increased surface charge/polarity with disturbance of hydrophobic patches even though resulting in a decreased inter-dimer association, i.e. lower stability of the native state, lead to the abolishment of the aggregational pathway with a critical gain in the kinetic stability of the enzyme. In an attempt to create an aggregation and thermodynamically-resistant variant we have introduced by site-directed mutagenesis the mutation A48P, present in the 2A1, into the 1B6 variant. In contrast to the 1B6, the resulting variant 1B6-A48P aggregates upon temperature raise while maintaining a similar $T_{opt}$ (data not shown). These results suggest the absence of a synergistic interaction between A48P mutation and the mutations present in 1B6 variant, in accordance with our DNA-shuffling results. In fact, none of the five variants selected in the 5th generation as a result of the shuffling among variant 2A1 and variants of the 4th generation, including A48P mutation (Table 1 and Figure 2). Most probably the surface mutations of 1B6, which are not present in 2A1, and A48P comprise an opposite effect in the stability of the native state of the variants since the endothermic transitions in the DSC assigned to dimer dissociation occur at ~55°C for 2A1 and at 39°C for 1B6 (i.e. 8°C higher and lower, respectively) than the wild type Tm ~47°C.

## Conclusions

Directed evolution is an effective and reliable approach to engineer proteins contributing simultaneously to our understanding of protein function and adaptation mechanisms. Two types of protein stability (thermodynamic and kinetic) are of interest from a fundamental and applied perspective. Increasing the thermodynamic thermostability is the main issue when an enzyme is used under denaturing conditions (i.e. high temperatures or organic solvents) [2]. Industrial applications require active enzymes rather than enzymes that are in a reversibly inactivated state. For other enzymes, for example those that can be utilized for diagnostic puroposes, to ensure a suitable shelf-life, it is often long term stability that needs to be improved. Most frequently both stabilities correlate since increasing the enzyme resistance to unfolding state also increases its resistance to inactivation. Remarkably in the present study we have identified a hit variant of PpAzoR, 1B6, with increased resistance to inactivation by aggregation. This was accomplished by drastically decreasing the rate of the irreversible step of denaturation (the aggregation of the unfolded monomers) upon thermal stress. Although variant 1B6 is less stable for thermal or chemically induced unfolding when compared to the wild type, lower amounts of the irreversibly denatured state accumulated in the long-term stability assay. The identification of this hit correlates well with the experimental screening strategy followed. In order to select for mutations that improve the stability of the catalytically competent native state a secondary screen strategy was implemented based on activity measurements at increasing temperatures i.e. by probing the dissociation of the dimer. Importantly our work shows that the resistance of the unfolded monomers to aggregation (resulting also in a destabilization of the dimeric folded state) is achieved through a delicate disturbance of hydrophobic patches and to a clearly increased surface net charge. On the other hand, the increase of robustness of the native structure is mainly an outcome of strengthening of solvent-exposed loops and inter-dimer interactions (Table 3). Overall, this work suggests that protein charge can be exploited to impart robust resistance to protein aggregation with implications on de novo protein design efforts, where unpredictable protein properties, including aggregation, remain a significant challenge. From the point of view of protein evolvability mutations that prevent the aggregation of the unfolded state, increasing thus the level of soluble protein, contribute to protein fitness [53], and therefore variant 1B6 is potentially a good candidate to be evolved for new functions of PpAzoR.

## Acknowledgments

The authors would like to thank Dr. Isabel Bento and Dr. Ana Maria Gonçalves for helpful discussions on structural aspects of PpAzoR.

## Author Contributions

Conceived and designed the experiments: LOM VB EPM. Performed the experiments: VB NG. Analyzed the data: LOM VB EPM. Contributed reagents/materials/analysis tools: LOM. Wrote the paper: LOM VB EPM.

## References

1. Volkin DB, Klibanov AM (1989) Minimizing protein inactivation In: CreightonTE, editors. Protein Function. A practical approach. Oxford: IRL press. pp. 1–24.
2. Bommarius AS, Paye MF (2013) Stabilizing biocatalysts. Chem Soc Rev 42: 6534–6565.
3. Sanchez-Ruiz JM (2010) Protein kinetic stability. Biophys Chem 148: 1–15.
4. Vieille C, Zeikus GJ (2001) Hyperthermophilic enzymes: sources, uses, and molecular mechanisms for thermostability. Microbiol Mol Biol Rev 65: 1–43.
5. Cherry JR, Fidantsef AL (2003) Directed evolution of industrial enzymes: an update. Curr Opin Biotechnol 14: 438–443.
6. Dalby PA (2011) Strategy and success for the directed evolution of enzymes. Curr Opin Struct Biol 21: 473–480.
7. Romero PA, Arnold FH (2009) Exploring protein fitness landscapes by directed evolution. Nat Rev Mol Cell Biol 10: 866–876.
8. Acharya P, Rajakumara E, Sankaranarayanan R, Rao NM (2004) Structural basis of selection and thermostability of laboratory evolved *Bacillus subtilis* lipase. J Mol Biol 341: 1271–1281.
9. Baik SH, Michel F, Aghajari N, Haser R, Harayama S (2005) Cooperative effect of two surface amino acid mutations (Q252L and E170K) in glucose dehydrogenase from *Bacillus megaterium* IWG3 on stabilization of its oligomeric state. Appl Environ Microbiol 71: 3285–93.
10. Flores H, Ellington AD (2002) Increasing the thermal stability of an oligomeric protein, beta-glucuronidase. J Mol Biol 315: 325–337.

11. Garcia-Ruiz E, Mate D, Ballesteros A, Martinez AT, Alcalde M (2010) Evolving thermostability in mutant libraries of ligninolytic oxidoreductases expressed in yeast. Microb Cell Fact 9: 17.
12. Miyazaki K, Takenouchi M, Kondo H, Noro N, Suzuki M, et al. (2006) Thermal stabilization of *Bacillus subtilis* family-11 xylanase by directed evolution. J Biol Chem 281: 10236–10242.
13. Salazar O, Cirino PC, Arnold FH (2003) Thermostabilization of a cytochrome p450 peroxygenase. Chembiochem 4: 891–893.
14. Shivange AV, Serwe A, Dennig A, Roccatano D, Haefner S, et al. (2012) Directed evolution of a highly active *Yersinia mollaretii* phytase. Appl Microbiol Biotechnol 95: 405–418.
15. Vazquez-Figueroa E, Yeh V, Broering JM, Chaparro-Riggers JF, Bommarius AS (2008) Thermostable variants constructed via the structure-guided consensus method also show increased stability in salts solutions and homogeneous aqueous-organic media. Protein Eng Des Sel 21: 673–80.
16. Ito K, Nakanishi M, Lee WC, Zhi Y, Sasaki H, et al. (2008) Expansion of substrate specificity and catalytic mechanism of azoreductase by X-ray crystallography and site-directed mutagenesis. J Biol Chem 283: 13889–13896.
17. Liu ZJ, Chen H, Shaw N, Hopper SL, Chen L, et al. (2007) Crystal structure of an aerobic FMN-dependent azoreductase (AzoA) from *Enterococcus faecalis*. Arch Biochem Biophys 463: 68–77.
18. Maier J, Kandelbauer A, Erlacher A, Cavaco-Paulo A, Gubitz GM (2004) A new alkali-thermostable azoreductase from *Bacillus* sp. strain SF. Appl Environ Microbiol 70: 837–844.
19. Suzuki Y, Yoda T, Ruhul A, Sugiura W (2001) Molecular cloning and characterization of the gene coding for azoreductase from *Bacillus* sp. OY1-2 isolated from soil. J Biol Chem 276: 9059–9065.
20. Wang CJ, Hagemeier C, Rahman N, Lowe E, Noble M, et al. (2007) Molecular cloning, characterisation and ligand-bound structure of an azoreductase from *Pseudomonas aeruginosa*. J Mol Biol 373: 1213–1228.
21. Cervantes FJ, Dos Santos AB (2011) Reduction of azo dyes by anaerobic bacteria: microbiological and biochemical aspects. Rev Environ Sci Bio 10: 125–137.
22. Dos Santos AB, Cervantes FJ, van Lier JB (2007) Review paper on current technologies for decolourisation of textile wastewaters: perspectives for anaerobic biotechnology. Bioresour Technol 98: 2369–2385.
23. Forgacs E, Cserhati T, Oros G (2004) Removal of synthetic dyes from wastewaters: a review. Environ Int 30: 953–971.
24. Deller S, Macheroux P, Sollner S (2008) Flavin-dependent quinone reductases. Cell Mol Life Sci 65: 141–160.
25. Leelakriangsak M, Huyen NT, Towe S, van Duy N, Becher D, et al. (2008) Regulation of quinone detoxification by the thiol stress sensing DUF24/MarR-like repressor, YodB in *Bacillus subtilis*. Mol Microbiol 67: 1108–1124.
26. Liu G, Zhou J, Fu QS, Wang J (2009) The *Escherichia coli* azoreductase AzoR is involved in resistance to thiol-specific stress caused by electrophilic quinones. J Bacteriol 191: 6394–6400.
27. Sollner S, Deller S, Macheroux P, Palfey BA (2009) Mechanism of flavin reduction and oxidation in the redox-sensing quinone reductase Lot6p from *Saccharomyces cerevisiae*. Biochemistry 48: 8636–8643.
28. Sollner S, Nebauer R, Ehammer H, Prem A, Deller S, et al. (2007) Lot6p from *Saccharomyces cerevisiae* is a FMN-dependent reductase with a potential role in quinone detoxification. Febs J 274: 1328–1339.
29. Mendes S, Pereira L, Batista C, Martins LO (2011) Molecular determinants of azo reduction activity in the strain *Pseudomonas putida* MET94. Appl Microbiol Biotechnol 92: 393–405.
30. Mendes S, Farinha A, Ramos CG, Leitao JH, Viegas CA, et al. (2011) Synergistic action of azoreductase and laccase leads to maximal decolourization and detoxification of model dye-containing wastewaters. Bioresour Technol 102: 9852–9859.
31. Gonçalves AMD, Mendes S, de Sanctis D, Martins LO, Bento I (2013) The crystal structure of *Pseudomonas putida* azoreductase: the active site revisited. FEBS J, 28D: 6643–6657.
32. Joern JM (2003) DNA Shuffling. In: Arnold FH, Georgiou G, editors. Directed evolution library creation: methods and protocols. New Jersey: Humana Press, Totowa. pp. 85–89.
33. Zhao H, Arnold FH (1997) Optimization of DNA shuffling for high fidelity recombination. Nucleic Acids Res 25: 1307–1308.
34. Martins LO, Soares CM, Pereira MM, Teixeira M, Costa T, et al. (2002) Molecular and biochemical characterization of a highly stable bacterial laccase that occurs as a structural component of the *Bacillus subtilis* endospore coat. J Biol Chem 277: 18849–18859.
35. Durão P, Bento I, Fernandes AT, Melo EP, Lindley PF, et al. (2006) Perturbations of the T1 copper site in the CotA laccase from *Bacillus subtilis*: structural, biochemical, enzymatic and stability studies. J Biol Inorg Chem 11: 514–526.
36. Fernandes AT, Martins LO, Melo EP (2009) The hyperthermophilic nature of the metallo-oxidase from *Aquifex aeolicus*. BBA-Proteins Proteom 1794: 75–83.
37. Cirino PC, Georgescu R (2003) Screening for thermostability. In: Arnold FH, Georgiou G, editors. Directed enzyme evolution: screening and selection methods. Totowa, New Jersey: Humana Press. pp. 117–125.
38. Salazar O, Sun S (2003) Evaluating a Screen and Analysis of Mutant Libraries. In: Arnold FH, Georgiou G, editors. Directed enzyme evolution: screening and selection methods. Totowa, New Jersey: Humana Press. pp. 85–97.
39. Jaenicke R, Bohm G (1998) The stability of proteins in extreme environments. Curr Opin Struc Biol 8: 738–748.
40. Lakowicz JR (1999) Principles of Fluorescence Spectroscopy. New York: Kluwer Academic/Plenum Publishers.
41. Correia B, Chen Z, Mendes S, Martins LO, Bento I (2011) Crystallization and preliminary X-ray diffraction analysis of the azoreductase PpAzoR from *Pseudomonas putida* MET94. Acta Crystallogr Sect F Struct Biol Cryst Commun 67: 121–123.
42. Eijsink VGH, Gaseidnes S, Borchert TV, van den Burg B (2005) Directed evolution of enzyme stability. Biomol Eng 22: 21–30.
43. Zhao HM, Arnold FH (1999) Directed evolution converts subtilisin E into a functional equivalent of thermitase. Protein Eng 12: 47–53.
44. Arnold FH, Wintrode PL, Miyazaki K, Gershenson A (2001) How enzymes adapt: lessons from directed evolution. Trends Biochem Sci 26: 100–106.
45. Spiller B, Gershenson A, Arnold FH, Stevens RC (1999) A structural view of evolutionary divergence. P Natl Acad Sci USA 96: 12305–12310.
46. Matthews BW (1993) Structural and Genetic-Analysis of Protein Stability. Annu Rev Biochem 62: 139–160.
47. Watanabe K, Kitamura K, Suzuki Y (1996) Analysis of the critical sites for protein thermostabilization by proline substitution in oligo-1,6-glucosidase from *Bacillus coagulans* ATCC 7050 and the evolutionary consideration of proline residues. Appl Environ Microbiol 62: 2066–2073.
48. Watanabe K, Masuda T, Ohashi H, Mihara H, Suzuki Y (1994) Multiple Proline Substitutions Cumulatively Thermostabilize *Bacillus cereus* Atcc7064 Oligo-1,6-Glucosidase - Irrefragable Proof Supporting the Proline Rule. Eur J Biochem 226: 277–283.
49. Lawrence MS, Phillips KJ, Liu DR (2007) Supercharging proteins can impart unusual resilience. J Am Chem Soc 129: 10110–10112.
50. Famm K, Hansen L, Christ D, Winter G (2008) Thermodynamically stable aggregation-resistant antibody domains through directed evolution. J Mol Biol 376: 926–931.
51. Zhang N, Suen WC, Windsor W, Xiao L, Madison V, et al. (2003) Improving tolerance of *Candida antarctica* lipase B towards irreversible thermal inactivation through directed evolution. Protein Eng 16: 599–605.
52. Kamal MZ, Ahmad S, Molugu TR, Vijayalakshmi A, Deshmukh MV, et al. (2011) *In vitro* evolved non-aggregating and thermostable lipase: structural and thermodynamic investigation. J Mol Biol 413: 726–741.
53. Tokuriki N, Tawfik DS (2009) Stability effects of mutations and protein evolvability. Curr Opin Struct Biol 19: 596–604.
54. Eijsink VG, Bjork A, Gaseidnes S, Sirevag R, Synstad B, et al. (2004) Rational engineering of enzyme stability. J Biotechnol 113: 105–120.
55. Fersht A (1999) Structure and mechanism in protein science: A guide to enzyme catalysis and protein folding. New York: W. H. Freeman and Company.
56. Lee B, Vasmatzis G (1997) Stabilization of protein structures. Curr Opin Biotechnol 8: 423–428.
57. van den Burg B, Eijsink VG (2002) Selection of mutations for increased protein stability. Curr Opin Biotechnol 13: 333–337.

# 7

# Deletional Protein Engineering Based on Stable Fold

**Govindan Raghunathan[1], Nagasundarapandian Soundrarajan[1], Sriram Sokalingam[1], Hyungdon Yun[2], Sun-Gu Lee[1]***

**1** Department of Chemical Engineering, Pusan National University, Busan, South Korea, **2** School of Biotechnology, Yeungnam University, Gyeongsan, South Korea

## Abstract

Diversification of protein sequence-structure space is a major concern in protein engineering. Deletion mutagenesis can generate a protein sequence-structure space different from substitution mutagenesis mediated space, but it has not been widely used in protein engineering compared to substitution mutagenesis, because it causes a relatively huge range of structural perturbations of target proteins which often inactivates the proteins. In this study, we demonstrate that, using green fluorescent protein (GFP) as a model system, the drawback of the deletional protein engineering can be overcome by employing the protein structure with high stability. The systematic dissection of N-terminal, C-terminal and internal sequences of GFPs with two different stabilities showed that GFP with high stability (s-GFP), was more tolerant to the elimination of amino acids compared to a GFP with normal stability (n-GFP). The deletion studies of s-GFP enabled us to achieve three interesting variants viz. s-DL4, s-N14, and s-C225, which could not been obtained from n-GFP. The deletion of 191–196 loop sequences led to the variant s-DL4 that was expressed predominantly as insoluble form but mostly active. The s-N14 and s-C225 are the variants without the amino acid residues involving secondary structures around N- and C-terminals of GFP fold respectively, exhibiting comparable biophysical properties of the n-GFP. Structural analysis of the variants through computational modeling study gave a few structural insights that can explain the spectral properties of the variants. Our study suggests that the protein sequence-structure space of deletion mutants can be more efficiently explored by employing the protein structure with higher stability.

**Editor:** Nediljko Budisa, Berlin Institute of Technology, Germany

**Funding:** This research was supported by Basic Science Program through the National Research Foundation of Korea (NRF) funded by the Ministry of Education, Science and Technology (NRF-2011-0021222). The funders had no role in study design, data collection and analysis, decision to publish, or preparation of the manuscript.

**Competing Interests:** The authors have declared that no competing interests exist.

* E-mail: sungulee@pusan.ac.kr

These authors contributed equally to this work.

## Introduction

Protein engineering tools have enabled not only the generation of valuable proteins with new functions and stabilities but also the various fundamental studies on protein structure, function, stability and evolution [1,2]. In general, the efficiency of a protein engineering method relies on the diversity of protein sequence-structure space, because the chance to find out new properties may be increased by exploring more diverse protein sequence-structure space [3]. Substitution mutagenesis is a major tool which has been employed in various protein engineering studies. The approach induces the diversification of target protein sequence-structure space by altering the amino acid residues in natural sequence with other amino acids, which allows for the modulation of target protein function and stability. Indeed, it has tremendously contributed to the development of various academic and industrial fields related to protein engineering and science.

Deletion mutagenesis is another type of mutation used in protein science and engineering. The deletion approach has been mainly employed in mapping the function of a target protein, which significantly contributed to elucidating the relation between protein sequence, structure and function [4]. On the other hand, the deletion mutagenesis has also been recognized as an important protein engineering tool. In particular, deletion mutagenesis is known to induce different structural changes in target proteins from those caused by substitution mutations, which provides important tools for altering protein structures and functions in directions not achievable with substitutions alone [4]. Protein properties such as stability, activity, interface, and size could be changed by the deletion mutagenesis such as N- or C-terminus deletions, loop deletions and random deletions [5,6,7,8,9,10,11]. In addition, protein structural database indicate that deletion mutations are also involved in protein sequence evolution in nature [11], suggesting that deletion mutations can be included in the mutagenesis step of directed evolution. Despite these potentials of deletion mutagenesis in protein sequence engineering and evolution, the approach has not been popularly used compared to substitution mutagenesis method because deletions lead to relatively large structural perturbations, and so the target proteins are prone to be inactivated [12,13,14].

Recently, there have been theoretical and experimental demonstrations that protein stability promotes the evolvability of proteins [15,16,17,18]. The concept is that a protein's capacity to evolve is enhanced by the mutational robustness of stable protein. Although the protein stability-evolvability relationship is not absolutely clear yet, the reports sufficiently provided the insight that the protein engineering and design tasks can be facilitated using stable protein folds with high stability. Indeed, the concept was successfully employed in some typical protein engineering studies based on canonical amino acid mediated substitution mutagenesis [19,20,21]. We also demonstrated that the concept could be applied to non-canonical amino acid mediated protein engineering approach [22], which permitted the generation of functional proteins with unnatural properties efficiently [23,24].

In this study, we show that the protein stability-evolvability concept can also be used to overcome the drawback of the deletion mutagenesis method mentioned above. Our idea is that the use of a more stable fold may allow for more deletions without perturbing the protein's native structure too much, which may confer us more chances to generate deletion mutants with functional and varied properties. To demonstrate this, green fluorescent proteins (GFP) with different stabilities were mutated by deletion mutations as a model system. We systematically investigated the effect of GFP stability on the tolerance to N-terminal, C-terminal, and internal loop sequence deletions. From the deletion study using a highly stable GFP fold, some interesting variants which cannot be obtained from GFP fold with normal stability could be generated, and their biophysical properties were studied. To understand the properties of generated variants, *in silico* studies on the structural changes of the variants were also performed.

## Materials and Methods

### Materials

The reagents for molecular cloning work were purchased from New England Biolabs (Tokyo, Japan) and chemicals were from Sigma chemicals (St. Louis, MO, USA). pET30b(+) from Novagen was used for the expression of target proteins. The cloning host *Escherichia coli* strain DH5α was used for plasmid DNA preparation in this study. *E. coli* BL21 (DE3) was used as the host for expression of target protein.

### Construction and Expression of s-GFP, n-GFP and Variants

Primers were designed for specific deletions and their sequence details were provided in **Supporting Information S1** along with target genes. The genes of n-GFP variants and s-GFP variants were respectively PCR amplified from pQE80-GFP [25] and pQE-80L-GFPhs1 [22] using the designed primers and cloned into Nde I and Xho I restriction sites of pET30b(+). The internal deletions were made using overlap extension PCR [26]. The constructs had N- or C-terminal hexahistidine tag for affinity purification. The PCR amplified genes were cloned into pET30b(+) and over-expression of s-GFP, n-GFP and deletion variants were carried out in *E. coli* BL21 (DE3). *E. coli* BL21 (DE3) harboring the variants were grown at 37°C and induced at 0.6–0.8 $O.D_{600}$ with 1 mM IPTG for 5 hrs. For SDS-PAGE analysis, 3 $O.D_{600\ nm}$ of over expressed cells were collected by centrifuging at a speed of 5000 rpm for 10 min at 4°C and lysed using BugBuster$^{TM}$ Protein Extraction Reagent (Novagen). The lysate was centrifuged at 13,000 rpm for 30 min to separate soluble and insoluble fractions, and analyzed on SDS-PAGE gel. For affinity column purification of protein, cells were grown at 37°C and induced at 0.6–0.8 $O.D_{600\ nm}$ for 18–48 hrs at 15°. The cells were collected by centrifugation at 5000 rpm for 10 min at 4°C after overexpression and lysed using French press (Thermo Scientific). The soluble protein fractions were purified by Ni-NTA column chromatography (GE Healthcare Bio-Sciences, Sweden) by standard protocol. Elution fractions were analyzed by SDS-PAGE and dialyzed against phosphate buffered saline (PBS) pH 7.4.

### Measurement of Fluorescence Intensity

The fluorescence of s-GFP, n-GFP and deletion variants were recorded on Perkin Elmer/Wallac Victor 2 Multilabel Counter (1420–011) by measuring the fluorescence intensity by exciting at 485 nm and emission at 515 nm with excitation/emission slits of 5.0 nm. The whole cell fluorescence assay was performed by suspending the cell pellet of 200 µl over expressed culture in PBS for measurement. The excitation/emission spectra of protein were recorded on PerkinElmer LS-55 Fluorescence spectrometer attached with FL WinLab$^{TM}$ software for spectral analysis. For excitation scan, the emission wavelength was set at 560 nm with slit of 2.5 nm except for s-C225 (10 nm). In emission scan, the emission wavelength was scanned by exciting at 440 nm, between 450–600 nm with excitation slit of 2.5 nm except for s-C225 (10 nm). The scan speed for both spectral scanning was 50 nm/min.

### Denaturation and Refolding of GFP Variants

Purified GFP variant proteins (30 µM) each were denatured under harsh condition in PBS (pH 7.4) containing 8 M urea and 5 mM DTT unfolded at 95°C for 5 min. Urea denatured samples were renatured at room temperature by 100 fold dilution into PBS containing 5 mM DTT without urea. The concentrations of denatured proteins were adjusted to 0.3 µM and recovered fluorescence was measured using PerkinElmer LS 55 Fluorescence spectrometer (490 nm excitation, 511 nm emission, 2.5 nm (10 nm for s-C225) excitation/emission slit) for 30 min with an interval of 3 sec. The recovered fluorescence was normalized by dividing with maximum fluorescence value of the spectra. The normalized values were fitted with Sigma Plot (Systat Software Inc., CA) using equations as described by previous report [27].

### Equilibrium Refolding Plot

Column purified proteins at a concentration of 100 µM in 1xPBS (pH 7.4) were diluted 5 fold in 10 M urea dissolved in refolding buffer (25 mM Tris pH 7.5, 0.2 M NaCl, 5% Glycerol, 1 mM DTT) to get final concentration of 8 M urea and unfolded at 95°C for 5 min. Fraction refolded was measured by diluting urea denatured proteins into refolding buffer containing 5 mM DTT to various concentration of urea (1 M–6 M) by allowing to refold for 24 hrs at 15°C. Fluorescence was measured by exciting at 485 nm and emission at 515 nm with excitation/emission slits of 5.0 nm and was recorded on Perkin Elmer/Wallac Victor 2 Multilabel Counter (1420–011). The concentration of urea at which the 50% of initial fluorescence was recovered after 24 hrs incubation is termed $C_{0.5}$. $C_{0.5}$ was determined by sigmoidal fit using sigma plot (Systat Software Inc., CA) [27].

### Fluorescence Microscope Imaging

Liquid culture of s-GFP and its variants of N-terminal and C-terminal deletions were spread with loop on clean glass slides and air dried for 5–10 min. The slides mounted with samples were observed under 40× magnification with 10× objective using Zeiss Axioplan 2 Fluorescence Microscope. The fluorescent cells were captured with digital imaging system and analyzed using Axiovision Rel. 4.8.2, software equipped with microscope.

### Measurement of Circular Dichroism Spectrum

Protein samples were buffer exchanged to 10 mM sodium phosphate (pH 7.4) and the concentration was adjusted to 0.2 mg/ml. CD experiments were performed on JASCO J-715 Spectropolarimeter with a Jasco PTC 348WI temperature controller. The Far-UV spectra (240–200 nm) were measured in

0.1 cm path length quartz cuvette filled with 0.2 mg/ml protein solution. The secondary structure content from the Far-UV spectra was estimated through DichroWeb [28,29] server with 7 set of reference in Contin program.

## Molecular Modeling of s-GFP and its Variants

The three dimensional structure of the GFP variant, s-GFP was modeled using the template structure of superfolder GFP (PDB ID 2B3P) [27]. The Build mutant protocol from the Protein modeling module of the Discovery Studio 2.0 (Accelrys Inc., San Diego, CA, USA) was used to model the s-GFP structure with the following steps. In the first step, the mutations required to represent the s-GFP were incorporated by replacing those amino acids of the template superfolder GFP structure. Second, the modified structure was optimized using the simulated annealing algorithm only for the mutated residues and its surrounding atoms within the radius of 4.5 Å so that the newly introduced amino acid positioned properly without any steric clash with the surrounding amino acids. Finally, energy minimization was performed for the entire s-GFP structure in order to optimize the overall structure for the introduced mutations. The conditions used for energy minimization were as follows; the atoms were typed with Momany & Rone CHARMm force field, non-bonded interactions cutoff were set to 10 to 12 Å, solvent conditions used were distance-dependent dielectrics with a dielectric constant set to 4 to mimic the screening effects of the solvent at the crude level, energy minimization cycles were carried out until the RMS gradient tolerance was ≤0.1 kcal mol $^{-1}$Å $^{-1}$. This modeled s-GFP structure having chromophore was used as the template for modeling the terminal deletion variants of GFP by deleting the corresponding residues mentioned in the Table 1 for each variant. For the modeling of internal deletion variants, the primary sequences after deleting the residues mentioned in the Table 1 for each variant were prepared and modeled using the s-GFP as the template structure with the help of Build homology module of the Discovery Studio 2.0. All the deletion variant structures were energy minimized using the same procedure mentioned above for the energy minimization of s-GFP. The solvent accessible surface area were predicted using the 1.4 Å probe radius, and the amino acid side chains with more than 25% accessibility to this probe radius were considered to be surface exposed residues in this study. The bond distance between N-O for an electrostatic pair to form a strong salt-bridge was considered to be ≤4 Å, and for the hydrogen bond, the distance between the donor and the acceptor atom was considered to be ≤2.5 Å [30,31].

## Molecular Dynamic Simulation of s-GFP and s-DL4

The three dimensional structure of s-GFP and s-DL4 were modeled with template crystal structure of GFP (PDB ID 1GFL) [32] which does not contain the chromophore using the same procedure mentioned above for modeling the s-GFP with chromophore. The MD simulation was performed using the GROMACS 4.5.5 with GROMOS96 53a6 as the force field [33]. The protein was solvated using the SPC/E water and the system was neutralized with $Na^+$ as counter ions. The system was energy minimized before simulation to avoid bad contacts and inappropriate geometry using the steepest descent algorithm for 2000 steps. Position restrained simulation was performed for equilibrating the system using the *NVT* and *NPT* ensemble, each for 100 ps. Unrestrained simulation was performed for 10 ns with Berendsen thermostat of 300 K and the pressure was maintained at 1.0 bar with Parrinello-Rahman pressure coupling factor. The trjconv, g_rms, and g_rmsf utilities of GROMACS 4.5.5 were used to analyze the MD results.

# Results

## GFP Folds Used in this Study

Two different GFP sequences, i.e. $GFP_{mut3.1b}$ and GFPhs1, with different stabilities were employed to compare the effect of tolerance to deletion mutation. $GFP_{mut3.1b}$, a S65G/S72A mutant of wild type GFP, is one of the rapid folding GFPs engineered by directed evolution [34] and has been used in various studies [35,36]. GFPhs1 is a derivative of superfolder GFP constructed by introducing 11 superfolder GFP mutations [27] and other 2 mutations known to stabilize GFP structure into $GFP_{mut3.1b}$ [22]. The GFPhs1 exhibited higher stability and mutational robustness compared to $GFP_{mut3.1b}$, and their detailed sequences and biophysical properties were described in our previous report [22]. In this study, the $GFP_{mut3.1b}$ and GFPhs1 were used as a GFP fold with normal stability (n-GFP) and a GFP fold with higher stability (s-GFP), respectively.

## Tolerance of s-GFP and n-GFP for Deletion

**Figure 1** shows the representative secondary structure of GFP, a single chain polypeptide of 238 amino acids with 7 α-helices, 11 β-strands and some connecting loop regions [27], and the primary sequence of s-GFP. The residues at position 65–67 in GFP form fluorophore on spontaneous oxidation and this characteristic spectral signal directly marks the structure and activity of the protein [37]. First, we generated various N-terminal, C-terminal, and internal sequence deletion variants of s-GFP using designed primers, expressed in *E. coli*, and measured their whole cell fluorescence. Based on the results, we constructed corresponding n-GFP deletion variants, and conducted same experiments.

**Table 1** presents the representative results comparing the relative whole cell fluorescence of s-GFP variants with those of n-GFP variants. The dissection of the N-terminal deletions of s-GFP allowed us to identify that the deletion of N-terminus up to 14$^{th}$ amino acid retains the fluorescent activity of s-GFP despite the decrease of fluorescence and the deletion of 2–15 amino acids completely abolished the activity. C-terminal deletion studies revealed that the C-terminal can be eliminated up to the 225$^{th}$ amino acid with fluorescence activity. On the other hand, the deletion variants of 2–11, 2–14 and 225–238 for n-GFP lost their activities completely. We didn't further dissect the N- and C-terminal deletions for n-GFP, but the results clearly indicate that the higher the protein stability is, the more we can delete the terminals of target proteins.

For the study on the elimination of internal sequences, we deleted four internal sequences (76–81, 83–88, 132–139, 191–196) of s-GFP which generated the mutants of s-DL1, s-DL2, s-DL3, and s-DL4. The 76–81 and 83–88 residues are the sequences that form small α-helices and the 132–139 and 191–196 are the sequences involving large loops of GFP. The sequences are known to be essential in the formation of proper structure of GFP and the deletions of these sequences led to the complete inactivation of GFP [6]. As shown in **Table 1**, among the s-GFP variants individually devoid of four internal loops, s-DL1, s-DL2 and s-DL3 were completely inactivated, but the deletion mutant of 191–196 residues (s-DL4) exhibited about 81% of fluorescent activity of s-GFP. On the other hand, the 191–196 deletion mutant of n-GFP (n-DL4) was completely inactivated. The combinatorial deletions of the internal loop (191–196), N-terminal (2–11 or 2–14), and C-terminal (227–238) for s-GFP were also carried out. Among the variants, only the mutant of 2–11 and 227–238 deletions (s-N11C227) showed about 36% fluorescence activity of s-GFP, and other mutants such as s-N11C227DL4 and s-N14C227 showed 4% and 7% of fluorescence activity respectively. s-N14C227DL4

**Table 1.** Comparison of relative whole cell fluorescence of s-GFP, n-GFP and deletion variants.

| Deleted sequences | Name & Relative fluorescence/au (%) of s-GFP variants[a] | | Name & Relative fluorescence/au (%) of n-GFP variants[b] | |
|---|---|---|---|---|
| No deletion | s-GFP | 100 | n-GFP | 100 |
| 2–11 | s-N11 | 80 | n-N11 | 0 |
| 2–14 | s-N14 | 35 | n-N14 | 0 |
| 2–15 | s-N15 | 0 | n-N15 | 0[c] |
| 225–238 | s-C225 | 18 | n-C225 | 0 |
| 224–238 | s-C224 | 5 | n-C224 | 0[c] |
| 223–238 | s-C223 | 0 | n-C223 | 0[c] |
| 76–81 | s-DL1 | 5 | n-DL1 | -[+] |
| 83–88 | s-DL2 | 0 | n-DL2 | -[+] |
| 132–139 | s-DL3 | 2 | n-DL3 | -[+] |
| 191–196 | s-DL4 | 81 | n-DL4 | 0 |
| 2–11, 227–238 | s-N11C227 | 36 | n-N11C227 | 0 |

The relative fluorescence (in arbitrary units) is defined as the whole cell fluorescence compared with fluorescence of cells expressing s-GFP[a] and n-GFP[b] respectively. All the fluorescence values were normalized by the $O.D_{600\ nm}$ of expressed cells. [c]Deletion of preceding residues in primary structure of n-GFP resulted in loss of fluorescence and experiments were not carried out. [+]Reported elsewhere [6] that these loop deletions makes the protein non-functional.

showed almost complete inactivation (**Figure S1**). On other hand, the deletion mutation of 2–11 and 227–238 for n-GFP (n-N11C227) showed no fluorescence. All these results also clearly suggest that s-GFP is more tolerant than n-GFP against various deletion mutations.

## Expression Analysis of the Deletion Variants

The SDS-PAGE analyses of the soluble and insoluble fractions for the above mutants were performed and the results are shown in **Figure S2**. In the analyses, we could observe some interesting points. First, most non-fluorescent variants of s-GFP and n-GFP showed insoluble expression of the target protein, but the inactive s-DL1 and s-DL2 mutants showed almost the same soluble expression level of s-GFP (**Fig. S2, (a)**). It is presumed that the deletions of 76–81 or 83–88 for s-GFP didn't significantly perturb the global structure of s-GFP fold despite the slight structural change that can inactivate fluorescence. Second, the N- or C- terminal deletion variants of s-GFP, i.e. s-N11, s-N14, s-C225, and s-C224, showing decreased fluorescence compared to s-GFP (**Table 1**) showed similar soluble expression levels to s-GFP (**Fig. S2, (b) and (c)**). These results suggest that the decreased fluorescent levels of the mutants were not from the decreased soluble expression levels, but from the decreased spectral intensities of the variants. Third, s-DL4, the 191–196 deletion mutant of s-GFP, with an 81% fluorescent level of s-GFP, showed a very intriguing result that the most expressed GFP was detected in insoluble fractions (**Figure 2a**). This suggests that the s-DL4 variant is expressed as insoluble form but functionally active, which was confirmed

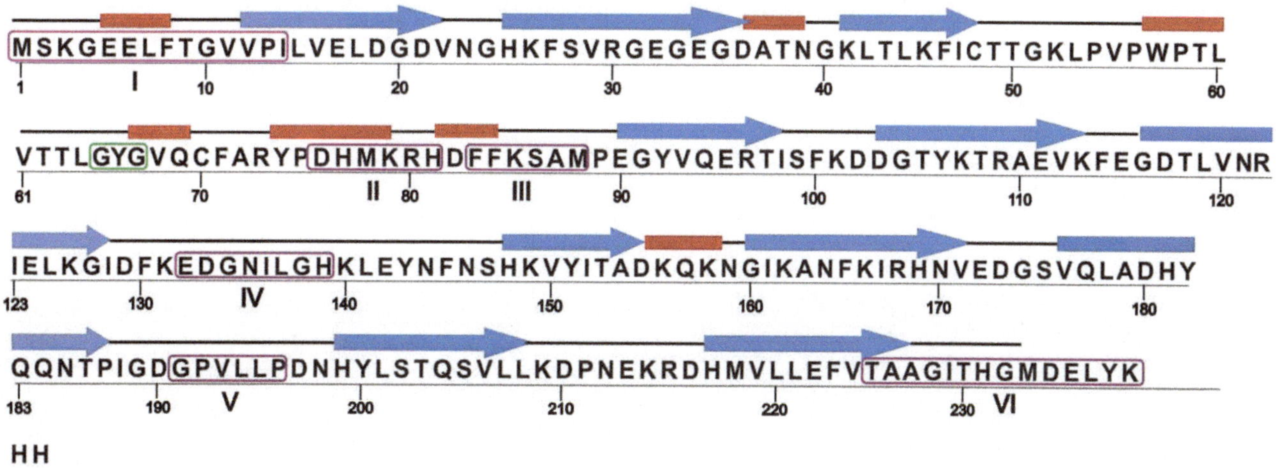


**Figure 1. The primary structure and secondary structure map of GFP [27].** Single-letter amino acid codes are used for representing the primary structure. Green color boxed residues 'GYG' represents the chromophore of GFP. Strands are represented in blue arrowhead and helices are indicated in red. The connecting black lines represent the turns and loops which conjoins the adjacent strands. The sequence of s-N14 is devoid of residues in box I. For s-C225, the residues in box VI are deleted. Residues in boxes II, III, IV and V are deleted in s-DL1, s-DL2, s-DL3 and s-DL4 respectively.

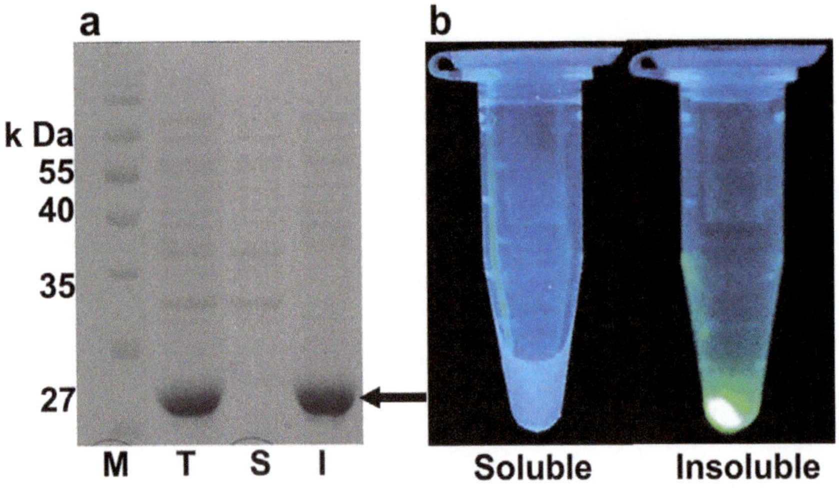


**Figure 2. Expression and fluorescence activity of s-DL4.** (a) Expression profiles of *Escherichia coli* BL21 (DE3) that produces s-DL4 variant which is devoid of internal loop (191–196). Lane M-Marker, Lane T-total cell protein, Lane S-Soluble fraction, Lane I-Insoluble Fraction. The expressed protein is indicated by arrow (b) UV irradiation of soluble and insoluble fraction of lysed s-DL4 variant to show the activity of the protein.

by detecting the fluorescence predominantly in the insoluble fraction (**Figure 2b**).

## Biophysical Properties of s-N14 and s-C225

Among the variants generated in above studies, the two deletion variants of s-GFP, s-N14 composed of 225aa and s-C225 composed of 224aa, were interesting because, as suggested in the previous reports [5,6], the N-terminal or C-terminal deletion mutants of GFP with reduced size can efficiently be utilized in some biochemical studies. To examine the possibility that s-N14 and s-C225 can be employed in biochemical studies, the biophysical properties of the two mutants were compared with those of n-GFP, one of the GFP variants which have been used in various studies as a folding reporter [35,36]. Specifically, we investigated the whole cell fluorescence, specific fluorescent activities, fluorescent spectral properties, refolding rates and tolerance to denaturant during refolding of the mutants. s-GFP was also characterized as a control. **Table 2** summarizes the biophysical properties of the selected mutants and the brief description of the results is followed.

The whole cell fluorescence activities of s-N14 and s-C225 were approximately 35% and 18% of s-GFP, but slightly higher and comparable to that of n-GFP, respectively. The specific activities of the purified s-N14 and s-C225 were about 70% and 22% of s-GFP, which corresponded to approximately 1.5 fold higher and two fold lower values of n-GFP, respectively. Here, the lower specific activities of s-N14 and s-C225 than s-GFP support the decreased whole cell fluorescence activities of the mutants despite the comparable soluble expressions described above. When the excitation and emission wavelengths were scanned, the s-N14 mutant showed almost the same properties as that of s-GFP, but the s-C225 exhibited the slight shift in wavelength of the ex/em spectra (**Figure 3**). An *in vitro* refolding study showed that the refolding rates of the two deletion mutants were almost the same as s-GFP and higher than n-GFP (**Figure 4**). This result is correlated with the SDS-PAGE analysis almost showing the similar soluble expression levels of the s-GFP, s-N14 and s-C225 described above. In order to test the tolerance of variants to urea during refolding, we performed the experiment of renaturation equilibrium plot against different concentrations of urea and measured the $C_{0.5}$'s of the variants (**Figure 5**). The refolding stability of s-N14 and s-C225 were superior to n-GFP but exhibited reduced stability than s-GFP.

**Table 2.** Comparison of biophysical properties of s-GFP, n-GFP and variants.

| Property | s-GFP | s-N14 | s-C225 | n-GFP |
|---|---|---|---|---|
| Relative whole cell fluorescence/au (%)[a] | 100 | 35 | 18 | 24 |
| Specific activity (%)[b] | 100 | 70 | 22 | 45 |
| $\lambda_{ex/em}$[c] | 490/511 | 490/511 | 467/506 | 501/511 |
| Refolding rate $\times 10^{-2} s^{-1}$[d] | 2.0 | 1.8 | 2.0 | 0.8 |
| $C_{0.5}$(M)[e] | 3.6 | 2.3 | 2.1 | 1.9 |

[a]The relative fluorescence (in arbitrary units) is defined as the whole cell fluorescence compared with fluorescence of cells expressing s-GFP. All the fluorescence values were normalized by the $O.D_{600\ nm}$ of expressed cells.
[b]Specific activity (in arbitrary units/μM of purified protein) is the fluorescence of purified protein compared with the fluorescence of purified protein of s-GFP.
[c]The excitation and emission maxima in units of nanometers.
[d]The refolding rate constant for the fast phase calculated by fitting refolding curve in sigma plot.
[e]Concentration of urea (M) at which 50% of the initial fluorescence is recovered during refolding of urea denatured protein.

Taken together, the biophysical properties of s-N14 and s-C225 were worse than s-GFP, but better or at least comparable to those of n-GFP. These results suggest that the variants generated by the truncation of terminal sequences can be employed in biochemical studies. To confirm the possibility of employment of the deletion variants in biochemical studies, we expressed the variants in *E. coli* and captured the images using fluorescence microscope. It was observed that the fluorescence levels of the deletion mutants were sufficient to perform *in vivo* imaging studies (**Figure 6**).

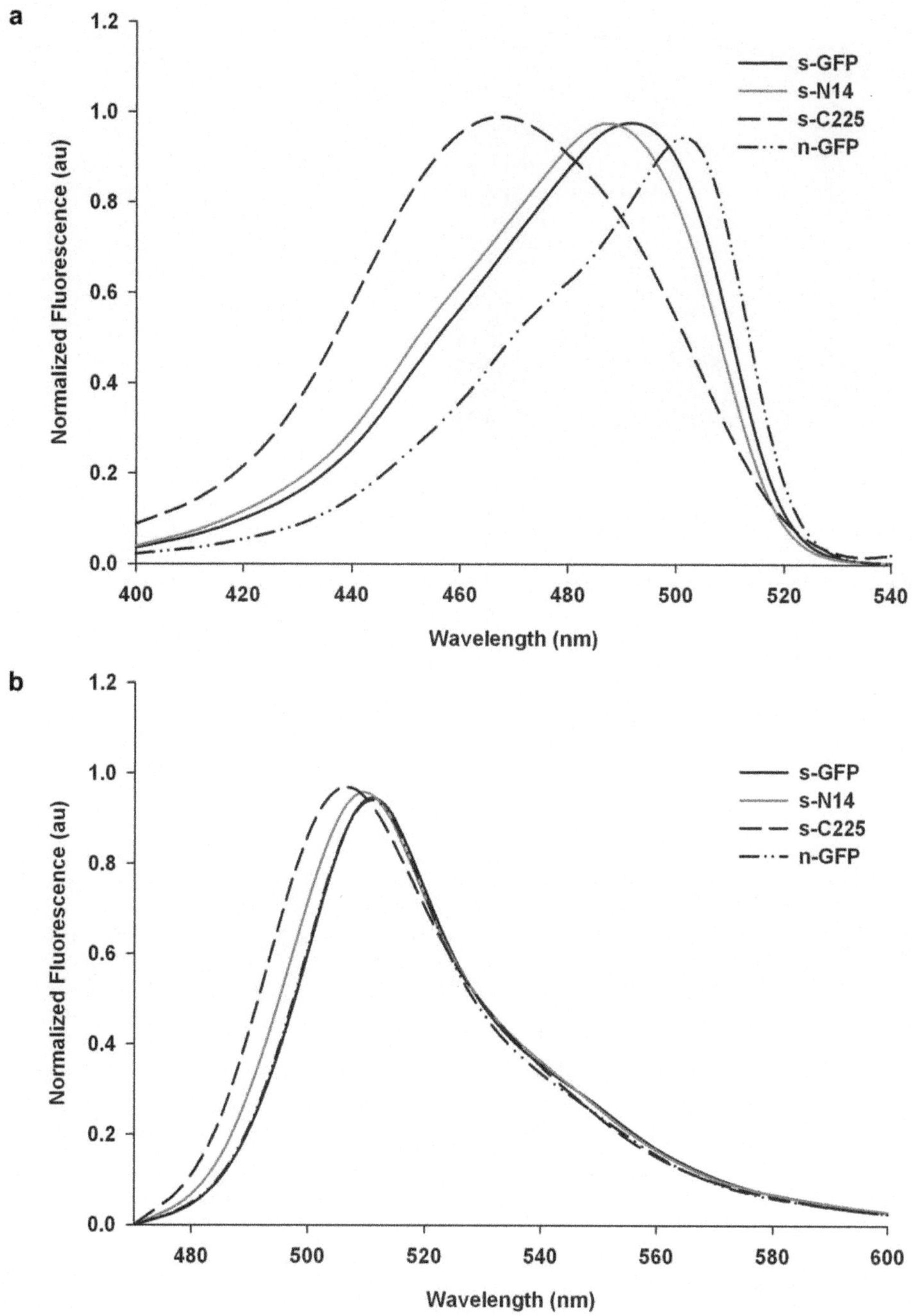


**Figure 3. Fluorescence excitation and emission spectra.** a) Excitation and b) Emission spectra of s-GFP, s-N14, s-C225 and n-GFP, measured using fluorescence spectrometer. All amplitudes were arbitrarily normalized to a maximum value of 1.0 to show the difference in spectral wavelength not the spectral intensity.

## Study on the Structural Perturbations of s-N14 and s-C225

To investigate whether or not there were any significant structural changes occurred due to the deletion of N-terminal and C-terminal of s-GFP, we performed CD spectral analyses of s-GFP, s-N14, and s-C225 (**Figure 7**). Although the analyses of the secondary structure content of the variants from the CD spectral data showed a slight variation of secondary structure amounts which may be caused by N- or C- terminal deletions (**Table S1**), we could not observe prominent structural changes between the s-GFP and the variants. These results indicate that the main β-barrel structure of s-GFP required for the fluorescent activity was almost maintained despite the elimination of the fractions of secondary structures in terminals.

## Structural Analysis of s-GFP and Deleted Variants by Molecular Modeling

As shown in the **Table 1** and **Figure S2**, the N- or C-terminal deletion mutants, i.e. s-N11, s-N14, s-C225, s-C224, s-N11C227, showed similar soluble expression levels but reduced whole cell fluorescence activities compared to s-GFP. In the case of loop deletion mutants, s-DL1 and s-DL2, significant fluorescent

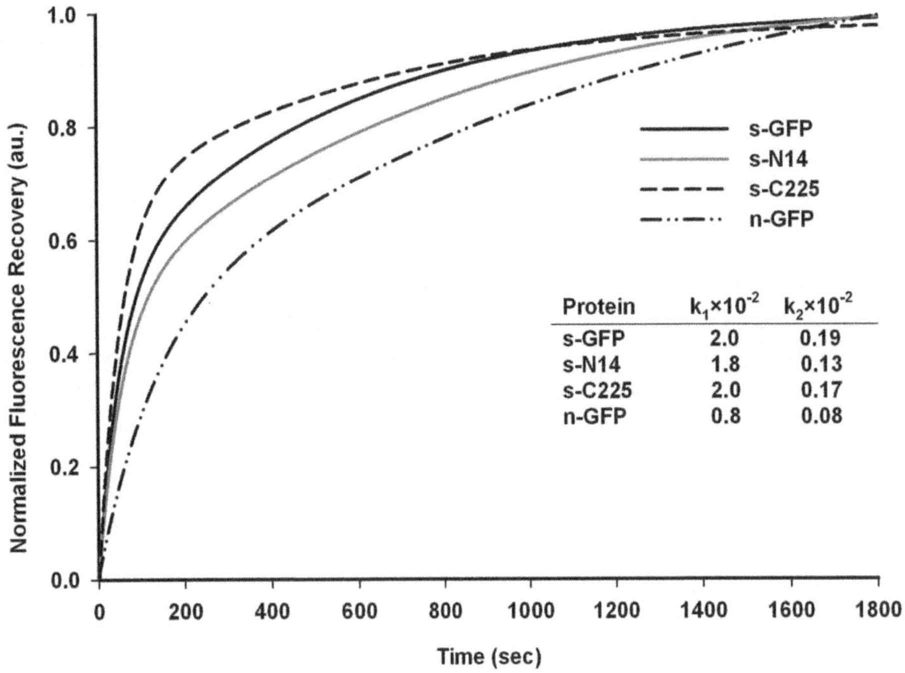


**Figure 4. Representative plot of refolding kinetics for s-GFP, s-N14, s-C225 and n-GFP.** Refolding kinetics was measured after denaturation in urea (8 M) followed by renaturation by dilution. Normalized fluorescence in arbitrary units (au) was plotted against time. Insert table showing the fast phase and slow phase rates of refolding process.

activities were not detected despite the soluble expression of the variants. Especially, s-C225 exhibited the shift in em/ex wavelength (**Figure 3**). These results suggest the possibility that the variants folded and formed proper β-barrel structure as demonstrated for s-N14 and s-C225 in the **Figure 7**, but their fluorescent properties were substantially changed. In general, the spectral properties of GFP are known to be sensitively influenced by the arrangement of amino acids and their interactions such as

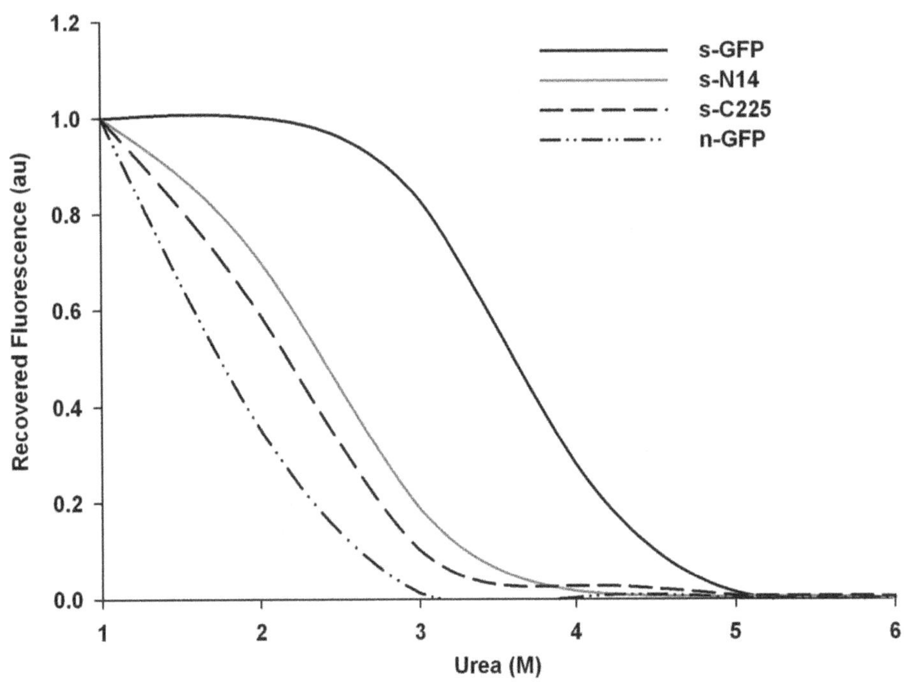


**Figure 5. Equilibrium refolding plot for s-GFP, n-GFP, s-N14 and s-C225.** Protein samples were denatured in 8 M urea and diluted to different concentration of urea in refolding buffer. Recovered fluorescence normalized by dividing by fluorescence of corresponding non-denatured samples diluted in similar fashion.

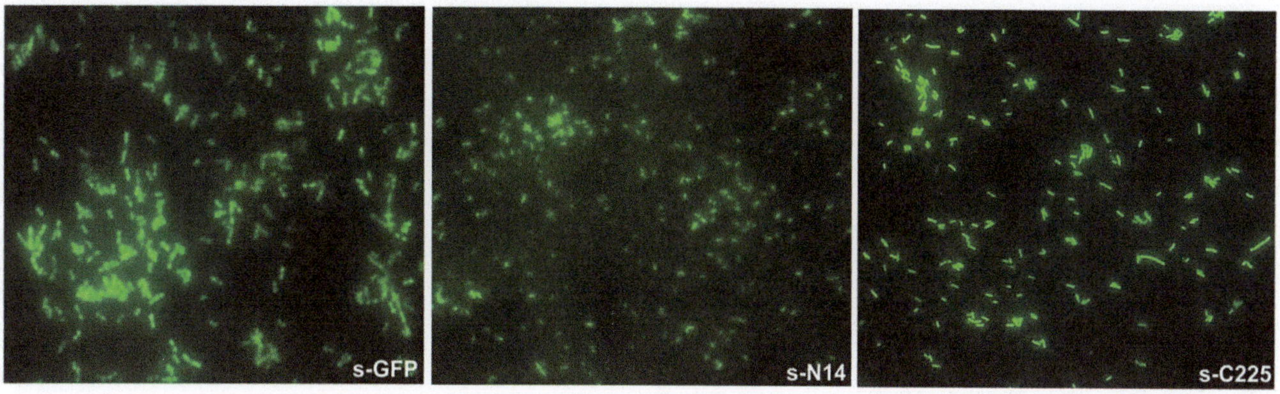


**Figure 6. Fluorescence microscope image of s-GFP and deletion variants.** s-GFP, s-N14 and s-C225 were employed in live cell imaging by expressing in *E. coli*. The images were captured using fluorescence microscope equipped with digital image analyzer.

hydrogen bonding and ion pair around chromophore [32,37,38,39]. Therefore, to understand the reduced or lost fluorescence activities of the variants, three dimensional structures of the deletion variants (s-N11, s-N14, s-C225, s-C224, s-N11C227, s-DL1 and s-DL2) were predicted by *in silico* modeling and their structural differences from s-GFP were analyzed. Superfolder GFP (2B3P) was used as the template structure to model the s-GFP because s-GFP is the variant that includes mostly superfolder GFP mutations. Detailed procedures for the structure modeling are described in *Materials and Methods* section. The root mean square deviations of the modeled structures of s-GFP and its variants were found to be less than 1Å compared to their corresponding template.

The lost and new electrostatic interactions of the N-terminal deletion variants (s-N11 and s-N14) were shown in **Table S2**. The modeled structure of s-N11 has one new hydrogen bond, but it lost 2 salt bridges and 12 hydrogen bonds. About 5 residues of the lost interactions were located within 10Å distance of chromophore. The modeled structure of s-N14 had 2 putative hydrogen bonds besides the loss of original interactions of s-GFP which includes 12 hydrogen bonds and 2 salt bridges. Among the 24 residues involved in the lost hydrogen bonds, 10 residues were found in the 10Å distance of chromophore, and I47 which forms putative hydrogen bonding with T49 was also located within 10Å distance. The interaction map of s-GFP, s-N11 and s-N14 showed that the Phe8 residue and F8–G4 interaction was lost due to the deletion in s-N11, and Phe8, Val12, Pro13, Ile14 residues and their related interactions were lost due to the deletion in s-N14 (**Figure 8(a)**).

The interactions of C-terminal deletion variants (s-C225 and s-C224) were also analyzed using the modeled structures and the changes in interactions were shown in **Table S3**. The energy minimized model of s-C225 was found to lose 9 parent hydrogen

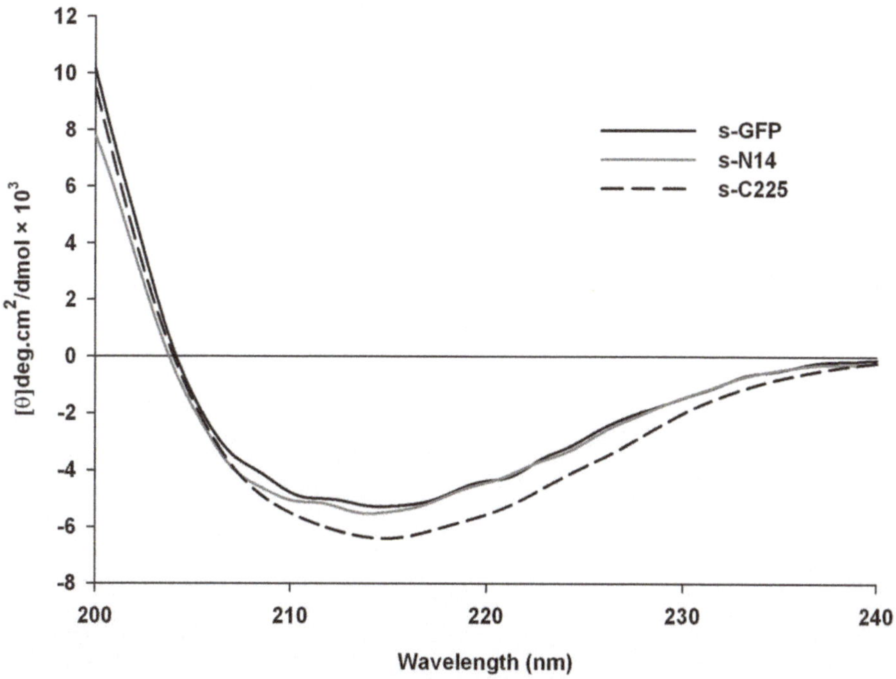


**Figure 7. Structural validation by Circular Dichroism spectroscopy.** CD spectral analysis of s-GFP, s-N14 and s-C225. Comparison of Far-UV (200–240 nm) CD spectra of s-GFP, s-N14 and s-C225 showed the major peak for secondary structure of β-sheet.

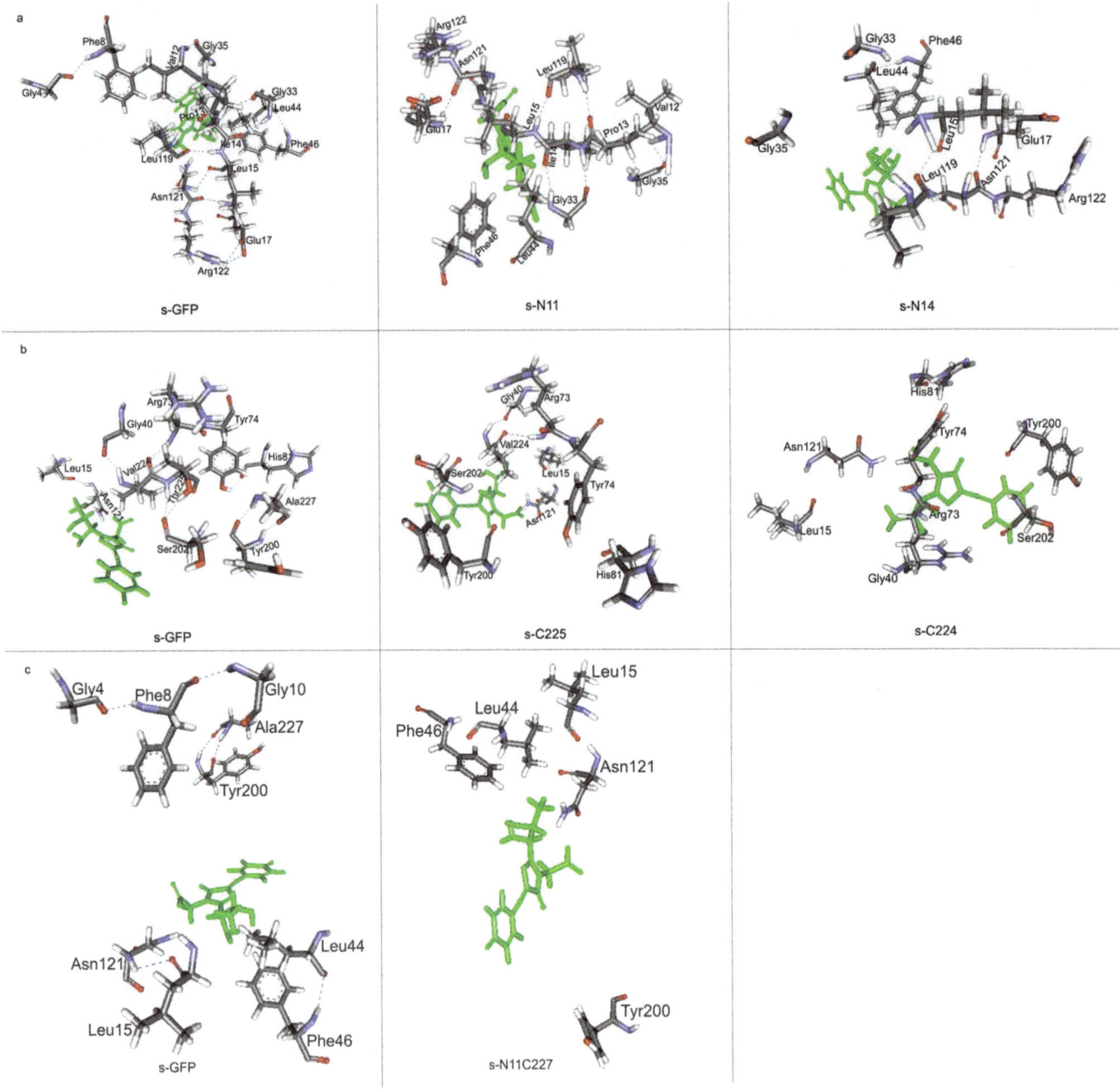


**Figure 8. Computational analysis of changes in structural interactions of s-GFP variants.** Comparison of lost hydrogen bonding network between s-GFP and deleted variants a) In s-N11 variant, deletion of F8 disrupted the hydrogen bond interaction with G4 along with loss of other interactions as compared with s-GFP. In s-N14 mutants, deletion of V12, P13 and I14 were shown clearly by comparing with s-GFP which resulted in loss of hydrogen bond interaction with G35, L119 and G33 respectivley. b) The interaction map of s-C225 showed the deleted residues T225 and A227, and the lost hydrogen bonds. Deletion of V224, T225 and A227 were shown in the interaction map of s-C224. c) The combined deletion variant s-N11C227 interaction map shows the deleted residues Phe8, Gly4, Gly10 and Ala227 and the interactions they were involved.

bonds and to have one putative salt bridge and 4 putative hydrogen bonds. Among the residues involved in the lost hydrogen bonds, newly generated hydrogen bonds, and newly generated salt-bridges, 9 residues, 5 residues, and 2 residues were found around the chromophore in 10Å distance, respectively. The modeled structure of s-C224 had 3 putative hydrogen bonds and 2 new salt bridges besides the loss of 8 original interactions. When the interactions within the 10Å area of chromophore of s-GFP, s-C225 and s-C224 were compared (**Figure 8(b)**), in addition to the lost and altered hydrogen bond network mentioned in **Table S3**, two residues Thr225 and Ala227 were eliminated in the interaction map of s-C225, and three residues Val224, Thr225 and Ala227 were deleted in s-C224.

**Table S4** shows the lost and new interactions of combined deletion variant s-N11C227. The s-N11C227 was found to lose 14 parent hydrogen bonds and 2 parent salt bridges besides 8 new interactions, which included 6 hydrogen bonds and 2 salt bridges. Among the residues involved in lost interactions, 6 of lost hydrogen bonds, 4 of newly generated hydrogen bonds and 4 of new salt bridges were located within 10Å distance of chromophore. In s-N11C227, the interactions of residues within 10Å distance of the chromophore were mapped and compared with s-

GFP, highlighting the elimination of Gly4, Phe8, Gly10 and Ala227 (**Figure 8(c)**).

**Table S5** shows the lost interactions of loop deleted variants (s-DL1 and s-DL2), and **Figure 9** presents their interaction maps around chromophore. The analysis informed that the residues of the deleted loops resulted in the loss of some amino acid residues and their related interactions around chromophore. In s-DL1, the residues Lys79 and His81 were missing along with their interactions with Tyr74 and Phe83 due to the loop deletion within 10Å distance of chromophore. In s-DL2, the missing of two amino acids (Phe83 and Met88) and loss of 3 hydrogen bonds were found within the 10Å distance of chromophore due to the deletion.

Above molecular modeling analysis gave following probable clues for the changed spectral properties of the variants. First, the structural analysis revealed that some hydrogen bonds, salt bridges and amino acid residues of the s-GFP were lost and reoriented due to the deletions. The loss of electrostatic interactions as well as some amino acid residues was also observed around chromophore in the deletion variants (**Figure 8 and 9**). As mentioned above, the chromophore of fluorescent proteins exhibits different fluorescent characteristics depending on the microenvironment around the fluorophore [32,37,38,39]. It is presumed that the deletions changed the packing interactions around the fluorophore environment and led to the reduced fluorescence of the variants. Second, above analysis informed us some correlations between spectral properties of variants and their structural changes around chromophore. An observed trend was that the variants showing almost complete loss of fluorescence lost more amino acid residues and electrostatic interactions around chromophore compared to the variants showing reduced fluorescence. In particular, deletion of C-terminal residues close to the catalytic residue E222 [40] would have affected the electrostatic microenvironment around chromophore more significantly, which might contribute to the shift of $\lambda_{em/ex}$ wavelengths in addition to the reduced fluorescent activity of s-C225.

## *In silico* Analysis of the Structure and Stability of s-DL4

An interesting result in our study was that the s-DL4 obtained from the internal loop deletion showed active but insoluble production (**Figure 2**). In general, the expression of GFP in *E. coli* leads to the soluble active expression and insoluble inactive expression. Although the insoluble fraction can include active GFP, the ratio is known to be very low [41]. On the other hand, in the case of s-DL4, approximately more than 95% of the expressed insoluble GFP was active when estimated based on the whole cell fluorescence and the amount of expressed protein. Although further detailed studies are needed, it is presumed that the protein sequences for s-DL4 folds into the proper structure of GFP for fluorescence, but the folded structure is prone to be aggregated or multimerized.

To understand the fluorescence activity of s-DL4 showing almost same activity with s-GFP, the structure of s-DL4 was modeled likewise the other variants. The modeled s-DL4 structure showed that only one hydrogen bond interaction was lost due to deletion of internal loop (191–196). Residue Leu195 had hydrogen bond interaction with Arg80, which was lost due to loop deletion (**Figure 10(a)**). The loss of only one hydrogen bond in the deleted sequences for s-DL4 might be the plausible explanation for retaining its fluorescence activity with less structural perturbation. In addition, the lost L195 and related hydrogen bond are not within 10Å distance of chromophore, which also supports the non-changed fluorescent activity of s-DL4.

To find the possible explanation for the aggregation behavior of s-DL4, the structure of s-DL4 was compared with the structure of s-GFP, and structural changes were examined. In this analysis, we focused on examining the structural changes of the exposed hydrophobic amino acids in GFP surface and the amino acids in the GFP dimer interface [32]. Any major structural differences between s-GFP and s-DL4 were not observed in the hydrophobic residues on surface and the residues in dimer interface, which might be caused by the current limitations in the computational structure modeling. To probe the structural perturbations of s-DL4 induced by the loop deletion further, molecular dynamic (MD) simulations study was performed. MD simulations of s-GFP and s-DL4 showed that the main chain of s-DL4 was more dynamically fluctuated than s-GFP (**Figure 10(b)**), which suggests that the loop deletion may destabilize the structure of the folded s-GFP. The analysis of the time-average fluctuations of the individual residues suggested that there were some specific regions showing high fluctuations in the primary sequence of s-DL4 (**Figure 10(c)**). In particular, it was identified that 30 residues of s-DL4 showed fluctuations with the difference of more than 0.5 Å compared to the corresponding residues of s-GFP (**Table S6).** In addition, 9 exposed residues which were reported to be in the dimer interface of GFP [32] and 17 other exposed hydrophobic amino acids of s-DL4 were showing higher fluctuations than s-GFP (**Table S6).** These results suggested the possibility that these 30 highly fluctuating residues along with some hydrophobic and dimer interface residues in surface exhibit dynamic nature by the deletion. A hypothesis is that the fluctuation of these unstable residues might induce the formation of locally unfolded states, which can initiate aggregation or multimerization of s-DL4.

## Discussion

In this study, it has been shown that the deletional protein engineering approach can be more efficiently performed by using a more stabilized protein structure. This was demonstrated by comparing two GFP sequences with different stabilities, n-GFP and s-GFP. In addition, the deletion mutagenesis of s-GFP with high stability also enabled us to generate deletion mutants such as s-N14, s-C225 and s-DL4, which cannot be obtained from normal GFP fold. s-N14 is the 2–14 residues deleted mutant of s-GFP without the small turn, one helix and a part of first β-strand in N-terminus of GFP. To our knowledge, s-N14 is the first GFP mutant showing fluorescent activity without the amino acid residues forming the first turn and helix. The deletion of 8 residues in the GFP N-terminus has been known to abolish the GFP activity completely so far [6]. The C-terminus deletion mutant of 225–238, s-C225, is also likely to be the first deletion mutant without a part of the last β-strand and complete tail sequence but showing fluorescence activity. In the previous study, the maximum deletion for C-terminus was the deletion of 229–238 residues while retaining its functional activity [6]. In the study, internal loops deletions were made but none of the loop deleted variants were found to be fluorescent, whereas deletion mutant of internal loop (191GPVLLP196), s-DL4, showed fluorescent activity in our study. Various types of split GFPs and truncated circularly permuted GFPs were reported to fluoresce [42,43]. However, in those studies, only excisions were made to prepare two fragments and the two fragments were reassembled to form complete β-barrel structure, whereas the variants were functionally active after true deletions in our study.

Though the tendency of a protein to form aggregate under physiological conditions is unpredictable yet, aggregation can happen through thermal fluctuations occurring under physiological conditions [44,45]. For example, an enzyme acylphosphatase from *Sulfolobus solfataricus* have been studied extensively

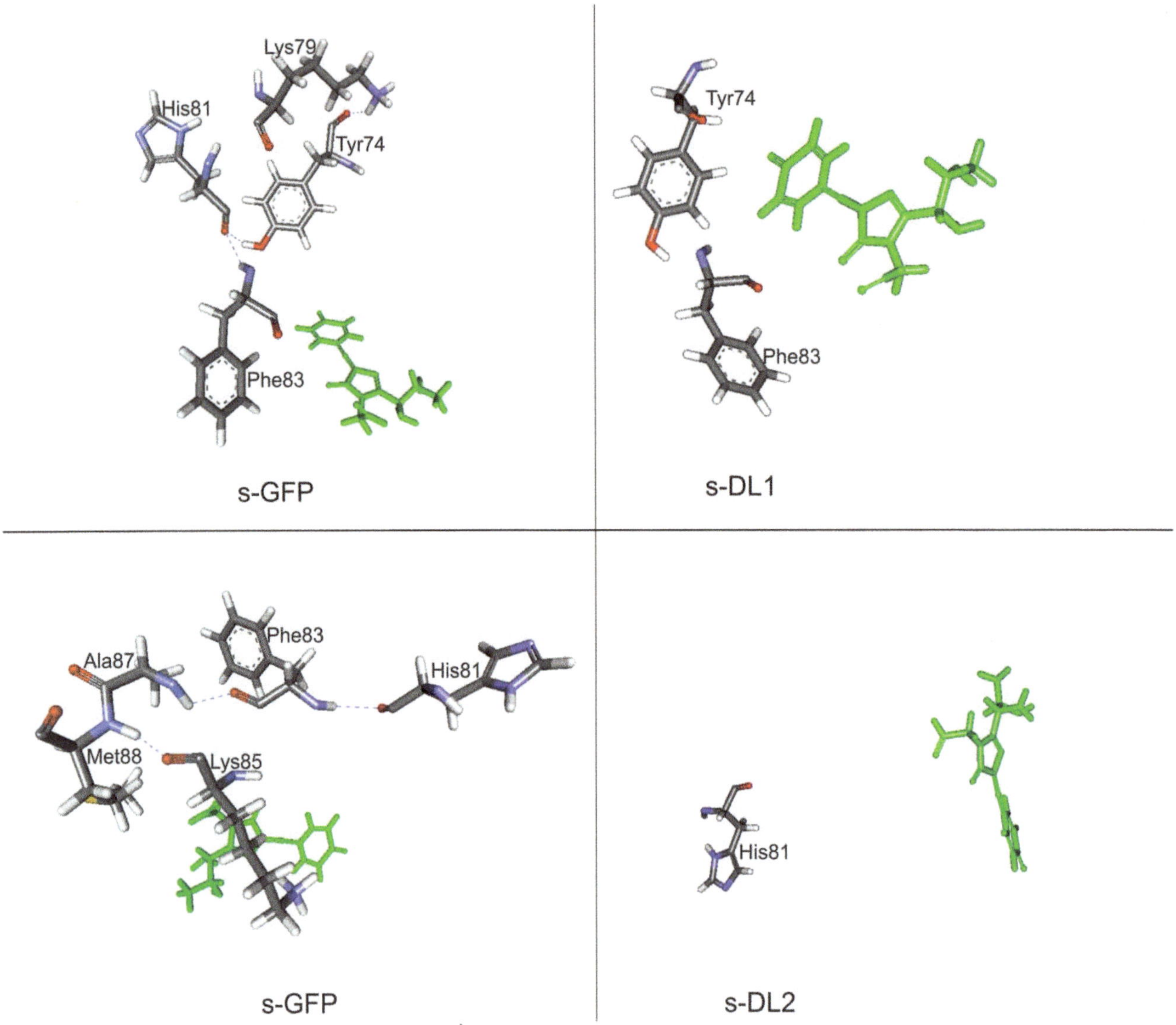


**Figure 9. Structural interactions of modeled structure of loop deleted variants.** a) In s-DL1, the loop deletion eliminated hydrogen bonds involved by the residues K79 and H81. b) The residues F83, K85, A87 and M88 were deleted in the loop sequence of s-DL2.

and found to form active native-like aggregates through partially unfolded protein molecules [46,47,48]. These results support our hypothesis for the aggregation of active s-DL4 variant that the dynamic fluctuations of some hydrophobic residues in exposed positions or dimer interface residues in GFP can induce such aggregation. However, detailed structural characterization of the s-DL4 by NMR spectroscopy or X-ray diffraction may be needed to give the exact reason for the aggregation of the s-DL4.

Our study focused on demonstrating that deletion mutagenesis can be more efficiently performed using stable fold by performing terminal deletions and loop deletions. We performed relatively systematic deletion studies for N-terminal and C-terminal residues, and found the important residues in respective terminals that are essential for retaining the native structure of stable GFP. On the other hand, the loop deletion mutagenesis strategy was not so systematic and just adapted from the previous work to compare the tolerance of deletion mutation between s-GFP and n-GFP. Further systematic deletion studies for the loop regions are needed to understand more exactly how such mutations have influence on the protein structure. Different type of deletion studies such as deletions in β-strands are also expected to provide chances to generate different interesting variants.

Although the engineering of physical properties such as protein size and protein aggregation were shown in our study, the approach using stabilized protein may also facilitate the engineering of catalytic proteins through deletion mutagenesis. For example, it has been reported that the aminopeptidase PepC was engineered by deleting four amino acids in C-terminal, which resulted in a new enzyme with altered property [49]. This demonstrated that deletion mutagenesis can be used for changing catalytic activities of protein. Therefore, in the case that the deletions for catalytic activity changes are related with protein structure formation, the improvement of protein stability may give us more chance to achieve active protein with altered catalytic properties.

In conclusion, this study demonstrated that the deletional protein engineering approach can be more efficiently performed

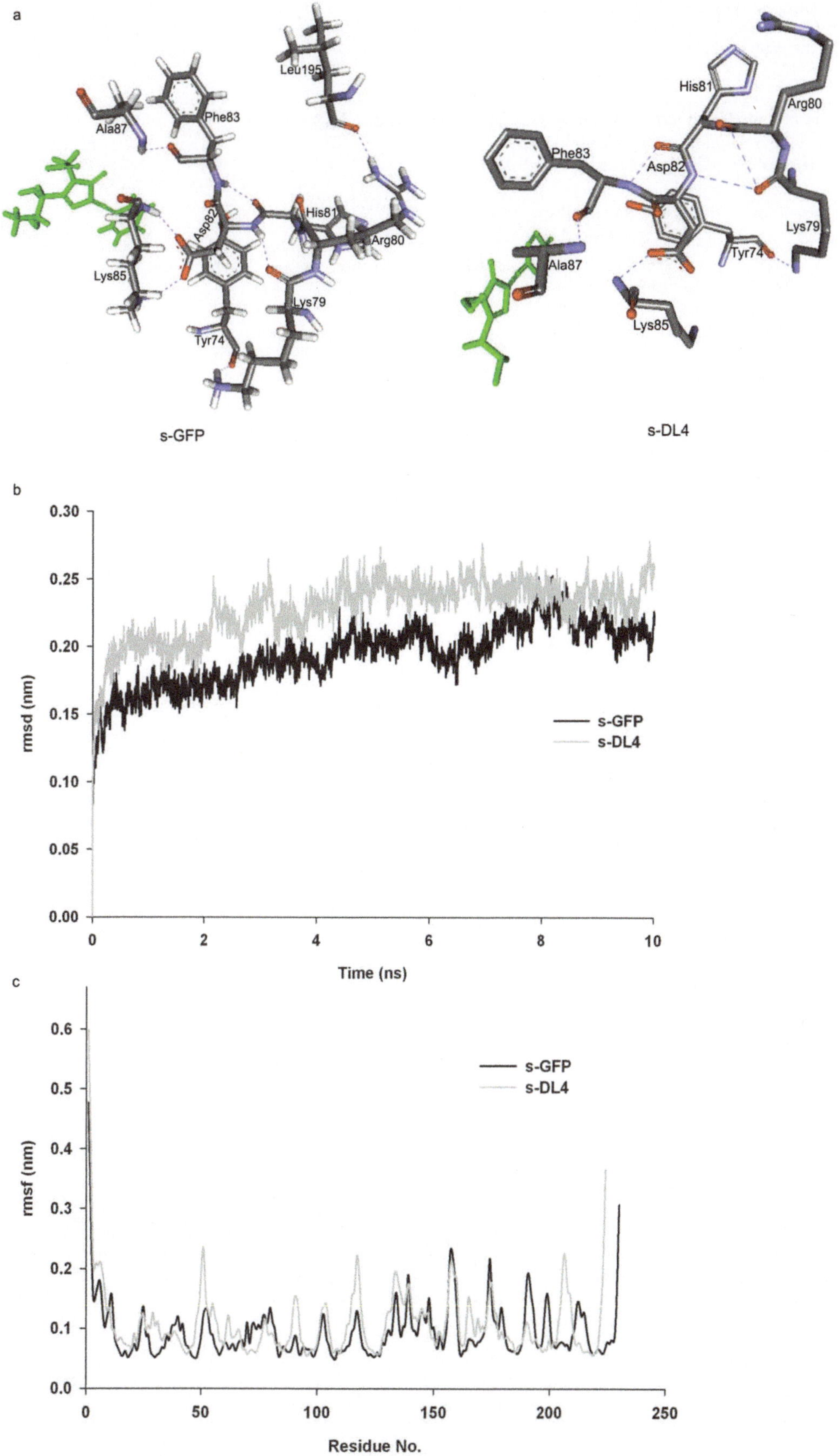
a
Leu195
Ala87
Phe83
Asp82
His81
Arg80
Lys85
Lys79
Tyr74
s-GFP
His81
Arg80
Phe83
Asp82
Lys79
Tyr74
Ala87
Lys85
s-DL4
b
0.30
0.25
0.20
0.15
0.10
0.05
0.00
rmsd (nm)
s-GFP
s-DL4
0
2
4
6
8
10
Time (ns)
c
0.6
0.5
0.4
0.3
0.2
0.1
0.0
rmsf (nm)
s-GFP
s-DL4
0
50
100
150
200
250
Residue No.

**Figure 10. Computational modeling and molecular dynamic simulation of s-DL4.** a) Lost hydrogen bonding between R80–L195 is shown in the interaction map with additional interaction surrounding the chromophore. Comparison of b) root mean square deviation (rmsd) and c) root mean square fluctuation (rmsf) of the main chain atoms of s-GFP and s-DL4.

by using a more stabilized protein structure. This indicates that the concept of protein stability-evolvability relationship can be applied to overcome the problem of deletional protein engineering. Our study also demonstrates that the deletion mutagenesis can generate a sequence-structure space different from the space derived by traditional substitution mutagenesis, which allows us to obtain the variants that cannot be obtained from the usual substitution mutagenesis. For instance, the functionally active insoluble s-DL4 looks like a unique GFP variant that has not been obtained from the tremendous previous GFP engineering studies mainly based on substitution mutagenesis. It is expected that our study can be helpful to facilitate the use of deletion method in protein engineering.

## Supporting Information

**Figure S1** Relative whole cell fluorescence activity of combinatorial deletion. Relative whole cell fluorescence (in arbitrary units) of combined deletion of N-terminal, C-terminal and internal loops (191–196) of s-GFP. The fluorescence of *E.coli* BL21 (DE3) expressing deletion variants was measured after the reached $O.D_{600}$ of 0.6 at 37°C was induced with 1 mM IPTG for 5 hrs. The relative fluorescence (in arbitrary units) is defined as the whole cell fluorescence compared with fluorescence of cells expressing s-GFP. All the fluorescence values were normalized by the $O.D_{600\ nm}$ of expressed cells.

**Figure S2** SDS-PAGE analysis of expressed protein. Expression profiles of designed s-GFP and n-GFP variants. Solid head arrow mark shows the expressed proteins in all images. T-Total cell protein, S-Soluble fraction, I-Insoluble fraction. Expression and solubility level of a) Internal loop deletions b) N-terminal deletions c) C-terminal deletions and d) Combined deletions of s-GFP. SDS-PAGE analysis and solubility level of e) N-terminal deletions f) combined deletions and internal loop deletion (n-DL4) of n-GFP variants.

**Supporting Information S1**

**Table S1** Estimated Secondary structure content recovered through DichroWeb server.

**Table S2** Lost and new interactions of N-terminal deletion mutants (s-N11, s-N14).

**Table S3** Change in interactions of C-terminal deletion mutants (s-C225, s-C224).

**Table S4** Lost and new electrostatic interactions in both terminals combined deletion mutants (s-N11C227).

**Table S5** Lost Hydrogen bonds and salt bridges of internal loop deletions s-DL1 and s-DL2.

**Table S6** Fluctuating amino acids in s-DL4.

## Acknowledgments

We thank Prof. Do Sik Min (Department of Molecular Biology, Pusan National University) for extending his help in providing fluorescence microscope.

## Author Contributions

Conceived and designed the experiments: GR NS SGL. Performed the experiments: GR NS SS. Analyzed the data: GR NS SS SGL. Contributed reagents/materials/analysis tools: HY SGL. Wrote the paper: GR NS SS SGL.

## References

1. DePristo MA, Weinreich DM, Hartl DL (2005) Missense meanderings in sequence space: a biophysical view of protein evolution. Nat Rev Genet 6: 678–687.
2. Camps M, Herman A, Loh E, Loeb LA (2007) Genetic constraints on protein evolution. Crit Rev Biochem Mol Biol 42: 313–326.
3. Povolotskaya IS, Kondrashov FA (2010) Sequence space and the ongoing expansion of the protein universe. Nature 465: 922–926.
4. Shortle D, Sondek J (1995) The emerging role of insertions and deletions in protein engineering. Curr Opin Biotechnol 6: 387–393.
5. Dopf J, Horiagon TM (1996) Deletion mapping of the Aequorea victoria green fluorescent protein. Gene 173: 39–44.
6. Li X, Zhang G, Ngo N, Zhao X, Kain SR, et al. (1997) Deletions of the Aequorea victoria green fluorescent protein define the minimal domain required for fluorescence. J Biol Chem 272: 28545–28549.
7. Santhoshkumar P, Murugesan R, Sharma KK (2009) Deletion of $^{54}$FLRAPSWF$^{61}$ residues decreases the oligomeric size and enhances the chaperone function of αB-crystallin. Biochemistry 48: 5066–5073.
8. Gong R, Wang Y, Feng Y, Zhao Q, Dimitrov DS (2011) Shortened engineered human antibody CH2 domains. J Biol Chem 286: 27288–27293.
9. Vamvaca K, Lansbury Jr PT, Stefanis L (2011) N-terminal deletion does not affect α-synuclein membrane binding, self-association and toxicity in human neuroblastoma cells, unlike yeast. J Neurochem 119: 389–397.
10. Haglund E, Danielsson J, Kadhirvel S, Lindberg MO, Logan DT, et al. (2012) Trimming down a protein structure to its bare foldons. J Biol Chem 287: 2731–2738.
11. Liu L, Zhang G, Zhang Z, Wang S, Chen H (2011) Terminal amino acids disturb xylanase thermostability and activity. J Biol Chem 286: 44710–44715.
12. Trevino RJ, Tsalkova T, Kramer G, Hardesty B, Chirgwin JM, et al. (1998) Truncations at the $NH_2$-terminus of rhodanese destabilize the enzyme and decrease its heterologous expression. J Biol Chem 273: 27841–27847.
13. Trevino RJ, Gliubich F, Berni R, Cianci M, Chirgwin JM, et al. (1999) $NH_2$-terminal sequence truncation decreases the stability of bovine rhodanese, minimally perturbs its crystal structure, and enhances interaction with GroEL under native conditions. J Biol Chem 274: 13938–13947.
14. Putnam-Evans C, Wu J, Bricker TM (1996) Site-directed mutagenesis of the CP 47 protein of photosystem II: alteration of conserved charged residues which lie within lethal deletions of the large extrinsic loop E. Plant Mol Biol 32: 1191–1195.
15. Bloom JD, Labthavikul ST, Otey CR, Arnold FH (2006) Protein stability promotes evolvability. Proc Natl Acad Sci U S A 103: 5869–5874.
16. Tokuriki N, Tawfik DS (2009) Stability effects of mutations and protein evolvability. Curr Opin Struct Biol 19: 596–604.
17. Caetano-Anollés G, Mittenthal J (2010) Exploring the interplay of stability and function in protein evolution. BioEssays 32: 655–658.
18. Bloom JD, Wilke CO, Arnold FH, Adami C (2004) Stability and the evolvability of function in a model protein. Biophys J 86: 2758–2764.
19. Müller MM, Kries H, Csuhai E, Kast P, Hilvert D (2010) Design, selection, and characterization of a split chorismate mutase. Protein Sci 19: 1000–1010.
20. Leconte AM, Patel MP, Sass LE, McInerney P, Jarosz M, et al. (2010) Directed evolution of DNA polymerases for next-generation sequencing. Angew Chem Int Ed Engl 49: 5921–5924.
21. Li Y, Drummond DA, Sawayama AM, Snow CD, Bloom JD, et al. (2007) A diverse family of thermostable cytochrome P450s created by recombination of stabilizing fragments. Nat Biotechnol 25: 1051–1056.

22. Nagasundarapandian S, Merkel L, Budisa N, Govindan R, Ayyadurai N, et al. (2010) Engineering protein sequence composition for folding robustness renders efficient noncanonical amino acid incorporations. ChemBioChem 11: 2521–2524.
23. Ayyadurai N, Saravanan Prabhu N, Deepankumar K, Lee S-G, Jeong H-H, et al. (2011) Development of a selective, sensitive, and reversible biosensor by the genetic incorporation of a metal-binding site into green fluorescent protein. Angew Chem Int Ed Engl 50: 6534–6537.
24. Ayyadurai N, Prabhu NS, Deepankumar K, Jang YJ, Chitrapriya N, et al. (2011) Bioconjugation of l-3,4-dihydroxyphenylalanine containing protein with a polysaccharide. Bioconjug Chem 22: 551–555.
25. Ayyadurai N, Neelamegam R, Nagasundarapandian S, Edwardraja S, Park H, et al. (2009) Importance of expression system in the production of unnatural recombinant proteins in Escherichia coli. Biotechnol Bioprocess Eng 14: 257–265.
26. Higuchi R, Krummel B, Saiki R (1988) A general method of in vitro preparation and specific mutagenesis of DNA fragments: study of protein and DNA interactions. Nucleic Acids Res 16: 7351–7367.
27. Pedelacq J-D, Cabantous S, Tran T, Terwilliger TC, Waldo GS (2006) Engineering and characterization of a superfolder green fluorescent protein. Nat Biotechnol 24: 79–88.
28. Whitmore L, Wallace BA (2008) Protein secondary structure analyses from circular dichroism spectroscopy: Methods and reference databases. Biopolymers 89: 392–400.
29. Whitmore L, Wallace BA (2004) DICHROWEB, an online server for protein secondary structure analyses from circular dichroism spectroscopic data. Nucleic Acids Res 32: W668–W673.
30. Barlow DJ, Thornton JM (1983) Ion-pairs in proteins. J Mol Biol 168: 867–885.
31. McDonald IK, Thornton JM (1994) Satisfying hydrogen bonding potential in proteins. J Mol Biol 238: 777–793.
32. Yang F, Moss LG, Phillips GN (1996) The molecular structure of green fluorescent protein. Nat Biotechnol 14: 1246–1251.
33. Hess B, Kutzner C, van der Spoel D, Lindahl E (2008) GROMACS 4: Algorithms for highly efficient, load-balanced, and scalable molecular simulation. J Chem Theory Comput 4: 435–447.
34. Cormack BP, Valdivia RH, Falkow S (1996) FACS-optimized mutants of the green fluorescent protein (GFP). Gene 173: 33–38.
35. Bulter T, Lee S-G, Wong WW, Fung E, Connor MR, et al. (2004) Design of artificial cell–cell communication using gene and metabolic networks. Proc Natl Acad Sci U S A 101: 2299–2304.
36. Fung E, Wong WW, Suen JK, Bulter T, Lee S-g, et al. (2005) A synthetic gene-metabolic oscillator. Nature 435: 118–122.
37. Ormö M, Cubitt AB, Kallio K, Gross LA, Tsien RY, et al. (1996) Crystal structure of the Aequorea victoria green fluorescent protein. Science 273: 1392–1395.
38. Sample V, Newman RH, Zhang J (2009) The structure and function of fluorescent proteins. Chem Soc Rev 38: 2852–2864.
39. Tsien RY (1998) The green fluorescent protein. Annu Rev Biochem 67: 509–544.
40. Ehrig T, O'Kane DJ, Prendergast FG (1995) Green fluorescent protein mutants with altered fluorescence excitation spectra. FEBS Letters 367: 163–166.
41. Peternel S, Grdadolnik J, Gaberc-Porekar V, Komel R (2008) Engineering inclusion bodies for non denaturing extraction of functional proteins. Microb Cell Fact 7: 34.
42. Huang Y-M, Nayak S, Bystroff C (2011) Quantitative in vivo solubility and reconstitution of truncated circular permutants of green fluorescent protein. Protein Sci 20: 1775–1780.
43. Kent KP, Oltrogge LM, Boxer SG (2009) Synthetic control of green fluorescent protein. J Am Chem Soc 131: 15988–15989.
44. Chiti F, Dobson CM (2009) Amyloid formation by globular proteins under native conditions. Nat Chem Biol 5: 15–22.
45. Ventura S, Villaverde A (2006) Protein quality in bacterial inclusion bodies. Trends Biotechnol 24: 179–185.
46. Bemporad F, Vannocci T, Varela L, Azuaga AI, Chiti F (2008) A model for the aggregation of the acylphosphatase from Sulfolobus solfataricus in its native-like state. Biochim Biophys Acta 1784: 1986–1996.
47. Pagano K, Bemporad F, Fogolari F, Esposito G, Viglino P, et al. (2010) Structural and dynamics characteristics of acylphosphatase from Sulfolobus solfataricus in the monomeric state and in the Initial native-like aggregates. J Biol Chem 285: 14689–14700.
48. Plakoutsi G, Bemporad F, Monti M, Pagnozzi D, Pucci P, et al. (2006) Exploring the mechanism of formation of native-like and precursor amyloid oligomers for the native acylphosphatase from Sulfolobus solfataricus. Structure 14: 993–1001.
49. Mata L, Gripon J-C, Mistou M-Y (1999) Deletion of the four C-terminal residues of PepC converts an aminopeptidase into an oligopeptidase. Protein Engineering 12: 681–686.

# Directed Evolution of the Transcriptional Regulator DntR: Isolation of Mutants with Improved DNT-Response

**Rosa Lönneborg*, Edina Varga, Peter Brzezinski***

Department of Biochemistry and Biophysics, Arrhenius Laboratories for Natural Sciences, Stockholm University, Stockholm, Sweden

## Abstract

The transcriptional regulator DntR, which previously has been isolated from bacterial strains capable of degrading 2,4-dinitrotoluene (DNT), was engineered in order to improve the ability to detect DNT. A directed evolution strategy was employed, where sequence diversity first was created by random mutagenesis in three subsequent rounds, followed by recombination of previously selected mutants. A *gfp* gene was used as a reporter for transcriptional activity mediated by DntR and cells with higher GFP expression after addition of DNT were sorted out using fluorescence-activated cell sorting (FACS). A DntR mutant, which displayed 10 times higher induction levels than wild-type DntR in response to DNT was isolated. This mutant still maintained low levels of *gfp* expression in the absence of DNT. The detection limit was ~10 μM, a 25-fold improvement compared to wild-type DntR. The functional role of some substitutions found in this clone is discussed in the framework of the structural changes observed when comparing the recently determined structures of DntR with and without bound inducer ligand.

**Editor:** Christopher B. Doering, Emory University School of Medicine, United States of America

**Funding:** Swedish Research Council (www.vr.se). The funders had no role in study design, data collection and analysis, decision to publish, or preparation of the manuscript.

**Competing Interests:** The authors have declared that no competing interests exist.

* E-mail: Rosa@dbb.su.se (RL); peterb@dbb.su.se (PB)

## Introduction

2,4-dinitrotoluene (DNT) is the main component in the vapor-phase derived from the explosive TNT [1]. The compound is one of the major contaminants found near TNT-producing facilities. Because it is toxic and presumably a human carcinogen there is an increasing need for sensitive detection of DNT. Several attempts have been made to develop biosensors for the detection of DNT, both *in vivo* systems with whole-cell biosensors [2,3] and *in vitro* systems using purified proteins/peptides [4,5]. The major challenge in developing a sensor for detection of DNT is to achieve both high sensitivity and high selectivity. Biological systems that have evolved over million of years can attain these properties, which has been taken advantage of in development of whole-cell biosensors for naturally occurring compounds such as mercury [6] and arsenite/arsenate [7], with detection limits reaching nanomolar levels. Biosensors for detection of xenobiotic man-made compounds pose a much greater challenge because the time-span to evolve the ability for binding and degradation of such compounds has been significantly shorter. The previously reported whole-cell biosensors for detecting DNT displayed detection limits in the range of 25 μM [3] to 125 μM [2], with a limited specificity [2].

Only very few nitro-aromatic compounds existed in Nature before the era of synthetic chemistry. Compounds such as nitrotoluenes and nitrobenzenes are man-made and used in the manufacturing of dyes, polymers, pigments and explosives. These compounds are often highly toxic and persistent because few pathways for complete degradation exist. Isolation of several species of bacteria and fungi capable of degrading nitro-aromatic compounds have been reported [8,9]. Investigation of the degradation pathways in several organisms has revealed evidence of recent and ongoing evolution. Remnants of genetic material from horizontal gene transfer are often seen, and there is usually a scattered organization of the genes encoding the degrading enzymes, which are still poorly adapted to the substrate in question. In addition, there is usually primitive and inefficient regulation of the enzyme synthesis [10].

DntR is an example of such an inefficient regulator. It is a LysR-type transcriptional regulator (LTTR) that was found in two different *Burkholderia* strains capable of degrading DNT [11,12]. DntR regulates the oxidative degradation pathway of DNT, but is poorly adapted to this task and DNT is a poor inducer for DntR-mediated transcriptional activity [13]. Instead, salicylate was found to act as a more efficient inducer. Although there is no functional role for salicylate in the DNT-degrading strains, it is an intermediary metabolite in the naphthalene-degrading pathways controlled by NahR [14] and NagR [15] and remnants of genes needed for naphthalene degradation were also found in the DNT degrading strains. NahR has 61% amino-acid sequence homology to DntR while NagR differs from DntR only by two amino-acid residues (T46A and N49K), clearly indicating recent divergence.

More recently NtdR/NbzR, another homologue to DntR was found in two different bacterial strains capable of degrading 2-nitrotoluene and nitrobenzene respectively [16]. NtdR was reported to react on DNT and several other nitro-aromatic compounds that were unable to function as inducers for DntR [13,16,17], although only seven amino acid residues differ between the two proteins.

DntR is functionally active as a homo-tetramer, where each monomer could be divided into an N-terminal DNA-binding domain (DBD) with a winged helix-turn-helix motif connected to a C-terminal inducer-binding domain (IBD) (**Figure 1**) via a linker region. The IBD could be further divided into two subdomains (RD1 and RD2) connected with a hinge region, with the inducer binding cavity (IBC) situated between RD1 and RD2. Structures of the full-length DntR have been determined with acetate or thiocyanate in the IBC [18]. Inducer binding is suggested to cause changes of the global protein conformation where re-positioning of the DBDs would relax the angle of bound promoter-DNA and thereby trigger transcriptional activation [13,18,19]. Very recently, we obtained structural data of a truncated DntR protein without the DBD-domain, with and without the inducer salicylate [20]. These data reveal two main conformations of the IBD, where salicylate binding induces a rotation of RD2 relative to RD1 such that several loops in RD2 move closer to loops in RD1, which results in wrapping of the salicylate between RD1 and RD2.

To this date, most efforts of altering the properties of transcription factors have relied on a rational-design strategy. However, it is commonly difficult to predict the effect of a single mutation at a functional level, especially when ligand binding is coupled to global structural changes [21]. A few studies have instead been based on the use of directed evolution to alter the properties of transcription factors [2,22,23]. In the present study we used the "NtdR" variant of DntR as a starting point and applied a directed evolution strategy to change the inducer specificity of the transcription factor DntR towards improved DNT response. A clone with a satisfactory repressor ability and a dramatic improvement of both the level of response and sensitivity for DNT was isolated. The effect of some substitutions found in this clone are also discussed in the framework of observed differences in the structures of DntR with and without the inducer salicylate [20].

## Results and Discussion

### Library construction, selection and screening

An "NtdR" mutant of DntR that showed broader inducer specificity than wild-type (wt) DntR (**Table S1**) was selected as a starting template in the search for DntR variants with improved response to DNT. The "NtdR" mutant was chosen due to this broadened inducer response, since a protein with a broad specificity is more likely to evolve its specificity towards a new ligand than a protein with a narrow specificity [2,24,25,26]. A *gfp* gene [27] was used as a reporter for DntR-mediated transcriptional activation, and cells displaying an increase in GFP fluorescence after exposure to DNT were sorted out from the constructed libraries using FACS.

The directed evolution process was performed in five consecutive rounds (see overview in **Figure 2**). The first three rounds employed mutagenic PCR with an average mutation frequency of ~2 mutations/gene fragment. Randomization was applied to residues 38–301, excluding the first part of the DBD. In round 4 and 5, libraries were created by recombination, using a variant of staggered PCR (StEP). An average increase in the DNT-response could be seen when analyzing randomly picked clones from the subsequent sorts (**Figure 3A**). This was in general also accompanied by an increase in the basal level of fluorescence (**Figure 3B**). To reduce the basal level of fluorescence, a switch of growth media was performed in round 5 from LB to M9*.

By switching from the rich LB medium to M9*, a subpopulation with lower basal level of fluorescence in M9* than in LB was observed. Upon addition of DNT, we observed a shift to higher fluorescence for this subpopulation. Consequently, this subpopulation was sorted out, thereby removing the constitutively active clones with high basal activities in both growth media and enriching the clones with lower basal level of fluorescence in M9*. Two additional rounds of sorting was performed, and from these

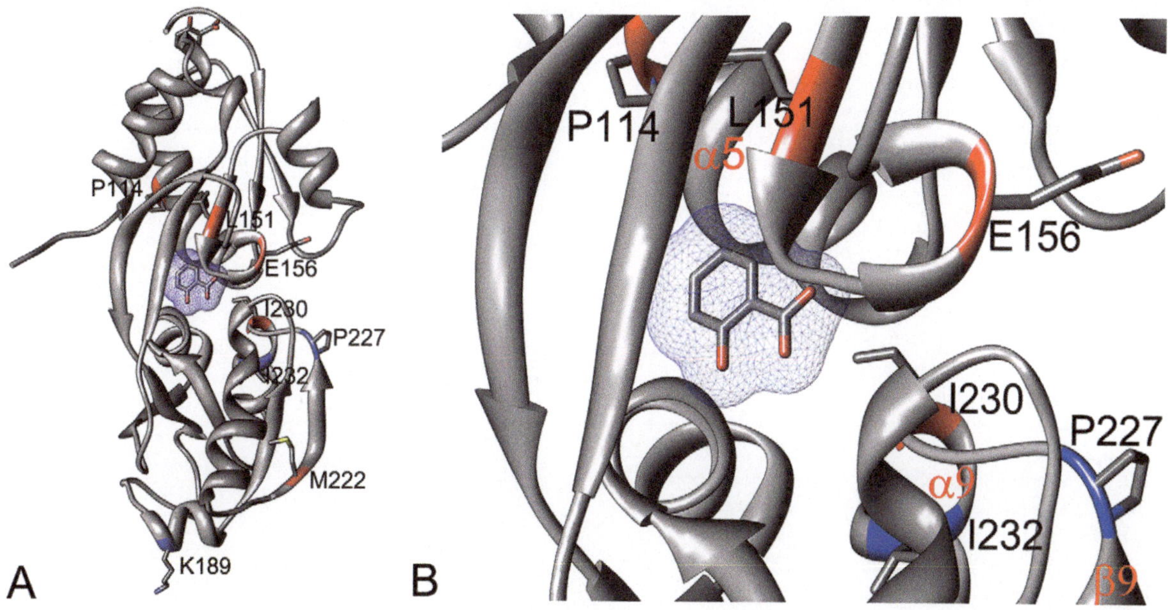


**Figure 1. Structure of DntR.** (A) Left: The ribbon structure of one inducer-binding domain of DntR with salicylate (salicylate is co-crystallized with the ΔN90-DntR construct in the fully closed state, PDB ID 2Y7P). The enriched mutations from the first four generations of libraries are shown colour-coded from red (occurring>10 times) to green (occurring 3–4 times). Right: Table listing the enriched mutations with the same colour-coding as in the picture above. The number of occurrences is calculated from all mutations occurring in randomly selected clones (a total of 82 clones) from library 1 to library 4. (B) Ribbon representation of the IBC of DntR with salicylate. The side-chains are shown for the residues where the enriched mutations are found. Also marked is the position of helix α5, helix α9 and β-strand β9 and the side-chain of D105.

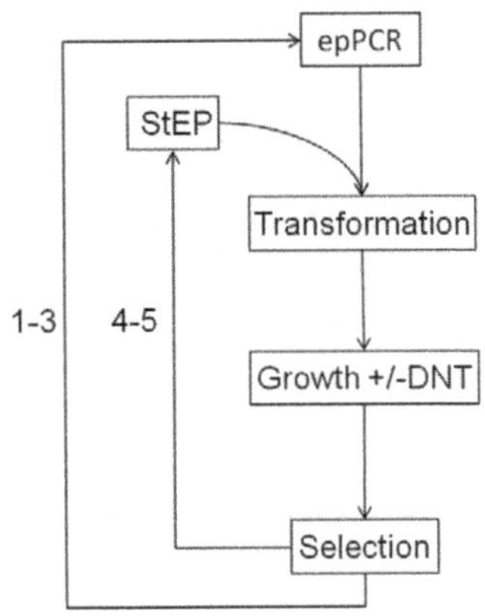


**Figure 2. Overview of the directed evolution process.** Five libraries were created in subsequent order, as indicated by the numbers 1–5. Library 1–3 were created using epPCR with Mutazyme II polymerase. The library size was in the range of $1–7\times10^7$ transformants/library. Library 4 was created using a modified variant of StEP using Mutazyme II polymerase, resulting in recombination of clones selected from library 3 with additional point mutations introduced by Mutazyme II. The best responding clones selected from library 4 were then recombined with StEP, this time using Pfu Polymerase. The success of recombination was confirmed by sequencing of 8 clones from each of library 4 and 5. Library 1–4 was grown in LB, while Library 5 was grown in both LB and M9*, with selection performed on cells grown in M9*. Each library was sorted 2–4 times (sorting and regrowth prior to another library generation indicated by dashed arrow).

sorts 132 clones were screened individually, and 11 unique clones with a DNT-response ranging from 3–12 fold induction with 500 µM DNT were identified (**Table 1**).

### Clones with improved DNT-response

Several clones with a 3–12 fold response to 500 µM DNT were isolated in the second and third sorting round of library 5 (**Table 1**). Also the 4.2c1 clone from the second sorting round of library 4 that was used as one of the templates for library 5, displayed a more than 3-fold induction with DNT in M9. Eight out of 11 of the best-responding clones in library 5 carried the I230N substitution derived from the 4.2c1 mutant, among them the best-responding clone 5.3p3c19. This clone showed an increase in fluorescence by a factor of ~12 upon addition of 500 µM DNT, compared to wt DntR. This mutant also had an increased response to salicylate, 2-nitrobenzoate, 4-nitrobenzoate and benzoate compared to both wt DntR and NtdR (**Table 2**). The response in the range of 250–1000 µM of both salicylate and DNT were at similar levels (13–15 fold induction) for this mutant (**Figure 4A–C**). The level of response to 250–1000 µM DNT for the 5.3p3c19 mutant was higher than that of wt DntR to salicylate in the same concentration range. This mutant apparently displayed both a large improvement in DNT response and a general improvement of the sensitivity both for the original inducer salicylate and for all tested compounds having a carboxy-group. The 5.3p3c19 mutant shows a general improvement in sensitivity, with a detection limit of ~10 µM for DNT and ~50 nM for salicylate, as compared to ~250 µM and ~250 nM, respectively for NtdR and ~250 µM and 1 µM respectively for wt DntR (**Figure 4C**).

### Structural significance of some substitutions obtained during the selection process

Sequencing of clones from each step of the sorting and selection of the first four libraries yielded a total of 81 clones carrying 291 mutations that altered the amino-acid composition. It was apparent that changes in some residues were reoccurring in several clones and were enriched during the sorting process, whereas ~45% of the unique point mutations occurred only once (~80% once or twice) and appear to be evenly distributed over the randomized *dntR* fragment.

The most frequently mutated residue was H169, but substitutions at this site were not found among any of the clones with improved DNT-response (the role of this residue is discussed in [28]).

The enriched substitution Arg165His (**Figure 1**) is found at about the same location as Arg156 in BenM, another member of the LysR type family induced by benzoate. In BenM, an R156H mutant was reported to have higher basal level of transcriptional activity than wild-type BenM [29].

When comparing the recently published structures of the IBD with and without salicylate bound [20], it is seen that inducer binding results in rotation of RD2 towards RD1, resulting in large movement of the α9 helix that also becomes partly unstructured. Also, the β8-α8 and β9-α9 loops in RD2 close towards the β4-α6 loop in RD1 (see Figure 4 in Devesse *et al.* for details [20]). In the closed conformation, these loops enclose the salicylate bound in the IBC, expelling water from the RD1–RD2 interface. In the best-responding clone, 5.3P3c19, an I230N substitution is found. I230 is situated in the β9-α9 loop (also the enriched substitution H228N is found in this loop). The I230N substitution would presumably result in a steric clash with the opposing D105 in α5 of RD1 upon transition of the IBD to the closed conformation (ligand bound). The side chains of Y110, F111 and I106 in this α5 helix constitute a hydrophobic wall in the IBC (**Figure 1B**). Repositioning of α5 due to the interactions between N230 and D105 could increase the volume of the hydrophobic part of the IBC which is expected to favor DNT-binding.

The combination of I230N with L151F is suggested to be the main reason for the improved DNT-response of the best-responding 5.3P3c19 clone compared to the clones with only 3–4 fold induction that carries the I230N substitution but not L151F (**Table 1**). The carbonyl oxygen of Leu151 forms a water-mediated interaction with salicylate in the IBC [20], and substitution with a Thr was previously suggested in order to reduce salicylate binding and facilitate DNT-binding [18]. However, the L151T substitution resulted in a lowered response to salicylate, but no change of inducer specificity [13]. The enriched L151F substitution found in the best responding mutant 5.3p3c19 is expected to change the position of the carbonyl oxygen of this residue, suggested to increase the volume of the hydrophobic part of RD1 close to the IBC, because the side-chain of the introduced F151 would have to move in order not to clash with F111.

In conclusion, by using several rounds of random mutagenesis followed by recombination of clones with a modest DNT-response, a DntR variant with greatly improved DNT-response was isolated. This variant also displayed a higher sensitivity to several other aromatic compounds, including the original inducer salicylate. Interestingly, the sensitivity observed for salicylate in this study is in fact higher than in studies aiming at improving the sensitivity of salicylate biosensors [30,31].

## Materials and Methods

### Chemical reagents and enzymes

Salicylic acid, 2-nitrotoluene, 2-nitrobenzoic acid, 2,4-dinitrotoluene and 2-hydroxytoluene (o-cresol) were all purchased from Sigma-Aldrich. Benzoic acid, 4-nitrobenzoic acid and 4-nitroto-

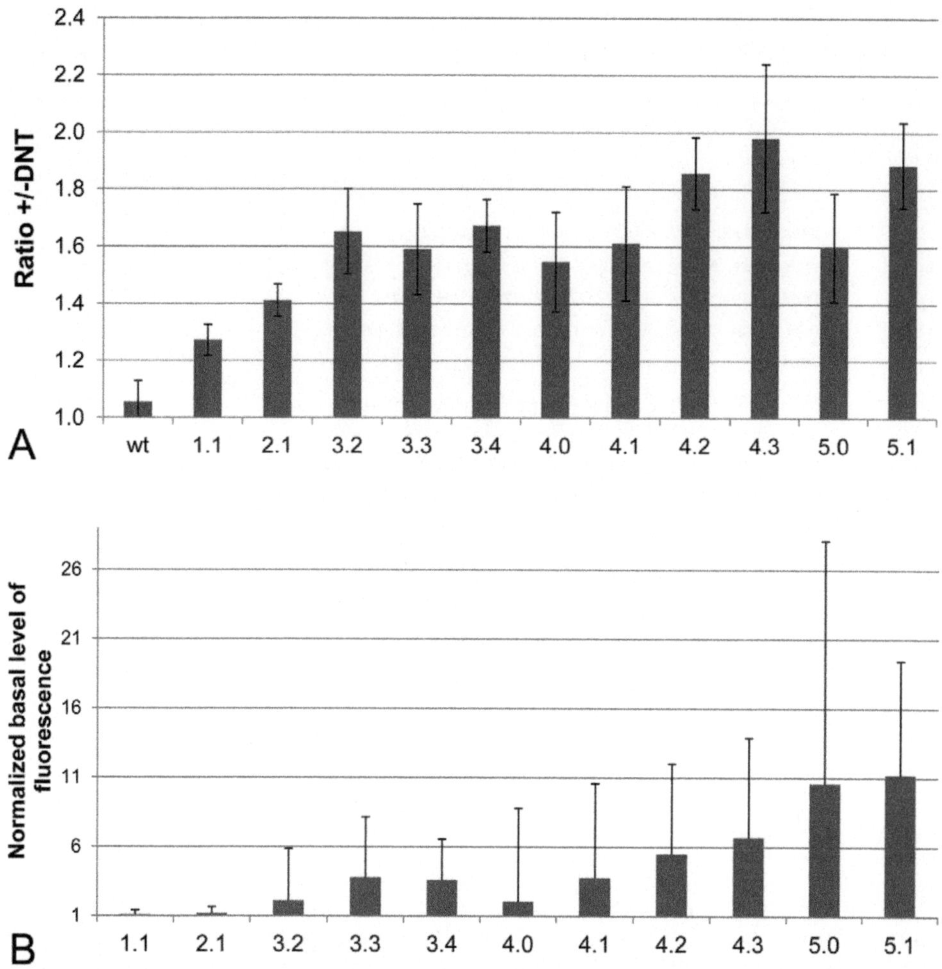


**Figure 3. The response to DNT and the basal level of fluorescence.** The diagram shows the mean response to DNT (A) and the mean basal level of fluorescence (B) for randomly selected clones from consecutive steps during the sorting procedure until the first sorting of library 5. The first number indicates the library, and the second number indicates the sorting step for that library. A) The response to DNT was measured for 5–10 clones for each sorting step. For each clone the response was calculated as the ratio between the mean fluorescence of 10000 cells grown in the presence of 500 μM DNT and the mean fluorescence of 10000 cells grown in the presence of the solvent DMSO. Shown in each bar is the mean for all the analyzed clones in that sorting step and in error bars are the standard deviation of the DNT response for these clones. As a comparison the mean response to DNT for wt DntR is shown with the standard deviation based on five independent data sets. B) The basal level of fluorescence is calculated as the ratio between the fluorescence of the population with DMSO added compared to the fluorescence of a WT DntR control with DMSO added for the same clones as in A). Clones were also analyzed for library 5, but since these were grown in M9* and first subjected to a screen on plate instead of randomly selected, there is no comparable data available.

luene were purchased from VWR International. All chemicals were of the highest grade available. For the flow cytometric analysis, 500 mM stock solutions of each compound were prepared in dimethylsulfoxide (DMSO). The stock solutions were kept frozen between usages. Restriction enzymes were purchased from Fermentas and T4 DNA ligase was purchased from New England Biolabs. For the random mutagenesis, the GeneMorph® II Random Mutagenesis kit (Stratagene, La Jolla) was used.

## Library construction, selection and screening

The introduction of mutations to create the starting template "NtdR" from wtDntR was done with the QuikChange Site-Directed-Mutagenesis kit (Stratagene, La Jolla). For an overview of the work process, see **Figure 2**.

For the first library, the primers GACATCGACTTGAAT-CTGCTGG and CATTGGGATATATCAACGGTGG were used together with MutazymeII polymerase (Stratagene, La Jolla), which amplifies an 1122 bp fragment containing the *dntR* gene from the pQE60DntR plasmid (construction of this plasmid has been described previously [13]). In each PCR tube, 90 ng of plasmid (corresponding to 45 ng of target DNA) carrying the "*ntdR*" gene was used. PCR was carried out using a program of 95°C for 5 min followed by 30 cycles of 95°C for 45 s, 55°C for 45 s and 72°C for 1 min 30 s. For the second and third library, the primers GACATCGACTTGAATCTGCTGG and GCCACCT-GACGTCTAAGAAACC were used to amplify a 1061 bp fragment containing the *dntR* gene library variants from the pQE*dntR*:$P_{DNT}$:*gfp* plasmid. PCR was performed as above, using 90 ng of plasmid in each tube.

For the fourth library, plasmids containing the clones from the third sorting of the third library were used as templates in a combination of staggered extension process PCR and epPCR by using the mutazyme and the following PCR programs: I) 95°C 7 min followed by 30 cycles of 95°C for 30 s and 60°C for 5 s. II) 95°C for 5 min followed by 40 cycles of 95°C for 45 s, 55°C for 45 s and 72°C for 1 min 30 s. For creation of the fifth library the

**Table 1.** The sequences (compared to the sequence of wt DntR) of the original templates and the clones with >3-fold induction in response to DNT when grown in M9 listed together with the response to 500 μM DNT and 500 μM salicylate expressed as the fold of induction.

| Protein variant | SAL | DNT | Sequence |
|---|---|---|---|
| wt | 10.4±0.9 | 1.2±0.1 | |
| NtdR | 12.6±0.6 | 1.9±0.2 | **T46A, N49K, I74V, H169L, K189R, P227S, I232V** |
| **5.3P3c19** | **12.0±0.9** | **11.6±1.3** | S40A, **T46A, N49K,** L54V, **I74V,** P114Q, L151F, E156K, **K189R,** M222I, **P227S,** I230N, **I232V** |
| 5.3P3c21 | 12.7±0.9 | 4.8±0.4 | **T46A, N49K,** L54V, P114Q, M222I, **P227S,** I230N, **I232V** |
| 5.2R3c24 | 10.1±0.5 | 4.1±0.3 | S40A, **T46A, N49K,** L54V, **I74V,** N203Y, M222I, **P227S, I232V** |
| 5.2R3c5 | 8.9±0.5 | 4.1±0.4 | S40A, **T46A, N49K, I74V,** R165H, **K189R, P227S,** I230N, **I232V** |
| 5.2R3c18 | 8.5±0.6 | 4.0±0.2 | S40A, **T46A, N49K, I74V,** R165H, **P227S,** I230N, **I232V** |
| 5.2R3c13 | 6.9±0.4 | 3.8±0.3 | **T46A, N49K,** L54V, **I74V,** T96R, N203Y, **P227S, I232V,** K280R |
| 5.3P3c27 | 6.7±0.6 | 3.7±0.2 | **T46A, N49K,** L54V, P114Q, R165H, **K189R, P227S,** I230N, **I232V** |
| 5.3P3c43 | 4.6±0.2 | 3.7±0.2 | **T46A, N49K,** L54V, A70S, **I74V,** P114Q, R165H, **P227S,** I230N, **I232V** |
| 5.2R3c11 | 7.4±0.3 | 3.5±0.3 | **T46A, N49K,** L54V, A70S, **I74V,** T79A, T96R, R165H, **P227S, I232V** |
| 4.2c1 | 5.4±0.1 | 3.6±0.4 | **T46A, N49K,** L54V, **I74V, P227S,** I230N, **I232V** |
| 5.3P2c31 | 4.4±0.7 | 3.2±0.1 | S40A, **T46A, N49K,** L54V, A70S, **I74V,** R165H, N203Y, M222I, **P227S,** I230N, **I232V** |

The mean fluorescence was measured for a cell population of 10 000 cells for each sample with the standard error based on at least three independent experiments shown in parenthesis. Clones are named after the library and sorting which they were isolated from followed by a unique clone label. The substitutions derived from the NtdR template are marked in bold.

same PCR programs were used with six selected clones from the fourth library as templates in an equimolar mixture, but this time Pfu polymerase was used in the process in order to achieve recombination without additional introduction of point mutations.

For all libraries the PCR products were purified from gel with the E.Z.N.A Gel Extraction Kit (Omega Bio-tek). The PCR products were digested with EagI and BglII. Digestion of the fragment resulted in a 775 bp fragment, corresponding to the *dntR* gene with the first 97 nucleotides excluded (corresponding to the 32 first N-terminal amino-acids). This fragment was purified from the gel, and then ligated into the EagI and BglII digested pQE*wtdntR*:$P_{DNT}$:*gfp* plasmid. Following ethanol precipitation, the ligated library (a total of 5–10 μg DNA) was transformed into *E. coli* DH5α cells through electroporation. The total number of transformants was estimated from plating a dilution series to be in the range of $1–7\times10^7$ transformants per library with a background of religated/undigested pQE*dntR*:$P_{DNT}$:*gfp* plasmid ranging from 3–30%.

For each library and each sorting of the library, 5–10 clones were sequenced to estimate the mutation frequency. For the first library the mutation frequency was estimated to 2.2 mutations/amplified gene fragment, based on sequencing of 10 clones. For the next libraries, more recombinations with the wt DntR sequence were observed, making an estimation of the mutation frequency difficult.

Since the fourth library was constructed from a pool of mutants derived from the third library using the MutazymeII polymerase that also introduces additional point mutations, the number of recombinations could not be estimated in this library. In the fifth library that was constructed using Pfu polymerase and where all of the sequences selected for recombination were known, sequencing of 8 clones in the unsorted library gave an average minimum number of recombinations of 3 per fragment, where fragments from all starting templates were represented, suggesting that the simplified (StEP) protocol used had been successful in achieving good recombination.

## Growth conditions

The first four libraries were grown in Luria-Bertani (LB) medium, while sort 1 to 3 of library 5 was grown in parallel in LB and a modified M9 medium, M9* (1×Difco M9 minimal salts, 1 mM $MgSO_4$, 0.1 mM $CaCl_2$, 0.2% glucose and 1%LB). For

**Table 2.** The response to a set of potential inducers for the original templates and the 5.3p3c19 clone grown in M9 22 h after induction.

| Protein variant | SAL | DNT | 2-NT | 2-NB | 4-NT | 4-NB | BEN | Basal F |
|---|---|---|---|---|---|---|---|---|
| wt DntR | 540±80 | 70±30 | 7±4 | 30±30 | 9±1 | 32±8 | −1±2 | 0 |
| "NtdR" | 820±30 | 110±20 | −15±15 | 40±30 | 0±27 | 260±70 | 150±60 | 60±35 |
| 5 5.3p3c19 | 1200±90 | 1200±130 | 21±16 | 820±120 | 65±35 | 530±70 | 1130±150 | 40±23 |

The response is expressed as the % increase in fluorescence when 500 μM of the potential inducer is added compared to the control where only solvent (DMSO) is added. The mean fluorescence was measured for a cell population of 10 000 cells for each sample with the standard error based on at least three independent experiments shown. The basal level of fluorescence (Basal F) is compared to that of WT DntR grown in M9 at the same signal amplification and is measured as % increase in fluorescence for mutant grown with DMSO/fluorescence for wt grown with DMSO. (Wt grown in LB gives a 100%(±30) increase in fluorescence compared to wt grown in M9). SAL: salicylate, DNT: 2,4-dinitrotoluene, 2-NT: 2-nitrotoluene, 2-NB:2-nitrobenzoate, 4-NT: 4-nitrotoluene, 4-NB: 4-nitrobenzoate, BEN: benzoate.

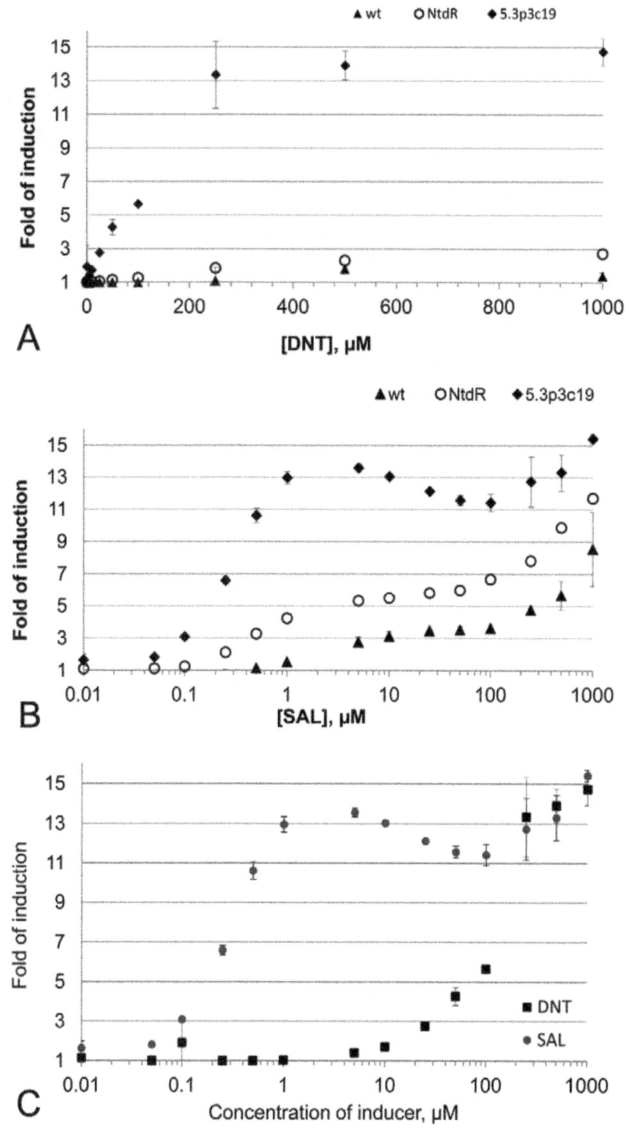


**Figure 4. The response to DNT and salicylate.** The measurements were done in the concentration range .10 nM-1000 µM for wt, NtdR and the 5.3p3c19 mutant grown in M9 medium, respectively (The water solubility limit of DNT is slightly above 1 mM). The response was measured as fold of induction and is calculated as the mean fluorescence of a cell population grown with the added concentration of inducer divided by the mean fluorescence of a cell population grown with addition of the solvent DMSO. Shown in error bars are the standard deviations. (A) The response to DNT measured as the fold of induction for wtDntR (filled triangles), NtdR (open circles) and 5.3p3c19 (filled diamonds). (B) The response to salicylate measured as the fold of induction for wtDntR (filled triangles), NtdR (open circles) and 5.3p3c19 (filled diamonds). (C) The response to DNT (black squares) and salicylate (grey circles) for the 5.3p3c19 clone (logarithmic scale).

later analysis of DNT-responding clones, M9 (1×Difco M9 minimal salts, 1 mM $MgSO_4$, 0.1 mM $CaCl_2$, 3 µM thiamine, 0.2% glucose and 0.7 mg/ml of each amino acid) was used. With all growth medias, the cells were grown at 30°C with shaking and supplied with 100 µg/ml ampecillin for maintenance of the pQE*dntR*:$P_{DNT}$:*gfp* plasmid. The transformed colonies containing the ligated library of *dntR* variants on the pQE*dntR*:$P_{DNT}$:*gfp* plasmid were pooled in 100 ml growth medium, grown for 2.5 h and pelleted. Freeze cultures were prepared from half of the culture, while the other half was used for plasmid preparation. Prior to analysis and sorting, a freeze culture of the library was grown in 25 ml of growth medium for 2 h. This culture was used to inoculate 2×100 ml culture so that the $OD_{600}$ at inoculation was 0.05. 2 h later one culture was induced with 500 µM DNT and DMSO was added to the other. The cultures were grown for 4 h in the presence or absence of DNT. The cells were pelletted down, and diluted in PBS prior to flow cytometric analysis and sorting. After sorting, cells were plated and regrown using the same protocol as above.

Individual clones from the fifth library were plated on M9*+agarose plates supplied with or without 500 µM DNT. Clones that appeared more fluorescent on the plate supplied with DNT were grown in liquid media as below.

Individual clones from all steps were grown overnight in 5 ml LB medium, and then inoculated to 1 ml growth media with a start $OD_{600}$ of 0.05. (The rest of the overnight cultures were used for plasmidpreparation followed by sequencing). After 2 h, 500 µM DNT was added, and 4 h/22 h later cells were taken out for FACS analysis. Selected clones were in addition also grown in the presence of 500 µM of a set of compounds with similarity to salicylate and/or DNT (2-nitrotoluene, 2-nitrobenzoate, 4-nitrotoluene, 4-nitrobenzoate and benzoate) with the same growth conditions as above.

The 5.2p3c19 clone, wt DntR and "NtdR" was also grown in M9 with concentrations of salicylate and DNT in the range of 10 nM-1000 µM, which were added from 1000× stock solutions for all concentrations to cell cultures grown as above.

## Flow cytometric analysis, sorting and screening

The flow cytometric analyses and sorting of libraries were performed on a FACS Vantage SE stream-in-air flow cytometry instrument (BD Biosciences, San Jose, CA, USA). The water-cooled argon ion laser was aligned prior to each analysis using flow cytometry beads for 488 nm excitation (Life technologies, Paisley, UK), and the same instrument settings were used for each time of analysis. Fluorescence was detected via a 530±15 nm (green) band pass filter. For each cell sample, data from 10 000 events were collected. For analysis of single clones, a FACS Calibur instrument (BD Biosciences, San Jose, CA, USA) was used with the same settings each time.

The cell population of the DNT-induced library was compared with the cell population of the non-induced library and with a control culture expressing wt DntR. A first gate was set to exclude cells with too high side and forward scatter. For the first four libraries, a gate was set to sort out cells with high fluorescence in the DNT-induced population while minimizing overlap with the population of the non-induced library. For the first three generations, 4–5% of the most fluorescent cells were sorted out in each sorting, with an overlap between the induced and non-induced population ranging from 15%–43%. In general, the overlap increased for each round of sorting due to enrichment of clones giving increased basal level of fluorescence. After regrowth of the sorted cells, the cells were sorted for 1–3 more rounds. For the fourth library where there was a prominent tail with higher fluorescence, 6–20% of the cells were sorted out with an overlap of 32–57%. In the first two rounds of sorting of the fifth library, cells with intermediary fluorescence (higher than the WT control+DNT, but lower than a large part of the population with mainly constitutively active mutants) were sorted out in the first round. In the following round, clones with higher fluorescence in response to DNT were again sorted out.

One particular challenge when selecting for DntR variants with increased response to DNT using GFP fluorescence as a sorting parameter, was the accumulation of clones with higher basal level of fluorescence in the absence of any added inducer. These clones were easily enriched after several rounds of cell sorting using high fluorescence as the major sorting parameter (**Figure 3B**). To reduce the number of false positives (i.e. inducer-independent phenotypes), another round of sorting was only done if reanalysis of the sorted cells grown with and without DNT showed that the ratio +/−DNT was increased compared to the previous round of sorting. A negative selection step where clones with low fluorescence in the absence of DNT were sorted from the second sort of the third library did lower the basal level of fluorescence in the sorted population, but the ratio +/−DNT was also reduced at this step. Later selection of clones in library 5 with lower fluorescence in M9 led to an increase in the ratio +/−DNT.

## Supporting Information

**Table S1 The response to a set of potential inducers for wt, NtdR and the 5.3p3c19 clone grown in LB.** The response is measured in the same way as in table 1. Data is based on at least three independent experiments with the standard error shown. The basal level of fluorescence is compared to that of WT DntR grown in LB at the same signal amplification.

## Acknowledgments

We would like to thank Per-Åke Nygren, the Royal Institute of Technology, for the access to the FACS Vantage SE instrument.

## Author Contributions

Conceived and designed the experiments: RL PB. Performed the experiments: RL EV. Analyzed the data: RL PB. Contributed reagents/materials/analysis tools: RL. Wrote the paper: RL PB.

## References

1. Jenkins TF, Leggett DC, Miyares PH, Walsh ME, Ranney TA, et al. (2001) Chemical signatures of TNT-filled land mines. Talanta 54: 501–513.
2. Galvao TC, Mencia M, de Lorenzo V (2007) Emergence of novel functions in transcriptional regulators by regression to stem protein types. Mol Microbiol 65: 907–919.
3. Radhika V, Proikas-Cezanne T, Jayaraman M, Onesime D, Ha JH, et al. (2007) Chemical sensing of DNT by engineered olfactory yeast strain. Nat Chem Biol 3: 325–330.
4. Nagatomo K, Kawaguchi T, Miura N, Toko K, Matsumoto K (2009) Development of a sensitive surface plasmon resonance immunosensor for detection of 2,4-dinitrotoluene with a novel oligo (ethylene glycol)-based sensor surface. Talanta 79: 1142–1148.
5. Jaworski JW, Raorane D, Huh JH, Majumdar A, Lee SW (2008) Evolutionary screening of biomimetic coatings for selective detection of explosives. Langmuir 24: 4938–4943.
6. Pellinen T, Huovinen T, Karp M (2004) A cell-free biosensor for the detection of transcriptional inducers using firefly luciferase as a reporter. Anal Biochem 330: 52–57.
7. Stocker J, Balluch D, Gsell M, Harms H, Feliciano J, et al. (2003) Development of a set of simple bacterial biosensors for quantitative and rapid measurements of arsenite and arsenate in potable water. Environ Sci Technol 37: 4743–4750.
8. Spain JC (1995) Biodegradation of nitroaromatic compounds. Annu Rev Microbiol 49: 523–555.
9. Crawford RL (1995) The Microbiology and Treatment of Nitroaromatic Compounds. Current Opinion in Biotechnology 6: 329–336.
10. Johnson GR, Spain JC (2003) Evolution of catabolic pathways for synthetic compounds: bacterial pathways for degradation of 2,4-dinitrotoluene and nitrobenzene. Appl Microbiol Biotechnol 62: 110–123.
11. Spanggord RJ, Spain JC, Nishino SF, Mortelmans KE (1991) Biodegradation of 2,4-dinitrotoluene by a Pseudomonas sp. Appl Environ Microbiol 57: 3200–3205.
12. Johnson GR, Jain RK, Spain JC (2002) Origins of the 2,4-dinitrotoluene pathway. J Bacteriol 184: 4219–4232.
13. Lönneborg R, Smirnova I, Dian C, Leonard GA, Brzezinski P (2007) In vivo and in vitro investigation of transcriptional regulation by DntR. J Mol Biol 372: 571–582.
14. Schell MA (1985) Transcriptional control of the nah and sal hydrocarbon-degradation operons by the nahR gene product. Gene 36: 301–309.
15. Jones RM, Britt-Compton B, Williams PA (2003) The naphthalene catabolic (nag) genes of Ralstonia sp. strain U2 are an operon that is regulated by NagR, a LysR-type transcriptional regulator. J Bacteriol 185: 5847–5853.
16. Lessner DJ, Parales RE, Narayan S, Gibson DT (2003) Expression of the nitroarene dioxygenase genes in Comamonas sp. strain JS765 and Acidovorax sp. strain JS42 is induced by multiple aromatic compounds. J Bacteriol 185: 3895–3904.
17. Ju KS, Parales JV, Parales RE (2009) Reconstructing the evolutionary history of nitrotoluene detection in the transcriptional regulator NtdR. Mol Microbiol 74: 826–843.
18. Smirnova IA, Dian C, Leonard GA, McSweeney S, Birse D, et al. (2004) Development of a bacterial biosensor for nitrotoluenes: the crystal structure of the transcriptional regulator DntR. J Mol Biol 340: 405–418.
19. Ezezika OC, Haddad S, Clark TJ, Neidle EL, Momany C (2007) Distinct effector-binding sites enable synergistic transcriptional activation by BenM, a LysR-type regulator. J Mol Biol 367: 616–629.
20. Devesse L, Smirnova I, Lonneborg R, Kapp U, Brzezinski P, et al. Crystal structures of DntR inducer binding domains in complex with salicylate offer insights into the activation of LysR-type transcriptional regulators. Mol Microbiol 81: 354–367.
21. Galvao TC, de Lorenzo V (2006) Transcriptional regulators a la carte: engineering new effector specificities in bacterial regulatory proteins. Curr Opin Biotechnol 17: 34–42.
22. Scholz O, Kostner M, Reich M, Gastiger S, Hillen W (2003) Teaching TetR to recognize a new inducer. J Mol Biol 329: 217–227.
23. Tang SY, Fazelinia H, Cirino PC (2008) AraC regulatory protein mutants with altered effector specificity. J Am Chem Soc 130: 5267–5271.
24. Jensen RA (1976) Enzyme recruitment in evolution of new function. Annu Rev Microbiol 30: 409–425.
25. Tokuriki N, Tawfik DS (2009) Protein dynamism and evolvability. Science 324: 203–207.
26. Bloom JD, Arnold FH (2009) In the light of directed evolution: pathways of adaptive protein evolution. Proc Natl Acad Sci U S A 106 Suppl 1: 9995–10000.
27. Cormack BP, Valdivia RH, Falkow S (1996) FACS-optimized mutants of the green fluorescent protein (GFP). Gene 173: 33–38.
28. Lönneborg R (2011) In search of a biosensor for DNT detection: studies of inducer response and specificity. Stockholm: Stockholm University. 68 p.
29. Ruangprasert A, Craven SH, Neidle EL, Momany C (2010) Full-Length Structures of BenM and Two Variants Reveal Different Oligomerization Schemes for LysR-Type Transcriptional Regulators. J Mol Biol 404: 568–586.
30. Huang WE, Wang H, Zheng H, Huang L, Singer AC, et al. (2005) Chromosomally located gene fusions constructed in Acinetobacter sp. ADP1 for the detection of salicylate. Environ Microbiol 7: 1339–1348.
31. Shin HJ (2010) Development of highly-sensitive microbial biosensors by mutation of the nahR regulatory gene. J Biotechnol 150: 246–250.

9

# Engineered Protein Nano-Compartments for Targeted Enzyme Localization

**Swati Choudhary[1], Maureen B. Quin[1], Mark A. Sanders[2], Ethan T. Johnson[1], Claudia Schmidt-Dannert[1]***

**1** Department of Biochemistry, Molecular Biology and Biophysics, University of Minnesota, St. Paul, Minnesota, United States of America, **2** University Imaging Centers, University of Minnesota, St. Paul, Minnesota, United States of America

## Abstract

Compartmentalized co-localization of enzymes and their substrates represents an attractive approach for multi-enzymatic synthesis in engineered cells and biocatalysis. Sequestration of enzymes and substrates would greatly increase reaction efficiency while also protecting engineered host cells from potentially toxic reaction intermediates. Several bacteria form protein-based polyhedral microcompartments which sequester functionally related enzymes and regulate their access to substrates and other small metabolites. Such bacterial microcompartments may be engineered into protein-based nano-bioreactors, provided that they can be assembled in a non-native host cell, and that heterologous enzymes and substrates can be targeted into the engineered compartments. Here, we report that recombinant expression of *Salmonella enterica* ethanolamine utilization (*eut*) bacterial microcompartment shell proteins in *E. coli* results in the formation of polyhedral protein shells. Purified recombinant shells are morphologically similar to the native Eut microcompartments purified from *S. enterica*. Surprisingly, recombinant expression of only one of the shell proteins (EutS) is sufficient and necessary for creating properly delimited compartments. Co-expression with EutS also facilitates the encapsulation of EGFP fused with a putative Eut shell-targeting signal sequence. We also demonstrate the functional localization of a heterologous enzyme (β-galactosidase) targeted to the recombinant shells. Together our results provide proof-of-concept for the engineering of protein nano-compartments for biosynthesis and biocatalysis.

**Editor:** Ashley M. Buckle, Monash University, Australia

**Funding:** This research was supported by the University of Minnesota's Institute on the Environment, Biocatalysis Initiative the BioTechnology Institute (University of Minnesota), and the Office of Naval Research (grant #N00014-10-1-0157). The funders had no role in study design, data collection and analysis, decision to publish, or preparation of the manuscript.

**Competing Interests:** The authors have declared that no competing interests exist.

* E-mail: schmi232@umn.edu

## Introduction

Engineering metabolic pathways into heterologous host cells to produce valuable chemical compounds and biofuels is a major goal of synthetic biology [1,2]. Factors like diffusion limitation, alternate metabolic routes, accumulation of toxic reaction intermediates and inhibitory products, however, frequently reduce the efficiency of such engineered pathways. In nature, cells often circumvent these issues by co-localizing metabolic enzymes [3]. Co-localization can be achieved, for example, by tethering enzymes to structures such as protein scaffolds or lipid membranes [4,5]. Drawing upon this approach, a synthetic protein scaffold was recently shown to dramatically increase flux through an engineered biosynthetic pathway [6]. Sequestration of enzymes into semi-permeable compartments or organelles is another strategy used by cells to spatially organize metabolic reactions. Unlike tethering, compartmentalization allows more stringent control over substrate and product transport to and from enzyme assemblies. It also protects the organism from harmful reaction intermediates. It is now known that several bacteria form proteinaceous shells that encapsulate functionally related enzymes. These are collectively referred to as bacterial microcompartments (BMCs) [7]. Recent advances in our understanding of BMC structure and function open up possibilities for engineering them into nano-bioreactors for biosynthesis and biocatalysis.

BMC-shell encoding genes are present in more than 400 sequenced bacterial genomes; and the encoded proteins are associated with enzymes involved in at least eight different metabolic pathways [7,8,9]. A subset of BMCs, called carboxysomes, plays an important role in $CO_2$ fixation [10,11]. Propanediol utilization (Pdu) BMCs catabolize 1,2-propanediol, and are found in *Salmonella*, *Citrobacter*, and some other bacteria [12,13,14]. *Salmonella* also forms BMCs during growth on the two-carbon substrate ethanolamine [7]. The membrane constituent phosphatidylethanolamine is hypothesized to serve as a major source of carbon, nitrogen and energy to enteric bacteria [7]. This view is supported by the attenuated behavior of *Salmonella* ethanolamine utilization (*eut*) loss-of-function mutants in the murine gut [15].

BMC shells have a viral capsid-like polyhedral structure with a diameter of 100–150 nm. While the shells of carboxysome are icosahedrons, those of Pdu and Eut BMCs appear to be semi-regular polyhedrons. Thin cell-section transmission electron micrographs of Eut BMCs indicate that Eut BMC shells have a rounder/less-sharp-edged morphology than either carboxysomes or Pdu BMCs [16,17]. BMC shells are formed by thousands of copies of a few proteins belonging to the BMC-domain family (Pfam family: Pf00936). BMC-domain proteins have been crystallized as flat cyclic hexamers (or pseudohexamers), and are believed to form the edges and facets of the microcompartment

[18]. The vertices of these shells are hypothesized to be capped by members of another BMC shell-associated protein family, some of which have been crystallized as pentamers (Pfam family: Pf03319) [7]. The multimeric shell protein structures have central pores of varying sizes and with different electrostatic properties. Crystallization of a few shell proteins with their central pores in either "open" or "closed" configuration suggests that they may function as gated transit points for cofactors and small metabolites [18].

To demonstrate the feasibility of engineering heterologous protein-based microcompartments in *E. coli*, we chose the Eut BMC shell proteins from *Salmonella enterica* LT2. While the exact composition of *S. enterica* Eut BMCs is unknown, they are believed to be made up only five shell proteins; and putative signal sequences that target enzymes to the interior of Eut BMCs can be inferred from BMC-targeting sequences recently reported by Bobik and colleagues in experiments with native Pdu BMCs in *S. enterica* [19]. The Eut BMC shell genes are encoded on the 17-gene *eut* operon, which also encodes for enzymes involved in the degradation of ethanolamine (**Fig. 1**) [20]. Evidence suggests that *S. enterica* Eut BMCs prevent dissipation of the volatile reaction intermediate acetaldehyde, and protect the cell from aldehyde toxicity [8,17,21]. A homologous *eut* operon is also present in *E. coli*, but is disrupted by a transposon in several common laboratory strains. Although there have been some reports of *eut* operon induction and Eut BMC formation in a few *E. coli* strains, these were observed only under very specific growth conditions [16,22]. Eut BMCs were not observed in our laboratory *E. coli* strains under either standard growth conditions or conditions reported to induce Eut BMC formation in *S. enterica*. To date, functional characterization of the *eut* operon has been conducted mainly using *S. enterica*; although crystal structures have recently been solved for putative Eut shell proteins cloned from *E. coli* [17,21,23,24,25,26,27,28].

While there is interest in engineering bacterial microcompartments to create intracellular protein compartments for biotechnological applications, the interactions between individual BMC shell proteins and enzymes targeted for encapsulation have not been studied in any heterologous system [10,19,29]. Previously, an attempt was made to produce empty *Citrobacter* Pdu BMCs by overexpressing the Pdu shell proteins in *E. coli*; however, purification of intact microcompartments was not reported [30]. Here, we show that *E. coli* can form polyhedral compartments with recombinantly expressed *S. enterica* Eut shell proteins. Purified recombinant Eut shells appear to be morphologically similar to the native *S. enterica* Eut compartments. We demonstrate that an N-terminal signal sequence targets Enhanced Green Fluorescent Protein (EGFP) and β-galactosidase to the recombinant shells. The ability to sequester catalytically active β-galactosidase indicates that the recombinant compartments may be engineered to encapsulate multi-enzymatic reactions. We also report the surprising discovery that one of the BMC-domain proteins, EutS, is necessary and sufficient for the formation of shells *in vivo*, and for targeting of heterologous proteins to these structures; thereby offering a simple strategy for the engineering of protein-based nano-bioreactors.

## Results

### Expression of *S. enterica* Eut shell proteins in *E. coli*

The *eut* operon in *S. enterica* encodes for a Pfam03319 protein (EutN) and four BMC-domain proteins (EutS, EutM, EutL and EutK, Pfam00936) which are homologs of Eut and Pdu BMC shell-associated proteins from *E. coli* and *Citrobacter* [31]. In order to investigate the role of Eut shell proteins in heterologous shell assembly, we cloned the *S. enterica* Eut shell genes into our in-house BioBrick$^{TM}$ expression vector pUCBB (**Fig. S1**) [32]. In 3-dimensional (3-D) crystals, wild type EutS displays a hexameric structure with a bend of approximately 40°, while the EutS-G39V mutant forms flat symmetric hexamers [28]. We hypothesized that the unusual bent structure formed by wild type EutS is important for its role in BMC shell function. In order to test this hypothesis, the EutS-G39V mutant was also functionally characterized in this study. As shown in **Fig. S2**, EutS, EutS-G39V, EutM, and EutK were overexpressed as soluble proteins in two different *E. coli* strains, while the expression of recombinant EutN and EutL varied between the two strains. The soluble expression of EutL and EutK was also affected by the co-expression of other Eut shell proteins. EutS and EutM showed aberrant SDS-PAGE migration ("smearing") typical of proteins that have hydrophobic peptide stretches [33].

### Formation of polyhedral microcompartments in *E. coli* co-expressing all five Eut shell proteins

Our next goal was to ascertain if the recombinant Eut shell proteins formed clearly demarcated protein shells in *E. coli*. For this purpose, cultures of *E. coli* harboring all five *S. enterica* Eut shell genes were grown overnight at 30°C. This lower temperature was expected to reduce the expression of individual recombinant proteins, and supply sufficient time for assembly of the 3-D shell. The cells were fixed, sectioned, and their internal structures examined by transmission electron microscopy (TEM). Expression of all five Eut shell proteins in *E. coli* resulted in the creation of clearly discernible protein shells in most of the examined cells (**Table S1**). The observed recombinant protein shells bore a strong resemblance to the BMCs produced by *S. enterica* control cells growing on ethanolamine (**Fig. 2A–D**, additional images in **Fig. S3**). The recombinant compartments were 100–200 nm in diameter and within range of the dimensions previously reported for native Eut BMCs [7]. However, while *S. enterica* cells displayed several Eut BMCs distributed through-out the cell, the recombinant EutSMNLK structures were restricted to only one or two per *E. coli* cell. Their intracellular localization was such that they were situated off-center, and close to the poles of the cell. These recombinant microcompartments were observed in two different strains of *E. coli* (C2566 and JM109). The structures formed in JM109 cells were larger in size than those observed in C2566 cells, a result likely arising from differences in shell protein expression rates between the two strains. The EutSMNLK shells were enveloped by an electron-transparent region which was also observed around many of the native *S. enterica* Eut BMCs. The Eut protein compartments may therefore be enveloped by an unknown matrix which is removed by detergent wash, suggesting that it may be of a lipophilic nature.

### EutS alone is able to form protein shells in *E. coli*

Based on the crystal structures of Eut shell proteins, a model has been proposed for their assembly to form the microcompartment [27,28]. According to this model, the edges and facets of the shell are formed by EutS and EutM respectively, while the central pore of the EutL pseudohexamer may facilitate gated transport into and out of the BMC. EutN is expected to 'cap' the vertexes of the icosahedral capsid, although unlike other Pfam03319 proteins which form pentamers, it crystallizes as a hexamer. The role of EutK in the Eut microcompartment assembly and function is at present unclear.

To explore the roles of *S. enterica* Eut shell proteins in the assembly of the Eut microcompartment, they were expressed in *E. coli* both individually and in different combinations. To our surprise, recombinant expression of EutS alone resulted in the

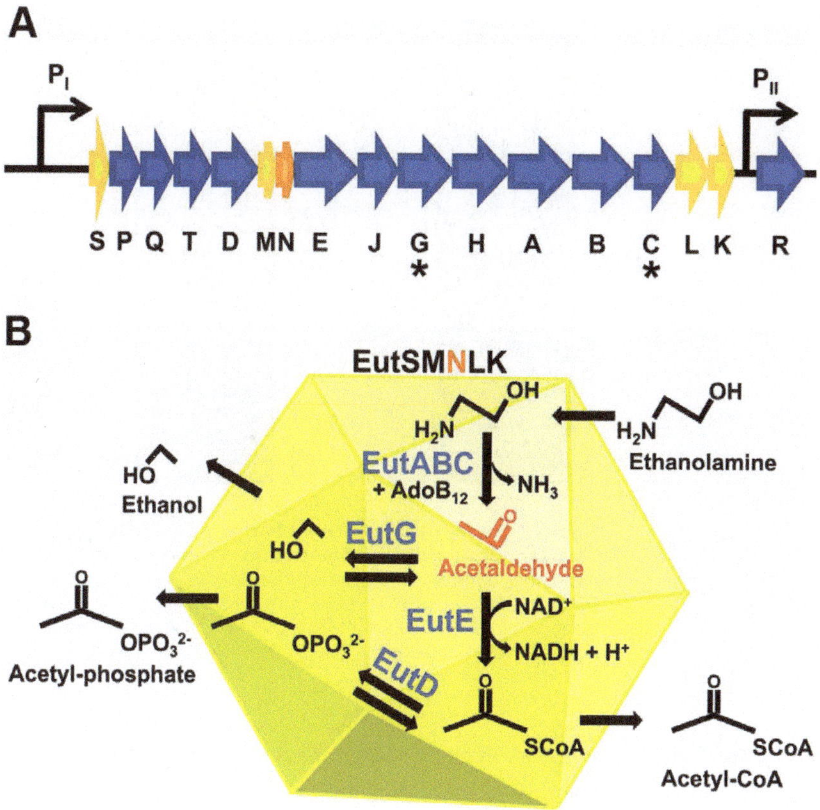


**Figure 1. Coenzyme-$B_{12}$-dependent ethanolamine utilization (*eut*) genes of *Salmonella enterica*.** (**A**) *eut* operon in *S. enterica*. *eutS*, *eutM*, *eutN*, *eutL* and *eutK* encode BMC shell proteins that are proposed to form the Eut microcompartment (yellow and orange) [20]. Asterisks indicate genes that encode for enzymes with predicted N-terminal signal sequences that target them to the BMC interior [19]. Transcription is induced from the $P_I$ promoter in the presence of both ethanolamine and vitamin $B_{12}$, while the promoter $P_{II}$ regulates weak constitutive expression of the transcription factor EutR [49]. (**B**) Model for catabolism of ethanolamine by the Eut BMC. Ethanolamine enters the microcompartment and is metabolized to ethanol, acetyl-phosphate and acetyl-CoA, which can enter the tricarboxylic acid cycle [7]. Eut BMC prevents dissipation of acetaldehyde, a volatile and toxic reaction intermediate (red) [21]. Enzymes assumed to reside in the BMC lumen include coenzyme-$B_{12}$-dependent ethanolamine ammonia lyase (EAL, EutBC), EAL reactivase (EutA), alcohol dehydrogenase (EutG), aldehyde dehydrogenase (EutE), and phosphotransacetylase (EutD).

formation of protein shells which were similar to those observed with EutSMNLK (**Fig. 2E–F**, additional images in **Fig. S3**). Engineered EutS protein shells appeared to be morphologically similar to the EutSMNLK compartments – they too were well-delimited and were surrounded by an electron-transparent region. One to two polyhedral bodies were observed per cell, and their appearance and localization matched that of compartments produced by EutSMNLK in *E. coli*. The ratio of 90 nm thin cell sections displaying recombinant compartments indicates that a majority of the *E. coli* cells expressed EutS compartments (**Table S1**). As shown in **Fig. 2G–H**, no defined structures were observed in cells co-expressing EutMNLK. These results indicate that engineering of *S. enterica* EutS is sufficient and necessary to form recombinant shells in *E. coli*.

Structures formed by over-expression of other Eut shell proteins are shown in **Fig. S3**. Recombinant expression of EutM alone created a thick axial filamentous structure which allowed cell division but interfered with separation. Similar fibers have been reported in *E. coli* expressing *Citrobacter freundii* PduAB, as well as the *S. enterica pduJ* deletion mutant [30,34]. Expression of EutK alone resulted in the formation of an electron-translucent region within the cell, while an electron-dense region was produced by co-expression of EutM and EutN. Comparable amorphous aggregates (albeit of a smaller size) are formed by *S. enterica pduBB'* loss-of-function mutants [34]. A small fraction of cells co-expressing EutLK displayed internal filaments similar to those observed in *E. coli* expressing *Citrobacter* PduABKN [30]. Apart from a few polar granules, *E. coli* expressing either EutN or EutL alone did not display clearly discernible structures.

## An N-terminal signal sequence targets EGFP to Eut shells

Recently, Fan *et al.* have demonstrated that an N-terminal signal sequence of PduP, a Pdu BMC-associated enzyme, targets heterologous proteins to the interior of native Pdu microcompartments in *S. enterica* [19]. Their sequence analysis also predicted N-terminal targeting sequences for other BMC-lumen associated proteins, including EutC and EutG (**Fig. 1**). In order to test if the predicted signal sequences (the first nineteen amino acids of EutC and EutG: $EutC^{1-19}$ and $EutG^{1-19}$, respectively) indeed function as Eut BMC-targeting sequences, we fused them to the N-terminus of EGFP; and expressed the fusion proteins in *S. enterica* grown in the presence of either glycerol or ethanolamine (to induce BMC formation). Bright fluorescent localization was observed in $EutC^{1-19}$-EGFP expressing cells grown on ethanolamine, suggesting that the tagged EGFP was being targeted to the Eut BMCs (**Fig. 3**). We noticed that the fluorescent loci were not stationary but were moving around within cells, suggesting interaction of the Eut BMCs with the *S. enterica* cytoskeleton (**Video S1**).

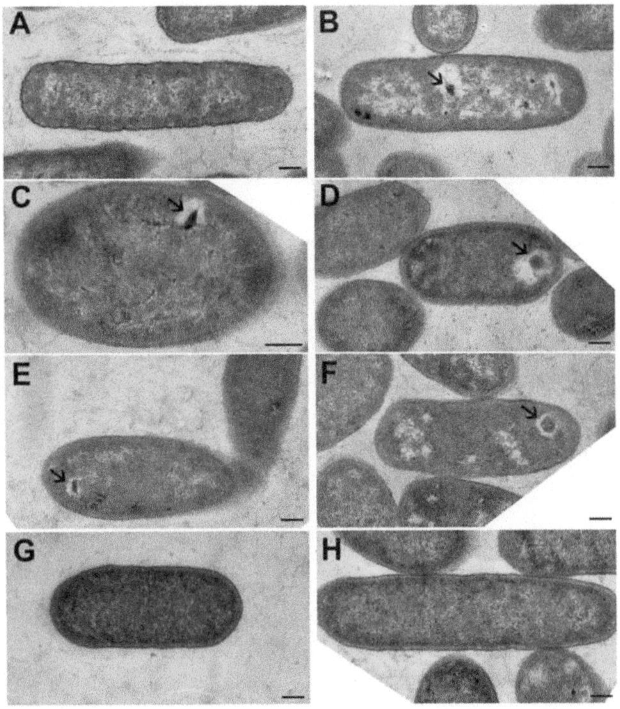


**Figure 2. Formation of engineered protein shells by expression of *S. enterica* Eut shell proteins in *E. coli*.** Transmission electron micrographs of thin sections of *S. enterica* and recombinant *E. coli*. (**A**) *S. enterica* grown on glycerol. (**B**) *S. enterica* grown on ethanolamine. (**C**) *E. coli* C2566 expressing recombinant EutSMNLK. (**D**) *E. coli* JM109 expressing recombinant EutSMNLK. (**E**) *E. coli* C2566 expressing recombinant EutS. (**F**) *E. coli* JM109 expressing recombinant EutS. (**G**) *E. coli* C2566 expressing recombinant EutMNLK. (**H**) *E. coli* JM109 expressing recombinant EutMNLK. Arrows indicate the location of recombinant BMCs. (Scale bar: 200 nm).

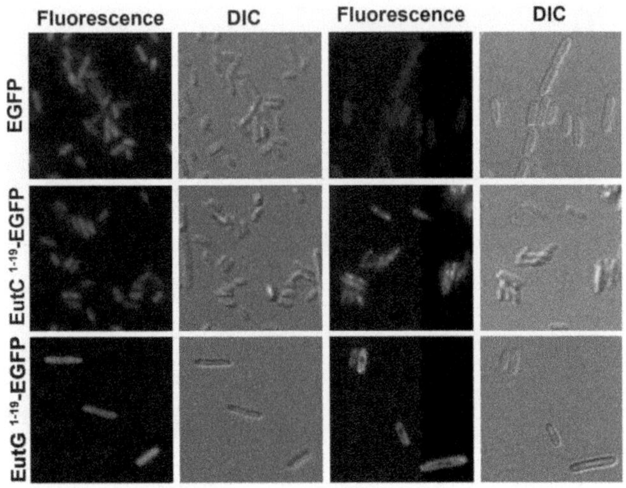


**Figure 3. Distribution of EGFP bearing putative N-terminal Eut BMC-targeting signal sequences in *S. enterica*.** *S. enterica* cells containing constructs for constitutive expression of EGFP, $EutC^{1-19}$-EGFP or $EutG^{1-19}$-EGFP were cultured with either glycerol or ethanolamine. Distribution of green fluorescence within the cells was observed by fluorescence microscopy. DIC images show the cell boundaries.

No fluorescent localization was observed when $EutC^{1-19}$-EGFP expressing cells were grown on glycerol, or with the EGFP control. Cells expressing $EutG^{1-19}$-EGFP also failed to display punctate green fluorescence when grown on ethanolamine, indicating that fusion with the first nineteen amino acids of EutG is not sufficient to target heterologous proteins to the Eut BMC.

After establishing that the first nineteen amino acids of EutC functioned as a BMC-targeting sequence in *S. enterica*, we sought to explore whether this sequence also localized heterologous proteins to the recombinant Eut protein shells engineered in *E. coli*. Therefore, $EutC^{1-19}$-EGFP was expressed in *E. coli* cells harboring the entire complement of Eut BMC shell proteins (EutSMNLK). Strong localized fluorescence was observed in 84% of the recombinant *E. coli* C2566 cells (**Fig. 4, Table S2**). Fluorescent green foci were also observed by co-expression of $EutC^{1-19}$-EGFP and EutSMNLK in *E. coli* JM109 cells; which indicates that the effect is not strain-specific (**Fig. S4, Table S2**). The number and location of these fluorescent foci (one to two per cell, near the poles) are consistent with the capsid structures observed by TEM. Transmission electron micrographs of thin cell sections indicated that similar structures were formed by *E. coli* cells expressing recombinant EutSMNLK alone or in combination with $EutC^{1-19}$-EGFP, supporting previous reports that formation of the bacterial microcompartment *in vivo* does not require scaffolding provided by a cargo protein (**Fig. S3**) [30,35]. Unlike in *S. enterica*, fluorescent loci in *E. coli* cells co-expressing EutSMNLK with $EutC^{1-19}$-EGFP appeared to be stationary.

## EutS is sufficient and necessary for targeting $EutC^{1-19}$-EGFP to recombinant Eut shells

To investigate whether one or a combination of Eut shell protein(s) is required for EGFP localization in *E. coli*, we next co-expressed $EutC^{1-19}$-EGFP with various Eut shell proteins. Remarkably, co-expression of only EutS and $EutC^{1-19}$-EGFP resulted in the formation of fluorescent foci within 87% of the *E. coli* C2566 cells and 84% of the *E. coli* JM109 cells, while cells expressing other Eut shell protein combinations without wild type EutS did not display discrete fluorescent localizations (**Fig. 4, Fig. S4, Fig. S5, Table S2**). Our results therefore indicate that EutS is necessary and sufficient for targeting $EutC^{1-19}$-EGFP to the engineered compartment.

## EutS and EutSMNLK shells are neither inclusion bodies nor are they enveloped by a hydrophobic matrix

The red fluorescent, lipophilic stain Nile Red was employed to examine the composition of the matrix surrounding the engineered proteins shells and to confirm that the shells were not inclusion bodies [36]. *E. coli* cells co-expressing $EutC^{1-19}$-EGFP and Eut shell proteins were stained with Nile Red, reasoning that co-localization of red and green fluorescent foci would indicate that the observed shells are either enveloped by a hydrophobic layer or are made up of aggregated, mis-folded proteins (aka. inclusion bodies). As shown in **Fig. S6**, no such co-localization was observed, indicating that the engineered protein shells formed by EutS alone or EutSMNLK are not inclusion bodies nor are they surrounded by a lipophilic matrix. As a control, we stained with Nile Red *E. coli* cells co-expressing the cyanobacterial carotenoid cleavage dioxygenase NSC-1, which we have previously shown to form inclusion bodies [37], and $EutC^{1-19}$-EGFP. As expected, clear Nile Red deposits were observed in *E. coli* expressing NSC1 inclusion bodies (**Fig. S7**). Furthermore, we observed dispersed green fluorescence, indicating that $EutC^{1-19}$-EGFP is not targeted to the NSC1 inclusion bodies. Therefore, our results indicate that overexpressed EutS or EutSMNLK do not form inclusion bodies in *E. coli* nor are the shells enveloped by a hydrophobic layer. The

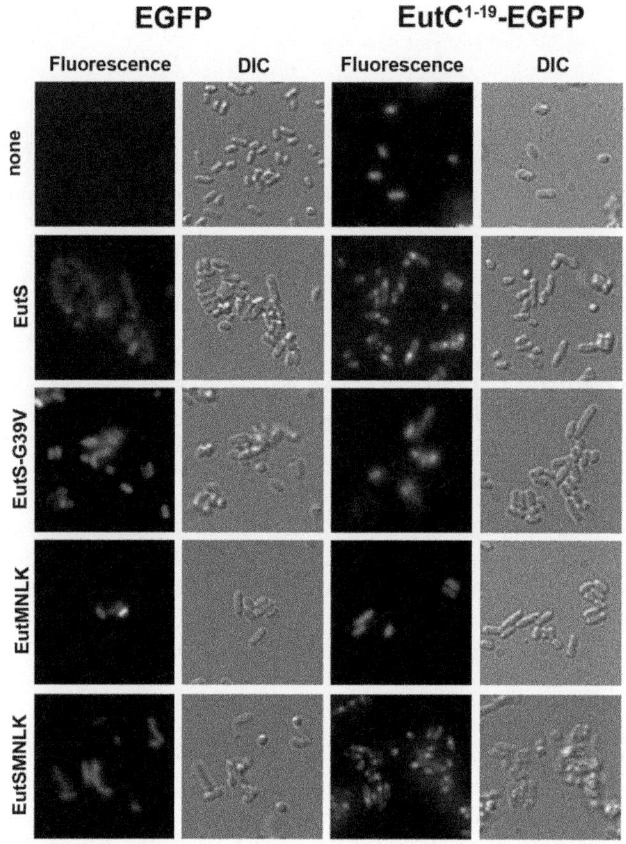


**Figure 4. Localization of EutC$^{1-19}$-EGFP in recombinant *E. coli* expressing *S. enterica* Eut shell proteins.** Fluorescence microscopy images of *E. coli* C2566 cells co-expressing EGFP or EutC$^{1-19}$-EGFP with EutS (wild type or the G39V mutant), EutMNLK or EutSMNLK. Cell boundaries are shown by the DIC images. (see **Table S2** for the quantification of EGFP localization in recombinant *E. coli*, and **Fig. S4** for the localization of EutC$^{1-19}$-EGFP in the *E. coli* JM109 strain).

composition of the matrix surrounding the shells (appearing as electro-transparent region after detergent wash) remains unclear.

The localization of EutC$^{1-19}$-EGFP but not EGFP to either EutS or EutSMNLK protein shells also shows that the recombinant protein compartments are made of functional, properly folded proteins; and are not merely aggregates of mis-folded proteins. Further evidence for the protein shells not being inclusion bodies is provided by the absence of EutC$^{1-19}$-EGFP puncta in *E. coli* cells co-expressing the EutS-G39V mutant (**Fig. 4, Fig. S4, Table S2**). Crystallographic studies have shown that this particular mutation causes a major change in the conformation adopted by EutS hexamers [28]. Our data links the altered 3-D crystal structure of EutS-G39V with an inability to sequester EutC$^{1-19}$-EGFP, suggesting that the conformation adopted by properly folded EutS-WT *in vivo* is essential for targeting EutC$^{1-19}$-EGFP to the engineered protein shell.

## Purification of Eut microcompartments

To further characterize Eut microcompartment formation in *E. coli*, we sought to purify the recombinant compartments. To the best of our knowledge, there have been no previous reports of successful purification of *S. enterica* Eut microcompartments. We therefore chose to first establish a protocol for purifying native Eut BMCs from *S. enterica* grown on ethanolamine. To assist in tracking fractions containing Eut BMCs during ultracentrifugation steps, we used *S. enterica* cells expressing EutC$^{1-19}$-EGFP, anticipating that encapsulation of EutC$^{1-19}$-EGFP by Eut BMCs would make them visible under UV light. Native Eut compartments were then isolated by modifying a previously established protocol reported for native *S. enterica* Pdu BMCs [13]. As a negative control we performed the same purification procedures with *E. coli* C2566 cells expressing only EutC$^{1-19}$-EGFP.

A white band that was fluorescent under UV light after sucrose gradient ultracentrifugation of *S. enterica* cell lysate was collected for further analysis. The *E. coli* EutC$^{1-19}$-EGFP control revealed a faint white band, but no fluorescent bands, in the sucrose gradient that was also collected. Isolated fractions were then analyzed by SDS-PAGE and TEM. SDS-PAGE analysis of the *Salmonella* sample (**Fig. 5A, lane 1**) showed a number of protein bands, including bands with molecular weights expected for recombinant Eut shell proteins and EGFP (**Fig. S2**). Aberrant protein migration ("smearing") of low molecular weight bands occurred similar to what was previously seen with the expression of EutS and EutM in *E. coli*. SDS-PAGE analysis of the control fraction from *E. coli* EutC$^{1-19}$-EGFP did not show any protein bands.

Negative stain TEM of the purified native Eut organelles revealed structures which were irregular in shape, with dimensions in the range of 100–150 nm (**Fig. 5B**). The purification was reproduced several times and negatively-stained structures of the same morphology were viewed on several different occasions,

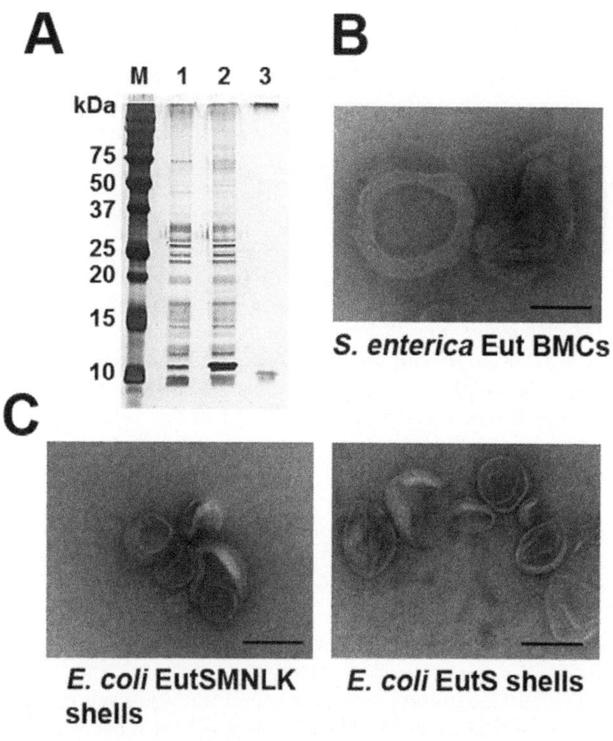


**Figure 5. Purification of Eut compartments.** (**A**) Silver stained SDS-PAGE gel showing purification of (lane 1) Eut BMCs from *S. enterica* cells harboring EutC$^{1-19}$-EGFP, (lane 2) recombinant EutSMNLK BMCs, and (lane 3) recombinant EutS BMCs from *E. coli* C2566 cells co-expressing EutC$^{1-19}$-EGFP. Calculated protein sizes are as follows: EutS (11.6 kDa), EutM (9.8 kDa), EutN (10.4 kDa), EutL (22.7 kDa), EutK (17.5 kDa), EutC$^{1-19}$-EGFP (29.1 kDa). (**B**) Transmission electron micrographs of isolated native and recombinant Eut compartments. From left to right: Eut BMCs from *S. enterica*, EutSMNLK shells from *E. coli* C2566, EutS shells from *E. coli* C2566. (Scale bar: 100 nm).

indicating that the structures isolated and viewed are the native Eut BMCs. The *E. coli* EutC$^{1-19}$-EGFP control sample did not contain any structures that could be visualized by TEM, confirming that the visualized *Salmonella* compartments are Eut BMCs and not membrane vesicles.

As an additional control, native Pdu BMCs from *S. enterica* were also purified and visualized following the published method (**Fig. S8A**) [13]. In our hands, the morphology of the purified Pdu and Eut BMCs is very similar – the structures are clearly discernible but somewhat deflated, perhaps as the result of the purification procedure and/or the destructive nature (dehydration) of negative stain EM.

Subsequently, we applied the same purification procedure developed for native Eut BMCs to the isolation of recombinant EutSMNLK and EutS compartments from *E. coli* C2566 co-expressing EutC$^{1-19}$-EGFP. As with the *S. enterica* Eut BMCs, a faint UV fluorescent band was detected after sucrose gradient ultracentrifugation and collected for further analysis. SDS-PAGE analysis of the isolated recombinant EutSMNLK shells (**Fig. 5A, lane 2**) showed a very similar protein pattern in the expected Eut shell protein size range when compared to the native Eut BMCs (**Fig. 5A, lane 1**). Protein patterns differ between the native BMC and recombinant protein shell preparations in the higher molecular weight range. The isolated EutS shell showed one smeared band at around 10 kDa as expected for EutS, but no band corresponding to EutC$^{1-19}$-EGFP. This might be due to a low protein concentration. We noticed that a large fraction of the loaded EutS protein does not migrate into the gel, suggesting that it may retain EutC$^{1-19}$-EGFP, thereby rendering its concentration in the gel too low for detection. Alternatively, lower levels of EutC$^{1-19}$-EGFP could be associated with engineered EutS protein shells which, however, is not supported by results from fluorescent microscopy (**Fig. 4**) and by the detection of a clear fluorescent band comparable to EutSMNLK BMCs during sucrose gradient centrifugation.

TEM analysis of the isolated recombinant EutSMNLK shells showed that they also appeared to be irregular structures (**Fig. 5C**). Purified recombinant EutSMNLK capsids were slightly smaller than the native Eut BMCs (the former are about 100 nm in diameter) (**Fig. 5B and 5C**). This difference in size may represent a difference in the number of different proteins constituting the native and recombinant compartments. While the native Eut shells are believed to be composed of only five shell proteins, a greater number of proteins may be involved in the formation of Eut BMCs in the native *S. enterica* host cell. For example, Pdu BMC shells are composed of seven proteins, one of which was only recently verified as a component of the shell (PduN) [12,34]. The morphology of purified EutS protein shells was similar to that of EutSMNLK compartments, (**Fig. 5C**), which supports the conclusion that EutS is capable of forming protein shells on its own.

The purification was reproducible and was also applied to the partial purification of native and recombinant Eut shells from cells that do not express EutC$^{1-19}$-EGFP (**Fig. S8B**). Negative stain TEM indicated that the size and morphology of the purified native Eut and recombinant EutSMNLK compartments were unaffected by the presence or absence of the cargo protein EutC$^{1-19}$-EGFP. However, empty EutS compartments (isolated from *E. coli* cells recombinantly expressing EutS only, and not the cargo protein EutC$^{1-19}$-EGFP) appeared to be less able to withstand purification. Isolated empty EutS shells were observed to be 50 nm in diameter, the samples were less homogeneous, and a number of shells appeared to be broken (**Fig. S8B**). Empty EutS shells also sedimented differently from the other protein compartments during sucrose gradient centrifugation.

## Eut BMC-targeted EGFP is sequestered within the Eut microcompartments

Anti-GFP immunogold TEM was performed to confirm the localization of EutC$^{1-19}$-EGFP to engineered EutSMNLK shells in permeabilized *E. coli* cells (**Fig. 6A**). As a control, immunofluorescence studies were performed to confirm that the anti-GFP antibody was able to access EutC$^{1-19}$-EGFP in permeabilized *E. coli* cells (**Fig. S9**). As shown in **Fig. 6A**, anti-GFP immunogold beads co-localize at a discrete polyhedral structure located near the pole of the recombinant *E. coli* cell, indicating that EutC$^{1-19}$-EGFP is specifically targeted to the EutSMNLK recombinant compartment. It should be noted that a new protocol (microwave-assisted low temperature processing followed by sectioning and antibody labeling of the permeabilized cells) was developed for our immunogold TEM experiment. Sample preparation for immunogold TEM favored optimal antigenicity but was less than optimal for preserving intracellular structures. The difference in sample preparation protocols can explain the variance in morphology of the shells observed via either immunogold or regular TEM (**Fig. 2**).

We also sought to determine whether EutC$^{1-19}$-EGFP is encapsulated within the Eut shells or if it interacts with the outer surface of the compartments. Eut BMCs isolated from *S. enterica* and Eut shells from *E. coli* (all co-expressing EutC$^{1-19}$-EGFP) were broken by sonication, and intact and broken shells were incubated with anti-GFP antibody using a modified version of the procedure recently described for protein localization into native Pdu BMCs [19]. Western blotting of samples run on a polyacrylamide native

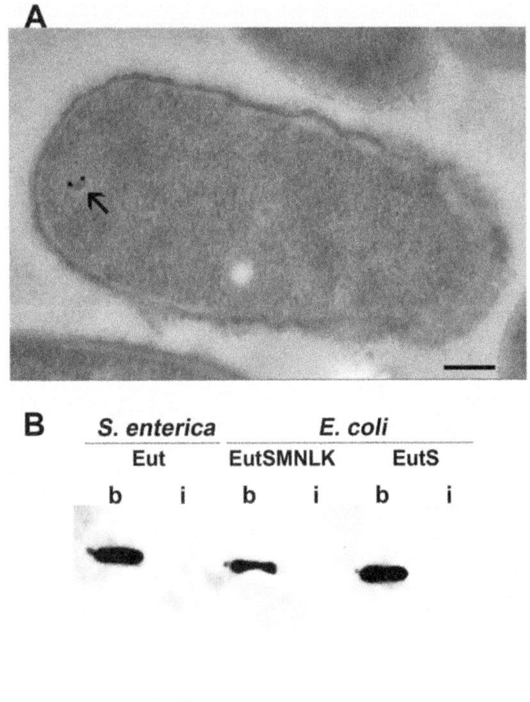


**Figure 6. EutC$^{1-19}$-EGFP is sequestered in the recombinant EutSMNLK compartment.** (**A**) Anti-GFP immunogold TEM of a thin section of *E. coli* JM109 cells co-expressing EutSMNLK and EutC$^{1-19}$-EGFP. Gold particles are localized to a protein shell. (Scale bar: 200 nm). (**B**) Native gel electrophoresis followed by anti-GFP western blot analysis of broken (b) and intact (i) Eut shells, harboring EutC$^{1-19}$-EGFP.

gel shows that GFP could be detected in the broken Eut shell preparation (*Salmonella* Eut BMCs as well as recombinant *E. coli* Eut compartments), but not in intact Eut shell preparation (no bands were visible in the high-molecular weight region of the gel) (**Fig. 6B and Fig. S10**). These findings indicate that $EutC^{1-19}EGFP$ molecules are encapsulated within both native and recombinant Eut compartments (preventing their interactions with anti-GFP antibodies), and are not localized on the outside surface of the Eut protein shells.

## Targeting of an enzyme to the recombinant Eut engineered protein shell

Finally, to explore the feasibility of developing recombinant Eut shells for enzyme sequestration and catalysis, we chose to study hydrolysis of the chromogenic substrate 5-bromo-4-chloro-3-indolyl-β-D-galactopyranoside (X-gal) by β-galactosidase fused N-terminally with the EutC BMC-targeting sequence. Hydrolysis of X-gal produces an insoluble colored indole (indigo), which we hypothesized would form discrete deposits within cells, confirming that $EutC^{1-19}$-ß-galactosidase was targeted to recombinant Eut compartments, and that X-gal could pass into the interior of the capsid. Overnight cultures of *E. coli* cells expressing ß-galactosidase or $EutC^{1-19}$-ß-galactosidase either alone or with different complements of Eut shell proteins (and grown in the presence of X-gal) were all blue, indicating that they possessed functional ß-galactosidase enzyme (data not shown). As shown in **Fig. 7**, *E. coli* co-expressing $EutC^{1-19}$-ß-galactosidase and either EutS or EutSMNLK displayed discrete accumulation of the colored X-gal cleavage product within the cell. The intracellular location of the colored precipitate was compatible with the localization of $EutC^{1-19}$-EGFP, as well as the location of capsid-like structures observed in TEM. No accumulation of the insoluble indigo product was observed within cells co-expressing $EutC^{1-19}$-ß-galactosidase with EutMNLK. Intracellular indole deposits were also absent in cells co-expressing untagged β-galactosidase and Eut shell proteins. Taken together, these data indicate that $EutC^{1-19}$-β-galactosidase is functionally sequestered within the engineered Eut protein shell. The indole deposits in *E. coli* cells expressing EutS appeared to be more diffuse; suggesting that the EutS compartments may be more accessible either due to the presence of only EutS-derived pores or gaps in the protein shell.

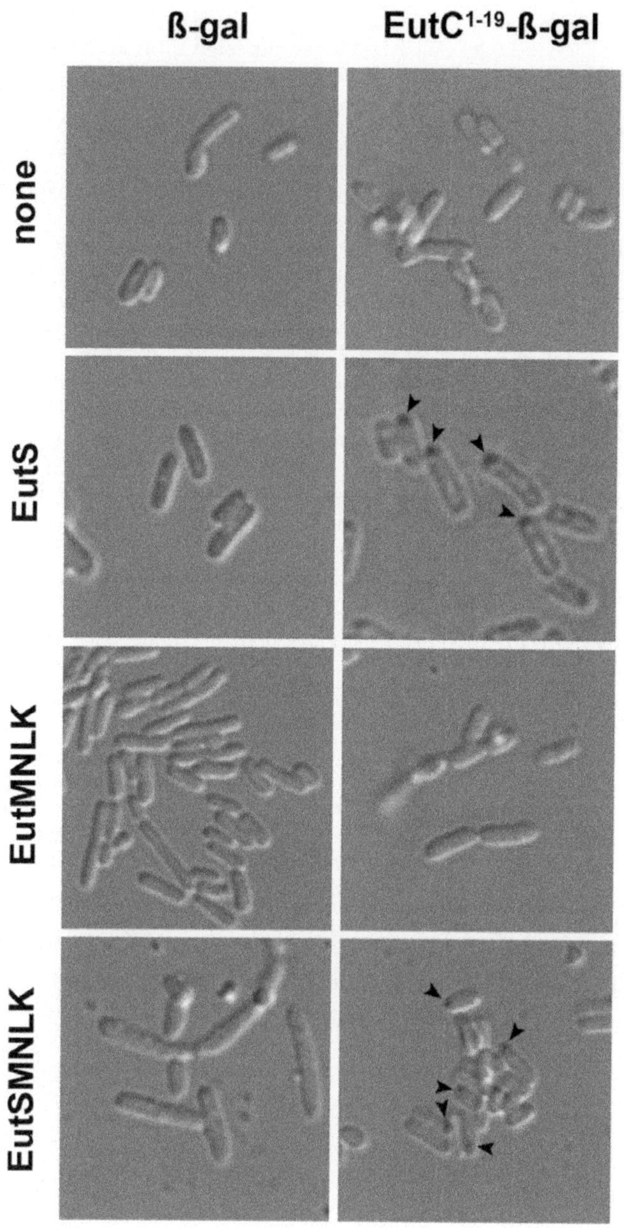


**Figure 7. Hydrolysis of X-gal by *E. coli* co-expressing $EutC^{1-19}$-β-galactosidase and recombinant Eut shell proteins.** *E. coli* C2566 cells with constructs for constitutive expression of β-galactosidase (β-gal) or $EutC^{1-19}$-β-gal and different combinations of Eut shell proteins were grown with the β-gal substrate X-gal. Intracellular accumulation of the insoluble X-gal cleavage product was observed by Differential Interference Contrast (DIC) microscopy. Arrows point to intracellular indole deposits.

## Discussion

Using the Eut BMC shell proteins from *S. enterica* as our model system, we demonstrate that proteinaceous compartments can be engineered in *E. coli*. We also show that heterologous proteins are efficiently targeted into the recombinant compartments; therefore enabling engineering of multi-step biocatalysis within tailored microcompartments as *in vivo* or *in vitro* nano-bioreactors.

In this study, we made the discovery that one of the Eut shell proteins, EutS, is necessary and sufficient for formation of engineered protein shells within *E. coli*. Considering that a 17 gene operon is implicated in the formation of native Eut BMCs, it is surprising that recombinant expression of a single Eut protein (EutS) results in the formation of well-defined compartments that resemble the native *S. enterica* Eut BMCs. Previously, the "bent" hexamer formed by wild type EutS in 3-D crystals had led to the hypothesis that it forms the edges of the BMC shell [28]. Our data indicates that EutS can also form the facets of the capsid, raising the possibility that it adopts more than one conformation *in vivo*. Alternatively, if EutS exists in only one conformation, it can be speculated that a strong reliance on EutS for shell formation may explain why the Eut BMCs are 'rounder' than Pdu BMCs or carboxysomes. Remarkably, EutS is also necessary and sufficient for targeting EutC-signal sequence tagged EGFP and β-galactosidase to the recombinant compartment. The EutS-G39V mutant (which forms flat symmetric hexamers) is unable to sequester targeted EGFP, suggesting that the unusual bent configuration adopted by EutS in 3-D crystal lattices is physiologically relevant to the role played by EutS in the engineered compartment [28].

Another significant observation was the successful hydrolysis of X-gal by β-galactosidase localized to the EutS and EutSMNLK compartments, which indicates that these shells allow the indole galactopyranoside to pass through to the interior. The scenarios in which X-gal may gain access to the interior of shells include entry through central pores in shell protein multimers, or via possible gaps between adjacent multimers that form the shell. A large but occluded central pore has been reported in the 3-D crystal structures of EutS and PduU (a close homolog of EutS) [28,38]. While a substantial rearrangement would be necessary to open the central pore of EutS, it is not without precedent – a recent report demonstrates that exposure to zinc ions causes the central pore of EutL to adopt an "open" conformation [39]. Further studies are required to determine if the central pore in EutS hexamers can also exist in "open" and "closed" conformations. Additionally, two-dimensional (2-D) crystal studies of EutM revealed a tiling pattern with apparent gaps between adjacent hexamers, a phenomenon not observed in the 3-D crystal structure [28,39]. This discrepancy raises the question if EutS hexamers, and indeed other Eut BMC proteins, can also form 2-D lattices which are not as tightly packed as those observed in their 3-D crystals.

The ability of EutS to self-assemble into compartments and sequester targeted proteins within its interior raises questions about the physiological roles of the other Eut shell proteins. It is possible that the EutM, EutN, EutL and EutK shell proteins impart selectivity in terms of the small molecules allowed to enter or exit the BMC shell. For example, EutL crystallizes as a pseudohexamer with a large, gated central pore that may play an important role in transporting substrates and bulky cofactors into the shells while preventing the loss of reaction intermediates [28,39]. Incorporation of EutMNLK may also increase the stability of the microcompartments. Other possible roles of EutMNLK in the native *S. enterica* Eut BMCs include interactions with the encapsulated ethanolamine utilization enzymes, thereby regulating their spatial organization within the compartment [18]. Further biochemical studies are required to elucidate the roles played by Eut shell proteins in BMC organization and function. The heterologous reconstitution of Eut compartments provides an important tool for such studies.

Bobik and colleagues had predicted the presence of N-terminal signal sequences on EutC and EutG enzymes [19]. Our experiments indicate that the first nineteen amino acids of EutC (but not EutG) are sufficient to target heterologous proteins to native Eut BMCs in *Salmonella* as well as recombinant Eut compartments in *E. coli*. The reasons behind the inability of $EutG^{1-19}$ to function as a BMC-targeting sequence are unclear. It is possible that steric hindrance prevents $EutG^{1-19}$ from functioning as a BMC-targeting sequence, and extending the length of the EutG sequence by a few amino acids may reduce the hindrance and allow it to function as a targeting signal. An additional example is the PduC enzyme, which was predicted to lack a BMC-targeting sequence [19]. However, a separate study showed that recombinant PduC-GFP fusion protein displays punctate localization when co-expressed with Pdu shell proteins, suggesting that it is targeted to recombinant Pdu compartments [30]. These observations suggest that there may be multiple mechanisms for targeting proteins for encapsulation within BMCs, and further experiments are required to understand them.

The functions of several genes in the *eut* operon are at present unknown. While *S. enterica* cells contained multiple Eut BMCs distributed throughout the cytoplasm, recombinant compartments in *E. coli* were restricted to only one or two per cell. Some of the non-BMC-shell *eut* genes may be required for the formation of more than one microcompartment per cell, through for instance, transcriptional regulation [40]. The off-center location of the recombinant compartments was also invariant. A recent report indicates that cyanobacterial carboxysomes are in motion within the cell and interact with the cytoskeleton, thereby ensuring their equitable distribution during cell division [41]. If one or more of the non-BMC-shell Eut proteins (for example EutP, an Era (*E. coli* Ras-like protein)-like GTPase) are required for interacting with the bacterial cytoskeleton, the intracellular location of the recombinant shell may simply result from aberrant interactions with the cytoskeleton leading to altered nucleation of the microcompartment. Our observation that native Eut BMCs exhibit movement in *S. enterica*, while recombinant Eut compartments appear to be stationary in *E. coli*, lends credence to this hypothesis. Further investigations into the role of as yet uncharacterized proteins encoded by BMC operons should provide important insights into BMC nucleation, assembly and cytoskeletal interaction, and may present a handle for controlling the number of recombinant compartments formed inside engineered cells.

While this manuscript was under review, heterologous expression of *Halothiobacillus neapolitanus* carboxysomes in *E. coli* was demonstrated by Bonacci *et al* [42]. Expression of several shell proteins appears to be required for the formation of carboxysomes, whereas EutS alone is sufficient to form heterologous protein shells. Compared to carboxysomes, a EutS-based system is less complex, and may find broader applicability in the engineering of metabolic pathways and multi-enzymatic biocatalysis. Our results show that Eut protein shells can be engineered to sequester heterologous enzymes for catalysis. A recent report outlines a different approach for enzyme encapsulation - addition of oppositely charged amino acids to a non-BMC shell protein (lumazine synthase) and the encapsulated enzyme [43]. Compared to this latter strategy, which involves optimization of electrostatic interactions between shell and cargo proteins while maintaining the catalytic efficiency of each sequestered enzyme, the *in vivo* system offered by EutS and the nineteen amino acid BMC targeting sequence is much simpler, and very specific.

Next steps in the engineering of microcompartments will include encapsulation of multiple enzymes for biosynthesis or biodegradation with engineered cells, and in addition, *in vitro* multi-step catalysis with isolated and immobilized compartments. The central pores formed by multimers of BMC shell proteins offer opportunities to engineer such protein nano-bioreactors with desired selectivities for substrates and products. Additionally, recombinant hybrid compartments may be formed by combining shell proteins from different BMC types and sources. In the future, synthetic biologists may be able to design protein nano-bioreactors to their desired specifications by selecting from a rapidly expanding library of native and engineered BMC shell proteins as well as targeting sequences. The results presented in this study bring us an important step closer to the rational design of protein compartments useful for a variety of *in vivo* and *in vitro* applications.

## Methods

### Microbiological methods

*E. coli* cultures were cultivated aerobically at 30°C for 15–18 h in Luria-Bertani (LB) medium supplemented with the appropriate antibiotic when required (ampicillin 100 $\mu g\ ml^{-1}$, chloramphenicol 50 $\mu g\ ml^{-1}$). For the *in vivo* β-galactosidase assay, cultures were grown overnight with X-gal (final concentration: 0.008% (w/v)). *S. enterica* cultures were grown aerobically at 37°C overnight in supplemented E medium with 150 nM vitamin $B_{12}$ (cyanocobalamin) and either 0.2% (v/v) glycerol or 30 mM ethanolamine [17]. 30 $\mu g\ ml^{-1}$ kanamycin was added to the growth media for culturing *S. enterica* strains containing pBBRBB-based plasmids.

### Gene cloning

Eut shell genes were amplified from *S. enterica* genomic DNA with gene specific primers containing suitable restriction sites.

Their accession numbers are as follows: EutS (NP_461405), EutM (NP_461400), EutN (NP_461399), EutL (NP_461391), and EutK (NP_461390). The *eut* genes were cloned into our in-house BioBrick$^{TM}$ expression vector pUCBB, and expressed from a constitutively active modified *lac* promoter described previously (**Fig. S1**) [44,45]. The EutS-G39V mutant was created by site-directed mutagenesis. Additional details of the cloning strategy are presented in **Methods S1**. β-galactosidase (amplified from *E. coli* MG1655 genomic DNA) and EGFP were cloned into our in-house low copy number BioBrick$^{TM}$ expression vector pACBB. Nucleotides coding for the putative EutC-signal sequence (EutC$^{1-19}$) from *S. enterica* LT2 (MDQKQIEEIVRSVMASMGQ) were added to N-terminus of EGFP and β-galactosidase through PCR [19]. Similarly, EGFP was also tagged with the predicted EutG-signal sequence (EutG$^{1-19}$) from *E. coli* K12 (MQNELQTALF-QAFDTLNLQ), which differs from the *S. enterica* LT2 EutG$^{1-19}$ by one amino acid. Both putative signal sequences were codon optimized for expression in *E. coli*.

## Transmission Electron Microscopy (TEM)

Bacterial cells were fixed in 2.5% glutaraldehyde in 0.1 M phosphate buffer, followed by three washes with 0.1 M phosphate buffer. Triton X-100 was added to the glutaraldehyde solution and rinse buffer to a final concentration of 0.1%. Subsequently, the pellets were post-fixed with 1% osmium tetroxide in 0.1 M phosphate buffer, washed with nanopure water, and embedded in 2% low melting agarose. The cell-agarose pellet was cut into 1 mm$^3$ cubes, and dehydrated using an ethanol gradient. The cell-agarose cubes were then incubated in 1:1 mixture of Embed 812 resin and 100% ethanol for 4 h, followed by 18 h incubation in 100% Embed 812 resin. Next, they were suspended in a fresh Embed 812 resin-N,N-dimethylbenzylamine (BDMA) solution and polymerized at 60°C for 48 h. 90 nm sections were sliced, placed on 200 mesh formvar-coated copper grids, and post-stained with 3% uranyl acetate and Triple lead stain. Specimens were observed and photographed with a Philips CM12 transmission electron microscope.

## Light Microscopy

Bacteria were viewed using a Nikon Eclipse E800 photomicroscope equipped with bright field, Differential Interference Contrast (DIC) and fluorescence optics including blue (excitation filter 470–490 nm, barrier 520–580 nm) and green (excitation filter 510–560 nm, barrier 570–620 nm) filter sets. The samples were viewed using a 100×, 1.4 n.a. plan apo objective. For fluorescence microscopy, 16-bit digital images were collected using a Roper CoolSnap HQ monochrome camera and captured using Image Pro Plus software. DIC microscopy was performed using a 1.4 n.a. oil condenser. Z-series images of cells were collected at 0.15 micron steps using a Ludl MAC 3000 controller interfaced with ImagePro Plus. The DIC images were deconvolved using the SharpStack Nearest Neighbor algorithm. A minimum projection of the resulting z-series was made and all images were identically adjusted for display using PhotoShop.

## Purification of protein compartments

*S. enterica* cells were made electrocompetent, and were transformed with pACBBEutC$^{1-19}$-EGFP. Cells harboring the plasmid were grown in 1 liter of NCE minimal medium supplemented with 50 μg ml$^{-1}$ chloramphenicol and 30 mM ethanolamine to induce Eut BMC production [46]. *E. coli* C2566 cells harboring pUCBBEutSMNLK or pUCBBEutS were also transformed with pACBBEutC$^{1-19}$-EGFP and were grown in 1 liter of LB medium supplemented with 100 μg ml$^{-1}$ ampicillin and 50 μg ml$^{-1}$ chloramphenicol. *E. coli* C2566 cells harboring pACBBEutC$^{1-19}$-EGFP were also grown as a control. Cultures were grown at 37°C with shaking at 275 rpm for 16 h. Cells were harvested by centrifugation at 10, 000×g for 30 min at 4°C. Purification of Eut shells was carried out using procedures previously described for Pdu BMCs, with the exception that the TEMP buffer was replaced by a TEME buffer (50 mM Tris.HCl, 10 mM $MgCl_2$, 2 mM EDTA, 30 mM ethanolamine (pH 8.0)), and that the shells were applied to two separate discontinuous sucrose gradients: firstly, a three step gradient of 20, 40 and 65% sucrose, and secondly a ten step gradient from 22 to 54% sucrose [13]. Native Eut BMCs from *S. enterica* harboring EutC$^{1-19}$-EGFP and recombinant EutSMNLK and EutS shells from *E. coli* harboring EutC$^{1-19}$-EGFP formed a white translucent band two thirds the way down the centrifuge tube, which was also checked for fluorescence under UV light due to the localization of EutC$^{1-19}$-EGFP into the BMCs. Following a final clarification step, the native Eut BMCs and recombinant EutSMNLK and EutS shells were pelleted, and sample purity was judged to be sufficient for TEM by SDS-PAGE electrophoresis. Purified shell samples were fixed and negatively stained according to previously published procedures, with the exception that 2% uranyl acetate was used for staining [13]. The purification and negative staining procedures were repeated on separate occasions to ensure consistency in results and reproducibility of the procedure.

## Anti-GFP Immunogold labeling and TEM

Microwave-assisted low temperature processing was used to prepare *E. coli* cells for anti-GFP immunofluorescence and immunogold labeling. Steps for primary fixation of the samples were adapted from procedures published previously [47,48]. The samples underwent dehydration/substitution in methanol/LR white. Thin sections were placed on formvar-coated nickel 200 mesh grids followed by anti-GFP immunogold labeling. Additional experimental details are included in **Methods S1**.

**Additional methods** including SDS-PAGE electrophoresis and western blotting are described in **Methods S1**.

## Supporting Information

**Figure S1 BioBrick$^{TM}$ vectors and strategy for stacking multiple genes into a single plasmid.** (**A**) Our in-house BioBrick$^{TM}$ vectors contain an expression cassette with a constitutive promoter ($P_{lac*}$) and an EGFP reporter. (**B**) Cloning of Eut BMC shell genes into pUCBB. (i) EutS, EutMN and EutLK were cloned downstream of the constitutive $P_{lac*}$ promoter (blue arrow) using BglII and NotI. (ii) and (iii) Expression cassettes for EutMNLK and EutSMNLK were created as described in **Methods S1**.

**Figure S2 SDS/PAGE analysis showing recombinant expression of *S. enterica* Eut shell proteins in *E. coli*.** (**A**) Overexpression of Eut shell proteins in the *E. coli* strain C2566. (**B**) Overexpression of Eut shell proteins in the *E. coli* strain JM109. (**c**) Overexpression of wild type EutS and the EutS-G39V mutant in *E. coli* strains C2566 and JM109. 15 μg soluble protein fraction was loaded in each lane. Expected protein sizes are as follows: EutS (11.6 kDa), EutM (9.8 kDa), EutN (10.4 kDa), EutL (22.7 kDa), EutK (17.5 kDa), and EGFP (26.9 kDa). Proteins were stained with Coomassie Blue.

**Figure S3 Transmission electron micrographs of thin sections of recombinant *E. coli* expressing *S. enterica***

**Eut shell proteins.** (**A–C**) *E. coli* expressing recombinant EutS contain properly delimited shells (*E. coli* strain used in **A**: C2566, and in **B**, **C**: JM109). (**D–F**) *E. coli* expressing recombinant EutM form thick axial filaments that interfere with separation after cell-division (*E. coli* strain used in **D, E**: C2566, and in **F**: JM109). (**G**) *E. coli* JM109 expressing recombinant EutN. (**H**) *E. coli* JM109 expressing recombinant EutL. (**I**) *E. coli* JM109 expressing recombinant EutK shows an electron translucent region in the middle of the cell. (**J**) An electron dense region is visible in *E. coli* JM109 co-expressing recombinant EutM and EutN. (**K**) Intracellular filaments are formed in *E. coli* JM109 co-expressing recombinant EutL and EutK. (**L–N**) Clearly defined shells are observed in *E. coli* JM109 expressing recombinant EutSMNLK. (**O–Q**) Co-expression of EutSMNLK and EutC$^{1-19}$-EGFP results in the formation of compartments that are morphologically similar to the shells observed *in vivo* by expression of either EutS or EutSMNLK alone. (*E. coli* strain used in **O**: C2566, and in **P, Q**: JM109). Arrows indicate the location of recombinant shells. (Scale bar: 200 nm).

**Figure S4 Localization of EutC$^{1-19}$-EGFP in recombinant *E. coli* JM109 cells expressing *S. enterica* Eut shell proteins.** Fluorescence microscopy images of *E. coli* JM109 cells co-expressing EGFP or EutC$^{1-19}$-EGFP with EutS (wild type and the G39V mutant), EutMNLK or EutSMNLK. See **Table S2** for the quantification of EGFP localization in recombinant *E. coli*, and **Fig. 4** for the localization of EutC$^{1-19}$-EGFP in the *E. coli* C2566 strain. Cell boundaries are shown by the DIC images.

**Figure S5 Localization of EutC$^{1-19}$-EGFP in recombinant *E. coli* C2566 cells expressing various combinations of *S. enterica* Eut shell proteins.** Fluorescence microscopy images of *E. coli* C2566 cells with constructs for constitutive expression of EGFP or EutC$^{1-19}$-EGFP with EutM, EutN, EutL, EutK, EutMN and EutLK. In the absence of EutS, there is no discrete fluorescent localization of EutC$^{1-19}$-EGFP, which indicates that EutS is required for targeting EutC$^{1-19}$-EGFP to the engineered microcompartments. Cell boundaries are shown by the DIC images.

**Figure S6 Nile Red staining of recombinant *E. coli* expressing EutC$^{1-19}$-EGFP.** *E. coli* C2566 cells co-expressing EutC$^{1-19}$-EGFP and EutS or EutSMNLK were stained with the fluorescent, lipophilic inclusion body stain Nile Red. Co-localization of red and green fluorescence was not observed, indicating that the recombinant Eut shells are not inclusion bodies nor are the surrounded by a hydrophobic matrix.

**Figure S7 Nile Red staining of recombinant *E. coli* expressing NSC1.** *E. coli* C2566 cells co-expressing the cyanobacterial carotenoid cleavage dioxygenase NSC1 either alone or with EutC$^{1-19}$-EGFP. While red fluorescent puncta corresponding to inclusion bodies were observed in the presence of NSC1, co-localization of red and green fluorescence was not seen, showing that EutC$^{1-19}$-EGFP is not targeted to NSC1 inclusion bodies.

**Figure S8 Transmission electron micrographs of partially purified protein compartments.** (**A**) Native Pdu BMCs isolated from *S. enterica*. (**B**) Native Eut BMCs and recombinant Eut protein shells isolated from cells not expressing the cargo protein EutC$^{1-19}$-EGFP. From left to right: Native Eut BMCs isolated from *S. enterica*, recombinant EutSMNLK shells isolated from *E. coli* C2566, and recombinant EutS shells isolated from *E. coli* C2566. Scale bar: 100 nm.

**Figure S9 Immunofluorescence analysis of EutC$^{1-19}$-EGFP localization in recombinant *E. coli* expressing Eut shell proteins.** EGFP, anti-GFP antibody (red) and merged EGFP-anti-GFP antibody fluorescence signals from *E. coli* cells with constructs for constitutive expression of EGFP or EutC$^{1-19}$-EGFP with EutS or EutSMNLK. (**A**) anti-GFP immunofluorescence studies in the *E. coli* strain C2566. (**B**) anti-GFP immunofluorescence studies in the *E. coli* strain JM109.

**Figure S10 Separation of EutC$^{1-19}$-EGFP from broken and intact Eut shells by native polyacrylamide electrophoresis.** Visualization of protein migration by silver stain of native gel. EGFP control is shown in lane 1, followed by broken (lane 2) and intact (lane 3) Eut BMCs from *S. enterica* cells harboring EutC$^{1-19}$-EGFP; broken (lane 4) and intact (lane 5) recombinant EutSMNLK BMCs co-expressing EutC$^{1-19}$-EGFP; and broken (lane 6) and intact (lane 7) recombinant EutS BMCs from *E. coli* C2566 cells co-expressing EutC$^{1-19}$-EGFP.

**Video S1 Dynamics of EutC$^{1-19}$-EGFP in *S. enterica* grown on ethanolamine.** Representative time-lapse movie of *S. enterica* cells harboring pBBRBB-EutC$^{1-19}$-EGFP, and grown in the presence of ethanolamine. Discrete fluorescent foci are observed to be in motion within the *S. enterica* cells, suggesting that Eut BMCs (which would be expected to encapsulate EutC$^{1-19}$-EGFP) are moving around within the cell. Time stamp on video indicates elapsed time. Preparations were viewed using a Nikon Eclipse E800 photomicroscope. Time-lapse images were collected at 15 second intervals. Shutters were opened only during camera exposure.

**Table S1 Quantification of the distribution of recombinant Eut shells in *E. coli*.** Thin cell sections of *E. coli* expressing EutS or EutSMNLK were observed by TEM. Assuming an average *E. coli* cell has a height of 2 μm and a diameter of 0.5 μm, about 20 thin sections (90 nm in width) perpendicular to the axis can be cut from each cell. An average recombinant Eut protein shell has a diameter of 100–200 nm. The average *E. coli* cell would have about 20 cross-sections parallel to the circular base, of which only two would pass through an engineered Eut shell. Even if 100% of *E. coli* had a recombinant Eut compartment, the actual fraction of cell cross-sections showing the phenotype would be around 10%. For sections parallel to the height of the cell, less than 40% would be expected to display the shell, and the number showing compartments at close to their maximum width will be even lower.

**Table S2 Quantification of the distribution of engineered Eut protein shells in *E. coli* by fluorescence microscopy.**

**Methods S1 Supporting Methods.**

## Acknowledgments

We thank Anthony Goering, Annie Kathuria and Ian Windsor (iGEM 2010 undergraduate student team members, College of Biological Sciences,

University of Minnesota) for their assistance in cloning the Eut BMC genes; Dr. Gail Celio, Alexander Cramer, Ashleigh Skaalen and Adam T. Vogel (University of Minnesota Imaging Centers) for help with microscopy and figure preparation; and Dr. Mark Held for critical reading of the manuscript. This manuscript is dedicated to Dr. Ethan T. Johnson, our dear friend, colleague and manuscript co-author, who died due to injuries suffered in a hit-and-run accident on September 21st, 2010.

## Author Contributions

Conceived and designed the experiments: SC MBQ ETJ CSD MAS. Performed the experiments: SC MBQ ETJ MAS. Analyzed the data: SC MBQ MAS ETJ CSD. Contributed reagents/materials/analysis tools: MAS CSD. Wrote the paper: SC MBQ CSD.

## References


1. Purnick PE, Weiss R (2009) The second wave of synthetic biology: from modules to systems. Nat Rev Mol Cell Biol 10: 410–422.
2. Keasling JD (2008) Synthetic biology for synthetic chemistry. ACS Chem Biol 3: 64–76.
3. Lopez-Gallego F, Schmidt-Dannert C (2010) Multi-enzymatic synthesis. Curr Opin Chem Biol 14: 174–183.
4. Fontes CM, Gilbert HJ (2010) Cellulosomes: highly efficient nanomachines designed to deconstruct plant cell wall complex carbohydrates. Annu Rev Biochem 79: 655–681.
5. Burbulis IE, Winkel-Shirley B (1999) Interactions among enzymes of the *Arabidopsis* flavonoid biosynthetic pathway. Proc Natl Acad Sci USA 96: 12929–12934.
6. Dueber JE, Wu GC, Malmirchegini GR, Moon TS, Petzold CJ, et al. (2009) Synthetic protein scaffolds provide modular control over metabolic flux. Nat Biotechnol 27: 753–759.
7. Kerfeld CA, Heinhorst S, Cannon GC (2010) Bacterial microcompartments. Annu Rev Microbiol 64: 391–408.
8. Cheng S, Liu Y, Crowley CS, Yeates TO, Bobik TA (2008) Bacterial microcompartments: their properties and paradoxes. Bioessays 30: 1084–1095.
9. Urano N, Kataoka M, Ishige T, Kita S, Sakamoto K, et al. (2010) Genetic analysis around aminoalcohol dehydrogenase gene of *Rhodococcus erythropolis* MAK154: a putative GntR transcription factor in transcriptional regulation. Appl Microbiol Biotechnol 89: 739–746.
10. Yeates TO, Kerfeld CA, Heinhorst S, Cannon GC, Shively JM (2008) Protein-based organelles in bacteria: carboxysomes and related microcompartments. Nat Rev Microbiol 6: 681–691.
11. Price GD, Badger MR, Woodger FJ, Long BM (2008) Advances in understanding the cyanobacterial $CO_2$-concentrating-mechanism (CCM): functional components, Ci transporters, diversity, genetic regulation and prospects for engineering into plants. J Exp Bot 59: 1441–1461.
12. Bobik TA, Havemann GD, Busch RJ, Williams DS, Aldrich HC (1999) The propanediol utilization (*pdu*) operon of *Salmonella enterica* serovar Typhimurium LT2 includes genes necessary for formation of polyhedral organelles involved in coenzyme $B_{12}$-dependent 1,2-propanediol degradation. J Bacteriol 181: 5967–5975.
13. Havemann GD, Bobik TA (2003) Protein content of polyhedral organelles involved in coenzyme $B_{12}$-dependent degradation of 1,2-propanediol in *Salmonella enterica* serovar Typhimurium LT2. J Bacteriol 185: 5086–5095.
14. Sriramulu DD, Liang M, Hernandez-Romero D, Raux-Deery E, Lunsdorf H, et al. (2008) *Lactobacillus reuteri* DSM 20016 produces cobalamin-dependent diol dehydratase in metabolosomes and metabolizes 1,2-propanediol by disproportionation. J Bacteriol 190: 4559–4567.
15. Stojiljkovic I, Baumler AJ, Heffron F (1995) Ethanolamine utilization in *Salmonella typhimurium*: nucleotide sequence, protein expression, and mutational analysis of the *cchA cchB eutE eutJ eutG eutH* gene cluster. J Bacteriol 177: 1357–1366.
16. Shively JM, Bradburne CE, Aldrich HC, Bobik TA, Mehlman JL, et al. (1998) Sequence homologs of the carboxysomal polypeptide CsoS1 of the thiobacilli are present in cyanobacteria and enteric bacteria that form carboxysomes – polyhedral bodies. Can J Bot 76: 906–916.
17. Brinsmade SR, Paldon T, Escalante-Semerena JC (2005) Minimal functions and physiological conditions required for growth of *Salmonella enterica* on ethanolamine in the absence of the metabolosome. J Bacteriol 187: 8039–8046.
18. Yeates TO, Thompson MC, Bobik TA (2011) The protein shells of bacterial microcompartment organelles. Curr Opin Struct Biol 21: 223–231.
19. Fan C, Cheng S, Liu Y, Escobar CM, Crowley CS, et al. (2010) Short N-terminal sequences package proteins into bacterial microcompartments. Proc Natl Acad Sci USA 107: 7509–7514.
20. Kofoid E, Rappleye C, Stojiljkovic I, Roth J (1999) The 17-gene ethanolamine (*eut*) operon of *Salmonella typhimurium* encodes five homologues of carboxysome shell proteins. J Bacteriol 181: 5317–5329.
21. Penrod JT, Roth JR (2006) Conserving a volatile metabolite: a role for carboxysome-like organelles in *Salmonella enterica*. J Bacteriol 188: 2865–2874.
22. Bertin Y, Girardeau JP, Chaucheyras-Durand F, Lyan B, Pujos-Guillot E, et al. (2011) Enterohaemorrhagic *Escherichia coli* gains a competitive advantage by using ethanolamine as a nitrogen source in the bovine intestinal content. Environ Microbiol 13: 365–377.
23. Penrod JT, Mace CC, Roth JR (2004) A pH-sensitive function and phenotype: evidence that EutH facilitates diffusion of uncharged ethanolamine in *Salmonella enterica*. J Bacteriol 186: 6885–6890.
24. Mori K, Bando R, Hieda N, Toraya T (2004) Identification of a reactivating factor for adenosylcobalamin-dependent ethanolamine ammonia lyase. J Bacteriol 186: 6845–6854.
25. Buan NR, Suh SJ, Escalante-Semerena JC (2004) The *eutT* gene of *Salmonella enterica* encodes an oxygen-labile, metal-containing ATP:corrinoid adenosyltransferase enzyme. J Bacteriol 186: 5708–5714.
26. Starai VJ, Garrity J, Escalante-Semerena JC (2005) Acetate excretion during growth of *Salmonella enterica* on ethanolamine requires phosphotransacetylase (EutD) activity, and acetate recapture requires acetyl-CoA synthetase (Acs) and phosphotransacetylase (Pta) activities. Microbiology 151: 3793–3801.
27. Tanaka S, Kerfeld CA, Sawaya MR, Cai F, Heinhorst S, et al. (2008) Atomic-level models of the bacterial carboxysome shell. Science 319: 1083–1086.
28. Tanaka S, Sawaya MR, Yeates TO (2010) Structure and mechanisms of a protein-based organelle in *Escherichia coli*. Science 327: 81–84.
29. Papapostolou D, Howorka S (2009) Engineering and exploiting protein assemblies in synthetic biology. Mol Biosyst 5: 723–732.
30. Parsons JB, Frank S, Bhella D, Liang M, Prentice MB, et al. (2010) Synthesis of empty bacterial microcompartments, directed organelle protein incorporation, and evidence of filament-associated organelle movement. Mol Cell 38: 305–315.
31. Yeates TO, Crowley CS, Tanaka S (2010) Bacterial microcompartment organelles: protein shell structure and evolution. Annu Rev Biophys 39: 185–205.
32. Shetty RP, Endy D, Knight TF, Jr. (2008) Engineering BioBrick vectors from BioBrick parts. J Biol Eng 2: 5.
33. Rath A, Glibowicka M, Nadeau VG, Chen G, Deber CM (2009) Detergent binding explains anomalous SDS-PAGE migration of membrane proteins. Proc Natl Acad Sci USA 106: 1760–1765.
34. Cheng S, Sinha S, Fan C, Liu Y, Bobik TA (2011) Genetic analysis of the protein shell of the microcompartments involved in coenzyme $B_{12}$-dependent 1,2-propanediol degradation by *Salmonella*. J Bacteriol 193: 1385–1392.
35. Menon BB, Dou Z, Heinhorst S, Shively JM, Cannon GC (2008) *Halothiobacillus neapolitanus* carboxysomes sequester heterologous and chimeric RubisCO species. PLoS One 3: e3570.
36. Steinmann B, Christmann A, Heiseler T, Fritz J, Kolmar H (2010) *In vivo* enzyme immobilization by inclusion body display. Appl Environ Microbiol 76: 5563–5569.
37. Marasco EK, Vay K, Schmidt-Dannert C (2006) Identification of carotenoid cleavage dioxygenases from *Nostoc* sp. PCC 7120 with different cleavage activities. J Biol Chem 281: 31583–31593.
38. Crowley CS, Sawaya MR, Bobik TA, Yeates TO (2008) Structure of the PduU shell protein from the Pdu microcompartment of *Salmonella*. Structure 16: 1324–1332.
39. Takenoya M, Nikolakakis K, Sagermann M (2010) Crystallographic insights into the pore structures and mechanisms of the EutL and EutM shell proteins of the ethanolamine-utilizing microcompartment of *Escherichia coli*. J Bacteriol 192: 6056–6063.
40. Sheppard DE, Roth JR (1994) A rationale for autoinduction of a transcriptional activator: ethanolamine ammonia-lyase (EutBC) and the operon activator (EutR) compete for adenosyl-cobalamin in *Salmonella typhimurium*. J Bacteriol 176: 1287–1296.
41. Savage DF, Afonso B, Chen AH, Silver PA (2010) Spatially ordered dynamics of the bacterial carbon fixation machinery. Science 327: 1258–1261.
42. Bonacci W, Teng PK, Afonso B, Niederholtmeyer H, Grob P, et al. (2011) Modularity of a carbon-fixing protein organelle. Proc Natl Acad Sci USA[Epub ahead of print] 2011 Dec 19. Pubmed PMID: 22184212.
43. Worsdorfer B, Woycechowsky KJ, Hilvert D (2011) Directed evolution of a protein container. Science 331: 589–592.
44. Schmidt-Dannert C, Umeno D, Arnold FH (2000) Molecular breeding of carotenoid biosynthetic pathways. Nat Biotechnol 18: 750–753.
45. Vick JE, Johnson ET, Choudhary S, Bloch SE, Lopez-Gallego F, et al. (2011) Optimized compatible set of BioBrick vectors for metabolic pathway engineering. Appl Microbiol Biotechnol 92: 1275–1286.
46. Vogel HJ, Bonner DM (1956) Acetylornithinase of *Escherichia coli*: partial purification and some properties. J Biol Chem 218: 97–106.
47. Munoz TE, Giberson RT, Demaree R, Day JR (2004) Microwave-assisted immunostaining: a new approach yields fast and consistent results. J Neurosci Methods 137: 133–139.
48. Webster P (2007) Microwave-assisted processing and embedding for transmission electron microscopy. Methods Mol Biol 369: 47–65.
49. Roof DM, Roth JR (1992) Autogenous regulation of ethanolamine utilization by a transcriptional activator of the *eut* operon in *Salmonella typhimurium*. J Bacteriol 174: 6634–6643.

# 10

# Rational Engineering of a Human Anti-Dengue Antibody through Experimentally Validated Computational Docking

**Luca Simonelli[1]**, **Mattia Pedotti[1]**, **Martina Beltramello[1]**, **Elsa Livoti[1]**, **Luigi Calzolai[2]**, **Federica Sallusto[1]**, **Antonio Lanzavecchia[1,3]**, **Luca Varani[1]***

**1** Institute for Research in Biomedicine, Bellinzona, Switzerland, **2** Institute for Health and Consumer Protection, Joint Research Centre, Ispra, Italy, **3** Institute of Microbiology, Eidgenössische Technische Hochschule (ETH) Zürich, Zürich, Switzerland

## Abstract

Antibodies play an increasing pivotal role in both basic research and the biopharmaceutical sector, therefore technology for characterizing and improving their properties through rational engineering is desirable. This is a difficult task thought to require high-resolution x-ray structures, which are not always available. We, instead, use a combination of solution NMR epitope mapping and computational docking to investigate the structure of a human antibody in complex with the four Dengue virus serotypes. Analysis of the resulting models allows us to design several antibody mutants altering its properties in a predictable manner, changing its binding selectivity and ultimately improving its ability to neutralize the virus by up to 40 fold. The successful rational design of antibody mutants is a testament to the accuracy achievable by combining experimental NMR epitope mapping with computational docking and to the possibility of applying it to study antibody/pathogen interactions.

**Editor:** Nikos Vasilakis, University of Texas Medical Branch, United States of America

**Funding:** Funding was provided by Canton Ticino, Swiss Vaccine Research Institute, and European Union (EU-NMR, Contract # RII3- 622 026145). The funders had no role in study design, data collection and analysis, decision to publish, or preparation of the manuscript.

**Competing Interests:** The authors have declared that no competing interests exist.

* E-mail: luca.varani@irb.usi.ch

These authors contributed equally to this work.

## Introduction

Improving our understanding of the structural rules governing antibody/antigen interactions is expected, in the long run, to accelerate vaccine development, since most modern vaccines aim to elicit an antibody response, and to help us design better antibodies for passive immunization or biotechnology applications such as the production of bio-recognition elements for target detection. As a proof of concept, we set forth to structurally characterize the binding of one antibody to the four existing Dengue Virus (DenV) serotypes and use this information to rationally alter its immunological properties, eliminating cross-reactivity and improving its ability to neutralize the virus.

DenV is responsible for 20,000 deaths and 500,000 hospitalizations annually [1], with economic impact rivaling that of malaria. Its epidemic activity and geographic expansion are increasing as climate changes, travel and urbanization create favorable conditions for the mosquito spreading it [2]. No cure or vaccine is currently available, mostly due to the presence of four serotypes and to a poorly understood process called Antibody Dependent Enhancement, where antibodies raised against a previous Dengue infection facilitate subsequent infection by another serotype [3]. In addition to its biomedical importance, the presence of related serotypes and the fact that they are structurally well characterized both at the protein and viral capsid level make DenV a good model for the study of antibody/antigen interactions. Although structural studies often concentrate on the complex between an antibody and a single serotype, usually the one against which the antibody is most effective, a comparison of the same antibody bound to antigens that it can and cannot neutralize may, in fact, teach us why it is only effective against some of them.

Having isolated a panel of human monoclonal antibodies from a donor recovered from infection from Dengue Virus serotype 2 (DenV2) [4], we selected and characterized one that would: *i*) bind all four DenV serotypes; *ii*) effectively neutralize only some of them and *iii*) bind to the so-called DIII, a small ig-like domain part of the E protein, whose homodimers are the main component of the viral surface [5,6,7] and a dominant target for the human antibody response against DenV [4,8,9,10].

We previously characterized the interaction between DV32.6, an antibody with the above mentioned properties, and DenV4 [11]. This alone however, cannot explain why DV32.6 can neutralize the other three serotypes given that the antibody binds stronger to its epitope on DenV4 rather than DenV1 or DenV3. If the antibody/antigen interaction were identical in all serotypes, then the antibody should fail to neutralize DenV1 and DenV3 just as it fails to neutralize DenV4.

Here we aim to elucidate the structure of DV32.6 in complex with all the remaining DenV serotypes and exploit the differences to rationally design mutated antibodies with i) selectively altered binding specificity and ii) improved ability to neutralize the virus.

We first use NMR epitope mapping to define the binding site of DV32.6 on DIII of all four DenV serotypes. We then use this information to filter computational predictions of the antibody/antigen complexes. Analysis of the resulting three dimensional structures proved sufficiently accurate for the rational design of antibody mutants with selectively altered binding specificity or improved neutralization properties.

## Results

### Antibody DV32.6 Binds to All Dengue Serotypes

DV32.6 is part of a panel of human monoclonal antibodies isolated from a donor recovered from infection by DenV2 [4]. It binds to DIII of all four DenV serotypes with KD 145±9 nM for DenV1; 7±0.2 nM for DenV2; 73±16 nM for DenV3; 34±7 nM for DenV4 according to SPR (Figure 1). The ability of DV32.6 to neutralize the virus was assessed by flow cytometry assays measuring the number of cells infected by DenV vaccine strains in the presence of different amounts of antibody. There is no direct correlation between DIII binding affinity and neutralization: the antibody is more efficient at neutralizing DenV2, DenV1 and DenV3 despite binding more strongly to DIII of DenV4 (Figure S1). Association and dissociation rates show no obvious correlation to the neutralizing activity, either. The approximate concentration of antibody required to neutralize 50% of the viral activity is 2 μg/ml for DenV2, 3 μg/ml for DenV3, 4 μg/ml for DenV1 and >74 μg/ml for DenV4. Incidentally, DV32.6 was isolated from a patient recovered from DenV2 and is most effective against this very serotype. Binding assays on isolated DIII allow us to compare the binding affinity of the antibody for its epitope but the natural target for the antibody is the full virus, where the surface of DIII is partially covered by neighbouring protein domains. In contrast to DIII binding, ELISA performed at 37°C on the full virus show that binding to DenV4 is not stronger than to other serotypes (Figure S1). No binding curve is obtained, instead, when performing ELISA on the full virus at 4°C; this will be later discussed in the context of the structural data.

### NMR Epitope Mapping

Solution NMR spectroscopy was used to characterize, at the residue level, the epitope (i.e. the region of the antigen that interacts with the antibody) of DV32.6 on DIII from each DenV serotype. In a so-called $^{15}$N-HSQC experiment, one of the simplest NMR experiments, each protein residue generates an individual signal. The position of these signals is sensitive to the local chemical environment, so much so that $^{15}$N-HSQCs are considered protein fingerprints. We exploit this property to determine the epitope: when DV32.6 binds to DIII, interface residues experience a different chemical environment and their NMR signal changes as a consequence. By comparing the NMR spectrum of DIII free or in complex with the antibody we can identify which signals change. Knowing the assignments, i.e. determining which signal belongs to which antigen residue, we can therefore determine the residues affected by antibody binding (Figure S2 and S3). NMR assignments are publicly available for DIII of three serotypes [12,13,14]. We assigned DIII of DenV3 according to standard techniques. We utilize a purely qualitative approach: if a signal changes position or disappears, then we know that the residue generating such signal is affected by antibody binding. In our experience, a quantitative analysis of the changes is neither required nor satisfactory in the case of large and poorly behaved complexes with low spectral quality. Nonetheless, the accuracy attainable by qualitative analysis even in such complicated cases is sufficient, as we have previously shown for a TCR/pMHC complex (similar in many ways to our DIII/DV32.6 complexes) that was subsequently validated by an x-ray structure [15] and in an RNA/protein complex later verified by an NMR structure [16]. Furthermore, by relying on simple and sensitive $^{15}$N-HSQC experiments, or their TROSY equivalent, we can adopt this approach even for the most difficult cases when more sophisticated NMR approaches are not suitable. Such NMR experiments may theoretically offer more information, but they fail to do so due to lack of sensitivity in the case of our DIII/antibody complexes.

The NMR signal of approximately 20% of the surface residues of DIII is perturbed upon binding of DV32.6. No unassigned peak shows chemical shift changes upon complex formation with the exception of one signal that presumably belongs to Q316, in the middle of the epitope. The epitope centers around residues 306–325 and the antibody footprint shows only slight variation amongst serotypes (Figure 2a and 2b). It includes residues conserved in all serotypes, explaining why DV32.6 binds all four of them, and residues that are not conserved, which are probably responsible for the different binding and neutralization properties. However, sequence or structural analysis of the epitope offers no information on the antibody residues important for interaction. To further characterize the binding interface we thus predicted the three-dimensional structure of the complex between DV32.6 and DIII by computational docking, guided and validated by the NMR epitope mapping data.

### Experimentally Validated Computational Docking

Computational docking predicts the structure of a multi-molecular complex starting from the separated structures of the individual components. Experimental structures of DIII from each serotype are available [6,13,14,17] and were used for docking. We predicted the structure of antibody DV32.6 by homology modeling according to the canonical structure method [18]. Although docking a model, thus making a model of a model, is known to generate problems in protein-protein docking, we and others have clearly shown that this is not the case for antibodies [11,18], since they can be modeled with high accuracy and precision in the vast majority of cases and exceptions are readily recognized [19]. Briefly, antibody antigen binding loops can adopt a limited set of conformations constrained by loop length and presence of specific amino acids at key positions [20]. Exploiting these constraints yields highly reliable antibody models, except for the H3 loop that can adopt highly variable conformations. We thus generated multiple models of DV32.6 [11], differing mainly in the conformation of the H3 and L3 loop and in the relative position of the six antigen binding loops. In particular, the H3 loop extends outwards in some of our models but presents a rather flat interacting surface in others. The purpose is two-fold: on one hand by using multiple models we increase the chances that at least one of them is accurate; on the other hand the ensemble of conformations may simulate the flexibility available to long protein loops such as the H3 of DV32.6. We then docked each antibody model independently to DIII with the program RosettaDock [21] and validated the results as described in the methods section. The difference between our approach and "ensemble docking" [22] protocols is that in the latter docks an ensemble of multiple conformations and uses a scoring function to select the best model. We, on the other hand, evaluate the docking of each conformation according to its agreement with the experimental data. We have previously shown that combining NMR mapping and docking in this way significantly increases the accuracy of computational docking [11]. In fact, the initial docking search we performed failed to generate a so-called "scoring

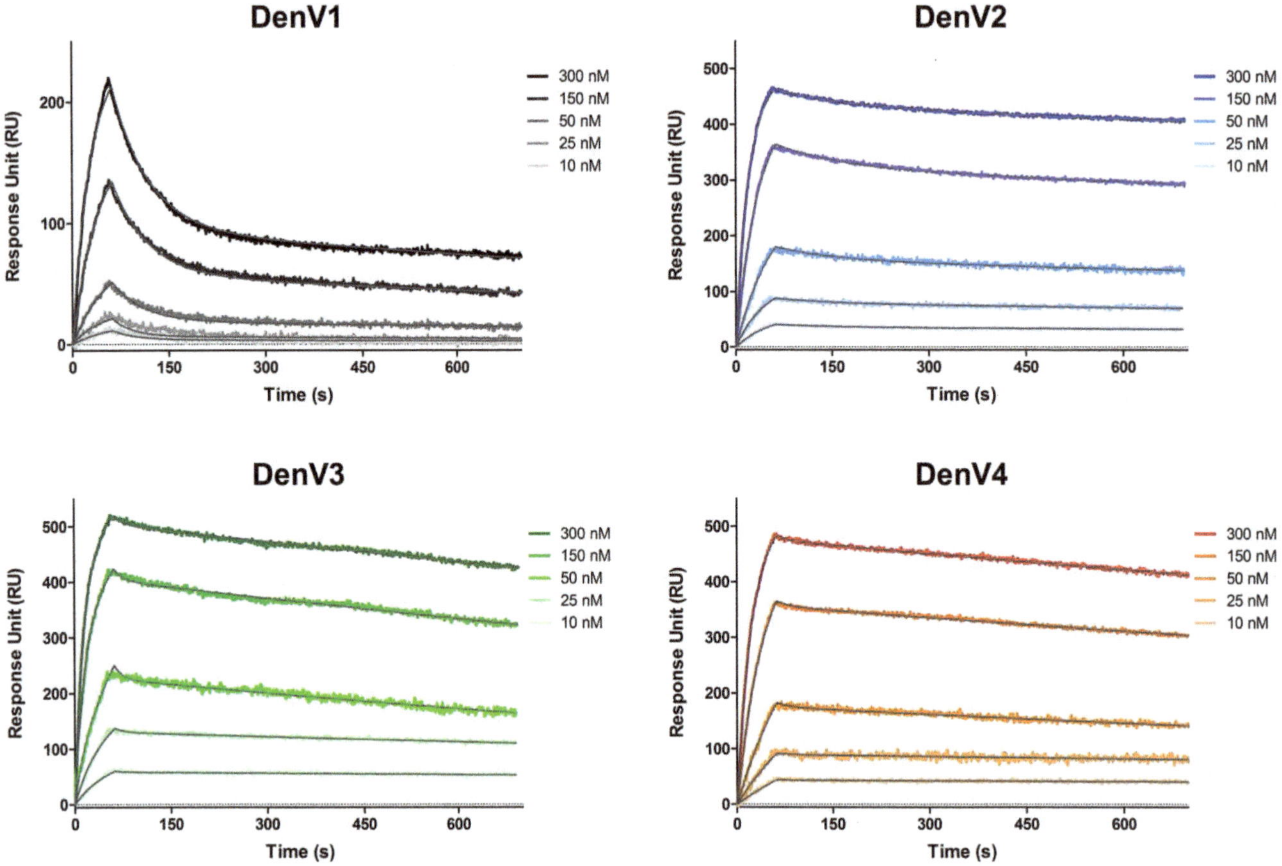


**Figure 1. Antibody DV32.6 binds the four Dengue serotypes with different affinity.** SPR sensorgrams showing association and dissociation of DV32.6 to DIII of each Dengue serotype. The antibody was immobilized on the sensor surface, followed by injection of DIII at the concentrations indicated in the figure. The line fitted to the experimental data and used to calculate the binding affinities is drawn in gray. KD values are 145±9 nM for DenV1; 7±0.2 nM for DenV2; 73±16 nM for DenV3; 34±7 nM for DenV4. Despite binding more weakly to DenV1 and DenV3 than DenV4, the antibody neutralizes those serotypes better than DenV4.

funnel", an ensemble of similar conformations significantly more accurate than the others according to the docking algorithm. Without our experimentally validated approach we would not be able to discriminate amongst a dozen or so solutions recognized as equally valid by the computational algorithm alone. Further refinement of the structure in better agreement with the NMR data (see methods) would often generate two scoring funnels, only one of which was in agreement with the experimental data.

## Interaction of Antibody DV32.6 with the Four Dengue Serotypes

According to our NMR validated docking predictions, DV32.6 primarily recognizes DIII residues located on adjacent beta strands, covering between 25 and 28 residues with a buried surface area between 684 Å$^2$ (DenV2) and 768 Å$^2$ (DenV1). These values are in line with those obtained from x-ray structures of other antibodies against DIII. Antibody 2H12 has buried surface area between 491 Å$^2$ (DenV4) and 652 Å$^2$ (DenV1). Antibody 1A1D-2 has a larger buried surface of 843 Å$^2$ and antibody 4E11 buries between 758 Å$^2$ (DenV3) and 894 Å$^2$ (DenV1) [23,24,25]. The predicted interface of antibody DV32.6 is dominated by electrostatic interactions and features several intermolecular hydrogen bonds and salt bridges. All DIII residues predicted to be at the interface by the selected docking solutions are also affected by complex formation in the NMR epitope mapping experiment.

K310, conserved in all serotypes, is at the center of the interface and involved in multiple intermolecular contacts. The epitope has several non-conserved residues, which are probably responsible for the different binding and neutralization properties of DV32.6. Indeed, residues 307, 309, 323, 325 and 327 differ among serotypes and appear to be involved in antibody binding. Residue 309, for instance, is Glu, Val, Lys or Asp in DenV1 to DenV4, respectively, and it may not be surprising that it has a different effect on the antibody partner. Perhaps surprisingly, some conserved DIII residues appear to have different antibody partners in different serotypes: E311, for instance, interacts with either the backbone of H-D102 (D102 of the antibody Heavy chain), L-S32 (S32 in the antibody Light chain) or L-S93. This may suggest conformational plasticity or may very well reflect a limit in the precision of the computational models. Even if the specific interactions are wrong, however, the models provide an indication of which residues are involved in intermolecular contacts. We may not know if E311 interacts with H-D102, L-S32 or L-S93, for instance, but we know that it is involved in electrostatic interactions with the antibody. This level of information was sufficient to successfully design antibody mutants.

Analyzing the predicted antibody interface (Figure 2c) shows differences that could be exploited to selectively alter the binding

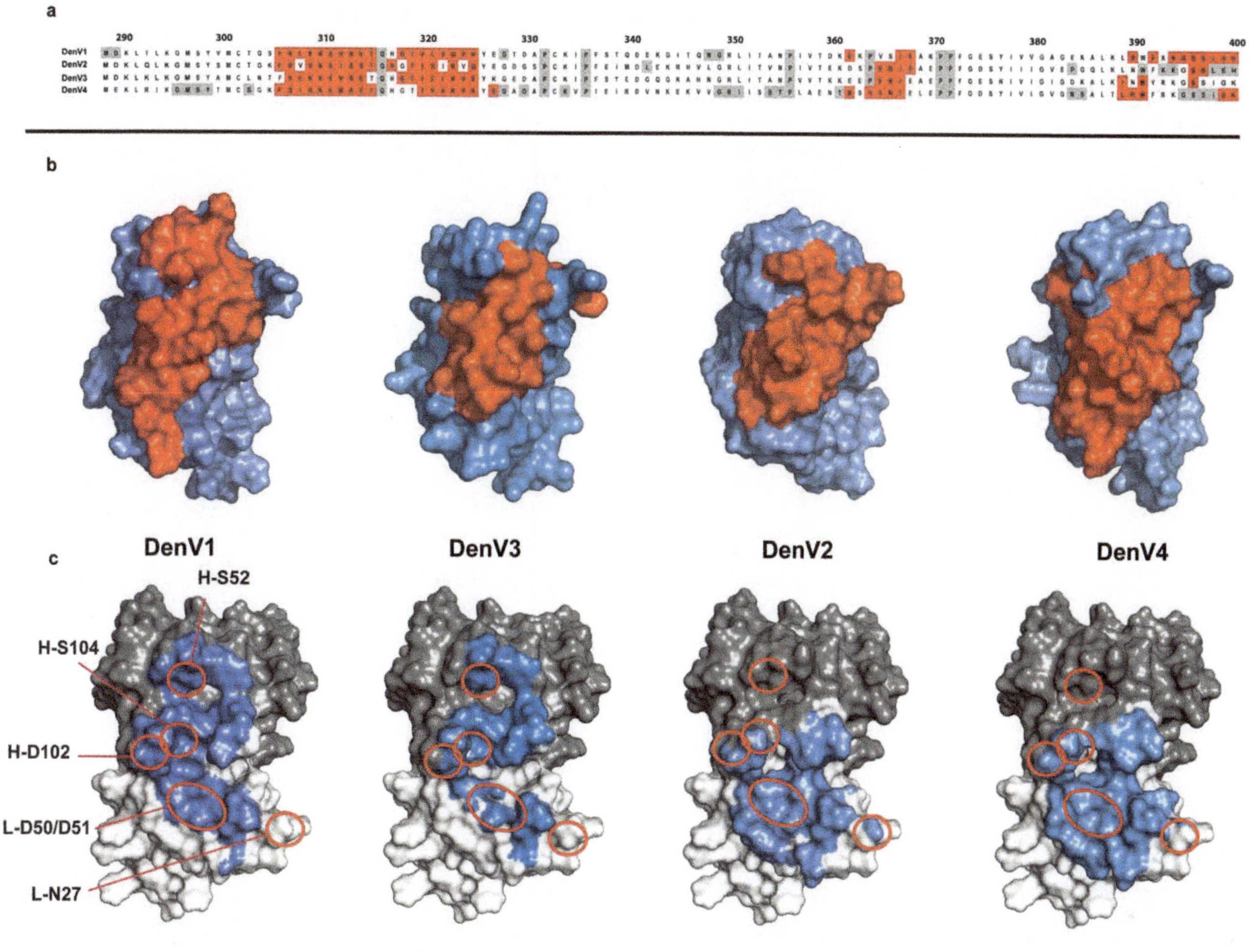


**Figure 2. Antibody DV32.6 interaction with the four Dengue serotypes.** NMR epitope mapping results (a, b): residues whose NMR signal is affected by antibody binding are indicated in red on the sequence (a) and on the surface representation (b) of DIII of each DenV serotype. Residues for which no NMR information is available are coloured gray in the sequence. The discontinuous epitope shows slight variations amongst serotypes both in sequence and structure, including some conserved residues and others that are not. The former explain why DV32.6 can bind to all four serotypes, the latter are likely responsible for the different binding and neutralization properties. Computational docking results (c). Surface representation of the antigen binding site of DV32.6. Antibody residues predicted to interact with DenV are shown in blue; light and heavy chains are in light and dark gray, respectively. Both antibody chains contribute to the binding interface in DenV1 and DenV3 whereas only the light chain and the H3 loop bind to DenV2 and DenV4. Some of the residues mutated to alter the antibody properties in a predictable way (see text) are indicated by red circles and labelled on DenV1.

properties: the light chain (L1, L2, L3) and H3 antigen binding loops interact with every serotype; the H2 loop interacts only with DenV1 and DenV3 and the H1 loop has no contact with any serotype. Although the total interface area is similar, the light and heavy chains equally contribute to it in DenV1 and DenV3, whereas 80% of the interface is formed by the light chain in DenV2 and DenV4.

## Rational Antibody Engineering to Selectively Alter its Binding Properties

By analyzing the three-dimensional models of the antibody/DIII complexes we designed several antibody mutants with the intent of further verifying our computational predictions, altering the binding properties of the antibody and ultimately improving its ability to neutralize the virus.

As a first test we aimed to modify the antibody without affecting its binding to DIII, proving that we can identify and avoid critical residues. Antibody sequence analysis can easily predict which residues belong to antigen binding loops and may, therefore, interact with the antigen. Our models go a step further and can identify which of these residues are not directly involved in intermolecular contacts and can thus be mutated without adverse consequences. We thus selected and mutated a subset of such residues in each of the three heavy chain antigen binding loops. As predicted, all the following mutations in the antigen binding loops didn't alter the antibody binding properties: H-S104A (Figure 3); H-T31A; H-S54Q; H-S103V; H-T106A. Binding assays for all the mutants described in the manuscript are included as figures S4 and S5.

To further validate our models we then aimed to abolish antibody binding to all serotypes. Residues L-D50 and L-D51 are predicted to be at the center of the antibody/DenV interface, forming an intermolecular network of hydrogen bonds and salt bridges in all serotypes. If our models are correct, disrupting this network should have a profound effect on the interface. Indeed, the double mutant L-D50A/D51A completely abolishes antibody binding (Figure 3). The same result is obtained when mutating the nearby H-S105D, since the

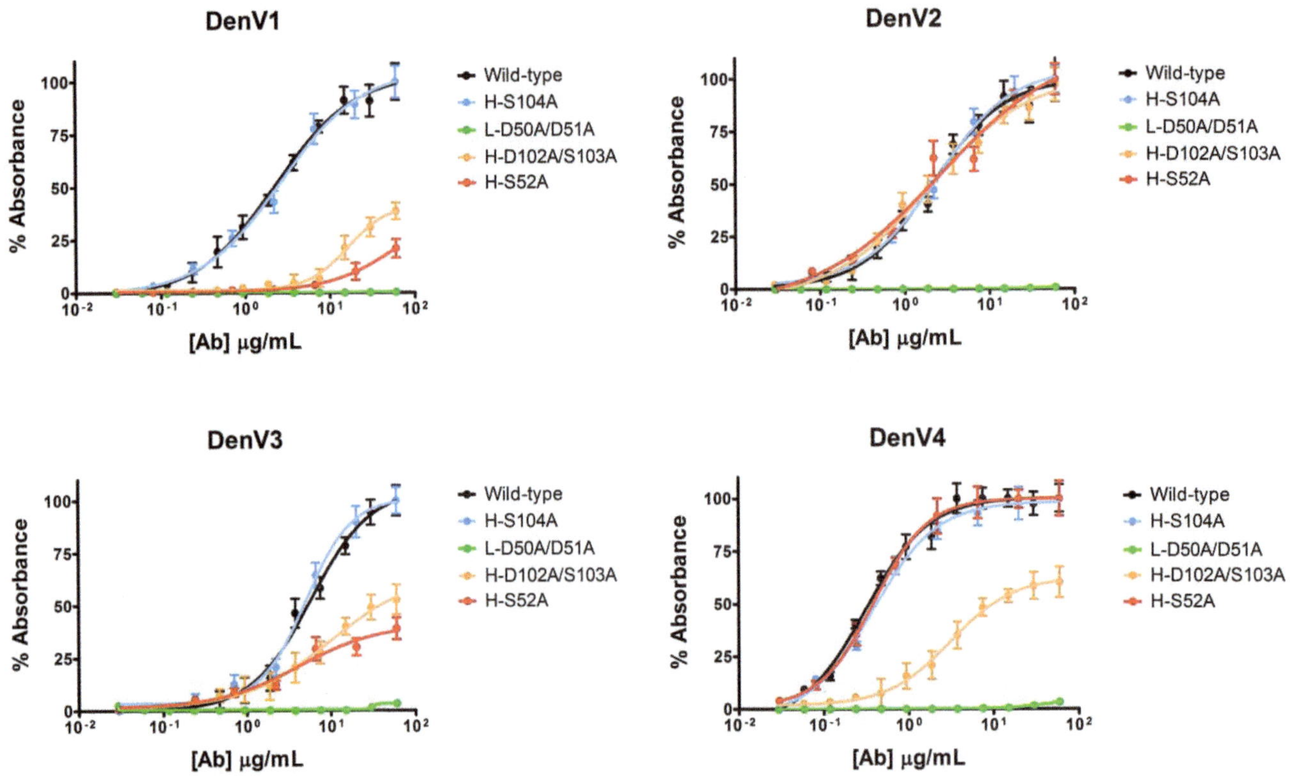


**Figure 3. Rational antibody engineering to selectively alter the binding specificity.** Binding assays (ELISA) of representative mutants for the four DenV serotypes are shown; higher y values correspond to increased antibody binding to DIII. The structural position of these mutants is indicated in Figure 2c. Wild-type DV32.6 (black) binds to all serotypes and H-S104A does not alter its properties. The L-D50A/D51A mutant cannot bind to any serotype. The H-D102A/S103A mutant binds to DenV2 and has severely limited binding to the other serotypes. The H-S52A mutant can only bind to DenV2 and DenV4.

negative charge introduced interferes with the negatively charged sidechain of L-D50.

Having shown that our approach can identify critical interface residues, we tackled the much more difficult task of improving the antibody properties. Generally speaking, increasing antibody selectivity is a useful exercise to eliminate cross-reactivity with undesired antigens or to design bio-recognition elements for specific antigen subtypes. As a proof of concept, we altered the interaction between DV32.6 and DenV, obtaining an antibody mutant specific for DenV2 and another that binds only DenV2 and DenV4.

According to our models, residues H-D102/S103, belonging to the H3 antigen binding loop, point away from the DenV2 antigen. Therefore, mutating them should have no effect on this serotype. They are, instead, in close proximity of DenV4 and DenV3. Finally, H-D102 is predicted to form an intermolecular salt bridge in DenV1. In agreement with this prediction, the H-D102A/S103A mutant binds DenV2 like the wild-type antibody but it has severely reduced binding to all other serotypes (Figure 3).

Since the antibody H2 loop is predicted to interact with DenV1 and DenV3 but not DenV2 and DenV4, we designed the H-S52A mutant to prevent antibody binding to DenV1 and DenV3 while leaving the other two serotypes unaltered (Figure 3). This mutant agrees with the computational prediction that the antibody uses different binding modes to interact with the serotypes as illustrated in Figure 2c.

## Rational Antibody Engineering to Improve the Neutralization Properties

The aforementioned mutations are a testimony to our ability to rationally alter the antibody binding properties and specificity. Improving the ability of therapeutic products to protect from infection is, however, the main goal of antibody engineering. As a proof of concept, analysis of the predicted antibody/antigen interfaces allowed us to design a mutated antibody 40 times more efficient than the wild-type in neutralizing DenV1 and another mutant more efficient against all serotypes, albeit to a lesser extent (Figure 4).

Our models show that H-S52 is in close proximity to a positively charged lysine sidechain in DenV1 (Figure 5). Introducing a nearby negative charge should favor the formation of intermolecular salt bridges, possibly resulting in improved neutralization properties. Indeed, the H-S52D antibody is 40 times more efficient in neutralizing DenV1; the estimated antibody concentration required to achieve 50% neutralization ($EC_{50}$) is 0.1 µg/ml for HS52D and 4.1 µg/ml for wild-type DV32.6 (Figure 4). SPR indicates that the $k_{off}$ is identical to the wild-type while the $k_{on}$ improves from $0.95\pm0.01\times10^{-4}$ to $4.8\pm0.4\times10^{-4}$ $nM^{-1}s^{-1}$, resulting in a KD of 28±2 nM versus 145±9 nM. DenV4, where H-S52D is also a slightly better neutralizer than the wild-type, has a similar improvement in $k_{on}$ ($1.5\pm0.1\times10^{-4}$ for the wild-type, $5.3\pm1.0\times10^{-4}$ $nM^{-1}s^{-1}$ for H-S52D) and the $k_{off}$ improves as well (from $5.1\pm1.3\times10^{-3}$ to $0.47\pm0.1\times10^{-3}$ $s^{-1}$) resulting in KD of 0.9±0.04 nM for the mutant versus 34±7 nM for the original antibody.

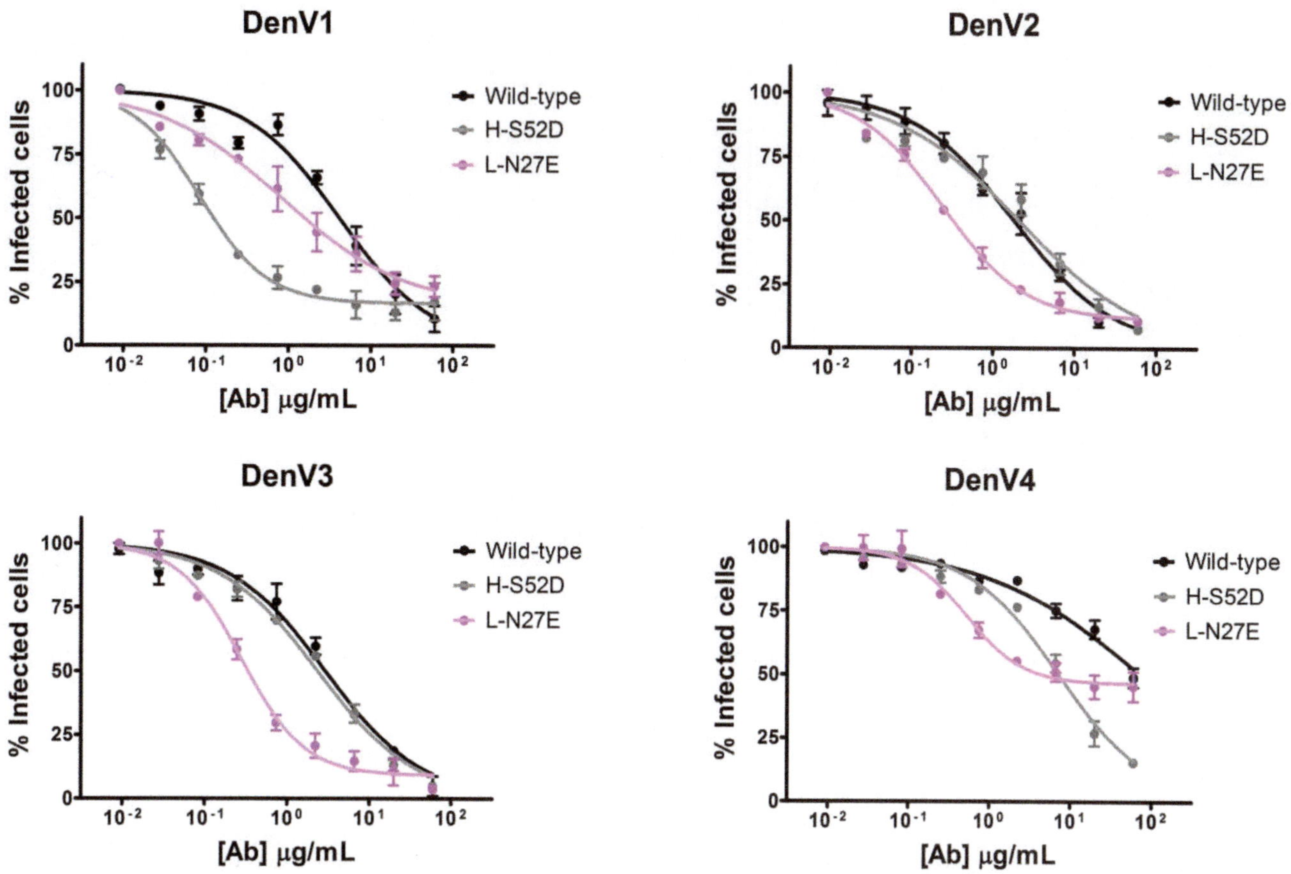


**Figure 4. Rational antibody engineering for improved virus neutralization properties.** We designed two antibody mutants with the intent of improving its neutralization properties. H-S52D (gray) neutralizes DenV1 40 times more efficiently than the wild-type (black) and L-N27E (violet) is better than the wild-type in all serotypes, albeit to a lesser extent. Viral neutralization assays are shown; the amount of infected cells (y axis) decreases at increasing antibody concentration (x axis). In comparison to the wild-type, a lower concentration of mutants is required to neutralize the same amount of virus.

L-N27, instead, is predicted to be close to a positively charged region in all serotypes (Figure 5) so introducing a negative charge should favor the interaction with all of them. An N>E rather than N>D mutation was introduced because a longer side chain would get closer to the antigen without creating clashes, according to the models. Indeed, the L-N27E mutant has improved neutralization properties for all serotypes (Figure 4). The L-N27E mutant is 3 times more effective on DenV1, 6 on DenV2, 9 on DenV3 and 17 on DenV4 (ratios between the $EC_{50}$ for wild-type and mutant antibody). KD is 5 times stronger for DenV1, 23 times stronger for DenV2, 4 times for DenV3 and 38 times for DenV4. The mutant has improved $K_{on}$ in all serotypes whereas the $K_{off}$ is equal to the wild type for DenV1 and DenV3. Unfortunately, the effects of the H-S52D and L-N27E mutations are not additive; the double mutant shows better neutralization only for DenV3. Apparently, the interaction cannot tolerate the simultaneous introduction of two negative charges on the other serotypes.

## Discussion

Strategies for optimizing and improving antibody properties are highly desirable, either to increase their efficacy or to alter their binding specificity. Generally speaking, abolishing or altering antibody binding to undesired antigens might confer selectivity to an otherwise broadly reactive antibody, with practical uses in avoiding cross-reactivity or designing specific biomarkers (or biosensors). On the other hand, antibodies that can recognize only selected serotypes on variable pathogens like Influenza or Dengue may be engineered to become more broadly reactive. Antibody optimization can be achieved by randomizing its sequence with display technologies [26] or by structural analysis and rational modification of the antibody/antigen interface. This latter strategy is often neglected and generally thought to require high resolution crystallographic structures [27], which are not always available. As a proof of concept, we here show that structure-based antibody engineering is feasible even without x-ray information and can be achieved by computational methods guided and validated by a limited set of rapidly obtained experimental results. We have previously shown that NMR mapping can provide an accurate representation of protein-protein interfaces [15] and that this can greatly increase the accuracy of computational docking by validating its results [11]. Here we move further ahead by showing that computational models of antibody/antigen complexes allowed us to rationally design antibody mutants that improve its properties by *i*) disrupting binding only to selected serotypes and *ii*) improving virus neutralization up to 40 fold.

Antibody DV32.6, isolated from the serum of a human donor recovered from DenV2 infection, can efficiently neutralize three of the four DenV serotypes but binds to all four in the so-called DIII, the most variable domain of the protein forming the viral surface

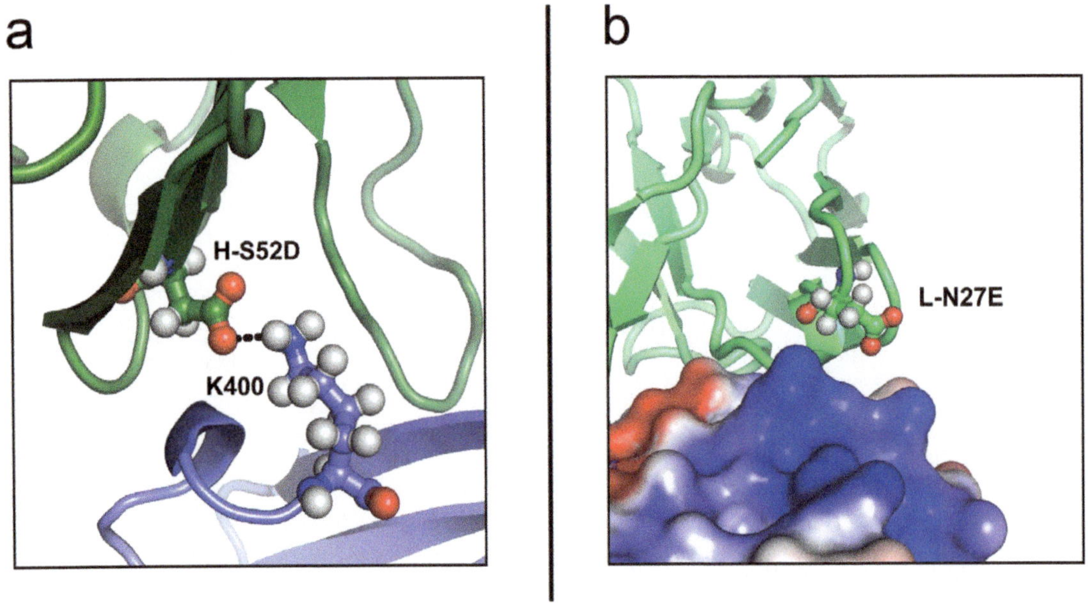


**Figure 5. Rational antibody engineering, a structural view.** a) H-S52 is close to the positive sidechain of K400 in DenV1. Mutating it to the negatively charged H-S52D favours the interaction and this mutant is 40 fold more effective than the wild type at neutralizing DenV1. The antibody is shown in green and Denv1 in blue. b) DV32.6 is shown as green cartoon over the electrostatic surface of DenV3; positively charged surfaces are in blue and negatively charged surfaces in red. L-N27 is close to a positive patch conserved in all serotypes. Introducing a negative charge (L-N27E) favours the interaction, resulting in improved neutralization properties. The L-N27E substitution was preferred to L-N27D since the longer sidechain was thought to bring the charged antibody moiety closer to DIII.

and the target of many potent antibodies described so far [4,23]. We identified the epitope of DV32.6 on each serotype at the residue level with solution NMR spectroscopy and used this information to guide and validate computational docking simulations yielding three-dimensional models of the antibody/antigen complexes. Visual analysis of these NMR validated computational structures allowed us to design a total of 22 antibody mutants, only 4 of which did not have the predicted effect: a testament to the accuracy of our approach that is unlikely to arise from a random process. The mutants that did not behave as desired involved: disruption of a predicted hydrogen bond that did not result in a detectable loss of binding; a Ser to Gly substitution that presumably altered the antibody structure and led to total loss of binding; double substitutions that did not yield the cumulative effect of the single mutants.

The Rosetta suite is probably the most successful software for protein folding and engineering [28]. We thus utilized it to complement our visual structural analysis. In particular, we attempted to design antibody mutants that would increase the binding affinity with the RosettaDesign [29] software. However, the algorithm is biased towards the introduction of hydrophobic residues at the intermolecular interface and fails to identify the electrostatic mutations that yield increased virus neutralization in our case. Unfortunately, we were not able to generate and test the hydrophobic mutations suggested by RosettaDesign. Hydrophobic antibodies are often problematic due to aggregation and lack of specificity ("sticky antibodies"); moreover, their refolding from inclusion bodies, if produced in *E.Coli*, tends to be very difficult.

Automated mutant design by the Rosetta software does not identify the electrostatic mutations that improve the antibody properties. Given the mutations, however, we wondered if a docking algorithm would be able to provide a confirmation of their efficacy. Judging this is not straightforward since a simple energetic comparison between wt and mutant by the empiric Rosetta scoring function is unlikely to provide meaningful results. We, thus, decided to look for the presence of so-called "scoring funnels". These are generated when the docking algorithm finds a large number of similar structures with a pseudo-energy score significantly better than the other structures. This results in a funnel shape in a plot of docking score versus structural similarity (RMSD from a reference structure). Scoring funnels are often considered an indication of an accurate docking solution, although exceptions are not uncommon, and in our experience are more likely to happen when the binding affinity between docking partners is high. In the case of the DV32.6 mutants there is no scoring funnel, nor we could detect any other indicator that would distinguish an effective mutation (e.g. L-N27E, which increases antibody binding) from an ineffective one (e.g. L-D50A/D51A, which abolishes binding). On the contrary, we had a scoring funnel only when docking the L-D50A/D51A mutant.

Software for protein design has leaped forward in recent years but it is not yet completely reliable. The advantage of human visual structural analysis, we believe, is that it can detect plausible intermolecular interactions those requirements may not be directly satisfied by the model but could be met with limited structural alterations. In an oversimplified example, an algorithm may suggest a mutation to form an intermolecular hydrogen bond if the two required chemical groups are within 3.5 Å; if the distance is 4.5 Å, however, it would fail to recognize the possibility. Human analysis, on the other hand, might recognize that a relatively minor structural rearrangement would bring the two chemical groups within bonding distance. This rearrangement might happen because proteins are flexible or, more simply, because the position of the chemical groups cannot be ascertained with high precision in the original structural models. In other words, human structural analysis can overcome the uncertainty inherent in computational models, provided that these are sufficiently accurate to reliably identify interface residues.

With this approach we produced DV32.6 antibody mutants that could *a*) leave binding unaltered, proving that we can identify residues that are not critical for interaction despite belonging to the antigen binding loops or *b*) abolish binding to all serotypes. More importantly, whereas wild-type DV32.6 binds to all four DenV serotypes, we could design antibody mutants binding only to DenV2 or to DenV2 and DenV4. Increasing antibody specificity is valuable to eliminate unwanted cross-reactivity or to design bio-recognition elements. If an antibody has therapeutic purposes, however, one would seek to improve its neutralization properties, which would have a beneficial effect on dosage and therapy. It may be argued that if nature was not able to generate a better antibody, then there is no reason why we should be. However, individuals generate antibodies against a specific antigen that may be different from the target of antibody engineering. In the case of Dengue, for example, an antibody generated against DenV2 might be rendered more effective against DenV1. Since the individual from which the antibody was isolated had not been exposed to DenV1, the immune system would not have been able to optimize the antibody against this viral strain. Furthermore, a better antibody may exist in the serum of an immunized individual but it may simply fail to be successfully isolated.

As a proof of concept, we were able to engineer antibody mutants up to 40 times more effective than the starting molecule at neutralizing DenV. These results prove that even when high resolution x-ray information cannot be easily obtained, antibody/antigen complexes can be structurally characterized with sufficient accuracy to allow a fine control of their binding specificity and immunological properties. Although computational algorithms by themselves often fail to discriminate inaccurate solutions [18], by validating them with rapidly obtained NMR epitope mapping results we could achieve accuracy sufficient for rational mutagenesis. It may be argued that sophisticated NMR experiments could provide more information than the simple $^{15}$N-HSQC experiments used in our approach. However, the very simplicity of such experiments provides reasonably rapid results even in large, poorly behaved biological molecules where other NMR experiments may fail due to lack of sensitivity. In fact, even relatively low affinity complexes that tend to escape successful crystallization can be investigated by NMR mapping through HSQCs.

In addition to providing the basis for rational mutagenesis, analysis of the structure of the DV32.6/DIII complexes may be used to investigate the immunological properties of the antibody. DV32.6, in fact, is a poor neutralizer of DenV4 despite binding stronger to its epitope on DIII of DenV4 rather than DenV1 or DenV3 (KD is 34 nM for DenV4, 74 nM for DenV3 and 146 nM for DenV1). It may be argued that $K_{on}$ or $K_{off}$, rather than KD, might be more relevant for neutralization. However, there is no linear correlation between those and neutralization, either. $K_{off}$ is better for DenV4 ($5.1\pm1.3\times10^{-3}$ s$^{-1}$, $23.0\pm1.1\times10^{-3}$ s$^{-1}$ and $13.9\pm2.0\times10^{-3}$ s$^{-1}$ for DenV4, DenV3 and DenV1) and $K_{on}$ is similar ($1.5\pm0.1\times10^{-4}$ nM$^{-1}$s$^{-1}$, $3.1\pm0.9\times10^{-4}$ nM$^{-1}$s$^{-1}$ and $0.95\pm0.01\times10^{-4}$ nM$^{-1}$s$^{-1}$ for DenV4, DenV3 and DenV1, respectively).

Given the same epitope and identical conditions, stronger binding should lead to stronger neutralization but this is not the case for DV32.6. One possibility, therefore, is that DenV4 has a different epitope from the other serotypes. NMR epitope mapping does not indicate this. The vast majority of the epitope residues are affected by antibody binding in all serotypes and the biological significance of those that aren't is not evident. The different immunological profile is thus likely to arise from differences not evident at the epitope level that manifest themselves on the full viral particle, which the antibody needs to recognize in order to neutralize the virus.

DV32.6 recognizes a discontinuous epitope centered around the so-called A strand of DIII of all four Dengue serotypes. This region is close to the intermolecular interface in the E protein homodimers that form the surface of mature Dengue virus, with DIII from one E protein contacting DII of the other unit [30]. On the viral surface, the DV32.6 epitope is partly covered by DII residues. Antibodies 1A1D-2 [23] and 4E11 [24] were similarly shown to bind to partially inaccessible epitopes close to the DV32.6 binding site. Cryo-EM data shows that1A1D-2 binds to a virus conformation different from the one determined in the absence of antibodies. It is likely that DV32.6 sports a similar mechanism. In fact, steric clashes between DV32.6 and neighbouring E proteins are present when superimposing our models to the mature structure of the full virus, whereas no clash is detected when superimposing our models to the structure of the virus in complex with 1A1D-2. Presumably, structural dynamic of the viral surface exposes the epitope for a time sufficient for antibody binding. This theory is in agreement with ELISAs data on DV32.6 binding to the full virus: at 37°C there is sufficient dynamic movement to uncover the epitope and allow DV32.6 binding; at 4°C, instead, the dynamic process is slowed or altogether halted, the epitope is not made accessible for a sufficient amount of time and DV32.6 cannot bind effectively anymore to the virus.

We previously noted that antibody binding to DenV4 is likely to cause steric clashes on the viral surface, preventing efficient neutralization. Why is this not the case for the other serotypes? One possibility is that DV32.6 has less steric clashes on the virion of DenV1 to 3 and that relieving them requires less extensive, thus more likely, viral movements than DenV4. Another possibility is that different viral strains have different extent of dynamic movements on the viral surface. Indeed, antibody 2H12 was recently shown to bind a partly inaccessible DIII epitope [25], although not in the same exact region of DV32.6. Similarly to DV32.6, 2H12 has the same binding affinity for DIII of all serotypes ($EC_{50}$ 0.4 nM in ELISA binding assays) but cannot neutralize DenV2. The authors, Midgley et al., suggest that the DenV2 virion might be less dynamic than other serotypes [25], resulting in less frequent exposure of the inaccessible antibody epitope and less efficient neutralization. In agreement with this interpretation, DV32.6 binds much stronger to DIII of DenV2 (KD 7 nM) than DenV1 (KD 145 nM), yet it neutralizes both strains equally well (EC50 is 1.8 μg/ml for DenV2 and 4.1 μg/ml for DenV1). The decreased binding affinity for DIII of DenV1 would be compensated by the additional rigidity of the virion of DenV2, resulting in similar neutralization. The same considerations hold true for the L-N27E antibody mutant, that neutralizes DenV2 and DenV3 equally well despite binding 100 times more strongly to DIII of DenV2 (KD 0.5 nM for DenV2 and 43 nM for DenV3, EC50 0.3 μg/ml in both cases).

However, in our case DenV4 would be the least dynamic virion, since the antibody binds strongly to DIII of this serotype but fails to neutralize the virus, suggesting that the epitope is seldom accessible on the viral surface. Midgley et al. do not describe a similar property for DenV4, although this serotype is less efficiently neutralized than DenV1 despite having similar binding affinity. One difference between the previously reported work and ours is that whereas both research teams use the same viral strains for DenV1 and DenV2, there are two aminoacid differences in DenV4 (L356F and N383D). Whether this is sufficient to justify a different viral flexibility is debatable.

Overall, the structural flexibility of Dengue viral strains remains largely unexplored. It is not clear, for instance, if it is exclusively an intrinsic property of the virion or if the antibodies can affect it. Antibody binding to one E protein on the viral surface, for instance, may distort the ordered structure of the virus and provoke a chain reaction making the epitopes on the other E proteins more readily accessible.

## Conclusions

Structural characterization of antibody/antigen complexes relies almost exclusively on x-ray information that may not always be available. Computational docking is emerging as an affordable alternative but it often fails to discriminate inaccurate solutions. We have shown that this problem can be alleviated by rejecting computational solutions that do not agree with a limited set of rapidly obtained experimental data. These may arise from viral escape mutants, peptide mapping or antigen mutagenesis; NMR epitope mapping, however, is a powerful tool for the detailed characterization of antibody binding sites with the potential to identify the epitope of several antibodies in a matter of hours, once assignment is available. We have shown that combining docking and NMR epitope mapping yields results sufficiently accurate for the rational modification of the antibody properties even in the absence of high resolution x-ray information. For antibody docking we have had good success with RosettaDock [21] but other programs like HADDOCK [31] are equally good. We believe that NMR validated computational docking can find widespread use in the structural characterization of antibody/antigen interactions, with implications for basic research, patenting purposes, screening and selection of candidate molecules as well as rational optimization of their binding and immunological properties.

## Materials and Methods

### Isolation of Human Monoclonal Antibody

Briefly, memory B cells were isolated from the blood serum of a donor recovered from DenV2 infection and immortalized with EBV as previously described [4]. The study protocol was approved by the Scientific and Ethical Committee of the Hospital for Tropical Diseases and the Oxford Tropical Research Ethical Committee. Written informed consent was provided. B cell clones were then tested for specificity by staining of Dengue infected cells and ELISA with recombinant E protein.

### Antigen and Antibodies Production and Purification

The sequence of DIII from each serotype was identical to that of the viral strains used for the binding and neutralization assays. DIII domains from E protein of each Dengue serotype were expressed in E. coli Rosetta cells with a pET21 vector (Novagen), induced at $OD_{600} = 0.7$ and harvested after 3 hours. After sonication and centrifugation, the pellet was repeatedly washed and centrifuged in sodium phosphate buffer pH 7.2, 1 M NaCl, 1 M urea and 1% Triton X-100. The pellet was finally resuspended in the same buffer with 8 M urea (buffer A). Following addition of 0.2% PEI and centrifugation, 65% ammonium sulphate was added to the supernatant. After centrifugation, the pellet was resuspended in buffer A and dialyzed for 3 days against 20 mM sodium phosphate buffer, 150 mM NaCl, 200 mM arginine, adding small amounts of pellet every 8 hours. DIIIs were finally purified on a superdex-75 size exclusion column (GE) with the buffer used for NMR and concentrated with Vivaspin concentrators (Sartorius). $^{15}N$ and $^{13}C$ labeled DIIIs, used for NMR experiments, were expressed in M9 minimal media using aptly labeled nutrients ($NH_4Cl$ and glucose) as sole source of nitrogen and carbon. DIIIs were proven to be correctly folded by NMR $^{15}N$-HSQC experiments.

DV32.6 from B-cell culture supernatant was purified by protein-A affinity chromatography followed by size-exclusion according to standard protocols. The $F_{ab}$ fragments utilized for NMR experiments were obtained by enzymatic digestion with immobilized papain (Pierce) for 8 hours at 37°C, followed by dialysis, protein-A affinity and finally size exclusion chromatography against the NMR buffer. Recombinant wild-type antibody and mutants were expressed in HEK293T cells and purified as before. All samples were freshly prepared and mixed immediately before the required experiments.

### NMR Epitope Mapping

NMR epitope mapping experiments were conducted on the complex between DIII of each serotype and either full DV32.6 (1:2 ratio to DIII) or its antigen binding fragment ($F_{ab}$, 1:1 ratio to DIII). Using the full antibody is more problematic for NMR due to its larger size but the enzymatic digestion required to obtain the $F_{ab}$ fragments comported a significant loss of material and, thus, sample dilution. Use of $F_{ab}$ fragments tended to provide better NMR results. TROSY versions or simple HSQCs experiments offered comparable sensitivity, they were both tested on each sample and the best one was chosen for the final experiment. Typical acquisition time was 30 minutes for free DIIIs and 18–24 hours for the complex at a temperature of 298 K. Typical concentration was around 0.25 mM in 20 mM NaPhosphate pH6, 50 mM NaCl. In a typical experiment, a $^{15}N$-HSQC spectrum of DIII from one serotype was recorded and a second spectrum was recorded after addition of the $F_{ab}$ of DV32.6. The two spectra were visually compared and residues with different signal in the free and bound form were identified. Knowing the assignment, i.e. which protein residue corresponds to each NMR signal, DIII residues affected by DV32.6 binding were localized. NMR assignments of DIII from DenV1, DenV2 and DenV4 are publically available [12,13,14]. We assigned the backbone atoms of DIII of DenV3 with HNCACB and $^{15}N$-NOESY-HSQC experiments according to standard techniques. Spectra were recorded on Bruker 750 MHz and 800 MHz instruments equipped with cryoprobe and analyzed with the program Sparky [32].

### Antibody Modeling and Docking

DV32.6 was modelled according to the canonical structure method with the programs PIGS [19] and RosettaAntibody [33] as previously described [18]. 11 models with different H3 loop conformations were generated and independently used for docking. The structure of DIII from all serotypes is publically available [6,13,14,17] and was used for docking.

Docking was performed using RosettaDock 2.3 as previously described [11]. In summary, a given antibody model was docked to DIII of one serotype; amongst the thousands of computationally generated complexes (typically 15,000), the one in better agreement with the NMR epitope mapping data was selected and further refined. If an antigen residue is in contact with the antibody in a docking decoy and is also affected by antibody binding in the NMR mapping experiment, then it is considered a valid contact. If it is not affected by antibody binding according to the NMR data then it is considered a violation. We select the docking decoys that maximize the number of valid contacts while avoiding violations. An antigen residue is defined "in contact" if either its N or H backbone atoms are within 7 Å of any antibody atom. This criterion proved to yield the most accurate results in extensive docking back-calculations of protein-protein complexes with known experimental structure and available NMR mapping information. By repeating the procedure for each of the 11 antibody starting models we obtained 11 putative complex structures. Not all of them equally satisfied the NMR epitope mapping data, so we finally retained only those complexes that had no significant disagreement with the experimental data and included all of them in the structural analysis. Computational models were discarded if the number of violations was greater than 1 or if the antibody showed extensive contacts with the antigen outside the canonical interaction regions (antigen binding loops and neighboring area). Amongst the remaining structures, those with a higher number of valid contacts were preferred. This was repeated for DIII of each Dengue serotypes. A table summarizing the docking and NMR epitope mapping results is offered as Table S1.

### Antibody Engineering

Antibody mutations were designed by visual analysis of the experimentally validated computational models. Single-point mutations were generated by site-directed mutagenesis of a vector encoding the wild-type antibody sequence using the QuikChange site-directed mutagenesis kit (Stratagene). The introduction of the desired mutations was confirmed by DNA sequencing. The PCR products were transformed into XL10-Blue supercompetent E. coli cells (Stratagene) and the plasmids purified according to standard techniques and transferred into HEK293T cells for antibody production.

### Binding and Neutralization Assays

All the neutralization and binding assays reported here were conducted on the wild-type antibody purified from immortalized B-cells isolated from the blood of a human donor, on the wild-type antibody cloned in HEK293T cells and on the mutated antibodies. Each measurement was conducted in duplicate, repeated 3 times and reported as the average of the resulting values. Results are normalized by considering as 100% the maximum point reached by the wild-type antibody, included as a reference in all experiments. $EC_{50}$ were calculated as the midpoint transition in the s shaped curves after line fitting according to standard procedures.

Binding assays: 96-well ELISA plates were coated at 4°C with recombinant DIII or purified virus from DenV1-4. After washing and blocking, antibodies were added for 1 hr at 37°C. After further wash, bound antibodies were revealed using goat anti-human IgG coupled to alkaline phosphatase (Jackson Immunoresearch).

The ability of DV32.6 to neutralize the virus was assessed by flow cytometry assays measuring the number of VERO or RAJI cells infected by DenV vaccine strains in the presence of different amounts of antibody [4]. Different antibody concentrations were pre-incubated with DenV attenuated virus and used to infect VERO or RAJI cells. The percentage of infected cells was determined by fluorescence-activated cell sorting using standard procedures, staining the cells with mouse mAb 4G2 (ATCC, D1-4G2-4-15).

### Surface Plasmon Resonance Binding Assays

The affinity of DV32.6 for DIII of the four Dengue serotypes was determined at 25°C with a ProteOn XPR-36 instrument (Bio-Rad). The antibodies (concentration 150 nM) were immobilized on the surface of a GLM sensor chip through standard amine coupling. DIII was injected at a flow rate of 100 µl/min (contact time, 1 min) at different concentrations (300, 150, 50, 25 and 10 nM, running in parallel on separate channels); dissociation was followed for 15 min. Analytes responses were corrected for responses from buffer-only injection both on a channel with antibody immobilized and on a channel with no antibody immobilized. Curve fitting and data analysis were performed with Bio-Rad ProteOn Manager software (version 3.1.0.6).

## Supporting Information

**Figure S1 DV32.6 has a stronger binding affinity for its epitope on DIII of DenV4 than for the other serotypes (panel a) but it is less efficient at neutralizing it (panel c).** a) Binding assay (ELISA) for wt DV32.6 on DIII. The antibody concentration is on the x axis and increased y values correspond to increased binding. DIII from each Dengue serotype was immobilized on a surface in the presence of different amounts of antibodies as described in the methods. Binding appears stronger for DenV4 than other serotypes. The experiment was done in duplicate and repeated 3 times. b) Binding assay (ELISA) for wt DV32.6 on the full virus at 37°C. The purified virus from each serotype was immobilized on a surface in the presence of different amounts of antibodies. In contrast to the results of isolated DIII, binding to DenV4 is not stronger than to other serotypes. In fact, the binding curve does not reach plateaux in DenV4 at the tested antibody concentrations. The experiment was done in triplicate. c) Viral neutralization assay; the amount of infected cells (y axis) decreases at increasing antibody concentration (x axis). A higher amount of antibody is required to neutralize DenV4.

**Figure S2 $^{15}$N HSQC spectra of DIII of the four Dengue serotypes free (blue) and in complex with antibody DV32.6 (red).** Residue V324 is affected by complex formation and shows chemical shift changes in DenV1, DenV3 and DenV4 but not DenV2.

**Figure S3 $^{15}$N HSQC spectra of DIII of the four Dengue serotypes free (blue) and in complex with antibody DV32.6 (red).** Residue K310 shows chemical shift changes upon complex formation in DenV4 and DenV1 (smaller changes). The peak corresponding to the bound state disappears in DenV2 and DenV3, revealing that the residue is affected by antibody binding.

**Figure S4 Binding assay (ELISA) for all the antibody mutants designed to alter its properties in a predictable manner and mentioned in the main text.** The antibody concentration is on the x axis; increased y values correspond to increased binding. DIII from each Dengue serotype was immobilized on a surface in the presence of different amounts of antibodies as described.

**Table S1 NMR validation and docking results for the DV32.6/DIII complexes chosen as representative for each different antibody homology model used for docking (Model1 to Model10 and PIGS).** Models in agreement with the NMR epitope mapping data and chosen as final result for the structural analysis (see main text) are highlighted in green. If an antigen residue is in contact with the antibody in the computational model and is affected by complex formation in the NMR epitope mapping experiments then it is considered a "valid contact". If it is in contact in the computational model but is not affected in the NMR epitope mapping experiment then it is considered a violation. The column "Valid-violations" indicates these numbers for each model. "Rosetta score" is the score calculated by the Rosetta-Dock scoring function, the lower the better; it is not sufficient by itself to discriminate incorrect models. "Rosetta ranking" indicates the rank, by score, assigned by Rosetta-Dock. A ranking of 1 means that the structure is the best scoring amongst the thousands of decoys generated in a typical docking run. Without NMR epitope mapping information one would have to rely entirely on the computational scoring function. In our case, instead, lower ranking models better satisfy the NMR epitope mapping information.

## Acknowledgments

We are grateful to Nico Callewaert and Maryse Letiembre for critical reading of the manuscript and Zinaida Yudina for technical assistance. We thank the CSCS Swiss Supercomputer Center for providing the computational resources; the MRC-NIMR and BMRZ NMR facilities for NMR time.

## Author Contributions

Conceived and designed the experiments: FS AL LV. Performed the experiments: LS MP EL MB LC. Analyzed the data: LS MP EL LV. Wrote the paper: LS MP LV.

## References

1. Gibbons RV, Vaughn DW (2002) Dengue: an escalating problem. BMJ 324: 1563–1566.
2. WHO (2009) Dengue and dengue haemorrhagic fever. World Health Organization.
3. Halstead SB (2003) Neutralization and antibody-dependent enhancement of dengue viruses. Adv Virus Res 60: 421–467.
4. Beltramello M, Williams KL, Simmons CP, Macagno A, Simonelli L, et al. (2010) The human immune response to Dengue virus is dominated by highly cross-reactive antibodies endowed with neutralizing and enhancing activity. Cell Host Microbe 8: 271–283.
5. Modis Y, Ogata S, Clements D, Harrison SC (2003) A ligand-binding pocket in the dengue virus envelope glycoprotein. Proc Natl Acad Sci U S A 100: 6986–6991.
6. Modis Y, Ogata S, Clements D, Harrison SC (2005) Variable surface epitopes in the crystal structure of dengue virus type 3 envelope glycoprotein. J Virol 79: 1223–1231.
7. Zhang Y, Zhang W, Ogata S, Clements D, Strauss JH, et al. (2004) Conformational changes of the flavivirus E glycoprotein. Structure 12: 1607–1618.
8. Crill WD, Roehrig JT (2001) Monoclonal Antibodies That Bind to Domain III of Dengue Virus E Glycoprotein Are the Most Efficient Blockers of Virus Adsorption to Vero Cells. Journal of Virology 75: 7769–7773.
9. Park M-S, Deng Y-Q, Dai J-X, Ji G-H, Jiang T, et al. (2011) A Broadly Flavivirus Cross-Neutralizing Monoclonal Antibody that Recognizes a Novel Epitope within the Fusion Loop of E Protein. PLoS One 6: e16059.
10. Wahala WM, Kraus AA, Haymore LB, Accavitti-Loper MA, de Silva AM (2009) Dengue virus neutralization by human immune sera: role of envelope protein domain III-reactive antibody. Virology 392: 103–113.
11. Simonelli L, Beltramello M, Yudina Z, Macagno A, Calzolai L, et al. (2010) Rapid structural characterization of human antibody-antigen complexes through experimentally validated computational docking. J Mol Biol 396: 1491–1507.
12. Volk DE, Lee YC, Li X, Barrett AD, Gorenstein DG (2006) NMR assignment of the dengue 4 virus envelope protein domain III. J Biomol NMR 36 Suppl 1: 62.
13. Volk DE, Anderson KM, Gandham SH, May FJ, Li L, et al. (2008) NMR assignments of the sylvatic dengue 1 virus envelope protein domain III. Biomol NMR Assign 2: 155–157.
14. Huang KC, Lee MC, Wu CW, Huang KJ, Lei HY, et al. (2008) Solution structure and neutralizing antibody binding studies of domain III of the dengue-2 virus envelope protein. Proteins 70: 1116–1119.
15. Varani L, Bankovich AJ, Liu CW, Colf LA, Jones LL, et al. (2007) Solution mapping of T cell receptor docking footprints on peptide-MHC. Proc Natl Acad Sci U S A 104: 13080–13085.
16. Varani L, Gunderson SI, Mattaj IW, Kay LE, Neuhaus D, et al. (2000) The NMR structure of the 38 kDa U1A protein - PIE RNA complex reveals the basis of cooperativity in regulation of polyadenylation by human U1A protein. Nat Struct Biol 7: 329–335.
17. Volk DE, Lee YC, Li X, Thiviyanathan V, Gromowski GD, et al. (2007) Solution structure of the envelope protein domain III of dengue-4 virus. Virology 364: 147–154.
18. Pedotti M, Simonelli L, Livoti E, Varani L (2011) Computational docking of antibody-antigen complexes, opportunities and pitfalls illustrated by influenza hemagglutinin. Int J Mol Sci 12: 226–251.
19. Marcatili P, Rosi A, Tramontano A (2008) PIGS: automatic prediction of antibody structures. Bioinformatics 24: 1953–1954.
20. Morea V. (2000) Antibody Modeling: Implications for Engineering and Design. Methods 20: 267–279.
21. Gray JJ, Moughon S, Wang C, Schueler-Furman O, Kuhlman B, et al. (2003) Protein–Protein Docking with Simultaneous Optimization of Rigid-body Displacement and Side-chain Conformations. Journal of Molecular Biology 331: 281–299.
22. Chaudhury S, Gray JJ (2008) Conformer Selection and Induced Fit in Flexible Backbone Protein–Protein Docking Using Computational and NMR Ensembles. Journal of Molecular Biology 381: 1068–1087.
23. Lok S-M, Kostyuchenko V, Nybakken GE, Holdaway HA, Battisti AJ, et al. (2008) Binding of a neutralizing antibody to dengue virus alters the arrangement of surface glycoproteins. Nature Structural & Molecular Biology 15: 312–317.
24. Cockburn JJ, Navarro Sanchez ME, Fretes N, Urvoas A, Staropoli I, et al. (2012) Mechanism of dengue virus broad cross-neutralization by a monoclonal antibody. Structure 20: 303–314.
25. Midgley CM, Flanagan A, Tran HB, Dejnirattisai W, Chawansuntati K, et al. (2012) Structural analysis of a dengue cross-reactive antibody complexed with envelope domain III reveals the molecular basis of cross-reactivity. J Immunol 188: 4971–4979.
26. Bradbury AR, Sidhu S, Dubel S, McCafferty J (2011) Beyond natural antibodies: the power of in vitro display technologies. Nat Biotechnol 29: 245–254.
27. Garcia-Rodriguez C, Levy R, Arndt JW, Forsyth CM, Razai A, et al. (2006) Molecular evolution of antibody cross-reactivity for two subtypes of type A botulinum neurotoxin. Nature Biotechnology 25: 107–116.
28. Leaver-Fay A, Tyka M, Lewis SM, Lange OF, Thompson J, et al. (2011) ROSETTA3: an object-oriented software suite for the simulation and design of macromolecules. Methods Enzymol 487: 545–574.
29. Liu Y, Kuhlman B (2006) RosettaDesign server for protein design. Nucleic Acids Res 34: W235–238.
30. Bressanelli S, Stiasny K, Allison SL, Stura EA, Duquerroy S, et al. (2004) Structure of a flavivirus envelope glycoprotein in its low-pH-induced membrane fusion conformation. EMBO J 23: 728–738.
31. Dominguez C, Boelens R, Bonvin AM (2003) HADDOCK: a protein-protein docking approach based on biochemical or biophysical information. J Am Chem Soc 125: 1731–1737.
32. Lee W, Westler WM, Bahrami A, Eghbalnia HR, Markley JL (2009) PINE-SPARKY: graphical interface for evaluating automated probabilistic peak assignments in protein NMR spectroscopy. Bioinformatics 25: 2085–2087.
33. Sircar A, Kim ET, Gray JJ (2009) RosettaAntibody: antibody variable region homology modeling server. Nucleic Acids Research 37: W474–W479.

# 11

# Seven N-terminal Residues of a Thermophilic Xylanase Are Sufficient to Confer Hyperthermostability on Its Mesophilic Counterpart

**Shan Zhang, Yongzhi He, Haiying Yu, Zhiyang Dong***

State Key Laboratory of Microbial Resources, Institute of Microbiology, Chinese Academy of Sciences, Beijing, P. R. China

**Abstract**

Xylanases, and especially thermostable xylanases, are increasingly of interest for the deconstruction of lignocellulosic biomass. In this paper, the termini of a pair of xylanases, mesophilic SoxB and thermophilic TfxA, were studied. Two regions in the N-terminus of TfxA were discovered to be potentially important for the thermostability. By focusing on Region 4, it was demonstrated that only two mutations, N32G and S33P cooperated to improve the thermostability of mesophilic SoxB. By introducing two potential regions into SoxB in combination, the most thermostable mutant, M2-N32G-S33P, was obtained. The M2-N32G-S33P had a melting temperature (Tm) that was 25.6°C higher than the Tm of SoxB. Moreover, M2-N32G-S33P was even three-fold more stable than TfxA and had a Tm value that was 9°C higher than the Tm of TfxA. Thus, for the first time, the mesophilic SoxB "pupil" outperformed its thermophilic TfxA "master" and acquired hyperthermostability simply by introducing seven thermostabilizing residues from the extreme N-terminus of TfxA. This work suggested that mutations in the extreme N-terminus were sufficient for the mesophilic xylanase SoxB to acquire hyperthermostability.

**Editor:** Danilo Roccatano, Jacobs University Bremen, Germany

**Funding:** This work was supported by the National Basic Research Program of China (2011CB707402), the National Natural Science Foundation of China (30970073) and the National High Technology Research and Development Program of China (2012AA092103). The funders had no role in study design, data collection and analysis, decision to publish, or preparation of the manuscript.

**Competing Interests:** The authors have declared that no competing interests exist.

* E-mail: dongzy@sun.im.ac.cn

## Introduction

Xylanases are increasingly recognized to be important for the deconstruction of lignocellulosic biomass [1,2]. Xylanases not only catalyze the hydrolysis of hemicelluloses but also render cellulose more accessible to enzymatic hydrolysis [3–6]. Thermostable xylanases possess obvious advantages over their counterparts because of the reduced associated costs and their ability to be manipulated for the deconstruction of lignocellulosic biomass [7,8]. It is not surprising, therefore, that widespread research endeavors have focused on developing thermostable xylanases by rational design or directed evolution.

The rational design strategy of replacing the N-terminus of mesophilic xylanases with the first 31 residues of the thermophilic xylanase TfxA from *Thermomonospora fusca* has successfully produced several thermostable hybrid xylanases, such as Stx15 [9], StxAB [10], Btx [11] and ATx [12]. These hybrid xylanases exhibited higher thermostabilities than their corresponding mesophilic parents. However, compared with TfxA, all of the hybrid xylanases displayed significantly inferior thermostabilities. Recently, another hybrid xylanase AEx11A was designed by substituting N-terminus of AoXyn11A with the corresponding region of a hyperthermostable xylanase, *Ev*Xyn11$^{TS}$ [13]. Likewise, the thermostability of the hybrid xylanase was higher than that of mesophilic parent AoXyn11A, but still significantly lower than that of thermophilic parent *Ev*Xyn11$^{TS}$ [13,14]. That is to say, all these hybrid xylanases containing the N-terminus of thermophilic xylanases displayed improved thermostabilities, but could not outperform their thermophilic parents to acquire hyperthermostabilities. Thus, a both industrially and biologically intriguing and important question arose: Are mutations in the extreme N-terminus sufficient for conferring hyperthermostability on mesophilic xylanases?

Directed evolution has been successfully applied to confer thermostability (or even hyperthermostability) on mesophilic xylanases. Using this technique, Ruller et al. [15] improved the melting temperature (Tm) of xylanase A from *Bacillus subtilis* from 59°C to 76.5°C. This mutant xylanase contained four mutations (*Q7H*, *G13R*, *S22P* and *S179C*), of which three were in the extreme N-terminus. Palackal et al. [16] also employed directed evolution and successfully obtained a thermostable variant (9X) with a high Tm (95.6°C). The 9X enzyme harbored nine mutations, of which four (*D8F*, *Q11H*, *N12L* and *G17I*) were in the extreme N-terminus. Likewise, Miyazaki et al. [17] created one thermostable variant, Xyl$^{st}$, by directed evolution. This mutant contained three amino acid substitutions, of which two were in the extreme N-terminus. In addition, Dumon et al. [14] successfully engineered a hyperthermostable GH11 xylanase, *Ev*Xyn11$^{TS}$. The Tm of *Ev*Xyn11$^{TS}$ was approximately 25°C higher than the Tm of the parent xylanase. It contained seven mutations, of which six (*S9P*, *T13F*, *N14H*, *Y18F*, *Q34L* and *S35E*) were in the extreme N-terminus. During the course of directed evolution, the importance of thermostabilizing mutations in the extreme N-terminus was highlighted. However, concomitant thermostabilizing mutations outside of the extreme N-terminus obscured the significance of the extreme N-terminal mutations in conferring thermostability. Thus, the same question remained: Would it be possible for mutations in the extreme N-terminus to confer hyperthermostability on mesophilic xylanase?

Moreover, it should be noted that several attempts have been made to obtain hyperthermostable xylanases by engineering thermophilic xylanases [18–22]. Using thermophilic xylanases, it is not difficult to obtain hyperthermostable mutants. However, for mesophilic xylanases, achieving hyperthermostability is a much bigger challenge.

In this work, by focusing on the extreme N-termini, a convenient and straightforward approach to obtain hyperthermostable xylanases was attained. Additionally, mutations in only the extreme N-terminus were confirmed to be sufficient to confer thermostability (or even hyperthermostability) on mesophilic SoxB.

## Materials and Methods

### Strains and Growth Conditions

*Escherichia coli* TOP10 and *E. coli* BL21 (DE3) were used for the cloning and heterologous expression, respectively, of xylanase genes. The *E. coli* was cultured in Luria-Bertani (LB) broth at 37°C.

*Streptomyces olivaceoviridis* (ATCC 23630) and *T. fusca* (ATCC 27730) were grown in GYM S*treptomyces* medium (0.4% glucose, 0.4% yeast extract, 1% malt extract and 0.2% $CaCO_3$) at 28°C and TYG medium (0.3% tryptone, 0.3% yeast extract, 0.3% glucose and 0.1% $K_2HPO_4$) at 50°C, respectively.

### DNA Manipulation and Plasmid Constructs

The genomic DNA of *S. olivaceoviridis* and *T. fusca* were used as polymerase chain reaction (PCR) templates to amplify the *SoxB* (GenBank accession number KF927165) and *TfxA* (GenBank accession number KF927166) genes. The region mutants of SoxB were constructed by overlap extension PCR [23]. The primers used in this study are listed in Table S1. All of the genes from the wildtype and region-mutant xylanases were digested with *Nde*I and *Eco*RI (TaKaRa) and ligated into the pET-28a (+) vector (Novagen). Each mutation was confirmed by sequencing of the entire xylanase gene.

### Expression and Purification of Recombinant Xylanases

The recombinant plasmids containing the wildtype and mutant xylanases were transformed into *E. coli* BL21 (DE3). Overnight cultures of *E. coli* BL21 (DE3) harboring the recombinant plasmids were diluted 1/100 in fresh LB containing 50 µg/ml kanamycin. The cultures were incubated at 37°C and shaken at 200 rpm until an absorbance between 0.6 and 0.8 at 600 nm was reached. The cultures were then induced with 0.5 mM isopropyl-β-D-thiogalactopyranoside (IPTG). Following 6 h of induction at 30°C, the cells were harvested by centrifugation at 6,000 rpm for 20 min at 4°C. After the pellet was resuspended in binding buffer (50 mM sodium phosphate [pH 7.4] and 500 mM sodium chloride), the cells were lysed ultrasonically. The insoluble material was removed by centrifugation at 12,000 rpm for 20 min at 4°C, and the supernatant was then applied to a Ni-NTA-agarose affinity column and purified by fast protein liquid chromatography (FPLC). The protein was washed with binding buffer containing 100 mM imidazole and eluted with 400 mM imidazole in binding buffer. The purified xylanase was pooled and dialyzed to remove the imidazole. For both the wildtype and all of the mutant xylanases, a single-step cation-exchange purification protocol yielded pure protein, as evaluated by Coomassie-stained sodium dodecyl sulphate-polyacrylamide gel electrophoresis (SDS-PAGE) [24]. The protein concentration was determined using the Pierce bicinchoninic acid (BCA) protein reagent (Thermo Scientific) and including bovine serum albumin (BSA) as the protein standard.

### Enzyme Activity Assay

Xylanase activity was determined using the dinitrosalicylic acid (DNS) method, with a few modifications [25]. The assay was performed in 50 mM sodium citrate buffer (pH 6.0), using birch wood xylan (Sigma-Aldrich) as the substrate. Initially, 40 µl of diluted enzyme was mixed with 360 µl of 10% birch wood xylan and incubated for 10 min at 50°C. Next, 600 µl of DNS reagent was added, and the solution was heated to 99°C for 15 min. The enzyme activity was determined from the increase in absorbance at 540 nm measured by a Sunrise Microplate Reader (TECAN). One international unit of enzyme activity was defined as the amount of enzyme necessary to liberate 1 µmol of reducing sugars per minute under the assay conditions. The optimum temperature for the enzyme activity was determined by a standard assay at various temperatures in 50 mM sodium citrate buffer (pH 6.0). The results were expressed as the activity relative to the value obtained from the optimum temperature assay.

### Thermal Inactivation Assay

Thermal inactivation assays were performed in 50 mM sodium citrate buffer (pH 6.0) at 70°C or 80°C in the absence of the substrate. At specific intervals, 40 µl aliquots were removed and cooled on ice. The residual activity was measured as described above for the enzyme activity measurement. The half-life curves were fitted to equation $y = A^*e^{-kt}$ and the half-life was then calculated from the thermal inactivation curve, as previously described [10].

### Differential Scanning Calorimetry (DSC) Experiments

DSC experiments were performed using a Nano DSC (TA Instruments) with sample concentrations of 1 mg/ml and a scan rate of 1°C/min, as previously described [10,26]. The instrumental baseline, with both cells filled with buffer, was also routinely recorded before experiments.

### Sequence Alignment and Homology Modelling

Sequence alignment was performed using the DNAman software package (Lynnon Biosoft). The homology model of SoxB and its mutants was developed using the software package modeller [27,28]. Three crystal structures (PDB codes: 1HIX, 1M4W and 3ZSE) were chosen as the templates. The structure visualisation was performed using Pymol [29] software.

## Results and Discussion

### Identification of Two Potential Regions in the Extreme N-terminus

The amino acid sequence of the mesophilic xylanase SoxB shares high-level identity (74%) with the sequence of the catalytic domain of the thermophilic homolog TfxA. However, this pair of xylanases exhibits only approximately 50% identity for the first 33 N-terminal residues (numbered by the SoxB sequence) (Fig. 1A and Fig. S1). As we demonstrated in our previous study [10], based on sequence alignment and structural analysis, the divergent positions of the 33 N-terminal residues were clustered in four regions (R1, R2, R3 and R4) (Fig. 1A). Accordingly, four region mutants of SoxB (M1, M2, M3 and M4) were produced by substituting the individual regional residues of SoxB with the corresponding regional residues in TfxA (Fig. 1A). Here, the thermostabilities of these four region mutants were compared with not only the thermostability of SoxB but also the thermostability of TfxA. Both DSC and thermal inactivation experiments demonstrated that the rank order of the thermostabilities was as follows: TfxA>M2> M4> SoxB>M3> M1 (Fig. 2). The importance of Region 2 for thermostability has been discussed in our previous study [10]. More specifically, we demonstrated that all five of the divergent residues in R2 of TfxA were essential and played a synergistic role in conferring thermostability. Here, the interest was focused on R4. The Tm of the region mutant M4 was 6.7°C higher than the Tm of SoxB (Fig. 2A). Moreover, at 70°C, the M4 displayed a nine-fold higher thermostability than SoxB (Table 1). Thus, the residues in R4 of TfxA were also important for conferring thermostability. However, it should be noticed that all the four region mutants displayed significantly lower thermostabilities than thermophilic TfxA. Even the M2 and M4 mutants exhibited four-fold and 70-fold lower thermostability at 80°C than thermophilic TfxA, respectively (Fig. 2B).

In summary, R2 and R4 of TfxA were confirmed to be two potential regions for conferring thermostability. Two region mutants, M2 and M4 mutants displayed higher thermostability than their mesophilic parent SoxB. However, compared with their thermophilic parent TfxA, all the region mutants displayed inferior thermostability.

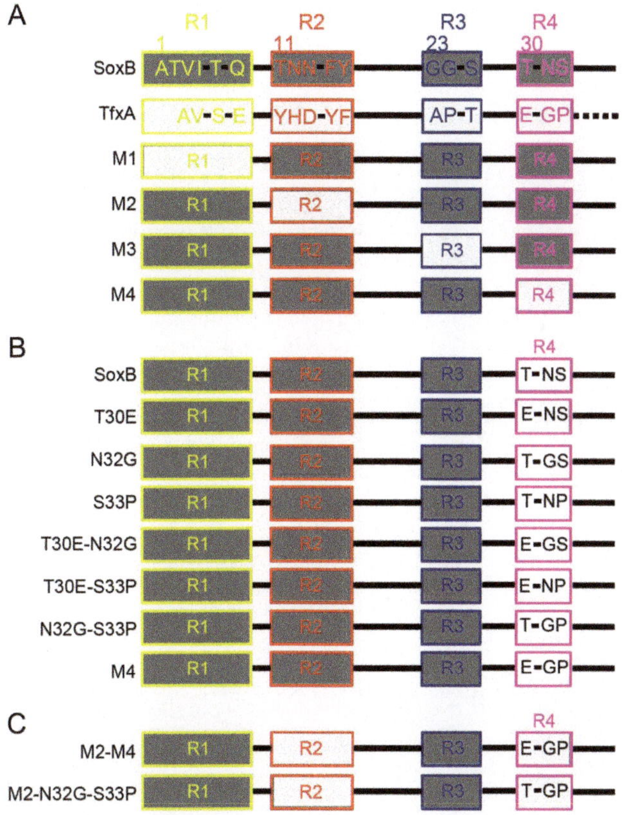


**Figure 1. Schematic representation of widetype xylanases and mutants.** (A) Region mutants were produced according to the divergent regions between the N-termini of SoxB and TfxA. (B) Focusing on R4, three mutants (T30E, N32G and S33P), each containing a single mutation and three double mutants (T30E-N32G, T30E-S33P and N32G-S33P) were created. (C) Two double region mutants were produced by introducing two potential regions into SoxB.

**Table 1.** Systematic analysis and comparison of the thermostabilities of mutants in R4.

| Xylanases | Mutations | $t_{1/2}$ at 70°C (min) | Tm (°C) | ΔTm |
|---|---|---|---|---|
| SoxB | – | 3.6±0.1 | 69.1 | – |
| M4 | *T30E, N32G* and *S33P* | 33±1.6 | 75.8 | 6.7 |
| T30E | *T30E* | 1.8±0.1 | 65.8 | -3.3 |
| N32G | *N32G* | 17±0.2 | 74.9 | 5.8 |
| S33P | *S33P* | 2.2±0.1 | 67.0 | -2.1 |
| T30E-N32G | *T30E* and *N32G* | 10±0.1 | 72.2 | 3.1 |
| N32G-S33P | *N32G* and *S33P* | 87±4.5 | 78.2 | 9.1 |
| T30E-S33P | *T30E* and *S33P* | 1.4±0.1 | 64.3 | -4.8 |

## The Effect of Each Mutation in Region 4 on Thermostability

As shown in Fig. 1A, three residues that diverged between SoxB and TfxA were found in R4. Therefore, three mutants of SoxB (T30E, N32G and S33P), each containing a single mutation, were generated (Fig. 1B). To our surprise, two of the three mutants, T30E and S33P, decreased the thermostability of SoxB rather than increasing this parameter. Only the N32G mutant displayed a higher thermostability than SoxB (Fig. 3A and Fig. 3B). Thus, the *N32G* mutation was of vital importance in conferring thermostability. However, compared with the M4 mutant, the N32G mutant still exhibited a lower thermostability (Fig. 3A and Fig. 3B), which suggested a potential synergistic effect between the three mutations (*N32G*, *T30E* and *S33P*).

## *N32G* and *S33P* in Region 4 Displayed Synergistic Effects

To discern the potential synergistic effect between the *N32G*, *T30E* and *S33P* mutations, three double mutants (T30E-N32G, T30E-S33P and N32G-S33P) were produced and their thermostabilities were analyzed (Fig. 1B).

First, the two double mutants harboring the *N32G* mutation, T30E-N32G and N32G-S33P, displayed a higher thermostability than the T30E and S33P mutants, respectively (Fig. 3A and Fig. 3B). This finding once again strongly indicated the importance of the *N32G* mutation in conferring thermostability. More importantly, the N32G-S33P mutant exhibited an approximately 40-fold and five-fold higher thermostability than S33P and N32G, respectively. And the Tm of the N32G-S33P mutant was also 11.2°C and 3.3°C higher than the Tm values of S33P and N32G, respectively (Table 1). Furthermore, to our surprise, the N32G-S33P mutant displayed even higher thermostability than M4 (Fig. 3A, Fig. 3B and Table 1). All these results clearly revealed a synergistic effect between N32G and S33P.

In contrast, the two double mutants harboring the *T30E* mutation, T30E-N32G and T30E-S33P, exhibited a lower thermostability than N32G and S33P, respectively (Fig. 3A and Fig. 3B and Table 1). This result agreed well with the negative effect of single mutation *T30E* on thermostability and demonstrated that the *T30E* mutation was unfavourable for the thermostability of SoxB.

As for the two double mutants containing the *S33P* mutation, the N32G-S33P mutant displayed a higher thermostability than the N32G mutant, whereas T30E-S33P exhibited a lower thermostability than the T30E mutant (Fig. 3A, Fig. 3B and Table 1). This finding appeared to be paradoxical at first but

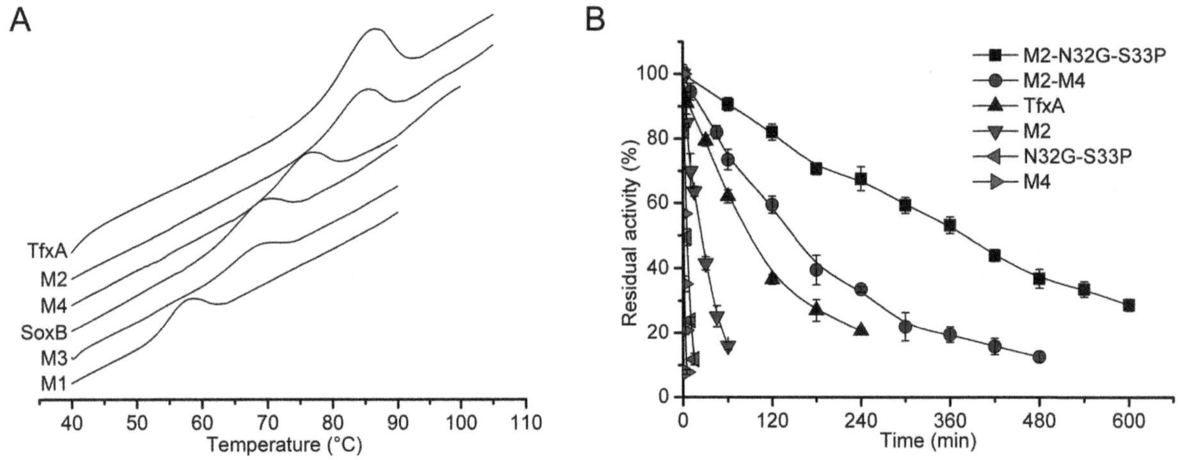


**Figure 2. Comparison of the thermostabilities of SoxB, TfxA and region mutants.** (A) The DSC profiles. (B) The thermal inactivation profiles at 80°C.

provided important information: the *S33P* was a thermostabilizing mutation only when combined with the *N32G* mutation, which further demonstrated the synergistic effect between *N32G* and *S33P*. Dumon [14] employed gene site saturation mutagenesis$^{TM}$ (GSSM) to discover all of the possible individual thermostabilizing mutations in *Ev*Xyn11$^{TS}$. Our result explains why the corresponding mutation of *S33P* was not detected as a thermostabilizing mutation in *Ev*Xyn11$^{TS}$, although the corresponding mutation of *N32G* (N30V) was discovered to be thermostabilizing.

To understand the structural basis of the synergistic effect between *N32G* and *S33P*, a structural model for N32G-S33P mutant was generated and analyzed (Fig. 4). The pyrrolidine ring restricts proline to fewer conformations than are available to the other amino acids [30,31]. Thus, proline is believed to strengthen the rigidity of proteins [32–34]. When the single *S33P* mutation was introduced into the sequence of SoxB, the Asn32 residue, which is immediately adjacent to Pro33, may have restrained the local configuration of Asn32 and Pro33. Thus, the single *S33P* mutation may have impaired the local configuration of SoxB and hence decreased the thermostability of SoxB. Unlike Asn32, glycine lacks a β-carbon and has more backbone conformational flexibility. Therefore, in N32G-S33P mutant, the *N32G* mutation would help Pro33 to form the desirable configuration, which would strengthen the rigidity of R4 of SoxB and improve the thermostability (Fig. 4A).

## Two Potential Regions in Combination Conferred Hyperthermostability on SoxB

To test whether the thermostability of SoxB was further enhanced by introducing mutations in both R2 and R4 simultaneously, two double region mutants, M2–M4 and M2–N32G-S33P, were produced (Fig. 1C). The M2–M4 mutant contained five mutations (*T11Y*, *N12H*, *N13D*, *F15Y* and *Y16F*) in Region 2 and three mutations (*T30E*, *N32G* and *S33P*) in Region 4, and M2-N32G-S33P harbored five mutations (*T11Y*, *N12H*, *N13D*, *F15Y* and *Y16F*) in Region 2 and the two mutations (*N32G* and *S33P*) in Region 4. Both the M2-M4 and M2-N32G-S33P mutants had a significantly higher thermostability than M2, M4 and N332G-S33P, suggesting that the thermostability of SoxB was significantly further enhanced. Due to the negative effect of the *T30E* mutation on thermostability, it was not surprising that the M2-M4 mutant displayed a lower thermostability than M2-N32G-S33P (Fig. 2B, Fig. 5A and Table 2).

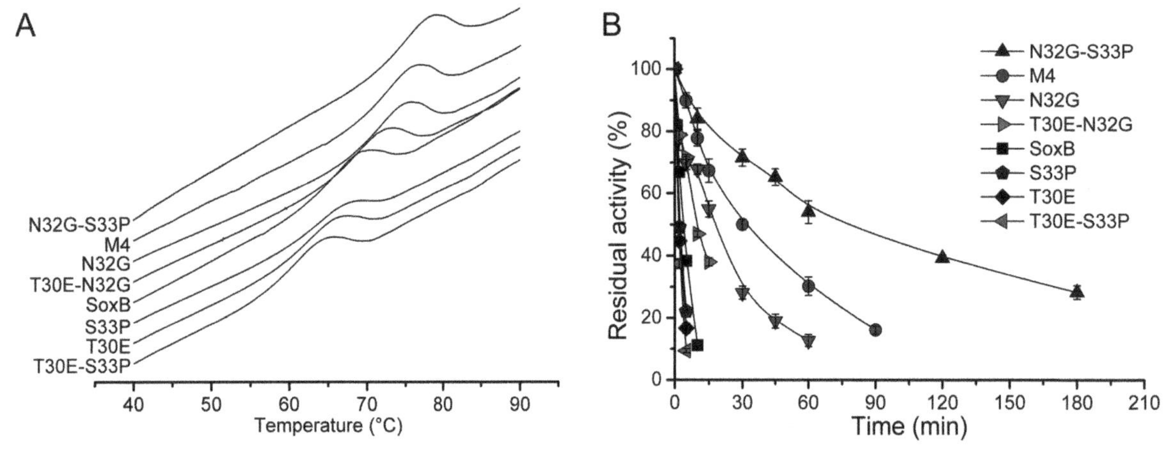


**Figure 3. Analysis of the effect of mutations in R4 on thermostability.** (A) DSC profiles. (B) The thermal inactivation profiles at 70°C for the mutants.

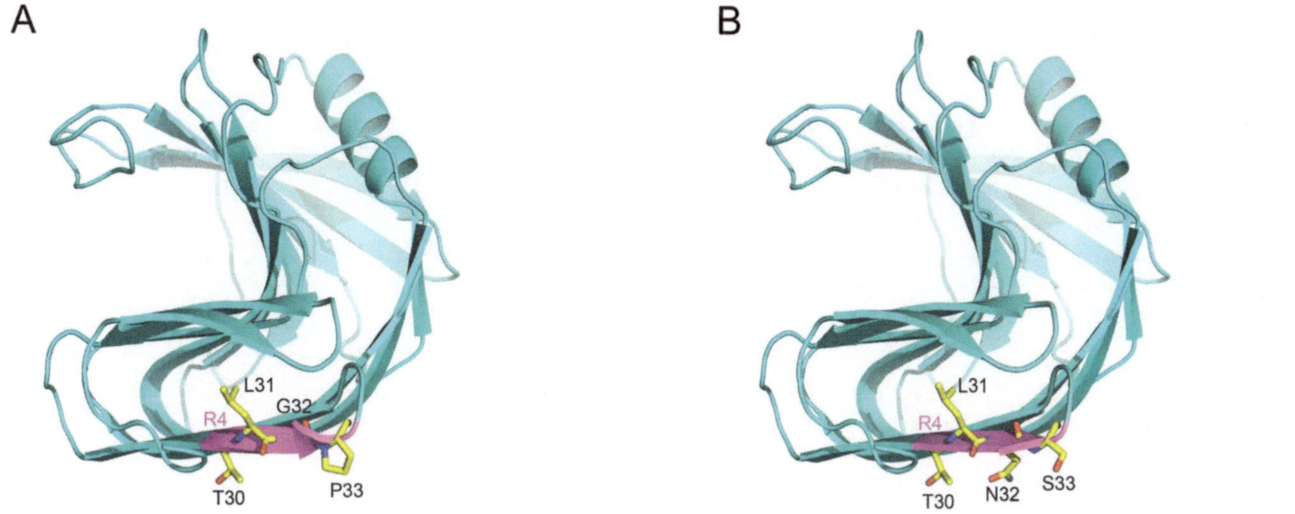


**Figure 4. Homology modelling and structure comparison of the N32G-S33P mutant (A) and wild type SoxB (B).** The Region 4 was in purple, and the residues in R4 were in yellow.

The M2-N32G-S33P mutant exhibited the highest thermostability among all of the mutants derived from SoxB. The Tm of M2-N32G-S33P was 94.7°C, which was 25.6°C higher than the Tm of SoxB. Furthermore, the M2-N32G-S33P mutant had an increased half-life at 80°C, which indicated 71-fold and 15-fold more stability than the half-life of N32G-S33P and M2, respectively (Fig. 2B, Fig. 5A and Table 2). In contrast, the SoxB lost its activity at 80°C so rapidly that the half-life could not be determined. Furthermore, the optimum temperature of M2-N32G-S33P was 80°C, which was 15°C higher than the optimum temperature of wildtype SoxB, and the specific activity of M2-N32G-S33P at this temperature (80°C) was 1.8-fold higher than the activity of wildtype SoxB at its optimum temperature (65°C) (Fig. 5B and Table 2). Thus, the seven mutations in the extreme N-terminus, which is located far from the catalytic centre of SoxB, did not compromise, and in fact improved, enzyme activity.

Because the seven or eight residues substituted in M2-N32G-S33P and M2-M4 were totally derived from TfxA, the thermostabilities of M2-N32G-S33P and M2-M4 were compared with the thermostability of TfxA. Intriguingly, both the M2-M4 and the M2-N32G-S33P mutants yielded an even higher thermostability than TfxA. In particular, M2-N32G-S33P was three-fold more stable than TfxA, and the Tm of M2-N32G-S33P was 9°C higher than the Tm of TfxA (Fig. 2B, Fig. 5A and Table 2). Thus, the mesophilic xylanase "pupil" outperformed its thermophilic "master" to acquire hyperthermostability simply by having seven mutations introduced into the extreme N-terminus. The specific activities of TfxA and M2-N32G-S33P were measured at 80°C. The M2-N32G-S33P had a quite comparable specific activity with that of TfxA (Fig. 5B and Table 2). The temperatures used for hydrolysis of heterologous xylans by xylanases may vary in different fields including food, animal feed, biofuel or paper production. In the field of biofuel, currently most plant cell wall polysaccharides-degrading enzymes including xylanases are derived from the mesophilic filamentous fungus *Trichoderma reesei*. Normally, these enzymes are used at around 50°C. It is expected that higher temperatures will be more beneficial at least for decreasing the viscosity of culture fluids, minimizing the possibility of microbial contamination, and facilitating the distillation of bioethanol. Therefore, there are tremendous efforts in discovering

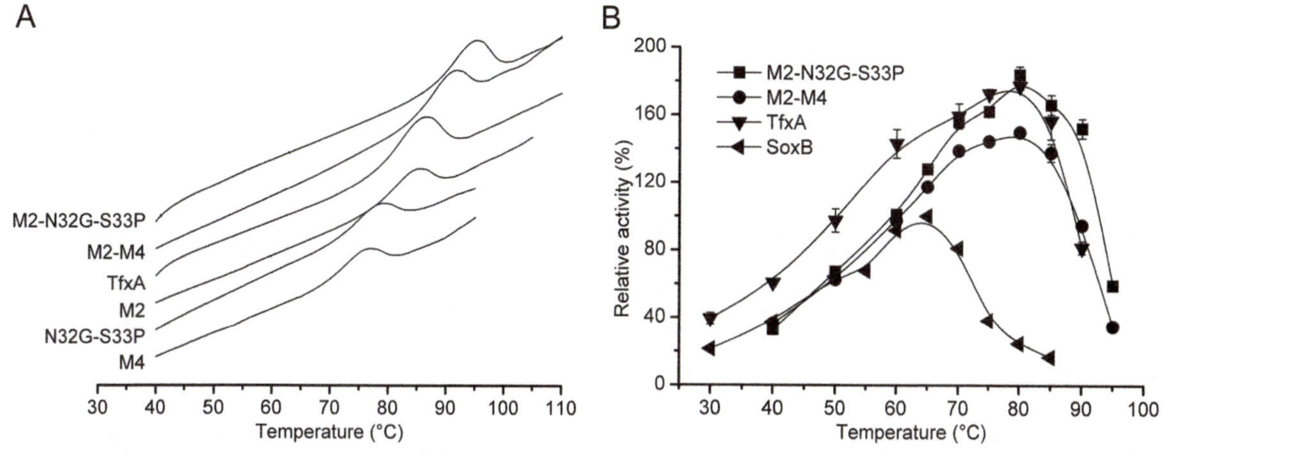


**Figure 5. Comparison of thermostabilities of double region mutants, M2-M4 and M2-N32G-S33P.** (A) DSC profiles. (B) The optimum temperature profiles.

**Table 2.** Comparsion of the thermostabilities of double region mutants with wildtype parents, SoxB and TfxA.

| Xylanases | $t_{1/2}$ at 80°C (min) | Tm (°C) | ΔTm | Optimum temperature (OT) (°C) | Specific activity at OT (IU/mg) |
|---|---|---|---|---|---|
| SoxB | –[a] | 69.1 | - | 65 | 801 |
| TfxA | 94±6 | 85.7 | 16.6 | 80 | 1419 |
| M2-M4 | 151±6 | 90.9 | 21.8 | 80 | 1200 |
| M2-N32G-S33P | 354±21 | 94.7 | 25.6 | 80 | 1472 |

[a]The xylanase lost all of its enzyme activity within only 1 min so that the $t_{1/2}$ could not be determined.

novel glycoside hydrolases [35] or engineering existing ones for better tolerance of higher temperatures [36]. This is also one of the purposes of current study. However, an industrial standard has not been settled yet for the reaction temperatures of all these thermophilic enzymes published previously. Here, the specific activities of TfxA and M2-N32G-S33P were also compared at a decreased temperature of 70°C, which is 10°C lower than the optimal temperatures of the M2-N32G-S33P and the TfxA and near the boiling temperature of the ethanol. The result showed at 70°C, the specific activity of M2-N32G-S33P (1249 IU/mg) was also comparable to that of TfxA (1276 IU/mg) (Fig. 5B). Our experiment was a proof-of-concept that the N-terminus, particularly the key residues in Region 2 and 4, of TfxA, can be used, but not limited, to improve the thermostability of the xylanase SoxB.

Many previous studies have been suggested that it was the combined effect of many mutations and certain structural features that conferred stability on proteins in high temperature [37]. Here, seven mutations in the extreme N-terminus in combination conferred hyperthermostability on mesophilic xylanase SoxB. In the Region 2, all five mutations were involved in conferring the thermostability on the SoxB. The five mutations seemed to confer the thermostability on SoxB through both additive (the thermostabilizing effect of the individual mutations T11Y, N12H and Y16F) and cooperative (the H12-D13 electrostatic interaction and Y11-F16 hydrophobic interaction) effects [10]. However, in the Region 4, not all but two mutations, N32G and S33P, cooperated to strengthen the structural stability of SoxB, thus improving the thermostability of this xylanase. On the other hand, the M2-N32G-S33P did not exhibit significant difference with SoxB regarding the interaction between the two thermostabilizing regions. Thus, it was likely that the two thermostabilizing regions conferred hyperthermostability on SoxB by additive rather than cooperative effect. Taken together, five beneficial mutations in Region 2 and two beneficial mutations in Region 4 strengthened the structural stability on Region 2 and Region 4, respectively. Then, these two thermostabilizing regions in combination strengthen the overall structure of N-terminus and improved the thermostability of SoxB.

## Conclusions

This work provided a convincing example that mutations in the extreme N-terminus could confer thermostability (or even hyperthermostability) on a mesophilic xylanase. Moreover, this study further clarified the potential mechanism of thermostabilizing effect of the N-terminus of TfxA. It was demonstrated that not all divergent regions but two thermostabilizing regions were essential and played a dominant role in conferring the thermostability. In addition, a series of xylanases with a wide range of thermostabilities were obtained in this work and would provide flexible options for the deconstruction of lignocellulosic biomass.

## Supporting Information

**Figure S1 Sequence alignment of TfxA, SoxB and the mutant M2-N32G-S33P.** The divergent regions in N-terminus were highlighted by square frames and seven substitutions of M2-N32G-S33P were marked by stars.

## Acknowledgments

We would like to thank Kai Zhang from the Institute of Biophysics, Chinese Academy of Sciences for technical support in homology modelling.

## Author Contributions

Conceived and designed the experiments: SZ ZD. Performed the experiments: SZ YH. Analyzed the data: SZ YH HY. Contributed reagents/materials/analysis tools: SZ. Wrote the paper: SZ ZD.

## References

1. Song L, Siguier B, Dumon C, Bozonnet S, O'Donohue MJ (2012) Engineering better biomass-degrading ability into a GH11 xylanase using a directed evolution strategy. Biotechnol Biofuels 5: 3.
2. Dodd D, Cann IK (2009) Enzymatic deconstruction of xylan for biofuel production. GCB Bioenergy 1: 2–17.
3. Penttilä PA, Várnai A, Pere J, Tammelin T, Salmén L, et al. (2013) Xylan as limiting factor in enzymatic hydrolysis of nanocellulose. Bioresour Technol 129: 135–141.
4. Hu J, Arantes V, Pribowo A, Saddler JN (2013) The synergistic action of accessory enzymes enhances the hydrolytic potential of a "cellulase mixture" but is highly substrate specific. Biotechnol Biofuels 6: 112.
5. Shen B, Sun X, Zuo X, Shilling T, Apgar J, et al. (2012) Engineering a thermoregulated intein-modified xylanase into maize for consolidated lignocellulosic biomass processing. Nat Biotechnol 30: 1131–1136.
6. Morais S, Barak Y, Caspi J, Hadar Y, Lamed R, et al. (2010) Cellulase-xylanase synergy in designer cellulosomes for enhanced degradation of a complex cellulosic substrate. MBio 1: 1–8.
7. Yeoman CJ, Han Y, Dodd D, Schroeder CM, Mackie RI, et al. (2010) Thermostable Enzymes as Biocatalysts in the Biofuel Industry. Adv Appl Microbiol 70: 1–55.
8. Viikari L, Alapuranen M, Puranen T, Vehmaanper J, Siika-Aho M (2007) Thermostable enzymes in lignocellulose hydrolysis. Adv Biochem Eng/ Biotechnol 108: 121–145.
9. Shibuya H, Kaneko S, Hayashi K (2000) Enhancement of the thermostability and hydrolytic activity of xylanase by random gene shuffling. Biochem J 349: 651–656.
10. Zhang S, Zhang K, Chen X, Chu X, Sun F, et al. (2010) Five mutations in N-terminus confer thermostability on mesophilic xylanase. Biochem Biophys Res Commun 395: 200–206.

11. Weng XY, Sun JY (2005) Construction, expression, and characterization of a thermostable xylanase. Curr Microbiol 51: 188–192.
12. Sun J-Y, Liu M-Q, Xu Y-L, Xu Z-R, Pan L, et al. (2005) Improvement of the thermostability and catalytic activity of a mesophilic family 11 xylanase by N-terminus replacement. Protein Expression Purif 42: 122–130.
13. Gao SJ, Wang JQ, Wu MC, Zhang HM, Yin X, et al. (2013) Engineering hyperthermostability into a mesophilic family 11 xylanase from *Aspergillus oryzae* by in silico design of N-terminus substitution. Biotechnol Bioeng 110: 1028–1038.
14. Dumon C, Varvak A, Wall MA, Flint JE, Lewis RJ, et al. (2008) Engineering hyperthermostability into a GH11 xylanase is mediated by subtle changes to protein structure. J Biol Chem 283: 22557–22564.
15. Ruller R, Deliberto L, Ferreira TL, Ward RJ (2008) Thermostable variants of the recombinant xylanase A from *Bacillus subtilis* produced by directed evolution show reduced heat capacity changes. Proteins: Structure, Function & Bioinformatics 70: 1280–1293.
16. Palackal N, Brennan Y, Callen WN, Dupree P, Frey G, et al. (2004) An evolutionary route to xylanase process fitness. Protein Sci 13: 494–503.
17. Miyazaki K, Takenouchi M, Kondo H, Noro N, Suzuki M, et al. (2006) Thermal stabilization of *Bacillus subtilis* family-11 xylanase by directed evolution. J Biol Chem 281: 10236–10242.
18. Wang Y, Fu Z, Huang H, Zhang H, Yao B, et al. (2012) Improved thermal performance of *Thermomyces lanuginosus* GH11 xylanase by engineering of an N-terminal disulfide bridge. Bioresour Technol 112: 275–279.
19. Zhang ZG, Yi ZL, Pei XQ, Wu ZL (2010) Improving the thermostability of *Geobacillus stearothermophilus* xylanase XT6 by directed evolution and site-directed mutagenesis. Bioresour Technol 101: 9272–9278.
20. Stephens DE, Rumbold K, Permaul K, Prior BA, Singh S (2007) Directed evolution of the thermostable xylanase from *Thermomyces lanuginosus*. J Biotechnol 127: 348–354.
21. Jeong MY, Kim S, Yun CW, Choi YJ, Cho SG (2007) Engineering a de novo internal disulfide bridge to improve the thermal stability of xylanase from *Bacillus stearothermophilus* No. 236. J Biotechnol 127: 300–309.
22. Paes G, O'Donohue MJ (2006) Engineering increased thermostability in the thermostable GH-11 xylanase from *Thermobacillus xylanilyticus*. J Biotechnol 125: 338–350.
23. Horton RM, Hunt HD, Ho SN, Pullen JK, Pease LR (1989) Engineering hybrid genes without the use of restriction enzymes: gene splicing by overlap extension. Gene 77: 61–68.
24. Laemmli UK (1970) Cleavage of structural proteins during the assembly of the head of bacteriophage T4. Nature 227: 680–685.
25. Bailey MJ, Biely P, Poutanen K (1992) Interlaboratory testing of methods for assay of xylanase activity. J Biotechnol 23: 257–270.
26. Sanchez-Ruiz JM (1992) Theoretical analysis of Lumry-Eyring models in differential scanning calorimetry. Biophys J 61: 921–935.
27. Eswar N, Webb B, Marti-Renom MA, Madhusudhan MS, Eramian D, et al. (2007) Comparative protein structure modeling using Modeller. Curr Protocols Protein Sci Chapter 2, Unit 29.
28. Sali A, Blundell TL (1993) Comparative protein modelling by satisfaction of spatial restraints. J Mol Biol 234: 779–815.
29. DeLano WL (2002) The PyMOL Molecular Graphics System The PyMOL Molecular Graphics System, DeLano Scientific, San Carlos, CA.
30. Suzuki Y, Hatagaki K, Oda H (1991) A hyperthermostable pullulanase produced by an extreme thermophile, *Bacillus flavocaldarius* KP 1228, and evidence for the proline theory of increasing protein thermostability. Appl Microbiol Biotechnol 34: 707–714.
31. Watanabe K, Masuda T, Ohashi H, Mihara H, Suzuki Y (1994) Multiple proline substitutions cumulatively thermostabilize *Bacillus Cereus* ATCC7064 oligo-1, 6-glucosidase. Eur J Biochem 226: 277–283.
32. Arnorsdottir J, Sigtryggsdottir AR, Thorbjarnardottir SH, Kristjansson MM (2009) Effect of proline substitutions on stability and kinetic properties of a cold adapted subtilase. J Biochem 145: 325–329.
33. Goihberg E, Dym O, Tel-Or S, Levin I, Peretz M, et al. (2007) A single proline substitution is critical for the thermostabilization of *Clostridium beijerinckii* alcohol dehydrogenase. Proteins 66: 196–204.
34. Tian J, Wang P, Gao S, Chu X, Wu N, et al. (2010) Enhanced thermostability of methyl parathion hydrolase from *Ochrobactrum* sp. M231 by rational engineering of a glycine to proline mutation. FEBS J 277: 4901–4908.
35. Su X, Han Y, Dodd D, Moon YH, Yoshida S, et al. (2013) Reconstitution of a thermostable xylan-degrading enzyme mixture from the bacterium *Caldicellulosiruptor bescii*. Appl Environ Microbiol 79: 1481–1490.
36. Dana CM, Saija P, Kal SM, Bryan MB, Blanch HW, et al. (2012) Biased clique shuffling reveals stabilizing mutations in cellulase Cel7A. Biotechnol Bioeng 109: 2710–2719.
37. Russell RJ, Taylor GL (1995) Engineering thermostability: lessons from thermophilic proteins. Curr Opin Biotechnol 6: 370–374.

# Circular Permutation in the Ω-Loop of TEM-1 β-Lactamase Results in Improved Activity and Altered Substrate Specificity

**Gurkan Guntas**[¤], **Manu Kanwar**, **Marc Ostermeier***

Department of Chemical and Biomolecular Engineering, Johns Hopkins University, Baltimore, Maryland, United States of America

## Abstract

Generating diverse protein libraries that contain improved variants at a sufficiently high frequency is critical for improving the properties of proteins using directed evolution. Many studies have illustrated how random mutagenesis, cassette mutagenesis, DNA shuffling and similar approaches are effective diversity generating methods for directed evolution. Very few studies have explored random circular permutation, the intramolecular relocation of the N- and C-termini of a protein, as a diversity-generating step for directed evolution. We subjected a library of random circular permutations of TEM-1 β-lactamase to selections on increasing concentrations of a variety of β-lactam antibiotics including cefotaxime. We identified two circularly permuted variants that conferred elevated resistance to cefotaxime but decreased resistance to other antibiotics. These variants were circularly permuted in the Ω-loop proximal to the active site. Remarkably, one variant was circularly permuted such that the key catalytic residue Glu166 was located at the N-terminus of the mature protein.

**Editor:** Mark Isalan, Center for Genomic Regulation, Spain

**Funding:** This work was supported by funding from the National Institute of General Medicine at the National Institutes of Health (R01 GM066972). The funder had no role in study design, data collection and analysis, decision to publish, or preparation of the manuscript.

**Competing Interests:** The authors have declared that no competing interests exist.

* E-mail: oster@jhu.edu

¤ Current address: Department of Biochemistry and Biophysics, University of North Carolina, Chapel Hill, North Carolina, United States of America

## Introduction

Directed evolution is a powerful technique used to improve protein properties. Directed evolution involves the generation of a protein library and subsequent rounds of selection or screening to identify improved protein variants. Generation of diverse protein libraries containing improved variants is therefore important for successful application of directed evolution techniques. Many studies have demonstrated the use of random mutagenesis, cassette mutagenesis, DNA shuffling and similar approaches as effective diversity generating methods for directed evolution. Circular permutation is the intramolecular relocation of the N- and C-termini of a protein [1]. Circular permutation is an atypical method for diversity generation, since it can be used to change the linear order of the primary sequence, but not the identity of the amino acids in the sequence. Circular permutation can alter the folding kinetics of a protein and create variants that could retain comparable wild-type functionality [2,3,4,5].

Very few directed evolution studies have used circular permutation as the diversity-generating step. However, there is growing evidence to suggest that the approach has merit for certain applications, including improving enzyme activity and altering substrate specificity. Although the goal was not to improve enzyme activity, a few circular permuted 5-aminolevulinate synthases were found to have higher catalytic efficiencies than the wild-type enzyme [6]. More recently, random circular permutation has been successfully applied as the diversity generation step for the directed evolution of *Candida antarctica* lipase B [7,8] and *Bacillus circulans* xylanase Bcx [9]. Circular permutation at select sites in lipase B was found to increase hydrolytic activity up to 175-fold on certain substrates [8,10]. Such improved variants often have the relocated termini proximal to the active site [7,9], which may alter activity through local, subtle conformational changes and increased backbone flexibility [11]. Circular permutation has also found use in directed evolution for creating diversity in the fusion geometry between two proteins in order to create protein switches [12,13].

TEM-1 β-lactamase (BLA) is a class A periplasmic bacterial enzyme that can hydrolyze a wide range of penicillin and cephalosporin antibiotics by a mechanism that involves the acylation and deacylation of a serine residue [14]. BLA cleaves the four atom β-lactam ring and deactivates the antibacterial properties of the antibiotic. This renders the antibiotics ineffective against gram-negative bacteria like *E. coli.* BLA's Ω-loop is considered to play an important role in the catalytic activity of BLA against β-lactam antibiotics [15,16,17]. The relative proximity of BLA's N- and C-termini makes the enzyme a good candidate for circular permutation; however, rationally designed circular permutants were found to be poor catalysts compared to the wild type protein [18].

We previously reported the creation of a hybrid of maltose binding protein (MBP) and BLA in which BLA was circularly permuted in the Ω-loop and inserted into MBP [19]. TEM-1 β-lactamase has poor activity against third-generation cephalosporin β-lactams such as cefotaxime (CFTX) in part due to steric reasons; however, many substitutions within and outside the Ω-loop are known to increase CFTX resistance and some of which are believed to act by repositioning the Ω-loop and enlarging the active site cavity [17]. The MBP-BLA hybrid gene conferred wild-type resistance to CFTX but reduced resistance to other β-lactam antibiotics. We speculated that the relatively high resistance to CFTX was a result of the circular permutation of BLA in the Ω-loop, since a change in substrate specificity due to mutations and insertions in or near the Ω-loop has been observed [15,16,17]. This result led us to wonder how effective circular permutation would be for altering the catalytic activities of BLA. We present here the results of a study to address this question.

## Results and Discussion

### Circular permutation of BLA and insertion into MBP alters BLA substrate specificity

We have previously used random circular permutation as a diversity-generating tool for the creation of protein switches. An iterative process involving random circular permutations of *bla* and random insertions of *bla* into the gene for maltose binding protein yielded hybrid proteins in which BLA enzyme activity was modulated by maltose [19]. One particular switch, MBP317-347, was identified from Library 7 in which random circular permutations of *bla* were inserted in place of the codon 317 in the gene for MBP. MBP317-347 has severely compromised catalytic activity in the absence of maltose that increases several hundred-fold in the presence of maltose. However, the catalytic activity in the presence of maltose was 40-fold less than wild-type BLA. Consistent with this, *E. coli* cells expressing MBP317-347 had a lower minimum inhibitory concentrations (MIC) for a variety of penicillins and cephalosporins than did cells expressing BLA (about 8–16 fold lower; Table 1). However, the MIC for cefotaxime for cells expressing either MBP317-347 or BLA were the same, suggesting that MBP317-347's substrate specificity was different than BLA and potentially, that MBP317-347 had higher activity than BLA for hydrolyzing cefotaxime.

### Random circular permutation of BLA

We wondered if the altered catalytic activity of the BLA domain of MBP317-347 might be a result of the circular permutation. The BLA domain in MBP317-347 is circular permuted within the Ω-loop that contains the key active site residue Glu166. Alternatively, the altered specificity could be a result of perturbations resulting from insertion of the BLA domain into MBP. We also wondered if circular permutation might be a route for improving enzyme activity in general. Systematic circular permutation and kinetic characterization of dihydrofolate reductase had earlier identified a few circular permutants with modest improvements in catalytic activity [5]. However, at the time we began these experiments it was unknown to what extent random circular permutation employed as a diversity-generating step in directed evolution would result in gain of function mutants, since such an experiment had not been reported. We thus examined whether random circular permutation of *bla* would result in enzymes with improved catalytic activity. Qian and Lutz have since demonstrated how random circular permutation employed for the directed evolution of *Candida antartica* lipase B can produce substantial improvements (up to 175-fold) in catalytic activity on some substrates [7,8].

**Table 1.** MIC of beta-lactam antibiotics conferred by BLA and variants of BLA.

| MIC for indicated antibiotic (μg/ml) | | | | |
|---|---|---|---|---|
| **Protein** | **Cefotaxime** | **Ampicillin** | **Cefazolin** | **Cephalothin** |
| **MBP[a] (Control)** | 0.02 | 2 | 1 | 4 |
| **MBP317-347[a]** | 0.04 | 512 | 2 | 8 |
| **BLA[b]** | 0.04 | 8192 | 16 | 64 |
| **CFX011[b]** | 0.18 | ≤2048 | ≤8 | ≤32 |
| **CFX019[b]** | 0.16 | ≤2048 | ≤8 | ≤32 |

[a]at 37°C.
[b]at 22°C.

### Construction and characterization of the library of circularly permuted BLA

The library of random circular permutations of *bla* contained in Library 7 [19] was amplified using primers designed to anneal to the MBP gene on either side of *bla*. The primers had BsgI sites such that digestion of the PCR product with BsgI followed by degradation of the 3′ overhangs by Klenow would result in a blunt ended product with no DNA originating from the MBP gene (Figure 1A). The original N- and C-termini of BLA are joined by a DKS linker in this library. The DKS linker has been previously identified as a beneficial tri-peptide linker for BLA circularly permuted at residue 216 [18]. The circularly permuted *bla* gene library was inserted into a vector such that it was fused to a sequence coding for the natural 23 amino acid *bla* signal sequence on the 5′ end and a series of three stop codons in all reading frames at the 3′ end (Figure 1B). The signal sequence is necessary to export BLA to the periplasm and to confer significant resistance to β-lactam containing antibiotics.

The number of transformants in the library that had a single insert was $5\times10^5$. This greatly exceeds the number of possible "perfect" circular permutations of the gene (798). However, the process of random circular permutation results in tandem duplications and deletions at the site of circular permutation in addition to perfect circular permutations. In addition, the insert can be either inserted in the correct orientation, or in the reverse orientation. If one estimates that at each position there could also be deletions and duplications of up to 50 bp each, the theoretical degeneracy of the library is $798\times(50+50)\times2 = 1.596\times10^5$. Assuming all variants are equally probable, the probability that our library contains the most active member of these possible variants is 87% [20].

Sequencing of 25 random members of the naive library indicated no obvious bias in the site of circular permutation (Figure 2A). Next, functional library members were selected by plating the naïve library on LB agar plates supplemented with 16 μg/ml ampicillin (this concentration is 8-fold higher than the MIC for cells not expressing any β-lactamase) or with 250 μg/ml ampicillin. Functional variants from each plating condition were chosen at random and sequenced (Figure 2B&C). Active variants identified from the 250 μg/ml ampicillin plates were heavily biased towards circular permutations very near the original N- and C-termini of the protein (Figure 2C). The site of circular permutation was more varied for those clones selected at the lower ampicillin concentration (Figure 2B).

### Selection of BLA variants that conferred improved antibiotic resistance

The above selection was for minimally functional β-lactamase activity. Selections were subsequently performed on the naïve library to identify library members that exhibited antibiotic resistance above that conferred by *bla*. The naïve library was

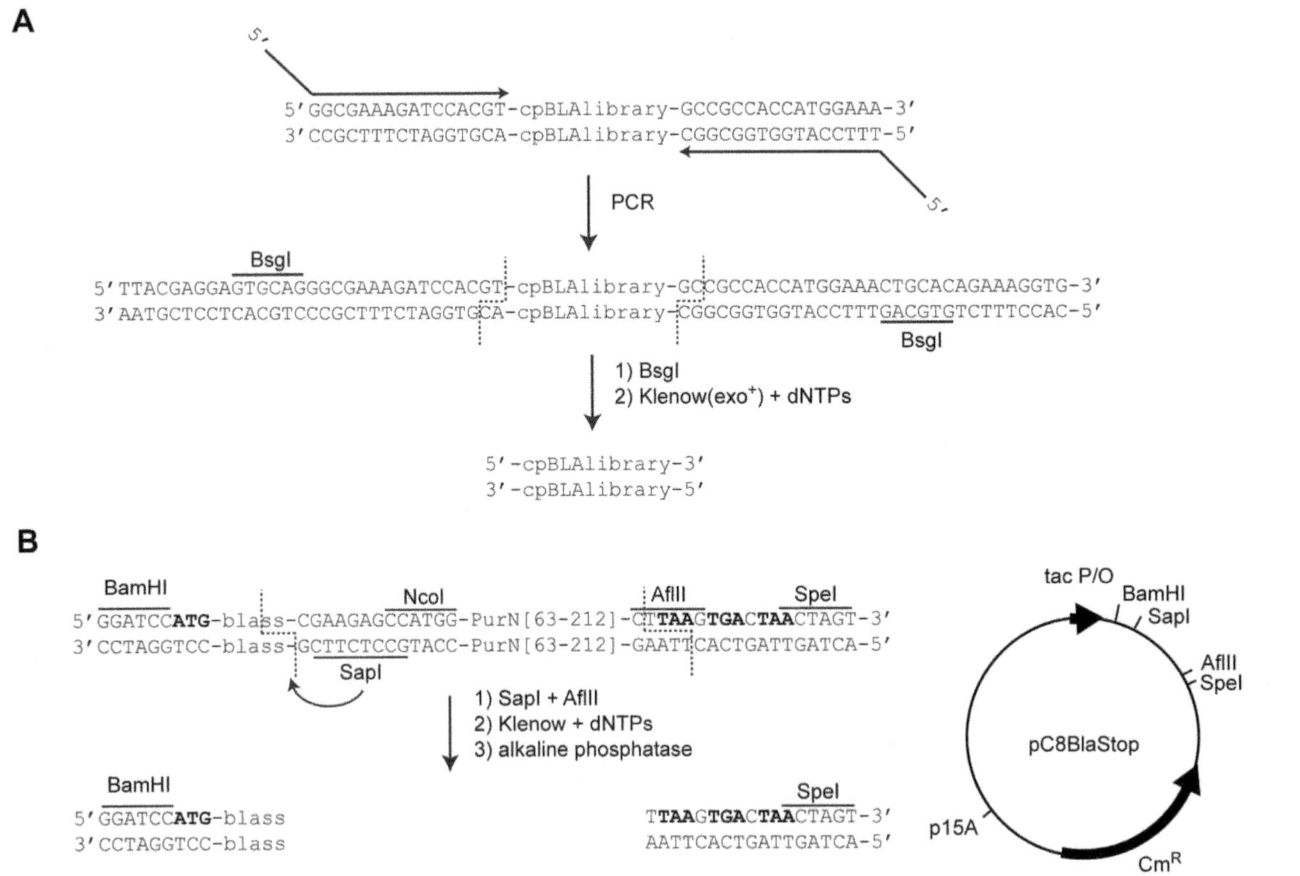


**Figure 1. Library construction.** (A) The previously described collection of circularly permuted *bla* genes [19] was PCR-amplified using primers designed to anneal just outside the circularly permuted *bla* DNA. Both primers contained an appropriately spaced BsgI restriction site (cuts and the indicated dashed lines) such that treatment of the digested product to remove the two-base 3′ overhand would result in the circularly permuted *bla* library without any "scars" from the surrounding DNA. (B) Plasmid pC8BlaStop is derived from pDIM-C8 [23] and contains appropriately place SapI and AflII sites such that fusion of the circularly permuted *bla* library can occur seamlessly to the *bla* signal sequence (blass) and a series of stop codons in all three reading frames (in bold).

plated (both at 37°C and at room temperature) on plates containing one of four antibiotics (ampicillin, cefazolin, cephalothin and cefotaxime) at or above the MIC for these antibiotics conferred by *bla*. Only in the case of cefotaxime were clones with elevated resistance identified. The number of colonies that grew indicated that about three library members exhibited resistance to at least 0.08 μg/ml cefotaxime and room temperature. The plasmid DNA from fourteen colonies that grew at room temperature on agar plates supplemented with 0.08 or 0.10 μg/ml cefotaxime were isolated and retransformed into fresh *E. coli* cells. Six colonies were false positives since the retransformed cells did not have resistance above background. The remaining eight library clones had an MIC for cefotaxime of at least 0.12 μg/ml (3-fold higher than the MIC of cells expressing BLA).

Sequencing revealed two library members with very similar sequences: CFX011 (BLA[166-286]-DKS-BLA[24-163]-VL) and CFX019 (BLA[157-286]-DKS-BLA[24-163]-VL). Like the BLA domain of MBP317-347, these two variants are circularly permuted in the Ω-loop. This change in substrate specificity due to alteration in or near the Ω-loop is consistent with previous reports [15,16]. Interestingly, CFX011 is permuted at exactly Glu166, which is known to be a key catalytic residue [21].

## Characterization of the improved circularly permuted variants

The MIC conferred by *CFX011* and *CFX019* were determined on a number of antibiotics and compared to that exhibited by *bla* (Table 1). Whereas the MIC for cefotaxime were 4-fold higher, the MIC's for ampicillin, cephalothin and cefazolin were all considerably lower than that conferred by *bla*. This indicates that the increased level of resistance provided by these two variants cannot solely be for reasons of increased expression of the protein, and that circular permutation has resulted in an alteration of the substrate specificity of the enzyme. Experiments designed to express CFX011 and CFX019 using the vectors on which they were isolated suffered from poor expression of the circular permutants (as judged by western blots and β-lactamase activity assays on lysates) and poor yields upon attempted purification using a phenylboronic acid agarose affinity column. This suggests that circular permutation compromised the production of CFX011 and CFX019 in *E. coli*. Thus, the most likely explanation for the increase MIC conferred by *CFX011* and *CFX019* is increased specific activity for hydrolyzing cefotaxime. Alternatively, these variants would have to be expressed at significantly higher level than BLA, retain TEM-1 levels of activity on cefotaxime, and

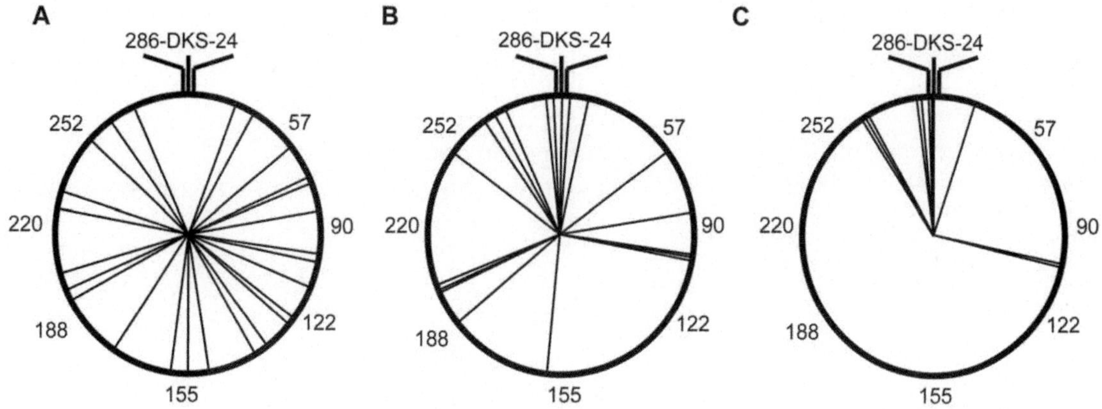


**Figure 2.** Sites of circular permutation BLA based on DNA sequencing of (A) 25 randomly selected members from the naïve library, (B) 20 randomly selected members capable of growing on plates containing 16 µg/ml ampicillin, and (C) 10 randomly selected members capable of growing on plates containing 250 µg/ml ampicillin. The first amino acid in the mature wild-type BLA is 24 (since the signal sequence of amino acids 1–23 is removed) and the last amino acid is 286. The linker joining the N- and C-termini has the amino acid sequence DKS.

selectively lose activity on ampicillin, cephalothin and cefazolin. This scenario seems unlikely.

## Conclusions

A directed evolution strategy involving random circular permutation was successfully applied to increasing the ability of TEM-1 β-lactamase to provide resistance to the antibiotic cefotaxime. The genes that provided increased resistance to CFTX were circularly permuted in the region corresponding to the Ω-loop of the protein. The genes conferred decreased activity against other cephalosporins and ampicillin. This finding indicates that circular permutation altered the substrate specificity of the enzyme. Since increased expression of the circularly permuted variants relative to wild-type would be unlikely, we proposed that the increased resistance derives from an increase in catalytic activity on cefotaxime, although confirmation will require characterization of the catalytic activity in vitro. It is interesting that insertions in the Ω-loop generated by pentapeptide mutagenesis also resulted in increased resistance to CFTX [15], indicating that different perturbations in this loop can result in similar changes in specificity, perhaps by removing steric constraints.

## Materials and Methods

### Materials

All enzymes were purchased from New England Biolabs. All DNA purification kits were purchased from Qiagen. Cefotaxime and ampicillin were purchased from Fisher Scientific. All oligonucleotides were purchased from Invitrogen. pGEM T-vector cloning kit and Taq polymerase were purchased from Promega (Madison, WI). All other chemicals were purchased from Sigma.

### Circular permutation library construction

The library was constructed in plasmid pC8BlaStop, a plasmid derived from pDIM-C8 [22] in which the section between the BamHI and SpeI sites is modified according to Figure 1A. Ten µg of pC8BlaStop was digested with 36 units of *AflII* and 25 units of *SapI.* The desalted digestion product was treated with Klenow and alkaline phosphatase according to the manufacturer's specifications. The large blunt fragment was isolated by agarose gel electrophoresis.

The insert was derived from a previously described library (called Library 7) of circularly permuted *bla* genes that was inserted into a specific location in the gene encoding maltose binding protein [19]. To create the library for the work described here we amplified this library of circularly permuted *bla* genes by PCR using the following primers: 5′-TTACGAGGAGTGCAGGGC-GAAAGATCCACGT-3′ and 5′-CACCTTTCTGTGCAGTT-TCCATGGTGGCGGC-3′ using 40 ng of template DNA. These primers located flanking BsgI sites an appropriate distance from the circularly permuted library such that digestion of the PCR product with BsgI and treatment with Klenow and dNTPs produced a blunt ended product without any extra DNA from the *malE* gene (Figure 1B). From the PCR reaction DNA of the correct size was isolated by agarose gel electrophoresis. Five µg of isolated DNA was digested with 50 units BsgI in the presence of 80 µM S-adenosylmethionine according to the manufacturer's instructions. BsgI cuts 16 base pairs downstream from its recognition sequence leaving behind the randomly circular permuted *bla* gene library with a 2-nucleotide 3′ overhang. These two overhang nucleotides were degraded by treating the isolated PCR product with Klenow(exo+) in the presence of dNTPs to create the blunt insert.

One hundred ng of linear, blunt pC8BlaStop vector was ligated to this 100 ng insert. The 10-µl reaction was carried out at 22°C overnight (for 24 hours) using 30,000 Weiss units of T4 DNA ligase. After ethanol precipitation, 5% of the precipitated product was electroporated into 75 µl DH5α *E. coli* cells using 0.2 cm cuvettes. Eight separate electroporations were performed and combined. The total number of transformants in the cpBLA library was about $0.5\times10^6$.

### Library characterization and selection

A frozen stock of the naïve library was diluted and plated on LB agar plates containing no antibiotics, 16 µg/ml ampicillin (Amp) or 250 µg/ml Amp. The plates were incubated overnight at room temperature. Plasmid DNA from inoculum prepared from randomly selected colonies were sequenced.

To select from genes that conferred increase resistance to β-lactam containing antibiotics, frozen stocks of the naïve library were diluted and plated on LB agar plates containing ampicillin, cefazolin, cephalothin or cefotaxime at concentrations at or above the MIC provided by *bla.* The plates were incubated overnight at

37°C and room temperature. Only on cefotaxime containing plates did any library members grow. The plasmid DNA was isolated and retransformed into DH5α *E.coli* to confirm that the reason for the elevated levels of resistance was plasmid-borne.

### Minimum inhibitory concentration determination

Minimum inhibitory concentrations (MICs) were determined by plating dilutions of overnight inoculums supplemented with varying concentrations of ampicillin, cefazolin, cephalothin or cefotaxime. Colonies that appeared on the LB plates after 24 hours incubation at room temperature were counted. The lowest concentration of the antibiotic required to prevent growth of at least 99% of the plated cells (compared to plates with no antibiotic) was taken to be the MIC.

## Author Contributions

Conceived and designed the experiments: MO GG. Performed the experiments: GG MK. Analyzed the data: GG MK MO. Wrote the paper: MO GG MK.

## References

1. Meister GE, Kanwar M, Ostermeier M (2009) Circular permutation of proteins. In: Lutz S, Bornscheuer U, eds. Protein Engineering Handbook: Wiley.
2. Baird GS, Zacharias DA, Tsien RY (1999) Circular permutation and receptor insertion within green fluorescent proteins. Proc Natl Acad Sci USA 96: 11241–11246.
3. Graf R, Schachman HK (1996) Random circular permutation of genes and expressed polypeptide chains: Application of the method to the catalytic chains of aspartate transcarbamoylase. Proc Natl Acad Sci USA 93: 11591–11596.
4. Topell S, Hennecke J, Glockshuber R (1999) Circularly permuted variants of the green fluorescent protein. FEBS Lett 457: 283–289.
5. Iwakura M, Nakamura T, Yamane C, Maki K (2000) Systematic circular permutation of an entire protein reveals essential folding elements. Nat Struct Biol 7: 580–585.
6. Cheltsov AV, Barber MJ, Ferreira GC (2001) Circular permutation of 5-aminolevulinate synthase. Mapping the polypeptide chain to its function. J Biol Chem 276: 19141–19149.
7. Qian Z, Lutz S (2005) Improving the catalytic activity of Candida antarctica lipase B by circular permutation. J Am Chem Soc 127: 13466–13467.
8. Qian Z, Fields CJ, Lutz S (2007) Investigating the structural and functional consequences of circular permutation on lipase B from Candida antarctica. Chembiochem 8: 1989–1996.
9. Reitinger S, Yu Y, Wicki J, Ludwiczek M, D'Angelo I, et al. (2010) Circular permutation of Bacillus circulans xylanase: a kinetic and structural study. Biochemistry 49: 2464–2474.
10. Yu Y, Lutz S (2009) Improved triglyceride transesterification by circular permuted Candida antarctica lipase B. Biotechnol Bioeng 105: 44–50.
11. Qian Z, Horton JR, Cheng X, Lutz S (2009) Structural redesign of lipase B from Candida antarctica by circular permutation and incremental truncation. J Mol Biol 393: 191–201.
12. Guntas G, Mitchell SF, Ostermeier M (2004) A molecular switch created by in vitro recombination of nonhomologous genes. Chem Biol 11: 1483–1487.
13. Ostermeier M (2005) Engineering allosteric protein switches by domain insertion. Protein Engineering Design and Selection 18: 359–364.
14. Fisher JF, Mobashery S (2009) Three decades of the class A beta-lactamase acyl-enzyme. Curr Protein Pept Sci 10: 401–407.
15. Hayes F, Hallet B, Cao Y (1997) Insertion mutagenesis as a tool in the modification of protein function. Extended substrate specificity conferred by pentapeptide insertions in the omega-loop of TEM-1 beta-lactamase. J Biol Chem 272: 28833–28836.
16. Petrosino JF, Palzkill T (1996) Systematic mutagenesis of the active site omega loop of TEM-1 beta-lactamase. J Bacteriol 178: 1821–1828.
17. Salverda ML, De Visser JA, Barlow M (2010) Natural evolution of TEM-1 beta-lactamase: experimental reconstruction and clinical relevance. FEMS Microbiol Rev 34: 1015–1036.
18. Osuna J, Pérez-Blancas A, Soberón X (2002) Improving a circularly permuted TEM-1 β-lactmase by directed evolution. Protein Eng 15: 463–470.
19. Guntas G, Mansell TJ, Kim JR, Ostermeier M (2005) Directed evolution of protein switches and their application to the creation of ligand-binding proteins. Proc Natl Acad Sci USA 102: 11224–11229.
20. Bosley AD, Ostermeier M (2005) Mathematical expressions useful in the construction, description and evaluation of protein libraries. Biomolecular Engineering 22: 57–61.
21. Minasov G, Wang X, Shoichet BK (2002) An ultrahigh resolution structure of TEM-1 beta-lactamase suggests a role for Glu166 as the general base in acylation. J Am Chem Soc 124: 5333–5340.
22. Ostermeier M, Nixon AE, Shim JH, Benkovic SJ (1999) Combinatorial protein engineering by incremental truncation. Proc Natl Acad Sci USA 96: 3562–3567.
23. Ostermeier M, Shim JH, Benkovic SJ (1999) A combinatorial approach to hybrid enzymes independent of DNA homology. Nat Biotechnol 17: 1205–1209.

# 13

# Engineering Agatoxin, a Cystine-Knot Peptide from Spider Venom, as a Molecular Probe for In Vivo Tumor Imaging

**Sarah J. Moore[1◑], Cheuk Lun Leung[2◑], Heidi K. Norton[1], Jennifer R. Cochran[1,2,3]***

**1** Department of Bioengineering, Stanford University, Stanford, California, United States of America, **2** Department of Chemical Engineering, Stanford University, Stanford, California, United States of America, **3** Stanford Cancer Institute and Bio-X Program, Stanford, California, United States of America

## Abstract

***Background:*** Cystine-knot miniproteins, also known as knottins, have shown great potential as molecular scaffolds for the development of targeted therapeutics and diagnostic agents. For this purpose, previous protein engineering efforts have focused on knottins based on the *Ecballium elaterium* trypsin inhibitor (EETI) from squash seeds, the Agouti-related protein (AgRP) neuropeptide from mammals, or the Kalata B1 uterotonic peptide from plants. Here, we demonstrate that Agatoxin (AgTx), an ion channel inhibitor found in spider venom, can be used as a molecular scaffold to engineer knottins that bind with high-affinity to a tumor-associated integrin receptor.

***Methodology/Principal Findings:*** We used a rational loop-grafting approach to engineer AgTx variants that bound to $\alpha_v\beta_3$ integrin with affinities in the low nM range. We showed that a disulfide-constrained loop from AgRP, a structurally-related knottin, can be substituted into AgTx to confer its high affinity binding properties. In parallel, we identified amino acid mutations required for efficient in vitro folding of engineered integrin-binding AgTx variants. Molecular imaging was used to evaluate in vivo tumor targeting and biodistribution of an engineered AgTx knottin compared to integrin-binding knottins based on AgRP and EETI. Knottin peptides were chemically synthesized and conjugated to a near-infrared fluorescent dye. Integrin-binding AgTx, AgRP, and EETI knottins all generated high tumor imaging contrast in U87MG glioblastoma xenograft models. Interestingly, EETI-based knottins generated significantly lower non-specific kidney imaging signals compared to AgTx and AgRP-based knottins.

***Conclusions/Significance:*** In this study, we demonstrate that AgTx, a knottin from spider venom, can be engineered to bind with high affinity to a tumor-associated receptor target. This work validates AgTx as a viable molecular scaffold for protein engineering, and further demonstrates the promise of using tumor-targeting knottins as probes for in vivo molecular imaging.

**Editor:** Matthew Bogyo, Stanford University, United States of America

**Funding:** This work was supported, in part, by a Developmental Project Grant from the Stanford In Vivo Cellular and Molecular Imaging Center NIH/NCI P50 CA114747, and NIH/NCI R21 CA143498. S.J.M. was supported by an NSF Graduate Research Fellowship, a Stanford Graduate Fellowship (Medtronic Fellow), a Siebel Scholars Fellowship, and a Gerald J. Lieberman Fellowship. C.L.L. was supported by an NSF Graduate Research Fellowship. H.K.N. was supported by a Stanford Bio-X Undergraduate Research Fellowship and by a Stanford Materials Science VPUE Summer Research Fellowship. The funders had no role in study design, data collection and analysis, decision to publish, or preparation of the manuscript.

**Competing Interests:** The authors have declared that no competing interests exist.

* E-mail: jennifer.cochran@stanford.edu

◑ These authors contributed equally to this work.

## Introduction

There is a critical need for in vivo molecular imaging agents that bind specifically and with high affinity to clinical targets of interest, while displaying desirable pharmacokinetics and tissue biodistribution properties [1,2]. For cancer, ideal molecular imaging agents are ones that exhibit robust tumor localization and rapid clearance from non-target tissues and organs [3,4]. Such attributes translate into high imaging contrast at early time points after probe injection, and low nonspecific or background imaging signals that otherwise obscure accurate identification of malignant tissue.

Recently, cystine-knot miniproteins, known as knottins, have emerged as promising agents for non-invasive molecular imaging of tumors in living subjects [5–7]. Knottins share a common disulfide-bonded framework, and contain loops of variable length and composition that are constrained to a core of anti-parallel beta-strands (Fig. 1) [8]. This structure confers high thermal, chemical, and proteolytic stability [9,10], which is desirable for in vivo biomedical applications. In addition, the small size of knottins (~30–60 amino acids) affords rapid blood clearance and the potential for chemical synthesis, allowing facile incorporation of a variety of imaging moieties [11,12].

Polypeptides containing cystine-knot motifs are found in myriad organisms such as plants, insects, and mammals, and carry out diverse functions including protease inhibition, ion channel blockade, and antimicrobial activity [13,14]. Although naturally-

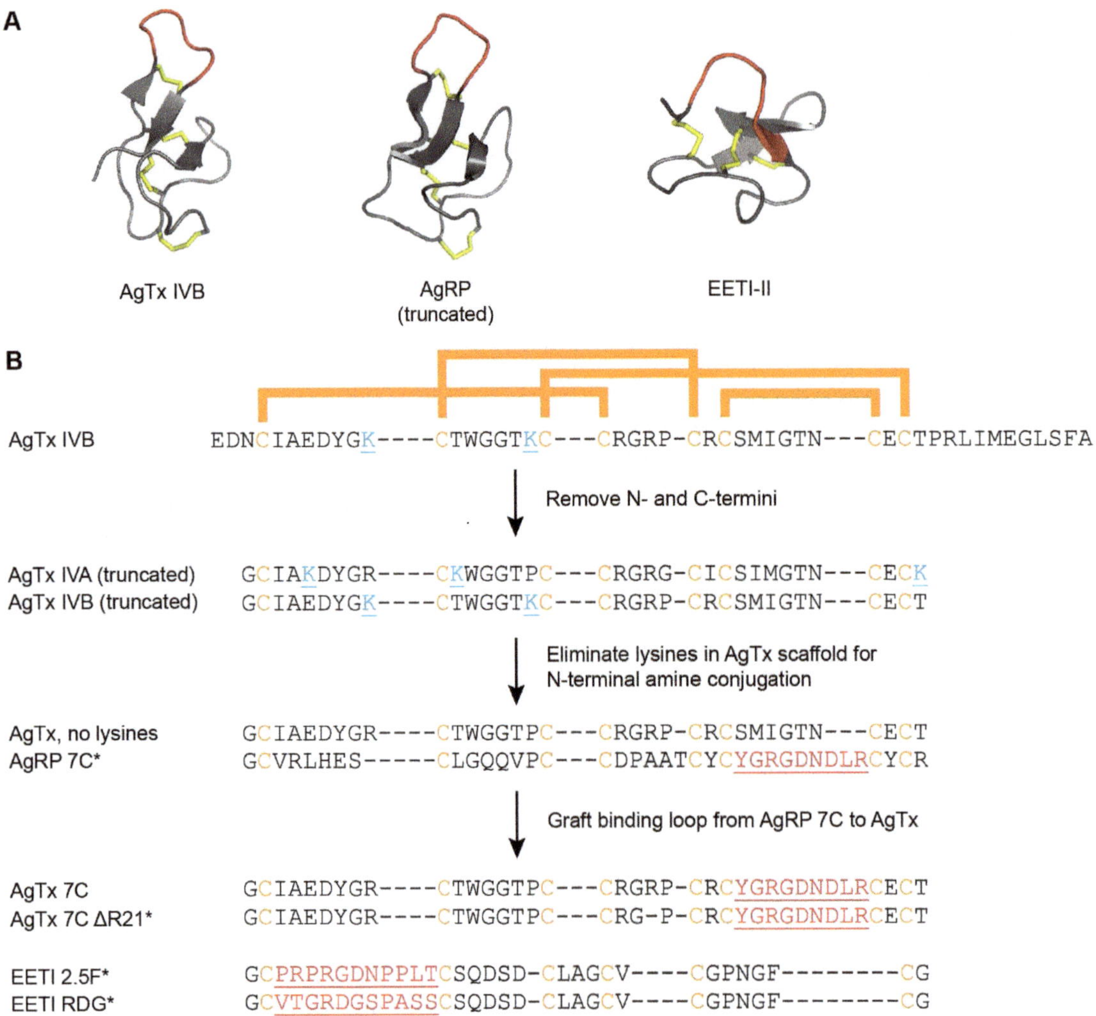


**Figure 1. AgTx, AgRP, and EETI knottins engineered to bind tumor-associated integrins.** (A) Native knottin structures. AgTx IVB (PDB 1OMB), truncated AgRP (PDB 1MR0), and EETI-II (PDB 2ETI), with disulfide bonds shown in gold, and native loops that were mutated to bind tumor-associated integrins shown in red. Structures were rendered in PyMOL. (B) Schematic of protein engineering strategy and sequences of native and engineered knottins used in this study. Conserved cysteine residues are shown in gold, and bars indicate disulfide bond connectivity. The N- and C-termini of AgTx were truncated and the sequences of isoforms IVA and IVB were combined to create a knottin scaffold with no lysine residues (cyan), allowing for site-specific conjugation of AF680 at the N-terminal amino group. The integrin-binding loop from AgRP 7C was grafted into the structurally analogous loop of this new scaffold to create AgTx 7C. Mutated loops are underlined and shown in red. * indicates knottins used for in vivo imaging. EETI RDG contains a scrambled sequence that does not bind integrins, and was used as a negative control.

occurring knottins have found important clinical applications [15,16], protein engineering is playing an increasing role in creating knottins that possess novel molecular recognition properties for use as therapeutics and diagnostics [17–20]. The disulfide-constrained loop regions of native knottins tolerate high levels of sequence diversity (Fig. 1B), providing a robust molecular framework for engineering proteins that recognize a variety of biomedical targets. Despite the large number of natural proteins with cystine knot motifs, engineering efforts have mainly utilized three knottins as molecular scaffolds: the *Ecballium elaterium* trypsin inhibitor-II (EETI), which is found in the seeds of the squirting cucumber [21,22]; a truncated version of the Agouti-related protein (AgRP), a neuropeptide that is involved in regulating metabolism and appetite [23,24]; and the cyclotide Kalata B1 from the African plant *Oldenlandia*, which has uterotonic activities [25].

We previously used yeast-surface display and high-throughput library screening to identify knottin variants, based on EETI and AgRP, that possess high affinity and specificity for integrin receptors expressed on tumor cells and their neovasculature (Fig. 1) [26,27]. Engineered integrin-binding EETI and AgRP knottins were labeled with a variety of contrast agents and used to non-invasively image tumors across multiple modalities, including positron emission tomography (PET) [28–34], single-photon emission computed tomography (SPECT) [35], ultrasound [36], and optical imaging [30,31]. In these studies, engineered EETI and AgRP knottins exhibited rapid tumor localization and blood clearance via the kidneys, resulting in robust tumor contrast compared to the surrounding tissue [28–34]. Given the critical roles that integrins play in tumor cell survival, invasion, metastasis, and angiogenesis [37–40], molecular imaging agents that selectively target tumor-associated integrins have potential diagnostic applications in disease staging and management, and monitoring response to therapy [41,42].

Spider venoms are a rich source of diverse peptides containing a cystine knot motif [43,44]. Here, we demonstrate the first use of Agatoxin (AgTx), a venom-derived knottin [45,46], as a molecular scaffold for protein engineering. AgRP, which is derived from mammals, is structurally homologous to AgTx, despite the fact that their sequences share only one identical residue beyond their conserved cysteines (Fig. 1). Using a rational approach, we grafted the integrin-binding loop from an engineered AgRP mutant into an AgTx scaffold, resulting in an AgTx variant that bound to tumor cells with low nanomolar affinity. Through this work, we identified scaffold mutations required for efficient in vitro folding of this engineered AgTx variant. We showed that engineered integrin-binding knottins based on AgTx, AgRP, and EETI can be conjugated to a near-infrared fluorescent dye and used for in vivo optical imaging of tumors in mice, but that EETI-based knottins had lower non-specific accumulation in the kidneys. This work validates AgTx as a molecular scaffold for protein engineering and in vivo imaging applications, expanding the repertoire of knottins that can be developed as potential therapeutic and diagnostic agents.

## Results

### Engineering a truncated AgTx scaffold that binds $\alpha_v\beta_3$ integrin

EETI 2.5F is an engineered knottin that binds to $\alpha_v\beta_3$, $\alpha_v\beta_5$, and $\alpha_5\beta_1$ integrins [26], while AgRP 7C is an engineered knottin that binds only to $\alpha_v\beta_3$ integrin [27]. To develop AgTx as a scaffold for molecular engineering, we first reduced its size to minimize complexity and allow for more facile peptide synthesis. There are two primary natural isoforms, AgTx IVA and IVB, which possess 71% sequence homology [45]. These isoforms each contain 48 amino acids, including disordered N- and C-terminal regions [46]. We defined the knottin core of AgTx IVA and IVB by sequence comparison with a truncated AgRP fragment [24] that was previously used as a scaffold to engineer integrin-binding knottins [27]. As the C-terminal region of AgTx is proposed to mediate ion channel binding and inhibition [46], we removed it to mitigate potential toxicity concerns of using full-length AgTx for in vivo studies. N- and C-terminal truncations resulted in peptides containing 35 amino acids for each AgTx isoform (Fig. 1B). AgTx IVA and IVB contain 3 and 2 lysine residues, respectively, at different positions throughout the polypeptide chain. We combined the sequences of the AgTx IVA and IVB isoforms to eliminate all lysine residues, so that a molecular imaging probe could be selectively conjugated to the N-terminal amino group. We define this combined sequence as "AgTx, no lysines" (Fig. 1B). To engineer this new AgTx scaffold to bind integrins we relied on the overall structural similarity between AgTx and AgRP (Fig. 1A), and substituted the integrin-binding loop from a previously engineered knottin variant, AgRP 7C, into the analogous location within AgTx to produce AgTx 7C (Fig. 1B).

### A specific arginine deletion in the integrin-binding AgTx variant promoted efficient in vitro folding

The linear precursor of the integrin-binding knottin AgTx 7C was chemically synthesized using solid-phase peptide synthesis. Crude peptide was folded in vitro using conditions previously established for disulfide bond formation with AgRP 7C [19], and the folded peptide was purified by reversed-phase high-performance liquid chromatography (RP-HPLC). Analytical-scale RP-HPLC was used to compare the crude peptide, folding reaction, and purified knottin (Fig. 2). Analysis of the folding reaction indicated the presence of a sharp elution peak (Fig. 2B), which is characteristic of folded knottin that can be separated from misfolded isomers. However, the molecular mass of this species was 156 Da less than expected, suggesting the deletion of an arginine residue (Fig. 2D). Upon further analysis, a peptide of the expected molecular mass was indeed present in the crude product of the synthesis (Fig. 2A), but efficient folding was only observed

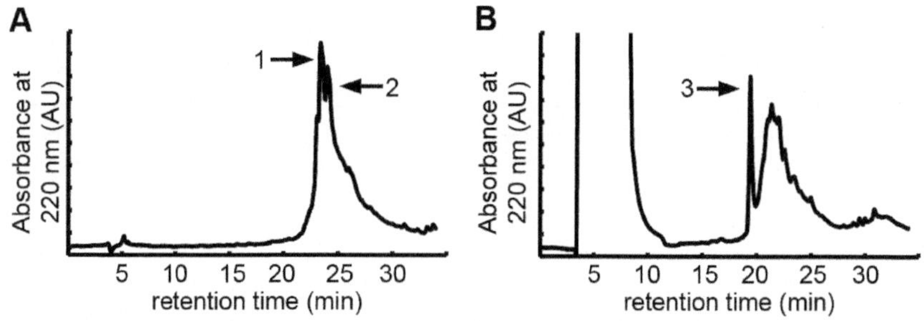


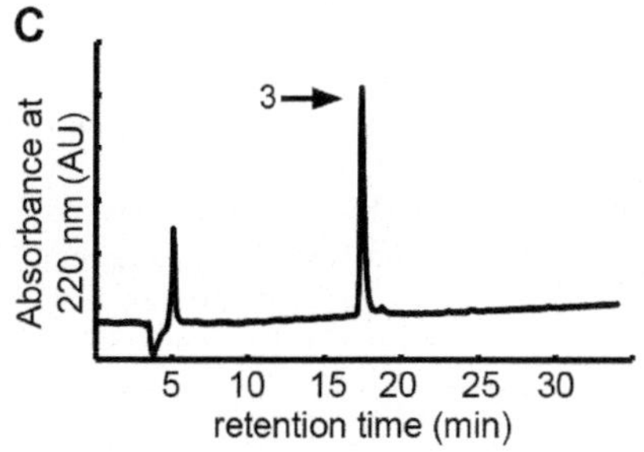


D

| Peak | Expected neutral mass (Da) | Observed mass (Da) |
|---|---|---|
| 1 | 4204 | 4205 |
| 2 | 4204 | 4048 |
| 3 | 4196 | 4040 |

**Figure 2. Synthesis and folding of AgTx 7C indicated a deletion product.** (A–C) RP-HPLC chromatograms. (A) The crude peptide from solid-phase peptide synthesis had two major peaks, one with the expected mass and the other with a loss of 156 Da, indicating possible deletion of an arginine residue. (B) Folding of crude peptide yielded a sharp peak which was 156 Da less than the expected mass. (C) Purified, folded peptide exhibited a single, sharp peak. (D) Expected and observed masses of indicated HPLC peaks as analyzed by MALDI-TOF mass spectrometry. Note that there is an 8 Da difference between unfolded and folded AgTx 7C due to the formation of 4 disulfide bonds.

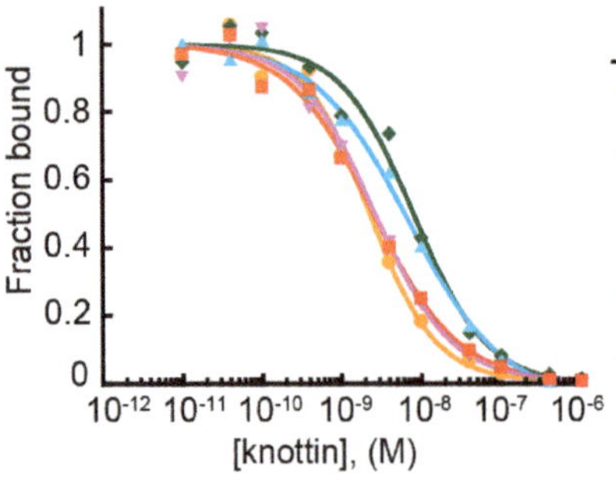


| Knottin | $IC_{50}$ (nM) |
|---|---|
| AgRP 7C | 2.2 ± 1.0 |
| AgTx 7C ΔR21 | 2.3 ± 0.5 |
| AgTx 7C P22G R24I | 8.1 ± 1.6 |
| AgTx 7C ΔR21 P22G R24I | 5.1 ± 1.6 |
| AgTx 7C ΔR9 | 2.5 ± 0.4 |

**Figure 3. Engineered AgTx 7C variants bind to K562-$\alpha_v\beta_3$ cells with similar $IC_{50}$ values.** Varying concentrations of AgRP 7C and AgTx 7C variants were incubated with FLAG-AgRP 7A and allowed to compete for binding to integrin receptors expressed on the surface of K562-$\alpha_v\beta_3$ cells. Representative competition binding curves are shown, and are plotted as knottin concentration versus the fraction of FLAG-AgRP 7A bound. $IC_{50}$ values reported as mean of three experiments ± SD.

for the AgTx 7C deletion variant. This serendipitous finding prompted us to determine the identity of this AgTx 7C deletion. Enzymatic digestion and tandem mass spectrometry analysis indicated the absence of an arginine at position 21 (Fig. S1). In solid-phase peptide synthesis, coupling of an arginine residue is sometimes problematic [47,48]. The side chain of arginine has two reactive nitrogen groups, and although one nitrogen is protected by 2,2,4,6,7-pentametyldihydrobenzofuran-5-sulfonyl (Pbf) during the synthesis, the unprotected nitrogen can form temporary intramolecular bonds, reducing coupling efficiency. The coupling efficiency is also reduced by the conformational rigidity of proline, whose unique side chain forms a covalent bond with the backbone of the growing peptide chain. Thus, the coupling of P22 followed by R21 would be expected to proceed at particularly low efficiency, potentially explaining the presence of high amounts of the AgTx 7C deletion product observed in our synthesis. To confirm these results, AgTx 7C was chemically synthesized without an arginine residue at position 21 (AgTx 7C ΔR21), resulting in a single species with the expected mass for both the purified linear precursor and the folded product (Fig. S2A, B). As further evidence that R21 was the problematic arginine residue for coupling, an AgTx variant was synthesized with an arginine deletion at position 9 (AgTx 7C ΔR9). The resulting peptide was again a mixture of two species, one with the correct mass, and one with a mass of 156 Da less than expected, presumably containing the additional R21 deletion from inefficient coupling. However, in this case we found that both AgTx peptides folded efficiently (data not shown). The cysteine adjacent to R9 forms a disulfide bond with the cysteine adjacent to P22, suggesting that the R9 deletion promotes folding through reduced steric constraints or reduced charge repulsion.

## Additional scaffold mutations promote efficient folding of AgTx 7C

To determine if other modifications to the AgTx scaffold influence folding, we synthesized additional AgTx 7C variants based on sequence variations of the AgTx isoforms. The R21 deletion is located in a region where the sequences of AgTx IVA and IVB differed. The "lysine-free" AgTx scaffold we defined included residues P22 and R24 from AgTx IVB (Fig. 1B); the corresponding residues in AgTx IVA are G22 and I24. A comparison of the AgTx IVA and AgTx IVB structures using the TM-align server [49] reveals that while overall the two structures are similar (TM-score of 0.70 and RMSD of 1.40), the region containing residues 22 and 24 has the largest variation compared to the rest of the structure. The inclusion of P22 in our combined AgTx scaffold, which directly precedes R21, might limit conformational flexibility required for efficient polypeptide folding, particularly when the 9-amino acid integrin-binding loop was incorporated into the construct. To test this hypothesis, we substituted G22 and I24 from AgTx IVA into the knottin scaffold, resulting in a variant denoted AgTx 7C P22G R24I. The mutations P22G and R24I resulted in efficient folding of AgTx 7C (Fig. S2). Moreover, only a single peptide species was produced from these syntheses, indicating that the substitution of proline with glycine at position 22 allows efficient coupling of the R21 residue. Finally, an AgTx variant incorporating P22G and R24I mutations and the ΔR21 deletion (denoted AgTx 7C ΔR21 P22G R24I) also demonstrated efficient folding (Fig S2). Folding conditions for each AgTx variant are described in the Supplemental Methods. Collectively, these results demonstrate that amino acid substitutions and deletions within the knottin scaffold can influence or promote more efficient folding.

## Engineered AgTx variants bind $\alpha_v\beta_3$ integrin with high affinity

Competition binding assays were used to measure the relative affinities of the AgTx 7C variants and AgRP 7C to K562 leukemia cells transfected to express high levels of $\alpha_v\beta_3$ integrin [50]. Recombinant FLAG-AgRP 7A, a related engineered knottin that binds specifically to $\alpha_v\beta_3$ integrin with high affinity [27], was used as the competitor. FLAG-AgRP 7A contains an N-terminal FLAG epitope tag (DYKDDDDK), which allows detection of cell surface binding by flow cytometry using a fluorescently labeled anti-FLAG antibody. All AgTx 7C variants showed similar relative binding affinities to K562-$\alpha_v\beta_3$ cells with half-maximal inhibitory concentration ($IC_{50}$) values in the single-digit nanomolar range (Fig. 3). AgRP 7C and AgTx 7C ΔR21 had nearly identical $IC_{50}$ values of 2.2±1.0 nM and 2.3±0.5 nM, respectively, confirming that loop grafting of a binding epitope from AgRP to the structurally similar AgTx scaffold was a successful protein engineering strategy. As the conformation of the Arg-Gly-Asp (RGD) integrin-binding motif is critically important for mediating high-affinity interactions [51,52], these results suggest that AgTx scaffold mutations at positions 9, 21, 22, or 24 do not significantly affect overall protein structure.

## Knottin peptides conjugated to AF680 dye retain high affinity integrin binding

For simplicity, we chose to move forward with AgTx 7C ΔR21 for in vivo molecular imaging studies. To visualize tumor targeting and tissue biodistribution, we conjugated the near-infrared dye

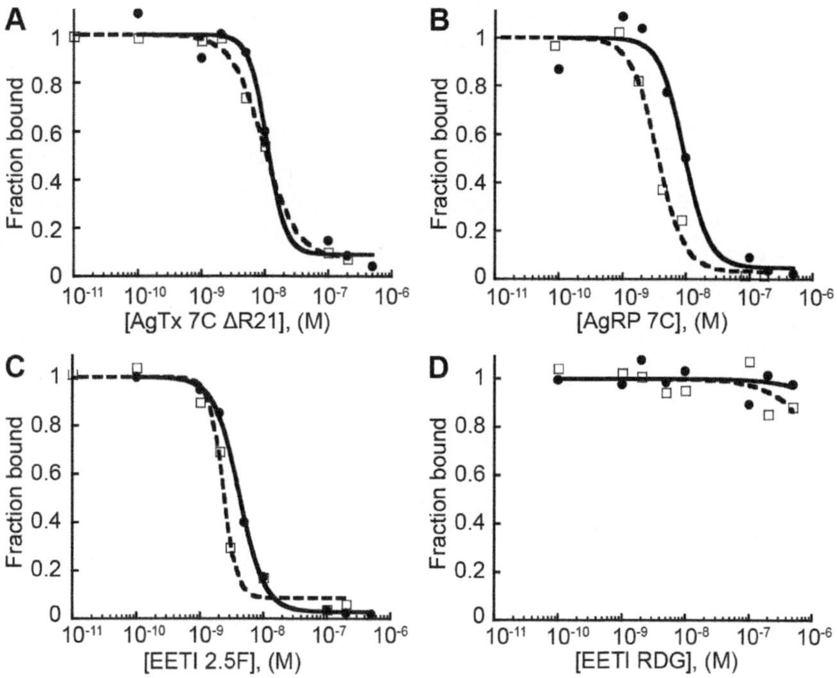


**Figure 4. Unlabeled and AF680-labeled knottins bind U87MG cells with high affinity.** Unlabeled (open squares, dashed line) and AF680-labeled (closed circles, solid line) knottins similarly compete off FLAG-AgRP 7A knottin binding to $\alpha_v\beta_3$ integrins expressed on U87MG glioblastoma cells. Representative competition binding curves shown for (A) AgTx 7C ΔR21, (B) AgRP 7C, (C) EETI 2.5F, and (D) EETI RDG control. $IC_{50}$ values reported as mean of three experiments ± SD.

Alexa Fluor 680 (AF680) to the N-terminal amino group of AgTx 7C ΔR21 (Fig. S3A). We also synthesized AF680-labeled versions of AgRP 7C and EETI 2.5F for comparison studies. AF680-labeled EETI RDG, which contains a scrambled integrin recognition sequence [26,30], served as a non-binding control. AF680-labeled knottin peptides were purified by RP-HPLC and had the expected masses for the addition of one dye molecule (Fig. S3B, C). Competition binding assays were used to determine if AF680 conjugation affected integrin recognition. FLAG-AgRP 7A was used to compete for knottin binding to U87MG human glioblastoma cells (Fig. 4). Unlabeled knottins and AF680-labeled knottins had similar $IC_{50}$ values in the low nanomolar range for each peptide tested, confirming that dye conjugation did not interfere with high-affinity integrin binding (Table 1).

## Integrin-binding knottins exhibit high tumor contrast in murine xenograft models

AF680-labeled knottins were evaluated as molecular imaging probes in subcutaneous U87MG tumor xenograft models. Non-invasive optical imaging was performed over a 24 hr time period after murine tail vein injection of 1.5 nmol AF680-labeled knottin peptide (Fig. 5). Whole-body fluorescent imaging signals were prominent at 1–2 hr post-injection for AgTx 7C ΔR21, AgRP 7C, and EETI 2.5F, in contrast to the EETI RDG control, which mainly showed kidney signal due to renal clearance (Fig. 5A). For all integrin-binding knottins, tumor signals steadily decreased over the 24 hr imaging experiment (Fig. 5B). AgTx 7C ΔR21 and AgRP 7C generated high kidney imaging signals at early time points, which decreased over time (Fig. 5C). In contrast, significantly lower kidney imaging signals were observed in mice injected with EETI 2.5F and EETI RDG at all time points. EETI 2.5F generated the greatest tumor imaging contrast amongst all the knottins (Fig. 5D), as defined by the ratio of the tumor signal to the normal flank tissue of the same mouse. A comparison of AgTx 7C ΔR21 versus AgRP 7C showed similar tumor contrast throughout the imaging time course, with levels significantly higher than the EETI RDG control. Maximum tumor-to-normal tissue contrast was observed at 6–8 hr post injection after clearance of probe from non-target tissue, with values of 10.0±1.1, 6.3±0.9, and 6.8±1.1 for EETI 2.5F, AgRP 7C, and AgTx 7C ΔR21, respectively (Fig. 5D).

## EETI-based knottins generate low kidney imaging signals compared to AgRP 7C and AgTx 7C ΔR21

To confirm tissue biodistribution observed with in vivo optical imaging experiments, mice were sacrificed at 4 hr and ex vivo imaging was performed on resected organs and tissue, including the tumor, kidney, liver, muscle, and blood (Fig. 6). Ex vivo imaging verified tumor-specific signals observed in mice injected with AF680-labeled AgTx 7C ΔR21, AgRP 7C, and EETI 2.5F

**Table 1.** Relative binding affinities of unlabeled and AF680-labeled knottins on U87MG cells, reported as $IC_{50}$ values.

| | $IC_{50}$ (nM) | |
|---|---|---|
| **Knottin** | **Unlabeled** | **AF680 labeled** |
| AgTx 7C ΔR21 | 10±1 | 11±3 |
| AgRP 7C | 4.0±0.4 | 9.2±0.2 |
| EETI 2.5F | 2.4±0.1 | 4.0±0.2 |
| EETI RDG | (-) | (-) |

(-) = no competition observed at highest concentration tested. $IC_{50}$ values reported as mean of three experiments ± SD.

compared to the EETI RDG control. In agreement with non-invasive in vivo optical imaging experiments, EETI 2.5F and EETI RDG generated low kidney signals at 4 hr post injection compared to AgTx 7C ΔR21 and AgRP 7C. In all animals, fluorescent signal was negligible in the muscle, blood, and liver, confirming efficient clearance from these organs and tissue.

## Discussion

Previous protein engineering studies have focused on the development and application of knottin scaffolds based on EETI, AgRP, and Kalata B1 [5–7,17,18,28]. Our current work expands the examples of knottins validated as molecular scaffolds to include AgTx. We first truncated the N- and C-termini of AgTx IVB to simplify the scaffold and abolish its native function. Based on structural similarities between AgRP and AgTx knottins, we next grafted a disulfide-bonded integrin-binding loop from an engineered AgRP variant into AgTx, and showed that high-affinity integrin binding was conferred with this new construct. A similar approach could potentially be used to incorporate disulfide-bonded cyclic peptides, such as those identified from phage display libraries [53], into knottin scaffolds.

Through this work, we identified an arginine deletion (ΔR21) that was required for efficient folding of the integrin-binding variant AgTx 7C. In contrast, native AgTx did not require this ΔR21 mutation for efficient folding (data not shown). The engineered integrin-binding loop we introduced into AgTx 7C is 9 amino acids in length, while the corresponding loop in native AgTx consists of 6 amino acids. This longer loop, together with the

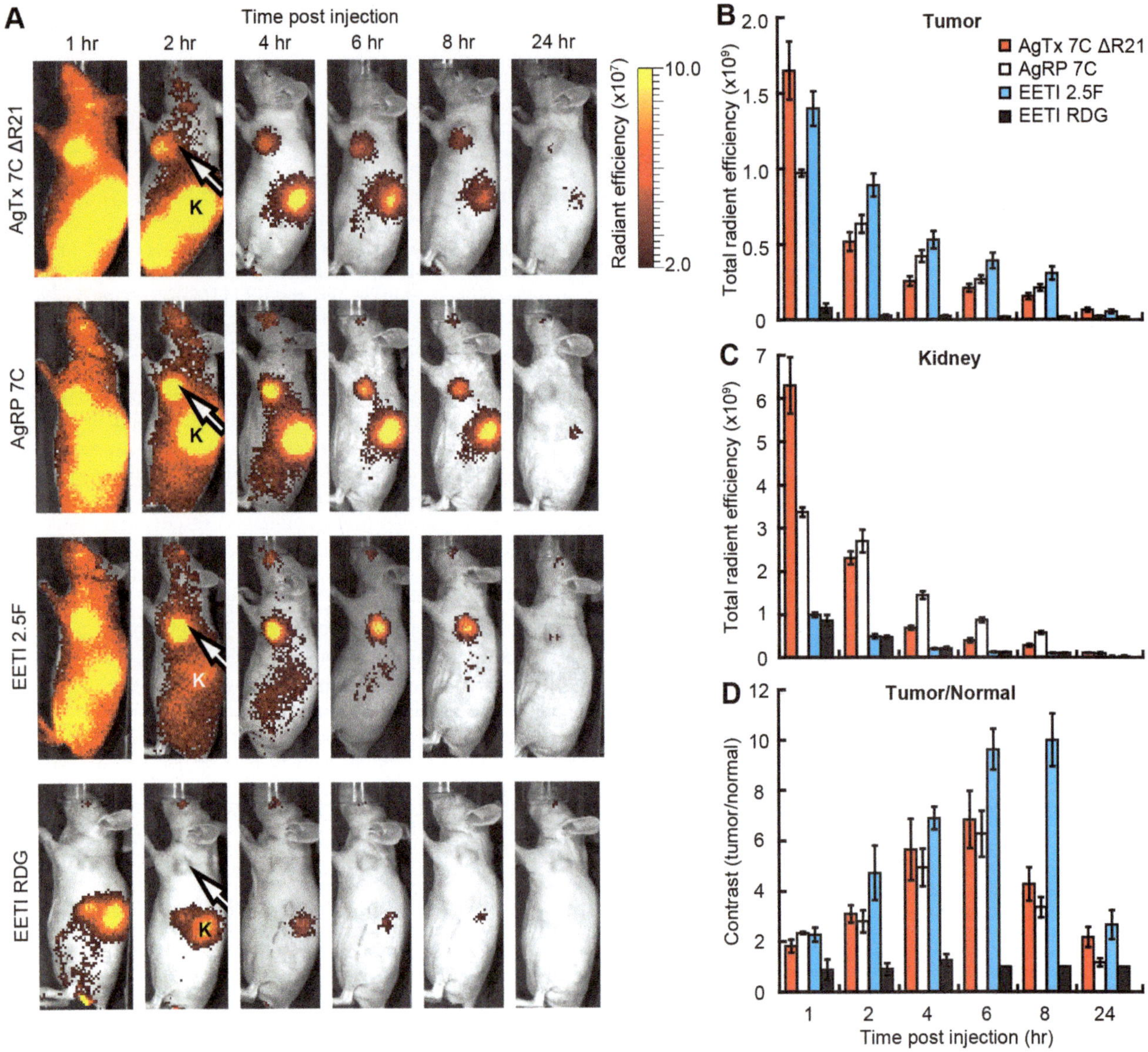


**Figure 5. Non-invasive in vivo imaging of AF680-labeled knottins in U87MG tumor xenografts.** (A) Representative whole-body fluorescent images of murine U87MG tumor xenografts injected via tail vein with 1.5 nmol AF680-labeled knottins AgTx 7C ΔR21, AgRP 7C, EETI 2.5F, and EETI RDG control. Tumors (white arrow) and kidneys (K) are indicated. Radiant efficiency [=] (p/s/cm$^2$/sr)/(μW/cm$^2$). (B–C) Quantification of imaging signals, reported as the total radiant efficiency, in the (B) tumor and (C) kidney over 24 hr. Total radiant efficiency [=] (p/s)/(μW/cm$^2$). (D) Imaging contrast, reported as the ratio of fluorescent signals for tumor versus normal tissue. There is no statistical difference in imaging contrast between AgTx 7C ΔR21 and AgRP 7C at all time points measured ($p>0.05$). Error bars represent ± SE, n = 4 for all knottins.

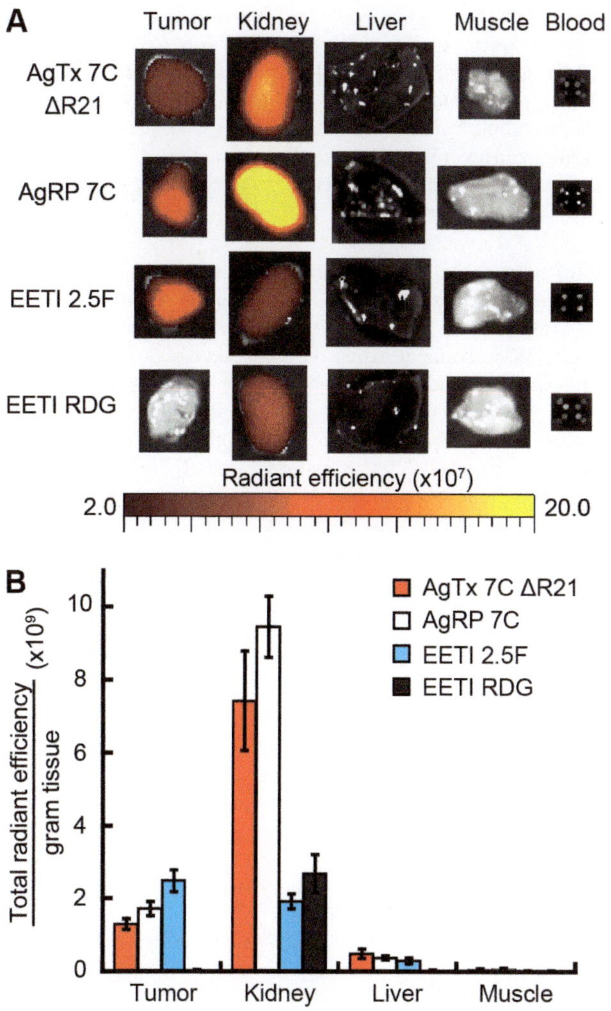


**Figure 6. Ex vivo imaging of tissue and organs from U87MG tumor xenografts.** (A) Representative ex vivo images of tumor, kidney, liver, muscle, and blood at 4 hr post injection of 1.5 nmol AF680-labeled knottins. (B) Quantification of total imaging signal per organ, normalized to organ mass, corroborates high kidney signals for AgTx 7C ΔR21 and AgRP 7C compared to EETI-based knottins. No fluorescent signal was detectable in blood samples at 4 hr post probe injection. Error bars represent ± SE. For all knottins, n = 9, except EETI RDG (n = 4).

inclusion of P22 which directly precedes the R21 mutation, may have disrupted an optimal balance of flexibility, size, and charge of amino acid side chains critical to folding amongst these isoforms. As a result of these changes, mutations immediately preceding the engineered integrin-binding loop were required for efficient folding. The AgTx mutations ΔR21 and/or P22G and R24I could potentially restore the thermodynamic favorability of knottin folding by removing steric clashes, improving backbone flexibility, or improving interaction of the beta strands in the knottin core. All of these AgTx 7C variants had similar relative binding affinities (Fig. 3), demonstrating that the scaffold modifications did not significantly affect the conformation of the engineered integrin-binding loop. AgRP 7C demonstrated efficient folding without the requirement for additional scaffold mutations. In contrast to the AgTx variants, AgRP has two Cys-Tyr-Cys motifs that flank the engineered integrin-binding loop. These Tyr residues have been proposed to participate in interactions that help mediate disulfide bond formation in native AgRP [54]. In addition, the Arg21 and Pro22 amino acid residues that were problematic for AgTx folding are Ala and Thr, respectively, in AgRP, which likely imparts conformational flexibility that better accommodates the engineered integrin-binding loop.

A myriad of small molecules and peptides containing an RGD integrin-binding motif have been developed, and several have advanced to human trials for diagnostic applications [55,56]. We showed that engineered integrin-binding knottins generate significantly higher levels of tumor contrast in subcutaneous U87MG xenograft models compared to imaging probes based on c(RGDfK) and c(RGDyK), two well-characterized cyclic pentapeptides that bind to $\alpha_v\beta_3$ and $\alpha_v\beta_5$ integrins. Maximum tumor-to-normal tissue contrast ratios for AF680-c(RGDfK) and AF680-c(RGDyK) were $3.2 \pm 0.5$ and $2.8 \pm 0.3$, respectively (Fig. S4), compared to imaging contrast ratios of 6-10 for AF680-labeled integrin-binding knottins (Fig. 5D). In addition to the potential for improved tumor contrast, molecular imaging probes can be directly conjugated to the knottin N-terminus without disrupting receptor binding interactions. The N-terminus of AgTx 7C ΔR21 is positioned on the opposite side of the knottin relative to the integrin-binding loop (Fig. 1). As such, conjugation of a fluorescent dye molecule to the N-terminus of AgTx 7C ΔR21 did not interfere with high affinity binding to cell surface integrin receptors. Similar results were obtained with imaging probe conjugation to the N-terminus of AgRP 7C and EETI 2.5F. In comparison, small molecules and disulfide-constrained cyclic peptides present binding epitopes within a more limited framework that can sometimes be disrupted upon imaging probe conjugation [53,55].

Another goal of our study was to compare the in vivo biodistribution profiles of engineered knottins derived from different organisms, including spiders, mammals, and plants. AF680-labeled AgTx 7C ΔR21, AgRP 7C, and EETI 2.5F all exhibited high tumor contrast in optical imaging experiments. The high tumor-to-normal imaging signals observed at early time points after knottin administration are consistent with theoretical and experimental studies examining the relationship between optimal molecular size and affinity needed for effective tumor targeting [57–60]. In particular, while small proteins accumulate rapidly in tumors, they must bind their targets with high affinity to be retained. Together, these results support further development of low molecular weight, high affinity tumor-targeting agents for molecular imaging applications.

Integrin-binding knottins based on the EETI scaffold exhibited remarkably low kidney signals, either due to more rapid clearance or reduced non-specific kidney accumulation compared to AgTx 7C ΔR21 and AgRP 7C. These results are consistent with our previous imaging studies using radiolabeled EETI and AgRP proteins [28–34]. In comparison, molecular imaging probes based on antibodies, antibody fragments, and other "alternative scaffolds" including affibodies, DARPins, and fibronectin domains, all exhibit high imaging signals in the liver and/or kidneys [57,61–64], which raises toxicity concerns and challenges for imaging the abdominal and thoracic regions. Such high kidney signals have been attributed to metabolites that accumulate in the kidneys through mechanisms such as renal tubular reabsorption [65,66]. We previously showed that AgRP 7C exhibits higher levels of metabolic breakdown in the liver, kidneys, and tumor compared to EETI 2.5F, although both probes were remarkably stable in serum and blood and were excreted intact in the urine [29,30].

The amino acid sequences of the knottin peptides could also play a role in their tissue biodistribution properties. In addition to having different scaffold sequences, the engineered loops of EETI 2.5F (PRPRGDNPPLT) and EETI RDG (VTGRDGSPASS) are markedly different from that of AgRP/AgTx 7C (YGRGDNDLR). In agreement with this hypothesis, our colleagues recently reported that highly charged residues, particularly arginine and glutamic acid, of *Momordica cochinchinensis* trypsin inhibitor-II knottins contribute greatly to non-specific kidney retention [67]. In addition, an $^{111}$In-labeled version of one of our alternative engineered integrin-binding AgRP knottins (AgRP 6E, engineered loop sequence: VERGDGNRR) had an approximately 50% reduction in kidney signal compared to $^{111}$In-labeled AgRP 7C [35], demonstrating the influence of the engineered loop on tissue biodistribution. Efforts to reduce undesirable non-target tissue accumulation have included: 1) introducing mutations that increase hydrophilicity and remove charged groups, 2) covalently attaching polymers such as polyethylene glycol, and 3) co-administration of the probe with charged amino acids or agents that block specific transporter receptors or alter overall kidney physiology [65,66,68]. The complexity of potential mechanisms for probe or metabolite retention in the kidneys highlights the need for polypeptides that do not require extensive efforts to block renal accumulation. Thus, in future studies it will be interesting to further explore the relationship between amino acid sequence, metabolic stability, and in vivo biodistribution properties of engineered knottins.

The methods we report for chemical synthesis, folding, dye conjugation, and purification of knottins as molecular imaging probes are compatible with production of clinical grade material, although yield has not yet been evaluated in a large-scale (>1 gram) production process. The homogeneity of the final product, with one fluorophore attached to the N-terminus of each knottin peptide, is desirable for clinical translation. Knottins are proposed to be non-immunogenic due to their high stability, which is thought to preclude presentation of peptide fragments to molecules that mediate immune function [69]. In addition, smaller quantities of knottin are used for in vivo molecular imaging compared to therapeutic dosage levels; however, immunogenicity and toxicity will need to be evaluated for each compound intended for human clinical trials.

In summary, we demonstrate that AgTx, an ion channel inhibitor found in spider venom, can be used as a pliable framework for constructing peptides that bind with high-affinity to a clinically-relevant integrin receptor. This work expands the repertoire of knottins that have been validated as molecular scaffolds for protein engineering, and provides insights for the further development of engineered knottins as in vivo molecular imaging agents.

## Materials and Methods

### Ethics statement

All animal procedures were in compliance with Protocol 22942 approved by the Stanford University Administrative Panels on Laboratory Animal Care. All procedures were conducted while animals were under general anesthesia with isofluorane, and all efforts were made to minimize suffering.

### Materials, reagents, and cell lines

Integrin binding buffer (IBB) was composed of 20 mM Tris (pH 7.5) with 1 mM $MgCl_2$, 1 mM $MnCl_2$, 2 mM $CaCl_2$, 100 mM NaCl, and 1 mg/mL bovine serum albumin (BSA). BPBS buffer was composed of phosphate-buffered saline (PBS) and 1 mg/mL BSA. 9-fluorenylmethyloxycarbonyl (Fmoc)-protected amino acids were purchased from Novabiochem/EMD Chemicals Inc. or CS Bio. The cyclic pentapeptides c(RGDfK) and c(RGDyK) were purchased from Peptides International. Human U87MG glioblastoma cells (ATCC) were cultured in DMEM (Gibco) supplemented with 10% fetal bovine serum (FBS) and 1% penicillin-streptomycin (Gibco). K562 cells transfected with $\alpha_v\beta_3$ integrins (courtesy of S. Blystone) were cultured in IMDM (Gibco) supplemented with 1 mg/ml active geneticin (Gibco), 10% FBS, and 1% penicillin-streptomycin [50].

### Peptide synthesis, folding, and purification

Knottin peptides were prepared as previously described in detail [19]. Briefly, linear precursor peptides were synthesized on a CS Bio CS336 instrument using Fmoc-based solid-phase peptide synthesis. After side-chain deprotection and resin cleavage, peptides were folded by promoting disulfide bond formation in oxidation buffers optimized for each peptide (Text S1). Folded knottins were purified by preparative-scale RP-HPLC using a Varian Prostar instrument and Vydac $C_{18}$ columns, where each folded peptide eluted as a sharp peak with an altered retention time from unfolded or misfolded precursors. Linear gradients of 90% acetonitrile in water containing 0.1% (v/v) trifluoroacetic acid were used for all peptide purifications, which were monitored at absorbances of 220 nm and 280 nm. Peptide purity was analyzed by analytical-scale RP-HPLC using a Vydac $C_{18}$ column. Molecular masses were determined by matrix-assisted laser desorption/ionization time-of-flight mass spectrometry (MALDI-TOF-MS; Stanford Protein and Nucleic Acid Facility) or electrospray ionization mass spectrometry (ESI-MS; Stanford Vincent Coates Foundation Mass Spectrometry Laboratory). Following purification, folded knottins were lyophilized and stored at room temperature until used. Purified knottins were dissolved in PBS, and concentrations were determined by amino acid analysis (UC Davis Proteomics Core Facility).

### AF680 dye conjugation

Pure, folded knottin (2 mg/ml) was incubated for 1 hr at room temperature and then at 4 °C overnight (with stirring) with Alexa Fluor 680 carboxylic acid, succinimidyl ester (Invitrogen) in a 0.1 M sodium bicarbonate solution, pH 8.0 at a 5:1 dye/peptide molar ratio (Fig. S3A). The resulting dye-conjugated knottins were purified by RP-HPLC (Fig. S3C). Masses were confirmed by MALDI-TOF mass spectrometry (Fig. S3B). AF680-labeled c(RGDfK) and c(RGDyK) were prepared in a similar manner. AF680-labeled compounds were lyophilized and resuspended in PBS, and concentrations were determined using UV-Vis spectroscopy, measuring dye absorption at 679 nm ($\varepsilon = 184{,}000\ cm^{-1}M^{-1}$). Alternatively, AF680-labeled knottins were purified by extensive buffer exchange with PBS using a centrifugal filter unit with a 3 kDa molecular weight cutoff (Amicon). In vivo imaging results were consistent between knottins purified through these two methods. AF680-labeled compounds, at a concentration of 15 μM in PBS, were passed through a 0.22 μm filter for animal experiments.

### Cell binding assays

Competition binding assays were performed on K562 leukemia cells transfected to express high levels of $\alpha_v\beta_3$ integrins [50]. Varying concentrations of AgRP 7C and AgTx variants were incubated with $5\times10^4$ K562-$\alpha_v\beta_3$ cells in IBB for 4 hr at 4 °C, along with a constant concentration (0.5 nM) of recombinantly expressed FLAG-AgRP 7A knottin as a competitor. FLAG-AgRP 7A binds with high affinity to $\alpha_v\beta_3$ integrin and contains an N-terminal epitope tag (DYKDDDDK), allowing cell surface binding to be detecting using an anti-FLAG antibody [27]. Care was taken to allow adequate time for equilibrium binding and to avoid ligand depleting conditions. After incubation with knottins, cells were washed with BPBS and resuspended with a 1:100 dilution of R-PE-conjugated anti-FLAG antibody (Prozyme) for 30 min on ice. Cells were washed with BPBS and analyzed by flow cytometry on a FACSCalibur instrument (BD Biosciences), and data was quantified using FlowJo software (TreeStar). $IC_{50}$ values were determined by nonlinear regression analysis using KaleidaGraph (Synergy Software). Similar competition binding assays were performed on U87MG glioblastoma cells, which express high levels of $\alpha_v\beta_3$ integrin receptors [70]. EETI 2.5F, AgRP 7C, AgTx 7C ΔR21, or EETI RDG, unlabeled or site-specifically labeled with one molecule of AF680 dye, were used for binding experiments. Varying concentrations of knottins were incubated with $5\times10^4$ U87MG cells in IBB for 4 hr at 4 °C with a constant concentration (5 nM) of FLAG-AgRP 7A as a competitor. Cell washes, secondary antibody incubation, and data collection and analysis were performed as for the K562-$\alpha_v\beta_3$ binding assays. $IC_{50}$ values are reported as the mean and standard deviation of at least three separate experiments.

## Mouse handling and generation of tumor xenograft models

Animal procedures were carried out on 4-week old female nude mice (Charles River Laboratory). $5\times10^6$ U87MG cells, suspended in 50 µl PBS along with 50 µl of Matrigel Basement Membrane Matrix (BD Biosciences, cat# 354234), were injected subcutaneously into the left shoulder of mice to generate human tumor xenografts. Mice were imaged when tumors reached 5–10 mm in diameter.

## In vivo and ex vivo optical imaging

Mice bearing U87MG tumor xenografts were anesthetized with isofluorane and injected via tail vein with 1.5 nmol AF680-labeled knottins, c(RGDfK), or c(RGDyK) in 100 µl of PBS. Whole-body in vivo fluorescence imaging was performed at the indicated times after probe injection using an IVIS 200 system (Caliper Life Sciences). The near-infrared fluorophore AF680 was excited at 615–665 nm and emission was analyzed at 695–770 nm. Background autofluorescence emission signal was also collected by exciting at 580–610 nm and analyzing at 695–770 nm. In each imaging set, a mouse injected with PBS alone (no knottin) was included to allow measurement of background signals for data processing. For ex vivo imaging, mice were sacrificed, and organs were excised and imaged using the same excitation and emission wavelengths as for in vivo imaging. Excised organs were weighed to determine the total fluorescent signal flux/gram of tissue.

## Imaging quantitation

All optical imaging quantification was performed using Living Image software (Caliper Life Sciences). In vivo signal was calculated as $[\text{Emission}_{615\text{–}665\text{ nm Excitation}}]-k*[\text{Emission}_{580\text{–}610\text{ nm Excitation}}]$, where the constant k was determined such that a background region of interest (ROI) would have no signal. This background ROI was drawn around the tumor of the control mouse in each experiment that received a PBS injection with no knottin. Contrast was calculated as total radiant efficiency (units of [photons/sec/cm$^2$/steradian]/[µW/cm$^2$]) for tumor tissue, divided by total radiant efficiency for an ROI on normal flank tissue on the same mouse. For ex vivo quantification, total radiant efficiency was measured for the entire organ, and normalized by the mass of the organ to determine flux/g of tissue. Two-tailed Student's t-tests were used to evaluate and assign statistical significance between data sets.

## Supporting Information

**Figure S1 Enzymatic digestion and tandem mass spectrometry analysis of folded AgTx 7C reveals an arginine deletion at position 21.** The modified AgTx 7C knottin (observed mass = 4040 Da) was reduced with dithiothreitol and alkylated with iodoacetamide. (A) Comparison of MALDI-TOF-MS of a chymotryptic digest to *in silico* chymotrypsin digestion using ExPASy PeptideMass revealed that the mass discrepancy is located in the sequence GGTPCCRGRPCRCY (position 13–26). (B) Comparison of tryptic digest to a Mascot search revealed that Arg19 is present and Arg24 is likely present due to the existence of fragment 25–34, indicating cleavage by trypsin after residue 24. This data suggests that the missing Arg is located at residue 21. (C) MS/MS analysis of the 1600 Da chymotryptic peptide further supports the sequence GGTPCCRG_PCRCY, with Arg21 as the most likely deletion, by the observation of $y_3$, $y_5$, $y_6$, $y_7$, and $y_8$ ions.

**Figure S2 Modifications to the AgTx scaffold promote in vitro folding of integrin-binding variants.** Analytical-scale RP-HPLC traces of linear, crude peptide (left), folding reaction (center), and purified, folded peptide (right) for AgTx 7C variants. Yield of purified, folded AgTx 7C was too low for further analysis. AgTx 7C P22G R24I and AgTx 7C ΔR21 P22G R24I were efficiently separated from misfolded isomers when folded from purified, linear precursor peptide, but not when folded from unpurified, crude peptide under the conditions tested. Thus, for these variants, crude linear peptide was first purified by preparatory-scale RP-HPLC using a Vydac $C_{18}$ column before folding. In contrast, purification of the AgTx 7C linear precursor prior to folding still resulted in very low folding efficiency. (B) Masses of folded, purified knottins were determined by ESI-MS or MALDI-TOF-MS.

**Figure S3 AF680 conjugation and characterization.** (A) The near infrared dye AF680 was site-specifically conjugated to knottins at their N-terminal amino group using succinimidyl ester chemistry. (B) Folded, purified knottins and AF680-labeled knottins were analyzed by mass spectrometry. Expected error in these measurements is 0.1%. (C) Analysis of purified AF680-labeled knottins by analytical-scale RP-HPLC. Purity was determined to be greater than 95%. Blue traces: absorbance at 220 nm by amide bonds, red traces: absorbance at 675 nm by AF680 fluorophore.

**Figure S4 Non-invasive in vivo imaging with AF680-labeled cyclic RGD peptidomimetics.** (A) Mice bearing U87MG tumor xenografts were injected with 1.5 nmol AF680-c(RGDfK) or AF680-c(RGDyK), which exhibited high tumor uptake but slow clearance from non-target tissues. Tumors (white arrow) and kidneys (K) are indicated. (B) Maximum tumor-to-normal tissue contrast ratios of 3.2±0.5 and 2.8±0.3 were measured for AF680-c(RGDfK) and AF680-c(RGDyK), respectively. Error bars represent ± SE, n = 3.

## Acknowledgments

We thank Dr. Scott Blystone (SUNY Upstate Medical University) for providing the K562-$\alpha_v\beta_3$ cell line, Dick Winant of the Stanford Protein and Nucleic Acids facility for assistance with enzymatic digestion and tandem mass spectrometry, and Tim Doyle and Frezghi Habte in the Stanford Small Animal Imaging Facility for assistance with in vivo imaging experiments and data quantification. We also acknowledge Frank Cochran for guidance with peptide synthesis, folding, and purification, Adam Silverman for initial work on knottin loop grafting, Kathleen Hershberger for technical assistance, and Jamie Bergen for critical comments on the manuscript.

## Author Contributions

Conceived and designed the experiments: SJM CLL HKN JRC. Performed the experiments: SJM CLL HKN. Analyzed the data: SJM CLL HKN JRC. Wrote the paper: SJM CLL JRC.

## References

1. Reynolds F, Kelly KA (2011) Techniques for molecular imaging probe design. Mol Imaging 10: 407–419.
2. Chen K, Chen X (2010) Design and development of molecular imaging probes. Curr Top Med Chem 10: 1227–1236.
3. Friedman M, Stahl S (2009) Engineered affinity proteins for tumour-targeting applications. Biotechnol Appl Biochem 53: 1–29.
4. Batra SK, Jain M, Wittel UA, Chauhan SC, Colcher D (2002) Pharmacokinetics and biodistribution of genetically engineered antibodies. Curr Opin Biotechnol 13: 603–608.
5. Daly NL, Craik DJ (2011) Bioactive cystine knot proteins. Curr Opin Chem Biol 15: 362–368.
6. Kolmar H (2011) Natural and engineered cystine knot miniproteins for diagnostic and therapeutic applications. Curr Pharm Des 17: 4329–4336.
7. Moore SJ, Leung CL, Cochran JR (2011) Knottins: Disulfide-bonded therapeutic and diagnostic peptides. Drug Discovery Today: Technologies 9: e3–e11.
8. Pallaghy PK, Nielsen KJ, Craik DJ, Norton RS (1994) A common structural motif incorporating a cystine knot and a triple-stranded beta-sheet in toxic and inhibitory polypeptides. Protein Sci 3: 1833–1839.
9. Colgrave ML, Craik DJ (2004) Thermal, chemical, and enzymatic stability of the cyclotide kalata B1: the importance of the cyclic cystine knot. Biochemistry 43: 5965–5975.
10. Werle M, Schmitz T, Huang HL, Wentzel A, Kolmar H, et al. (2006) The potential of cystine-knot microproteins as novel pharmacophoric scaffolds in oral peptide drug delivery. J Drug Target 14: 137–146.
11. Miao Z, Levi J, Cheng Z (2010) Protein scaffold-based molecular probes for cancer molecular imaging. Amino Acids epub Feb 12.
12. Reinwarth M, Nasu D, Kolmar H, Avrutina O (2012) Chemical Synthesis, Backbone Cyclization and Oxidative Folding of Cystine-knot Peptides; Promising Scaffolds for Applications in Drug Design. Molecules 17: 12533–12552.
13. Zhu S, Darbon H, Dyason K, Verdonck F, Tytgat J (2003) Evolutionary origin of inhibitor cystine knot peptides. FASEB J 17: 1765–1767.
14. Gracy J, Le-Nguyen D, Gelly JC, Kaas Q, Heitz A, et al. (2008) KNOTTIN: the knottin or inhibitor cystine knot scaffold in 2007. Nucleic Acids Res 36: D314–319.
15. Williams JA, Day M, Heavner JE (2008) Ziconotide: an update and review. Expert Opin Pharmacother 9: 1575–1583.
16. Veiseh M, Gabikian P, Bahrami SB, Veiseh O, Zhang M, et al. (2007) Tumor paint: a chlorotoxin:Cy5.5 bioconjugate for intraoperative visualization of cancer foci. Cancer Res 67: 6882–6888.
17. Craik DJ, Swedberg JE, Mylne JS, Cemazar M (2012) Cyclotides as a basis for drug design. Expert Opin Drug Discov 7: 179–194.
18. Kolmar H (2010) Engineered cystine-knot miniproteins for diagnostic applications. Expert Rev Mol Diagn 10: 361–368.
19. Moore SJ, Cochran JR (2012) Engineering knottins as novel binding agents. Methods in Enzymology 503: 223–251.
20. Getz JA, Rice JJ, Daugherty PS (2011) Protease-resistant peptide ligands from a knottin scaffold library. ACS Chem Biol 6: 837–844.
21. Favel A, Mattras H, Coletti-Previero MA, Zwilling R, Robinson EA, et al. (1989) Protease inhibitors from *Ecballium elaterium* seeds. Int J Pept Protein Res 33: 202–208.
22. Le Nguyen D, Heitz A, Chiche L, Castro B, Boigegrain RA, et al. (1990) Molecular recognition between serine proteases and new bioactive microproteins with a knotted structure. Biochimie 72: 431–435.
23. Ollmann MM, Wilson BD, Yang YK, Kerns JA, Chen Y, et al. (1997) Antagonism of central melanocortin receptors in vitro and in vivo by agouti-related protein. Science 278: 135–138.
24. Jackson PJ, McNulty JC, Yang YK, Thompson DA, Chai B, et al. (2002) Design, pharmacology, and NMR structure of a minimized cystine knot with agouti-related protein activity. Biochemistry 41: 7565–7572.
25. Craik DJ, Daly NL, Bond T, Waine C (1999) Plant cyclotides: A unique family of cyclic and knotted proteins that defines the cyclic cystine knot structural motif. J Mol Biol 294: 1327–1336.
26. Kimura RH, Levin AM, Cochran FV, Cochran JR (2009) Engineered cystine knot peptides that bind alphav beta3, alphav beta5, and alpha5 beta1 integrins with low-nanomolar affinity. Proteins 77: 359–369.
27. Silverman AP, Levin AM, Lahti JL, Cochran JR (2009) Engineered cystine-knot peptides that bind alphav beta3 integrin with antibody-like affinities. J Mol Biol 385: 1064–1075.
28. Jiang H, Moore SJ, Liu S, Liu H, Miao Z, et al. (2012) A novel radiofluorinated agouti-related protein for tumor angiogenesis imaging. Amino Acids: doi:10.1007/s00726-00012-01391-y.
29. Jiang L, Kimura RH, Miao Z, Silverman AP, Ren G, et al. (2010) Evaluation of a $^{64}$Cu-labeled cystine-knot peptide based on agouti-related protein for PET of tumors expressing alphav beta3 integrin. J Nucl Med 51: 251–258.
30. Kimura RH, Cheng Z, Gambhir SS, Cochran JR (2009) Engineered knottin peptides: a new class of agents for imaging integrin expression in living subjects. Cancer Res 69: 2435–2442.
31. Kimura RH, Miao Z, Cheng Z, Gambhir SS, Cochran JR (2010) A dual-labeled knottin peptide for PET and near-infrared fluorescence imaging of integrin expression in living subjects. Bioconjug Chem 21: 436–444.
32. Liu S, Liu H, Ren G, Kimura RH, Cochran JR, et al. (2012) PET imaging of integrin positive tumors using $^{18}$F-labeled knottin peptides. Theranostics 1: 403–412.
33. Miao Z, Ren G, Liu H, Kimura RH, Jiang L, et al. (2009) An engineered knottin peptide labeled with $^{18}$F for PET imaging of integrin expression. Bioconjug Chem 20: 2342–2347.
34. Nielsen CH, Kimura RH, Withofs N, Tran PT, Miao Z, et al. (2010) PET imaging of tumor neovascularization in a transgenic mouse model with a novel $^{64}$Cu-DOTA-knottin peptide. Cancer Res 70: 9022–9030.
35. Jiang L, Miao Z, Kimura RH, Silverman AP, Ren G, et al. (2012) $^{111}$In-labeled cystine-knot peptides based on the Agouti-related protein for targeting tumor angiogenesis. J Biomed Biotechnol 2012: Article ID 368075.
36. Willmann JK, Kimura RH, Deshpande N, Lutz AM, Cochran JR, et al. (2010) Targeted contrast-enhanced ultrasound imaging of tumor angiogenesis with contrast microbubbles conjugated to integrin-binding knottin peptides. J Nucl Med 51: 433–440.
37. Brooks PC, Clark RA, Cheresh DA (1994) Requirement of vascular integrin alpha v beta 3 for angiogenesis. Science 264: 569–571.
38. Kim S, Bell K, Mousa SA, Varner JA (2000) Regulation of angiogenesis in vivo by ligation of integrin alpha5beta1 with the central cell-binding domain of fibronectin. Am J Pathol 156: 1345–1362.
39. Brooks PC, Montgomery AM, Rosenfeld M, Reisfeld RA, Hu T, et al. (1994) Integrin alpha v beta 3 antagonists promote tumor regression by inducing apoptosis of angiogenic blood vessels. Cell 79: 1157–1164.
40. Friedlander M, Brooks PC, Shaffer RW, Kincaid CM, Varner JA, et al. (1995) Definition of two angiogenic pathways by distinct alpha v integrins. Science 270: 1500–1502.
41. Gaertner FC, Schwaiger M, Beer AJ (2010) Molecular imaging of avb3 expression in cancer patients. Q J Nucl Med Mol Imaging.
42. Cai W, Rao J, Gambhir SS, Chen X (2006) How molecular imaging is speeding up antiangiogenic drug development. Mol Cancer Ther 5: 2624–2633.
43. Escoubas P, Rash L (2004) Tarantulas: eight-legged pharmacists and combinatorial chemists. Toxicon 43: 555–574.
44. Wood DL, Miljenovic T, Cai S, Raven RJ, Kaas Q, et al. (2009) ArachnoServer: a database of protein toxins from spiders. BMC Genomics 10: 375.
45. Adams ME, Mintz IM, Reily MD, Thanabal V, Bean BP (1993) Structure and properties of omega-agatoxin IVB, a new antagonist of P-type calcium channels. Mol Pharmacol 44: 681–688.
46. Kim JI, Konishi S, Iwai H, Kohno T, Gouda H, et al. (1995) Three-dimensional solution structure of the calcium channel antagonist omega-agatoxin IVA: consensus molecular folding of calcium channel blockers. J Mol Biol 250: 659–671.
47. Rink H, Sieber P, Raschdorf F (1984) Coversion of NG-urethane protected arginine to ornithine in peptide solid phase synthesis. Tetrahedron Letters 25: 621–624.
48. Rubina A, Bespalova Zh D, Bushuev VN (2000) The solid-phase synthesis of peptides containing an arginine residue with an unprotected guanidine group. Russian Journal of Bioorganic Chemistry 26: 235–244.
49. Zhang Y, Skolnick J (2005) TM-align: a protein structure alignment algorithm based on the TM-score. Nucleic Acids Res 33: 2302–2309.
50. Blystone SD, Graham IL, Lindberg FP, Brown EJ (1994) Integrin alphav beta3 differentially regulates adhesive and phagocytic functions of the fibronectin receptor alpha5 beta1. J Cell Biol 127: 1129–1137.
51. Xiong JP, Stehle T, Goodman SL, Arnaout MA (2003) Integrins, cations and ligands: making the connection. J Thromb Haemost 1: 1642–1654.
52. Haubner R, Finsinger D, Kessler H (1997) Stereoisomeric peptide libaries and peptidomimetics for designing selective inhibitors of the avb3 integrin for a new cancer therapy. Angew Chem Int Ed 36: 1374–1389.

53. Cochran FV, Cochran JR (2010) Phage display and molecular imaging: expanding fields of vision in living subjects. Biotechnology and Genetic Engineering Reviews 27: 57–94.
54. McNulty JC, Jackson PJ, Thompson DA, Chai B, Gantz I, et al. (2005) Structures of the agouti signaling protein. J Mol Biol 346: 1059–1070.
55. Haubner R (2006) Alphavbeta3-integrin imaging: a new approach to characterise angiogenesis? Eur J Nucl Med Mol Imaging 33 Suppl 1: 54–63.
56. Beer AJ, Schwaiger M (2008) Imaging of integrin alphavbeta3 expression. Cancer Metastasis Rev 27: 631–644.
57. Zahnd C, Kawe M, Stumpp MT, de Pasquale C, Tamaskovic R, et al. (2010) Efficient tumor targeting with high-affinity designed ankyrin repeat proteins: effects of affinity and molecular size. Cancer Res 70: 1595–1605.
58. Orlova A, Magnusson M, Eriksson TL, Nilsson M, Larsson B, et al. (2006) Tumor imaging using a picomolar affinity HER2 binding affibody molecule. Cancer Res 66: 4339–4348.
59. Schmidt MM, Wittrup KD (2009) A modeling analysis of the effects of molecular size and binding affinity on tumor targeting. Mol Cancer Ther 8: 2861–2871.
60. Wittrup KD, Thurber GM, Schmidt MM, Rhoden JJ (2012) Practical theoretic guidance for the design of tumor-targeting agents. Methods Enzymol 503: 255–268.
61. Hackel BJ, Kimura RH, Gambhir SS (2012) Use of $^{64}$Cu-labeled fibronectin domain with EGFR-overexpressing tumor xenograft: molecular imaging. Radiology 263: 179–188.
62. Kaur S, Venktaraman G, Jain M, Senapati S, Garg PK, et al. (2012) Recent trends in antibody-based oncologic imaging. Cancer Lett 315: 97–111.
63. Lofblom J, Feldwisch J, Tolmachev V, Carlsson J, Stahl S, et al. (2010) Affibody molecules: engineered proteins for therapeutic, diagnostic and biotechnological applications. FEBS Lett 584: 2670–2680.
64. Wu AM, Olafsen T (2008) Antibodies for molecular imaging of cancer. Cancer J 14: 191–197.
65. Akizawa H, Uehara T, Arano Y (2008) Renal uptake and metabolism of radiopharmaceuticals derived from peptides and proteins. Adv Drug Deliv Rev 60: 1319–1328.
66. Vegt E, de Jong M, Wetzels JF, Masereeuw R, Melis M, et al. (2010) Renal toxicity of radiolabeled peptides and antibody fragments: mechanisms, impact on radionuclide therapy, and strategies for prevention. J Nucl Med 51: 1049–1058.
67. Kimura RH, Teed R, Hackel BJ, Pysz MA, Chuang CZ, et al. (2012) Pharmacokinetically stabilized cystine knot peptides that bind alphav beta6 integrin with single-digit nanomolar affinities for detection of pancreatic cancer. Clin Cancer Res 18: 839–849.
68. Hackel BJ, Sathirachinda A, Gambhir SS (2012) Designed hydrophilic and charge mutations of the fibronectin domain: towards tailored protein biodistribution. Protein Eng Des Sel 25: 639–648.
69. Maillere B, Mourier G, Herve M, Cotton J, Leroy S, et al. (1995) Immunogenicity of a disulphide-containing neurotoxin: presentation to T-cells requires a reduction step. Toxicon 33: 475–482.
70. Zhang X, Xiong Z, Wu Y, Cai W, Tseng JR, et al. (2006) Quantitative PET imaging of tumor integrin alphav beta3 expression with $^{18}$F-FRGD2. J Nucl Med 47: 113–121.

# 14

# Computational Design of Binding Proteins to EGFR Domain II

**Yoon Sup Choi[1,2,8] ❡, Soomin Yoon[1,3] ❡, Kyung-Lock Kim[4], Jiho Yoo[5], Parkyong Song[4], Minsoo Kim[1,6], Young-Eun Shin[4], Won Jun Yang[1,3], Jung-eun Noh[4], Hyun-soo Cho[5], Sanguk Kim[2,7]*, Junho Chung[1,3]*, Sung Ho Ryu[2,4]***

**1** Cancer Research Institute, Seoul National University School of Medicine, Seoul, Republic of Korea, **2** School of Interdisciplinary Bioscience and Bioengineering (I-Bio), Pohang University of Science and Technology (POSTECH), Pohang, Republic of Korea, **3** Department of Biochemistry and Molecular Biology, Seoul National University School of Medicine, Seoul, Republic of Korea, **4** Division of Integrative Bioscience and Biotechnology, Pohang University of Science and Technology (POSTECH), Pohang, Republic of Korea, **5** Department of Systems Biology, College of Life Science and Biotechnology, Yonsei University, Seoul, Republic of Korea, **6** Scripps Korea Antibody Institute, Chuncheon, Republic of Korea, **7** Division of IT Convergence Engineering, Pohang University of Science and Technology (POSTECH), Pohang, Republic of Korea, **8** KT Institute of Convergence Technology, Seocho-gu, Seoul, Korea

## Abstract

We developed a process to produce novel interactions between two previously unrelated proteins. This process selects protein scaffolds and designs protein interfaces that bind to a surface patch of interest on a target protein. Scaffolds with shapes complementary to the target surface patch were screened using an exhaustive computational search of the human proteome and optimized by directed evolution using phage display. This method was applied to successfully design scaffolds that bind to epidermal growth factor receptor (EGFR) domain II, the interface of EGFR dimerization, with high reactivity toward the target surface patch of EGFR domain II. One potential application of these tailor-made protein interactions is the development of therapeutic agents against specific protein targets.

**Editor:** Ashley M. Buckle, Monash University, Australia

**Funding:** This study was supported by a research grant from the Cancer Research Institute, Seoul National University (cri-12-1), a National Research Foundation of Korea (NRF) grant funded by the Korean government (MEST) (No. 2013R1A2A1A03010110). This work was also supported by the National Research Foundation of Korea (NRF) grant funded by the Korean government (No.2010-0028684). The funders had no role in study design, data collection and analysis, decision to publish, or preparation of the manuscript.

**Competing Interests:** The authors have declared that no competing interests exist.

* E-mail: sukim@postech.ac.kr (SK); junhochung@me.com (JC); sungho@postech.ac.kr (SHR)

❡ These authors contributed equally to this work.

## Introduction

Protein interaction networks evolve over time to create new protein interactions, which results in the dynamic rewiring of links among pre-existing nodes [1–3]. The current approaches to develop novel protein interactions are based on gene duplication and gene modification [1]. Gene duplication results in the addition of both a network node (i.e., protein) and links (i.e., interactions) to the protein interaction network [2,4,5]. Gene modification, which usually involves point mutations, results in the addition of links to the network [1,6].

Recent attempts to develop artificial binding proteins, which are based on a single protein framework, have been successful [7–9]. In these studies, a large number of random mutations have been introduced into predefined structural regions of protein frameworks, such as fibronectins [10–13], lipocalins [14–16], and the ankyrin repeat protein motif [17–19]. However, although the scaffolds constructed in these studies have shown affinity to various targets, the selection of different protein frameworks specific to a predetermined target surface patch has not been successful except in a recent study that developed protein binders for influenza hemagglutinin [20].

To mimic the evolutionary process by which protein networks evolve, we adopted the basic mechanism by which antibodies are produced against antigens. When animals are exposed to an antigen, B cells that express a low-affinity surface immunoglobulin are selected. During rapid B-cell proliferation, random mutations are introduced into the immunoglobulin sequences, and clones that express antibodies with high affinities are preferentially selected. To bind a specific antigen with high specificity and affinity, antibodies form a complementary shape to the target surface patch of the antigen using complementarity determining regions (CDRs). The amino acids in CDRs can produce extremely diverse structures, each of which forms the complement shape that recognizes a specific epitope (Figure 1A).

We have developed a strategy using protein docking simulation that imitates this process of antibody generation to select human protein scaffolds with complementary shapes (Figure 1B). This procedure designs novel protein interactions by selecting human protein scaffolds with shapes that complement a predetermined surface patch on a target protein (Figure 1C). In this procedure, key residues are optimized by using an amino acid residue randomization and phage display. The successful implementation of this strategy enables the reproduction of novel protein–protein interactions in the laboratory setting.

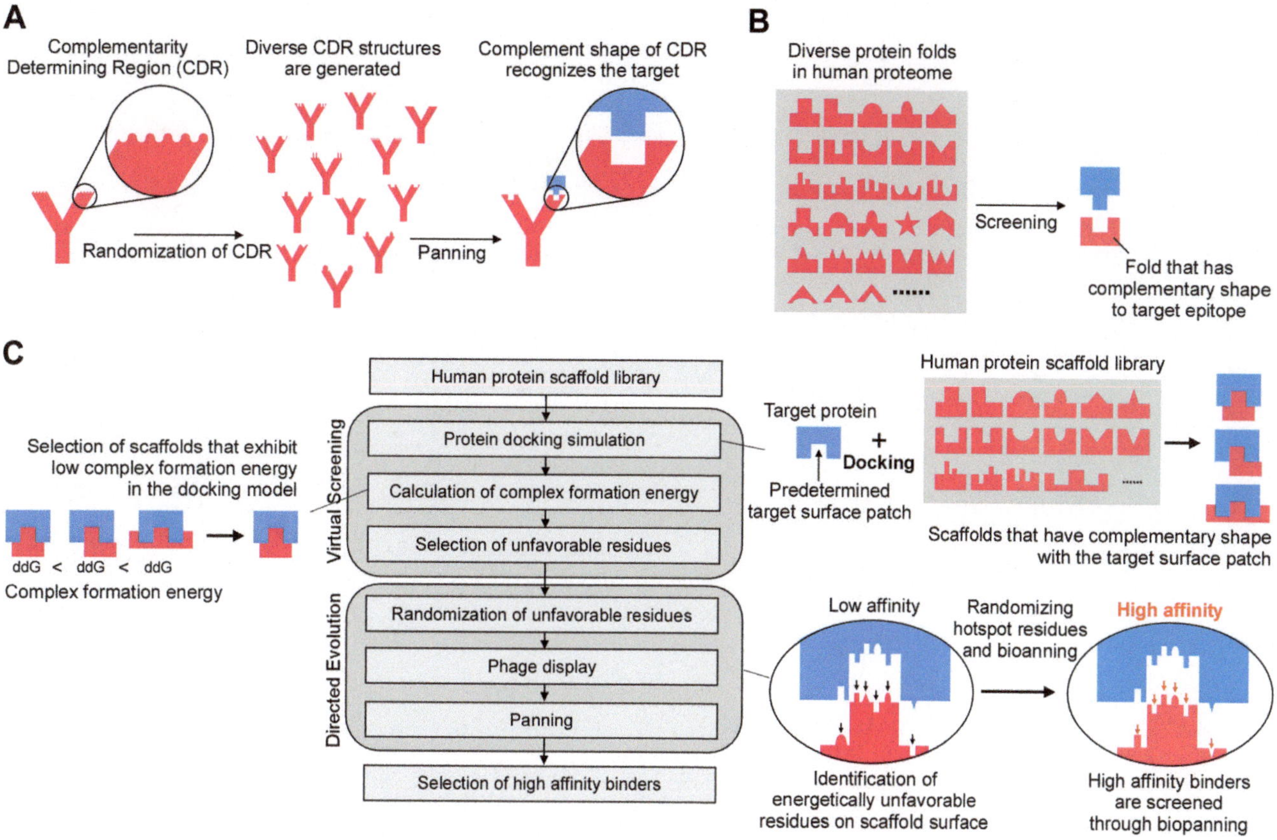


**Figure 1. Design scheme of target-specific scaffolds.** (A) Synthetic antibodies can achieve extremely diverse structures through sequence randomization of the complementarity determining region (CDR). Among diverse structures, only antibodies with complementary shapes are able to recognize and bind to a particular epitope. (B) By imitating synthetic antibody generation, we devised a strategy to select target-specific scaffolds from the human proteome with shapes that are complementary to the target surface patch. (C) The flow chart shows a two-step strategy to obtain target-specific scaffolds (middle). In the first step, a virtual screening of a human protein scaffold library is conducted to determine a framework specific to the surface patch of interest. Target specific-scaffolds with shapes complementary to the surface patch of interest are selected from the scaffold library through protein docking simulations (upper right). The scaffold–target docking structures with the most favorable complex formation energies are further evaluated (left). In the second step, the scaffold interface in the selected scaffold–target model is optimized by sequence randomization and phage display using directed evolution (lower right).

We have applied this method to the development of proteins that bind epidermal growth factor receptor (EGFR) domain II. EGFR, which is also known as ErbB1 and HER1, is one of the most extensively studied proteins, and plays key roles in many cancers, including colorectal and lung cancer [21–24]. EGFR undergoes a dramatic conformational change when activated to form homodimers or heterodimers with other receptors in the EGFR family [25,26]. In the absence of the EGF ligand, monomeric EGFR exists in a conformational equilibrium of tethered and untethered states (Figure 2A) [27]. The binding of EGF stabilizes EGFR in the untethered conformation and exposes domain II, which is otherwise occluded by intramolecular interactions, to form the homodimeric interface (Figure 2A) [25]. Binding EGFR domain II when EGFR is in the transition state inhibits EGFR activation. For example, monoclonal antibody 806 primarily binds domain II of an EGFR mutant [28–30].

Here, we generated six different protein binders from two different frameworks that have been computationally screened from the human proteome. We show that these binders have a high affinity for EGFR and EGFR fragments that harbor domain II. Moreover, we show that treatment with exogenous EGF further increases the reactivity of these binders to EGFR, which might suggest that these protein binders react with EGFR in its transition state. This ability to develop artificial protein binding pairs provides us with an experimental tool to create new protein interactions in a laboratory setting. Our results suggest that the non-interface surface of a protein can be changed into a *de novo* protein interface.

## Results

### Computational design scheme

We hypothesized that new protein interactions can be created in nature when two proteins share surface patches with complementary structures, and critical hotspot residues on these complementary surfaces are changed to more energetically favorable residues. We therefore devised a two-step strategy to create novel protein-protein interactions in a laboratory setting (Figure 1C). The first step involved virtual screening of protein scaffolds to find those with structural complementarity to the target surface patch and with favorable complex formation energy. In the second step, we improved the reactivity between the scaffold and target proteins using a directed evolution approach (second part of flowchart in Figure 1C). In this two-step strategy, we used the computational components to assist the directed evolution by selecting protein frameworks and narrowing down the potential interfacial residues

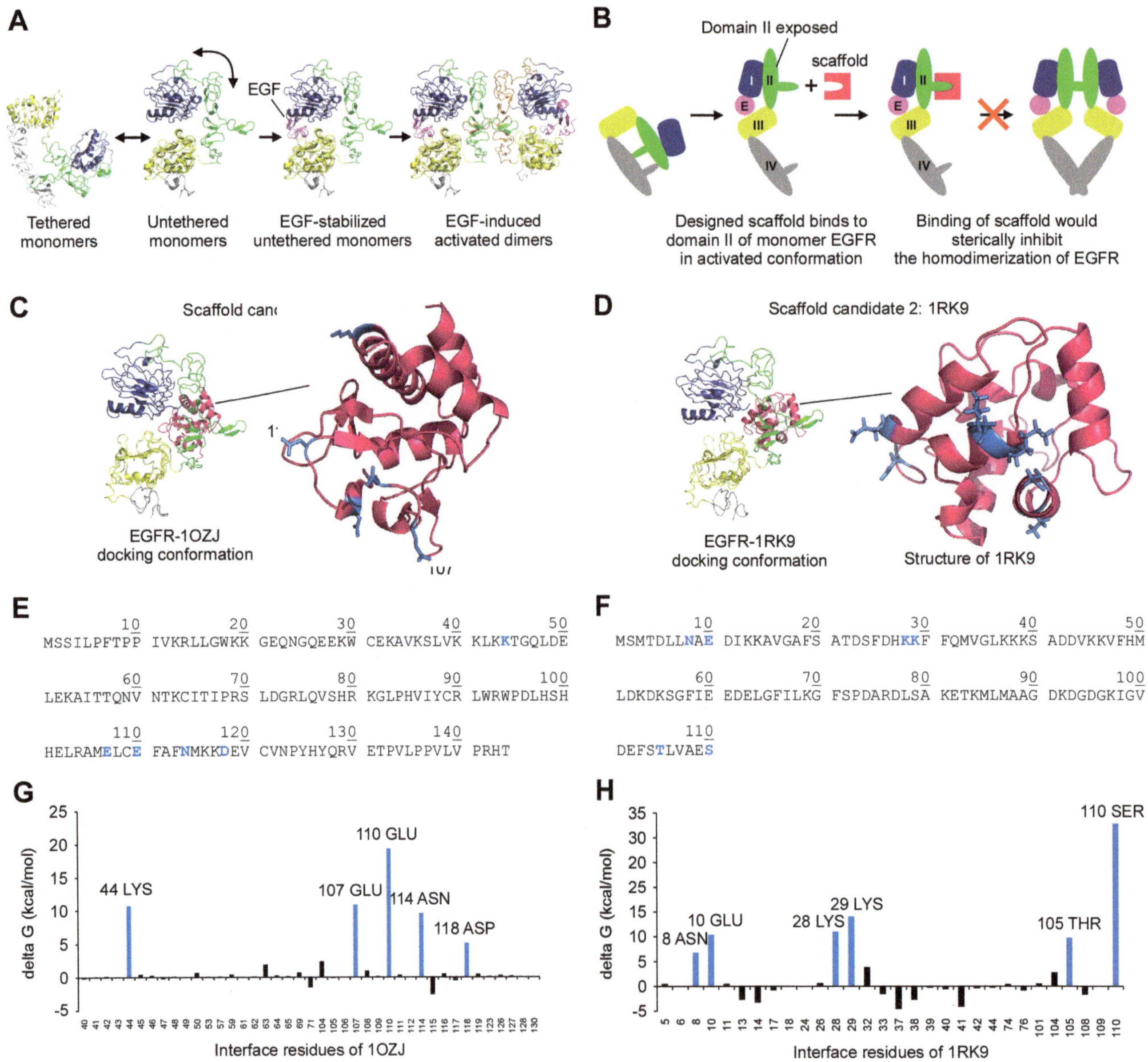


**Figure 2. Scaffolds 1OZJ and 1RK9 have shapes complementary to EGFR domain II.** (A) The inactivated EGFR monomer exists in equilibrium between the tethered and untethered conformations. The binding of EGF (pink) stabilizes the untethered monomers, which exposes the dimerization arm of domain II (green) and activates EGFR to form homodimers at the domain II dimeric interface (domain I, blue; domain II, green; domain III, yellow; domain IV, gray). For clarity, domain II on the right-hand EGFR in the homodimer is shown in orange. (B) EGFR activation can be blocked by binding domain II, which is exposed in the untethered conformation, with the designed scaffold (magenta), which sterically interferes with EGFR dimerization. (C) The docking conformation of 1OZJ–EGFR, which was selected from the virtual screening procedure (left), and the enlarged 1OZJ structure (magenta). (D) The docking conformation of 1RK9–EGFR, which was selected from the virtual screening procedure (left), and the enlarged 1RK9 structure (magenta). Energetically unfavorable residues that were selected for optimization are depicted in blue. (E–F) Amino acid sequences for 1OZJ and 1RK9, respectively. The blue residues represent amino acid residues that were energetically unfavorable for EGFR complex formation. (G–H) The complex formation energy of the scaffold interface residues upon the formation of the 1OZJ–EGFR and 1RK9–EGFR docking complexes, respectively. The blue bars depict the energetically unfavorable residues that were selected for optimization. The horizontal axis shows each amino acid in the scaffold interface, and the vertical axis represents the energy contribution (delta G) to the complex formation.

based on the protein structure and the calculation of the binding energy.

## Virtual screening of scaffolds complementary to EGFR domain II

We first constructed a massive *in silico* library of human protein scaffolds covering virtually all known human protein folds (see the Materials and Methods). The resulting library consisted of 716 semi-redundant human protein structures (Dataset S1 in File S1), which we used to screen for scaffolds that bind the target surface patch.

Subsequently, we conducted a protein docking simulation using the untethered monomer structure of EGFR, specifically chain A of the crystal structure of EGFR (Protein Data Bank [PDB] ID: 1IVO), to select scaffolds from the scaffold library with a surface shape complementary to EGFR domain II. All 716 scaffolds were evaluated using PatchDock [31,32]. For this rigid-body docking simulation, we assumed that the proteins are not undergoing large conformational changes upon binding.

Each docking attempt generated multiple binding modes, which were scored and ranked by PatchDock based on the complementarity of the protein–protein interface of the docking structures. We found that 708 out of 716 scaffolds (98.9%) generated more than 10 docking models. PatchDock determines the number of binding modes based on the number of complementary protein surface patches between the two input proteins. To select docking conformations with good shape complementarity, we chose the top 10 models for each scaffold for a total of 7,080 docking conformations. We examined the interface residues of each complex to select models in which EGFR domain II was involved in the complex formation. For this step, we chose docking structures in which at least 10 of the 28 residues of EGFR domain II were buried in the interface, which resulted in a total of 1,680 docking complexes selected from 340 scaffolds.

## Selection of scaffolds with favorable binding energy and identification of unfavorable key interface residues

From the 1,680 selected docking complexes, the two scaffolds with the most favorable binding energies (Figure 1C, left) were selected: the B chain of 1OZJ (Smad3 MH1 domain) [33] (Figure 2C) and the A chain of 1RK9 (human alpha-parvalbumin) [34] (Figure 2D). The binding energies were evaluated with the EGAD program, which was developed to calculate the residue interaction energies [35]. The amino acid sequences of 1OZJ and 1RK9 are shown in Figure 2E and 2F, respectively. The energy contributions of each interface residue of 1OZJ (Figure 2G) and 1RK9 (Figure 2H) were assessed by EGAD to select residues unfavorable to complex formation. Five unfavorable interface residues from 1OZJ and six unfavorable interface residues from 1RK9 were selected for randomization (shown in dark blue on the scaffold structures in Figure 2C and 2D, respectively, and in the amino acid sequences in Figure 2E and 2F, respectively).

## Interface refinement through directed evolution using sequence randomization and phage display

We optimized unfavorable interaction residues on the surface of scaffolds 1OZJ and 1RK9 through residue-specific sequence randomization (Figure 1C). High-complexity phage display libraries for 1OZJ and 1RK9 were constructed, and the five critical residues of 1OZJ and six residues of 1RK9 were fully randomized to cover all possible nucleotide combinations. The final size of the 1OZJ library ($4.0 \times 10^8$ combinations) was sufficient to cover all possible codon combinations. The 1RK9 library, which theoretically needed $1.1 \times 10^9$ combinations to cover all possible combinations, contained $8.7 \times 10^8$ combinations.

The libraries were subjected to five rounds of biopanning against EGFR-crosslinked magnetic beads to enrich for binders. Clones from the output titer plate of the final round of biopanning were subjected to phage-ELISA (Figure 3A). Five clones of 1OZJ and one clone of 1RK9 were selected for further analysis because their reactivity was significantly higher than that of wild-type 1OZJ and 1RK9 phages. Altered amino acid sequences of the selected 1OZJ and 1RK9 mutant clones are shown in Figure 3B and 3C, respectively. No consensus sequences were found among the high-reactivity binders of 1OZJ. However, we found that bulky amino acids (e.g., Glu, Lys, Asp) were replaced by smaller amino acids (e.g., Ala, Val, Ser).

## Binders showed reactivity against EGFR fragments containing domain II

We confirmed that the binders obtained from our directed evolution approach were specific to the EGFR domain II by using fragments of the protein. As shown in Figure 4, all six binders showed reactivity to an EGFR fragment containing domains I–IV (i.e., extracellular region of EGFR, dark blue) and an EGFR fragment containing domains I–II (light blue). Human $IgG_1$ Fc fragments and bovine serum albumin (BSA) were used as controls because the EGFR fragments were expressed and purified as Fc fusion proteins. We were unable to express a sufficient amount of an EGFR fragment containing domain II only.

To further test the specificity of selected binders, we generated chimeric EGFR fragments with the EGFR domain II replaced with domain II of ErbB2 or ErbB4 (Figure 5A). While all six binders showed reactivity to the fragments containing EGFR domain II (green bars), they did not show reactivity to the chimeric fragments containing ErbB2 domain II (orange bars) or ErbB4 domain II (dark blue bars, Figure 5B). Cetuximab, which binds to EGFR domain III [36], retained its reactivity to the chimeric EGFR fragments. Thus, the reactivity of the binders to EGFR depended on the presence of EGFR domain II.

We evaluated whether the scaffold binders showed higher reactivity to EGFR in the presence of the ligand EGF, which stabilizes EGFR in the tethered form to expose domain II (Figure 6A), using phage-ELISA with or without the addition of EGF. All six clones showed higher reactivity in the presence of EGF (pink bars) than in its absence (black bars; Figure 6B). This increased reactivity was not observed for wild-type 1OZJ, wild-type 1RK9, or the irrelevant control phage. To confirm these results, we performed phage-ELISA with eight-fold replication (i.e., eight samples for each phage clone) and found that the increased reactivity of the clones with the EGF treatment was significant (P value $= 1.5 \times 10^{-4}$ by a Wilcoxon rank-sum test; Table S1 in File S1).

We also evaluated whether the 1OZJ and 1RK9 clones showed a similar increase in reactivity to the extracellular domain of EGFR in the presence of EGF (Figure 6C). These results were similar to those obtained with whole EGFR: clones showed a higher affinity to EGFR domain I–IV in the presence of EGF (orange bars) than in its absence (black bars; Table S1 in File S1).

Interestingly, EGF treatment decreased the reactivity of cetuximab (i.e., a monoclonal antibody that targets EGFR) against both the whole EGFR protein (P value $= 1.5 \times 10^{-4}$; Figure 6B) and the EGFR extracellular domain (P value $= 0.01$; Figure 6C; Table S1 in File S1). This result is consistent with a previous observation that cetuximab binds to EGFR domain III, which is occluded by the binding of EGF [27].

## Discussion

In this study, we demonstrated a process to create protein interactions between EGFR domain II and computationally designed protein scaffolds. The two scaffolds, 1OZJ and 1RK9, previously had no common features except shape complementarity and favorable interaction energies with the target structure, EGFR domain II. By mutating several *in silico*-selected residues that had contributed to the unfavorable binding energies, we were able to produce novel protein–protein interactions for both 1OZJ and 1RK9 with EGFR domain II. We showed that the selected scaffolds specifically bound to EGFR domain II by showing that they did not bind to chimeric EGFR in which domain II was replaced by domain II of ErbB2 or ErbB4. We also confirmed that the designed scaffolds had a higher affinity to whole EGFR protein

A

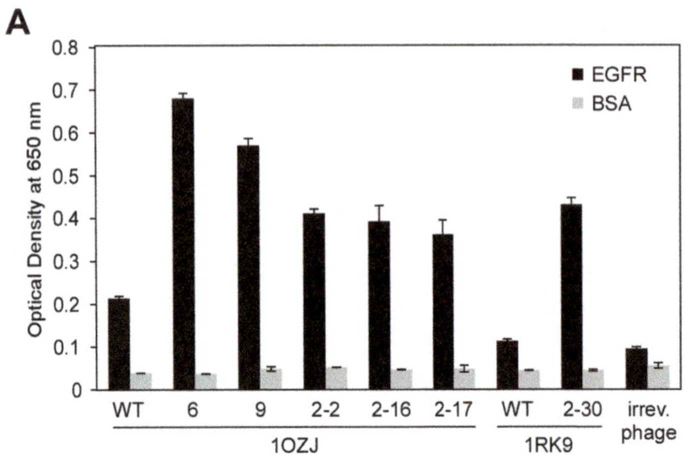


B

| Residue: | 44 | 107 | 110 | 114 | 118 |
|---|---|---|---|---|---|
| WT | K | E | E | N | D |
| 6 | S | G | S | N | L |
| 9 | A | Q | I | L | P |
| 2-2 | G | R | L | L | S |
| 2-16 | V | G | N | P | G |
| 2-17 | V | G | V | R | P |

C

| Residue: | 8 | 10 | 28 | 29 | 105 | 110 |
|---|---|---|---|---|---|---|
| WT | N | E | K | K | T | S |
| 2-30 | G | A | K | K | G | L |

**Figure 3. Strong binders were enriched by directed evolution of the 1OZJ and 1RK9 scaffolds.** (A) The reactivity of the protein binders (wild-type and mutant clones) against EGFR (black) and BSA (gray) as assessed by phage-ELISA. Error bars represent the standard deviation of triplicate measures. (B) The sequence of the mutant clones generated from 1OZJ. (C) The sequence of the mutant clone generated from 1RK9.

and to the extracellular domain of EGFR in the presence of the ligand EGF, which stabilizes EGFR in the untethered conformation to expose domain II [25,27]. Our results demonstrate how non-interface protein surfaces can be changed to protein interfaces in a process that is independent from gene duplication.

Interestingly, the replaced residues of the high-reactivity binding scaffolds did not share properties of charge or hydrophobicity (Figure 3B). However, in both scaffolds, large amino acids in the binding interface were replaced with smaller amino acids. This result indicates that, in a protein pair with overall shape complementarity and favorable residue–residue interactions, the creation of a *de novo* interface may involve the elimination of steric hindrance on the surface patch. In nature, this change is likely to occur through accumulated point mutations.

This study has implications for the production of docking decoys. Protein docking simulation between two arbitrary proteins can generate hundreds or thousands of docking models. Non-native docking structures are usually considered artifacts. However, these structures may be considered candidates for *de novo* protein–protein interactions if these structures can be optimized to improve binding affinity.

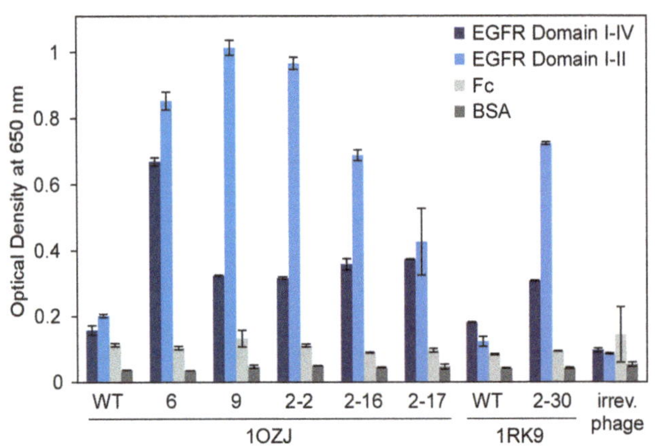


**Figure 4. Binders generated from the 1OZJ and 1RK9 scaffolds bind to EGFR fragments I–IV and I–II.** The reactivity of the protein binders (wild-type and mutant clones) against EGFR domain I–IV (dark blue) and EGFR domain I–II (light blue) as assessed by phage-ELISA. Error bars represent the standard deviation of triplicate phage-ELISA experiments.

We believe that computationally designed scaffolds have advantages over antibodies and conventional scaffolds. For example, our strategy enables a predetermined surface patch to be targeted at the start of the screening process. In contrast, the binding patch of conventional antibodies and single framework-dependent scaffolds is randomly determined. In addition, the process introduced in this paper selects epitope-specific frameworks. The structures that can be generated from this method are therefore more diverse than antibodies and single-framework-dependent scaffolds, because diversity in antibodies and single-framework-dependent scaffolds is inherently limited to their original template.

One limitation of our system is the need to produce a high-resolution structure of the target protein during the screening of the scaffold structure. Our method therefore cannot be used for targets with unknown structures. However, the scaffolds that can be generated using our computational method would make useful probes to screen for antibodies that bind to an epitope of interest, particularly because current epitope mapping methods are limited, and the identification of an epitope is not always easy. Current techniques, such as western-blot analysis, site-directed mutagenesis, and tandem mass spectroscopy, all have significant limitations. The antibody–antigen binding mode can be directly visualized at the atomic level by X-ray crystallography, which is probably the most definitive approach for epitope mapping [27,36,37]. X-ray crystallography is costly and time-consuming [38], however, and it is not always possible to crystallize the protein complex. Our computationally designed scaffolds can be used to screen for epitope-specific antibodies for use in competitive binding assays.

In summary, we used computational design and directed evolution to induce protein interactions between two previously unrelated proteins. Our results suggest that this method could be successfully applied to develop epitope-specific binder proteins for experimental or therapeutic use.

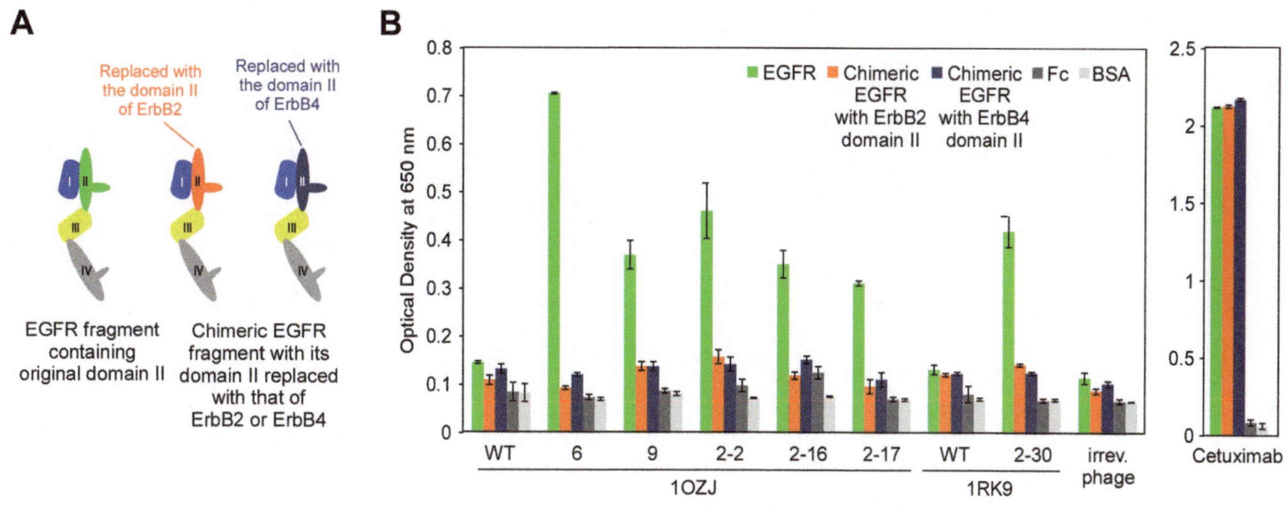


**Figure 5. Reactivity of the binders to EGFR depended on the presence of EGFR domain II.** (A) Chimeric EGFR fragments were generated with the EGFR domain II replaced with domain II of either ErbB2 (orange) or ErbB4 (dark blue). (B) The reactivity of the protein binders to the EGFR fragments containing EGFR domain II (green), the chimeric EGFR fragments with ErbB2 domain II (orange), and the chimeric EGFR fragments with ErbB4 domain II (dark blue), as assessed by phage-ELISA. Error bars represent the standard deviation of triplicate phage-ELISA experiments.

## Materials and Methods

### Construction of the human protein scaffold library

To construct a library that covers virtually all known human protein folds, we selected at most three representative protein folds from each structural class in the Structural Classification of Proteins database [39]. In structural classes consisting of fewer than three protein folds, we selected every available protein fold. However, only soluble and non-antibody proteins (molecular weight 10–40 kDa) were selected because low-molecular-weight proteins are likely to be overexpressed [7]. We eliminated membrane protein structures using annotations from the Protein Data Bank of Transmembrane Proteins [40,41]. We eliminated antibodies using a keyword search in the Protein Data Bank [42]. To avoid promiscuous binding, we excluded proteins with >10 known interactions in the annotated Human Protein Reference Database [43]. We also excluded proteins that form homomultimers (other than homodimers and homotrimers) based on the SUBUNIT section of the Swiss-Prot database [44]. All the databases used in this study were downloaded in March of 2007.

### Screening scaffolds with shapes complementary to EGFR domain II

To screen the scaffolds for shapes complementary to EGFR domain II, docking simulations between each scaffold in the library and the A chain of the human EGFR extracellular domain (RCSB PDB ID: 1IVO) [26] were conducted using PatchDock (version 1.3) [31,32]. For all other parameters in PatchDock, default values were used. The top 10 docking models were selected from each scaffold–EGFR docking result, and their binding modes were analyzed to find docking structures in which EGFR domain II was involved in the complex formation. A scaffold was considered to contact EGFR domain II in the docking complex if more than 10 of the following 28 residues of domain II were involved in the complex formation: 229, 230, 239, 242, 244, 245, 246, 248, 249, 250, 251, 252, 253, 262, 263, 264, 265, 275, 278, 279, 280, 282, 283, 284, 285, 286, 303, and 304. This residue numbering is consistent with the RCSB PDB structure of EGFR (1IVO). A residue was considered to be involved in binding if the residue underwent a change in solvent-accessible surface area >1 Å$^2$ upon the formation of the docking complex between the two chains [45,46]. The solvent-accessible surface area of each residue was calculated using NACCESS [47].

### Energy calculations for complex formation

The complex-formation energy and the energy contribution of each contact residue of the scaffolds were calculated using the EGAD program (April 2005 version) [35]. Using a docking structure generated by PatchDock as the template, JOBTYPE complex_formation_energy was used to calculate the binding energy; for all other options, default values were used. From the calculation result, the pseudo_DELTA_G_complex_formation value was used to evaluate complex formation energy, and dG values of level_1 interface residues were collected to determine unfavorable residues. Level_2 residues indirectly involved in the complex formation were excluded.

### Random library construction

The genes encoding the EGFR-binding scaffolds 1OZJ and 1RK9 were chemically synthesized (Genscript, Piscataway, NJ). Using these DNA molecules as templates, the target residues were randomized using NNK primers (Genotech Corporation, Daejeon, Korea). The following sequences of primers were used to randomize the five specific residues of the 1OZJ scaffold: primer 1: 5′ GG GCC CAG GCG GCC ATG TCG TCC 3′; primer 2: 5′ CTT GAG TTT CTT GAC CAG GCT CTT GAC CGC CTT CTC GCA 3′; primer 3: 5′ AGC CTG GTC AAG AAA CTC AAG NNK ACG GGG CAG CTG GAC GAG 3′; primer 4: 5′ GGG ATT CAC GCA GAC CTC MNN CTT CTT CAT MNN GAA GGC GAA MNN ACA CAG MNN CAT GGC CCG TAG CTC GTG 3′; and primer 5: 5′ GGC CGG CCT GGC CTG TGT GGC GTG GCA CCAACA CAG GAG GTA GAA CTG GTG TCT CTA CTC TCT GGT AGG GAT TCA CGC AGA CCT 3′. The following sequences of primers were used to randomize the six specific residues of 1RK9: primer 1: 5′ G GCC CAG GCG GCC ATG CGA ATG ACA GAC TTG CTG NNK GCT NNK GAC 3′; primer 2: 5′ CAG GCC GAC CAT TTG GAA GAA MNN MNN GTG GTC GAA GG 3′; primer 3: 5′

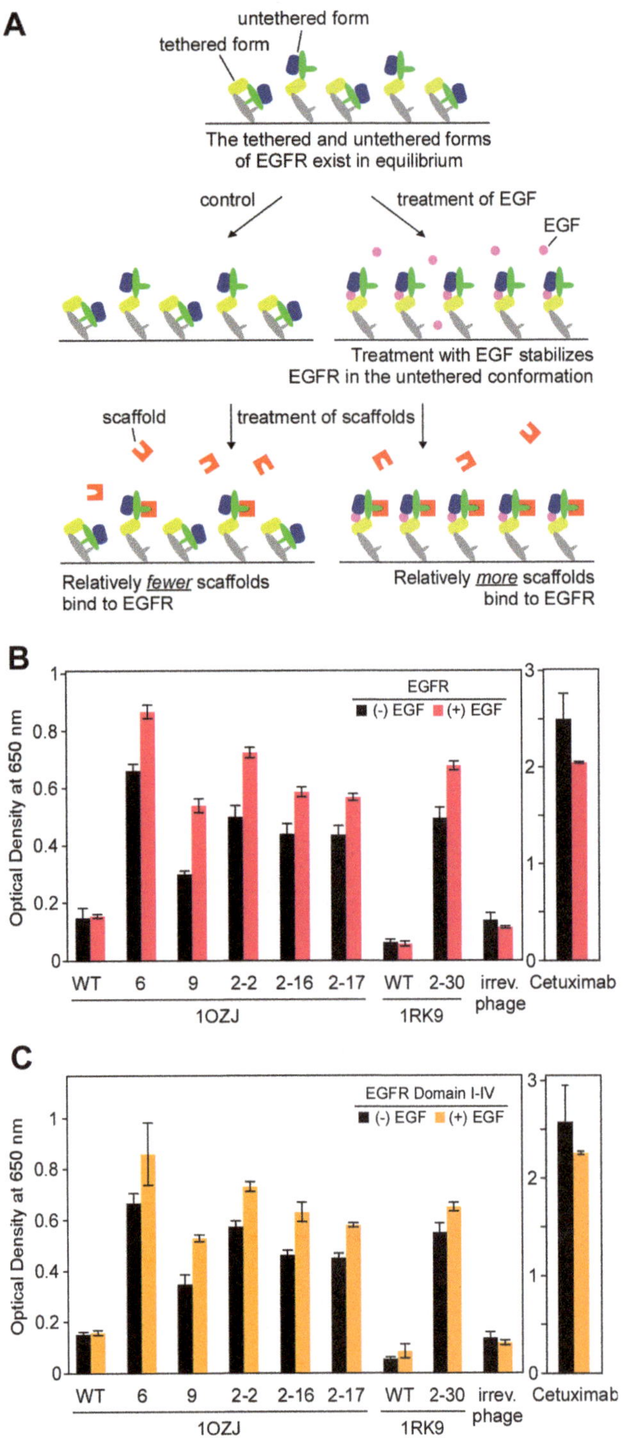


**Figure 6. Binders have a higher reactivity to EGFR in the presence of ligand EGF.** (A) Ligand EGF stabilizes the untethered form of EGFR to expose domain II and increase scaffold binding (domain I, blue; domain II, green; domain III, yellow; domain IV, gray; EGF ligand, pink; scaffold, red). (B) The reactivity to EGFR in the presence of EGF (pink) or the absence of EGF (black). (C) The reactivity to EGFR fragment containing domains I–IV in the presence of EGF (orange) or the absence of EGF (black) as assessed by phage-ELISA. Error bars represent the standard deviation of phage-ELISA experiments performed with eight-fold replication.

CC TTC GAC CAC NNK NNK TTC TTC CAA ATG GTC GGC CTG 3′; and primer 4: 5′ GGCC GGC CTG GCC TTA MNN TTC AGC CAC CAG MNN GG 3′.

The first polymerase chain reaction (PCR) amplification was performed in a 50-μl reaction volume containing 10 pmol of primers 1 and 2 or primers 3 and 4, 100 ng of the template DNA, 4 μl of 2.5 nM dNTP mix (Promega, Madison, WI), 10× reaction buffer, and 0.5 μl high-fidelity Taq polymerase (Roche, Indianapolis, IN). The PCR protocol consisted of 25 cycles of denaturation at 94°C for 15 s, annealing at 56°C for 30 s, and extension at 72°C for 90 s. Following 2% agarose gel electrophoresis, the specific DNA band was excised and purified using the QIAEX II Gel Extraction Kit (Qiagen).

To link the DNA fragment that was amplified with primers 1 and 2 with the DNA fragment that was amplified with primers 3 and 4, a second PCR amplification was performed in a 50-μl reaction volume containing 100 ng each of the two PCR products, 10 pmol of primers 1 and 4, 4 μl of 2.5 nM dNTP mix (Promega), 10× reaction buffer, and 0.5 μl high-fidelity Taq polymerase (Roche) using the same PCR protocol as described above. Linked DNA was confirmed and purified as described above. For the 1OZJ scaffold, one more step of PCR was performed with 10 pmol of primers 1 and 5 and the second PCR product as the template, using the same PCR protocol.

The PCR resulted in genes encoding the 1OZJ and 1RK9 scaffolds randomized at the selected residues and containing *Sfi* I restriction sites at the 5′ and 3′ ends. The scaffold DNA and phagemid vector pComb3X (Barbas III, 2001) were digested with *Sfi*I (Roche) and purified using the QIAEX II Gel Extraction Kit (Qiagen). The scaffold DNA and vector were ligated at 16°C overnight using T4 DNA ligase (Roche). The phage display library was constructed as previously described (Barbas III, 2001) the next day.

## Biopanning

Dynabeads (M-270 Epoxy, Invitrogen, Carlsbad, CA) were washed twice with 100 μl of sodium phosphate (0.1 M), and $5\times10^7$ beads were conjugated to 15 μg of EGFR (Sigma, St. Louis, MO) according to the manufacturer's instructions. On the following day, the beads were washed four times with phosphate buffered saline (PBS; 137 mM sodium chloride, 10 mM phosphate, 2.7 mM potassium chloride, pH 7.4), incubated with 3% BSA in PBS for 1 h, and washed with 0.5% Tween-20 in PBS (PBST).

To enrich for specific EGFR binders, five rounds of biopanning were performed as described previously (Barbas III, 2001) with some modifications. Briefly, $10^7$ beads were used for the first round of biopanning, and $5\times10^6$ beads were used for consecutive rounds of biopanning. EGFR-coated beads were incubated with 500 μl of randomized phage library for 2 h at room temperature in the rotator. A total of $10^{12}$ to $10^{13}$ phages were subjected to the first round of biopanning. Following an incubation period, the beads were washed with 0.05% TBST (v/v) to remove unbound phage one time during the first round of biopanning, three times during the second and third rounds of biopanning, and five times during the final two rounds of biopanning. Phages bound to the beads were eluted by adding 30 μl of glycine-HCl (0.1 M, pH 2.2) twice with 5-min incubations and neutralized with 2 M Tris-HCl (pH 9.1). Freshly grown *Escherichia coli* ER 2738 cells were infected with the eluted phages, and the phages were rescued by adding $10^{12}$ VCSM13 helper phage. Following an overnight culture, the rescued phages were precipitated from the culture supernatant for use in the next round of biopanning using polyethyleneglycol.

During each round of biopanning, the numbers of input and output phages were titrated.

### Expression and purification of EGFR fragments

EGFR cDNA was obtained from the Korea Human Gene Bank (Daejeon, Korea). Primers were synthesized (Genotech) to produce DNA fragments of EGFR domains I–II and I–IV with *Sfi* I restriction enzyme sites at the 5′ and 3′ ends. The following primer sequences were used for EGFR domain I–II: (forward) 5′ GGC CCA GGC GGC CCT GGA GGA AAA GAA AGT TTGC 3′ and (reverse) 5′ GGC CGG CCT GGC C GCG GCA AGG CCC TTC GC 3′. The following primer sequences were used for EGFR domain I–IV: (forward) 5′ GGC CCA GGC GGC CCT GGA GGA AAA GAA AGT TTGC 3′ and (reverse) 5′ GGC CGG CCT GGC C GCG GCA AGG CCC TTC GC 3′.

The EGFR fragments were amplified by PCR in a 50-µl reaction volume containing 100 ng of EGFR cDNA, 10 pmol of sense and reverse primers (specific to either domain I–II or I–IV), 4 µl of 2.5 nM dNTP mix (Promega), 10× reaction buffer, and 0.5 µl high-fidelity Taq polymerase (Roche). The PCR protocol consisted of 20 cycles of denaturation at 94°C for 15 s, annealing at 56°C for 30 s, and extension at 72°C for 90 s. Following 1% agarose gel electrophoresis, the specific DNA band was excised and purified using the QIAEX II Gel Extraction Kit (Qiagen). A modified pCEP4 vector containing the human IgG1 Fc fragment (Park et al, 2010) and the purified PCR products of EGFR domain I–II and EGFR domain I–IV were simultaneously digested with *Hind*III and *BamH*I (New England Biolabs, Ipswich, MA) at 37°C overnight. The inserts were ligated into the pCEP4 vector and transformed into the *E. coli* strain DH5α (Invitrogen). The plasmid DNA was prepared using the MG™ Plasmid SV kit (Macrogen, Daejeon, Korea) for transfection into human embryonal kidney (HEK) 293F cells (Invitrogen).

Gene constructs encoding the chimeric EGFR domain I–IV fragments in which the EGFR domain II was replaced with domain II of ErbB2 and ErbB4, respectively, were synthesized (Genscript) with *Sfi* I restriction sites at the 5′ and 3′ ends. The constructs were digested and ligated with pCEP4 vector.

The HEK 293F cells were cultured in GIBCO FreeStyle 293 Expression medium (Invitrogen) in Erlenmeyer tissue culture flasks (Corning Inc., Corning, NY) at 135 rpm at 37°C with 8% $CO_2$ on an orbital shaking incubator (Minitron, INFORS HT, Switzerland). Fresh medium was added the day before transfection to achieve a density of $1.0\times10^6$ cells/ml, which resulted in a density of $2.0\times10^6$ cells/ml on the day of transfection. HEK293F cells were transfected with the expression vectors using Lipofectamine™ 2000 (Invitrogen) according to the manufacturer's instructions. The transfected cells were cultured for 3 more days, and then the culture supernatants were harvested. Following filtration through a membrane (Advantec, Toyo Roshi Kaisha, Ltd., Japan), the culture supernatant was subjected to gel affinity chromatography (IPA 300 Protein A Affinity Resin, RepliGen, Waltham, MA). The purified EGFR fragments were analyzed by SDS-PAGE using NuPAGE 4%–12% Bis-Tris gel (Invitrogen).

### Phage-ELISA

Phage-ELISA was performed as described previously (Barbas III, 2001) with some modifications. Individual colonies were randomly selected from the output titration plate on the last round of biopanning and inoculated into 1 mL of super broth (3% tryptone, 2% yeast extract, 1% 3-[N-Morpholino] propanesulfonic acid, pH 7.0). The phages were rescued by adding $10^{12}$ VCSM13 helper phage. Following overnight growth at 37°C, the culture supernatant was analyzed by ELISA. A 96-well microtiter plate was coated overnight in 20 µl of PBS at 4°C with 4 µg/mL of EGFR (Sigma), the EGFR domain I–II fragment, the EGFR domain I–IV fragment, or the chimeric EGFR domain I–IV fragment with EGFR domain II replaced by ErbB2 domain II and ErbB4 domain II, respectively. The plate was washed twice with PBST and incubated with 150 µl of BSA (3%) in PBS for 1 h at 37°C. The phage-containing culture supernatant (25 µl) was mixed with 25 µl of BSA (6%) in PBS and added to each well. Following a 1-h incubation at 37°C, the plate was washed three times with PBST. Horseradish peroxidase-conjugated anti-M13 antibody (50 µl; Sigma) diluted with 3% BSA in PBS (1:5,000) was added to the plate, which was then incubated for 1 h at 37°C and washed with 0.05% PBST. For cetuximab, horseradish peroxidase-conjugated anti-Human Fc antibody (Sigma) was used. A substrate solution (3,3′,5,5′-tetramethylbenzidine; 50 µl; Pierce, Rockford, IL) was added to each well, and the optical density (OD) was measured at 650 nm. Phage titers in the culture supernatant were determined as described previously (Barbas III, 2001) to compare the reactivity of positive phage clones, and the same number of phages was used. All phage-ELISA experiments were performed in triplicate.

### ELISA with the addition of EGF

Selected phage clones expressing EGFR binders were amplified as described above. A 96-well microtiter plate was coated with 4 µg/ml of EGFR (Sigma) or EGFR domain I–IV in 20 µl of PBS at 4°C overnight. Following a 1-hr incubation with 150 µl of BSA (3%) in PBS at 37°C, a mixture of 25 µl of phage supernatant and 25 µl of EGF (160 ng/ml) in 6% BSA in PBS was added, and the assay was conducted as described above. Eight wells were used for each phage clone for maximum clarity. P-values for each template were calculated using the Wilcoxon rank-sum test.

## Supporting Information

**File S1 Supporting Dataset and Table. Dataset S1.** Human scaffold library. Research Collaboratory for Structural Bioinformatics Protein Data Bank (PDB) accession numbers of 717 proteins in human scaffold library. The first four characters indicate the PDB accession number. For proteins with more than one chain, the chain we used was specified after the colon. **Table S1.** Statistical significance of the difference in binding affinity in the presence of exogenous EGF, as determined by the Wilcoxon rank-sum test.

## Author Contributions

Conceived and designed the experiments: YSC HC SK JC SHR. Performed the experiments: YSC SY KK JY PS MK YS WJY JN. Analyzed the data: YSC SY HC SK JC SHR. Contributed reagents/materials/analysis tools: YSC SY KK JY PS MK YS WJY JN. Wrote the paper: YSC SY HC SK JC SHR.

## References

1. Yamada T, Bork P (2009) Evolution of biomolecular networks: lessons from metabolic and protein interactions. Nat Rev Mol Cell Biol 10: 791–803.
2. Berg J, Lassig M, Wagner A (2004) Structure and evolution of protein interaction networks: a statistical model for link dynamics and gene duplications. BMC Evol Biol 4: 51.

3. Levy ED, Pereira-Leal JB (2008) Evolution and dynamics of protein interactions and networks. Curr Opin Struct Biol 18: 349–357.
4. Wagner A (2003) How the global structure of protein interaction networks evolves. Proc Biol Sci 270: 457–466.
5. Kim PM, Lu LJ, Xia Y, Gerstein MB (2006) Relating three-dimensional structures to protein networks provides evolutionary insights. Science 314: 1938–1941.
6. van Noort V, Snel B, Huynen MA (2004) The yeast coexpression network has a small-world, scale-free architecture and can be explained by a simple model. EMBO Rep 5: 280–284.
7. Binz HK, Amstutz P, Pluckthun A (2005) Engineering novel binding proteins from nonimmunoglobulin domains. Nat Biotechnol 23: 1257–1268.
8. Skerra A (2007) Alternative non-antibody scaffolds for molecular recognition. Curr Opin Biotechnol 18: 295–304.
9. Gebauer M, Skerra A (2009) Engineered protein scaffolds as next-generation antibody therapeutics. Curr Opin Chem Biol 13: 245–255.
10. Koide A, Bailey CW, Huang X, Koide S (1998) The fibronectin type III domain as a scaffold for novel binding proteins. J Mol Biol 284: 1141–1151.
11. Xu L, Aha P, Gu K, Kuimelis RG, Kurz M, et al. (2002) Directed evolution of high-affinity antibody mimics using mRNA display. Chem Biol 9: 933–942.
12. Getmanova EV, Chen Y, Bloom L, Gokemeijer J, Shamah S, et al. (2006) Antagonists to human and mouse vascular endothelial growth factor receptor 2 generated by directed protein evolution in vitro. Chem Biol 13: 549–556.
13. Tolcher AW, Sweeney CJ, Papadopoulos K, Patnaik A, Chiorean EG, et al. (2011) Phase I and pharmacokinetic study of CT-322 (BMS-844203), a targeted Adnectin inhibitor of VEGFR-2 based on a domain of human fibronectin. Clin Cancer Res 17: 363–371.
14. Schlehuber S, Skerra A (2005) Lipocalins in drug discovery: from natural ligand-binding proteins to "anticalins". Drug Discov Today 10: 23–33.
15. Schlehuber S, Skerra A (2005) Anticalins as an alternative to antibody technology. Expert Opin Biol Ther 5: 1453–1462.
16. Skerra A (2007) Anticalins as alternative binding proteins for therapeutic use. Curr Opin Mol Ther 9: 336–344.
17. Stumpp MT, Binz HK, Amstutz P (2008) DARPins: a new generation of protein therapeutics. Drug Discov Today 13: 695–701.
18. Stumpp MT, Amstutz P (2007) DARPins: a true alternative to antibodies. Curr Opin Drug Discov Devel 10: 153–159.
19. Binz HK, Amstutz P, Kohl A, Stumpp MT, Briand C, et al. (2004) High-affinity binders selected from designed ankyrin repeat protein libraries. Nat Biotechnol 22: 575–582.
20. Fleishman SJ, Whitehead TA, Ekiert DC, Dreyfus C, Corn JE, et al. (2011) Computational design of proteins targeting the conserved stem region of influenza hemagglutinin. Science 332: 816–821.
21. Bianco R, Gelardi T, Damiano V, Ciardiello F, Tortora G (2007) Rational bases for the development of EGFR inhibitors for cancer treatment. Int J Biochem Cell Biol 39: 1416–1431.
22. Lockhart AC (2005) The epidermal growth factor receptor as a target for colorectal cancer therapy (vol 32, pg 52, 2005). Seminars in Oncology 32: 432–432.
23. Mendelsohn J, Baselga J (2000) The EGF receptor family as targets for cancer therapy. Oncogene 19: 6550–6565.
24. Woodburn JR (1999) The epidermal growth factor receptor and its inhibition in cancer therapy. Pharmacol Ther 82: 241–250.
25. Ferguson KM, Berger MB, Mendrola JM, Cho HS, Leahy DJ, et al. (2003) EGF activates its receptor by removing interactions that autoinhibit ectodomain dimerization. Mol Cell 11: 507–517.
26. Ogiso H, Ishitani R, Nureki O, Fukai S, Yamanaka M, et al. (2002) Crystal structure of the complex of human epidermal growth factor and receptor extracellular domains. Cell 110: 775–787.
27. Li S, Schmitz KR, Jeffrey PD, Wiltzius JJ, Kussie P, et al. (2005) Structural basis for inhibition of the epidermal growth factor receptor by cetuximab. Cancer Cell 7: 301–311.
28. Sivasubramanian A, Chao G, Pressler HM, Wittrup KD, Gray JJ (2006) Structural model of the mAb 806-EGFR complex using computational docking followed by computational and experimental mutagenesis. Structure 14: 401–414.
29. Garrett TP, Burgess AW, Gan HK, Luwor RB, Cartwright G, et al. (2009) Antibodies specifically targeting a locally misfolded region of tumor associated EGFR. Proc Natl Acad Sci U S A 106: 5082–5087.
30. Scott AM, Lee FT, Tebbutt N, Herbertson R, Gill SS, et al. (2007) A phase I clinical trial with monoclonal antibody ch806 targeting transitional state and mutant epidermal growth factor receptors. Proc Natl Acad Sci U S A 104: 4071–4076.
31. Schneidman-Duhovny D, Inbar Y, Nussinov R, Wolfson HJ (2005) PatchDock and SymmDock: servers for rigid and symmetric docking. Nucleic Acids Res 33: W363–367.
32. Schneidman-Duhovny D, Inbar Y, Polak V, Shatsky M, Halperin I, et al. (2003) Taking geometry to its edge: fast unbound rigid (and hinge-bent) docking. Proteins 52: 107–112.
33. Chai J, Wu JW, Yan N, Massague J, Pavletich NP, et al. (2003) Features of a Smad3 MH1-DNA complex. Roles of water and zinc in DNA binding. J Biol Chem 278: 20327–20331.
34. Baig I, Bertini I, Del Bianco C, Gupta YK, Lee YM, et al. (2004) Paramagnetism-based refinement strategy for the solution structure of human alpha-parvalbumin. Biochemistry 43: 5562–5573.
35. Pokala N, Handel TM (2005) Energy functions for protein design: adjustment with protein-protein complex affinities, models for the unfolded state, and negative design of solubility and specificity. J Mol Biol 347: 203–227.
36. Sheriff S, Silverton EW, Padlan EA, Cohen GH, Smith-Gill SJ, et al. (1987) Three-dimensional structure of an antibody-antigen complex. Proc Natl Acad Sci U S A 84: 8075–8079.
37. Cohen GH, Silverton EW, Padlan EA, Dyda F, Wibbenmeyer JA, et al. (2005) Water molecules in the antibody-antigen interface of the structure of the Fab HyHEL-5-lysozyme complex at 1.7 A resolution: comparison with results from isothermal titration calorimetry. Acta Crystallogr D Biol Crystallogr 61: 628–633.
38. Ladner RC (2007) Mapping the epitopes of antibodies. Biotechnol Genet Eng Rev 24: 1–30.
39. Lo Conte L, Ailey B, Hubbard TJ, Brenner SE, Murzin AG, et al. (2000) SCOP: a structural classification of proteins database. Nucleic Acids Res 28: 257–259.
40. Tusnady GE, Dosztanyi Z, Simon I (2005) PDB_TM: selection and membrane localization of transmembrane proteins in the protein data bank. Nucleic Acids Res 33: D275–278.
41. Tusnady GE, Dosztanyi Z, Simon I (2004) Transmembrane proteins in the Protein Data Bank: identification and classification. Bioinformatics 20: 2964–2972.
42. Sussman JL, Lin D, Jiang J, Manning NO, Prilusky J, et al. (1998) Protein Data Bank (PDB): database of three-dimensional structural information of biological macromolecules. Acta Crystallogr D Biol Crystallogr 54: 1078–1084.
43. Goel R, Muthusamy B, Pandey A, Prasad TS (2011) Human protein reference database and human proteinpedia as discovery resources for molecular biotechnology. Mol Biotechnol 48: 87–95.
44. Bairoch A, Boeckmann B, Ferro S, Gasteiger E (2004) Swiss-Prot: juggling between evolution and stability. Brief Bioinform 5: 39–55.
45. Choi YS, Yang JS, Choi Y, Ryu SH, Kim S (2009) Evolutionary conservation in multiple faces of protein interaction. Proteins 77: 14–25.
46. Choi YS, Han SK, Kim J, Yang JS, Jeon J, et al. (2010) ConPlex: a server for the evolutionary conservation analysis of protein complex structures. Nucleic Acids Res 38: W450–456.
47. Hubbard SJ, Thornton JM (1993) NACCESS. Computer Program, University College, London.

# 15

# Identification and Multidimensional Optimization of an Asymmetric Bispecific IgG Antibody Mimicking the Function of Factor VIII Cofactor Activity

**Zenjiro Sampei, Tomoyuki Igawa*, Tetsuhiro Soeda, Yukiko Okuyama-Nishida, Chifumi Moriyama, Tetsuya Wakabayashi, Eriko Tanaka, Atsushi Muto, Tetsuo Kojima, Takehisa Kitazawa, Kazutaka Yoshihashi, Aya Harada, Miho Funaki, Kenta Haraya, Tatsuhiko Tachibana, Sachiyo Suzuki, Keiko Esaki, Yoshiaki Nabuchi, Kunihiro Hattori**

Research Division, Chugai Pharmaceutical Co., Ltd., Gotemba, Shizuoka, Japan

**Abstract**

In hemophilia A, routine prophylaxis with exogenous factor VIII (FVIII) requires frequent intravenous injections and can lead to the development of anti-FVIII alloantibodies (FVIII inhibitors). To overcome these drawbacks, we screened asymmetric bispecific IgG antibodies to factor IXa (FIXa) and factor X (FX), mimicking the FVIII cofactor function. Since the therapeutic potential of the lead bispecific antibody was marginal, FVIII-mimetic activity was improved by modifying its binding properties to FIXa and FX, and the pharmacokinetics was improved by engineering the charge properties of the variable region. Difficulties in manufacturing the bispecific antibody were overcome by identifying a common light chain for the anti-FIXa and anti-FX heavy chains through framework/complementarity determining region shuffling, and by pI engineering of the two heavy chains to facilitate ion exchange chromatographic purification of the bispecific antibody from the mixture of byproducts. Engineering to overcome low solubility and deamidation was also performed. The multidimensionally optimized bispecific antibody hBS910 exhibited potent FVIII-mimetic activity in human FVIII-deficient plasma, and had a half-life of 3 weeks and high subcutaneous bioavailability in cynomolgus monkeys. Importantly, the activity of hBS910 was not affected by FVIII inhibitors, while anti-hBS910 antibodies did not inhibit FVIII activity, allowing the use of hBS910 without considering the development or presence of FVIII inhibitors. Furthermore, hBS910 could be purified on a large manufacturing scale and formulated into a subcutaneously injectable liquid formulation for clinical use. These features of hBS910 enable routine prophylaxis by subcutaneous delivery at a long dosing interval without considering the development or presence of FVIII inhibitors. We expect that hBS910 (investigational drug name: ACE910) will provide significant benefit for severe hemophilia A patients.

**Editor:** Peter J. Lenting, Institut National de la Santé et de la Recherche Médicale, France

**Funding:** Chugai Pharmaceutical Co., Ltd. supported and funded the research described in this report. The funders authorized our decision to publish, but had no role in study design, data collection and analysis, or preparation of the manuscript.

**Competing Interests:** All authors are employees of Chugai Pharmaceutical Co., Ltd. which is conducting the clinical study of hBS910 (ACE910). TS, T. Kojima and KH are inventors of the patents and patent applications which claim FVIII-mimetic bispecific antibodies to FIXa and FX. ZS, TI, TS, YON, CM, AM, T. Kojima, T. Kitazawa and KY are inventors of the patent application which claims hBS910, the investigational drug.

* E-mail: igawatmy@chugai-pharm.co.jp

## Introduction

Hemophilia A is caused by an X-linked inherited dysfunction of coagulation factor VIII (FVIII). Patients with severe hemophilia A, who have plasma FVIII levels of less than 1% of normal, typically experience bleeding events several times a month [1]. Routine supplementation with exogenous human FVIII to maintain FVIII levels at 1% of normal or above is effective for reducing joint bleeding events and improving joint status and health-related quality of life in hemophilia A patients [2]. However, there are two major drawbacks to this prophylactic usage of exogenous FVIII. The first drawback is the necessity of frequent intravenous administration: three intravenous injections weekly of FVIII are necessary because of its low subcutaneous bioavailability and its short plasma half-life [3,4,5]. The second drawback is the development of inhibitory anti-FVIII alloantibodies, known as "inhibitors" [6]. Once FVIII inhibitors have developed, routine supplementation with exogenous FVIII will be no longer effective and the usage of exogenous FVIII for treating on-going bleeds is restricted. In such cases, alternative agents, such as activated factor VII and activated prothrombin complex concentrate, which are more expensive and have less stable hemostatic effects, need to be used to control bleeding [7,8]. Therefore, a new agent that resolves these drawbacks of exogenous FVIII is awaited in the field of the bleeding prophylaxis of severe hemophilia A.

Monoclonal antibodies have become an important therapeutic option in numerous diseases and are expected to play a greater role in the future of disease treatment [9,10]. Various monoclonal antibodies have been generated [11]; these not only include those with antagonistic activity but also those with agonistic activity [12], catalytic activity [13], and allosteric activity [14]. Antibody engineering technologies to generate bispecific antibodies have been extensively studied due to the huge potential of these antibodies for therapeutic applications [15]. Bispecific antibodies can be applied to simultaneously target two disease related antigens, retarget effector cells against the target cell [16], and co-ligate two different antigens on the same cell [17].

FVIII is cleaved by thrombin or factor Xa (FXa), and the resultant factor VIIIa (FVIIIa) presents a heterotrimeric structure consisting of the A1 subunit, the A2 subunit, and the light chain [18]. Simultaneous binding of FVIIIa to FIXa and FX by the light chain and the A2 subunit, and by the A1 subunit, respectively, contributes to FVIII cofactor activity which places FIXa and FX into proximity, and also allosterically enhances the catalytic rate constant of FIXa [19,20,21,22,23] (Fig. 1A).

Considering this function of FVIII and the versatilities of antibodies, we hypothesized that a bispecific antibody recognizing FIXa with one arm and FX with the other arm could mimic the FVIII cofactor activity by placing FIXa and FX in spatially appropriate positions, and by allosterically enhancing the catalytic activity of FIXa (Fig. 1B) [23]. We have recently reported a recombinant humanized bispecific antibody to FIXa and FX, termed hBS23, which exerted coagulation activity in FVIII-deficient plasma, even in the presence of FVIII inhibitors, and showed *in vivo* hemostatic activity in a cynomolgus monkey model of acquired hemophilia A, and had high subcutaneous bioavailability and a 2-week half-life in cynomolgus monkey [23]. Although the pharmacological concept of FVIII-mimetic bispecific antibody was clearly demonstrated in this report, the detail of this anti-FIXa/FX asymmetric bispecific IgG antibody identification was not described, and moreover, it required further optimization in several ways before the clinical use of such an agent in humans.

For therapeutic development, optimization of the bispecific antibody by molecular engineering to enable large-scale manufacturing of the bispecific antibody at clinical grade would be required. Although a variety of molecular formats for bispecific antibodies have been studied, including single-chain diabody, tandem scFv, IgG-scFv, DVD-Ig, CrossMab, and dual-binding Fab [24], we selected an asymmetric bispecific IgG format because it is the only format that can recognize FIXa and FX with each arm and has a long half-life and a native IgG structure. However, recombinant production of this format is challenging in comparison to the other formats because it consists of two different heavy chains and two different light chains which would result in the secretion of a mixture of ten different combinations of heavy and light chains [25]. Purification of one desired bispecific antibody from a mixture including nine miss-paired byproducts is nearly impossible. Engineering technologies to resolve such a difficulty have been previously reported. First, identification of a common light chain by phage display as the partner of the two heavy chains can reduce the number of heavy and light combinations to three: one heterodimeric bispecific antibody and two homodimeric

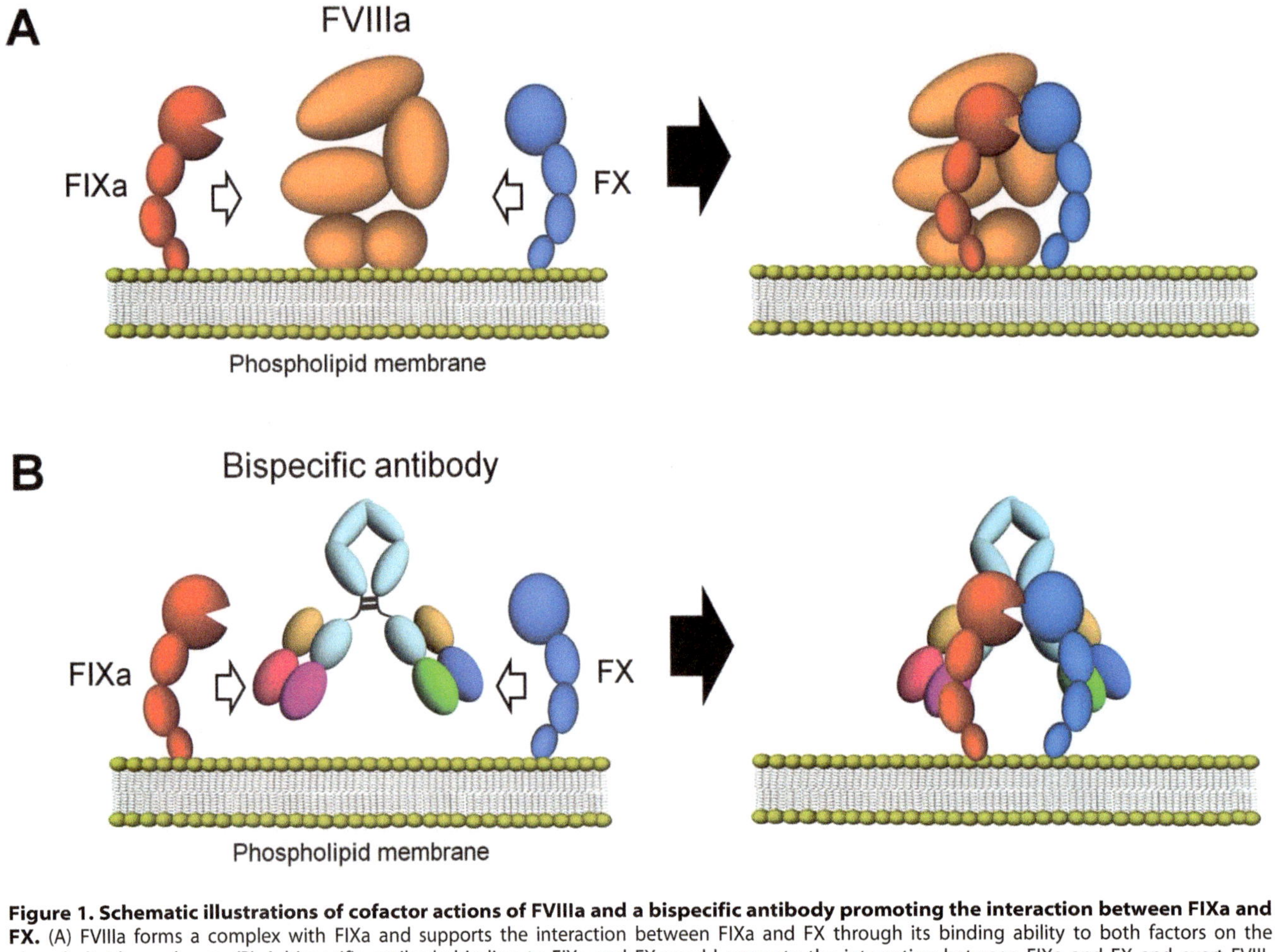


**Figure 1. Schematic illustrations of cofactor actions of FVIIIa and a bispecific antibody promoting the interaction between FIXa and FX.** (A) FVIIIa forms a complex with FIXa and supports the interaction between FIXa and FX through its binding ability to both factors on the phospholipid membrane. (B) A bispecific antibody binding to FIXa and FX would promote the interaction between FIXa and FX and exert FVIII-mimetic activity on the phospholipid membrane.

monospecific antibodies [26]. However, selection of a common light chain with potent FVIII-mimetic activity based on the binding affinity by phage display was not feasible for our bispecific antibody, because higher binding affinities was considered not to lead to higher activity. To date, alternative approach to obtain common light chain has not been reported. Second, engineering the $C_H3$ domain to facilitate Fc heterodimerization can minimize the amount of homodimeric byproducts [25,26]. Nevertheless, the efficiency of heterodimerization is not complete; therefore, a small amount of homodimeric byproduct is formed, which needs to be removed in the downstream purification process. However, because the biophysical properties of the homodimeric byproducts are often similar to those of the target bispecific antibody, purifying the target bispecific antibody on a large scale for clinical applications is still challenging. To date, technologies to address this issue have not been reported. Therefore, improving the manufacturability of this type of FVIII-mimetic antibody by molecular engineering is required for therapeutic application.

Moreover, for maximizing the therapeutic potential of FVIII-mimetic bispecific antibody, optimization to increase the FVIII-mimetic activity of the bispecific antibody, prolong the half-life, improve the physicochemical properties of the antibody and reduce the immunogenicity of the humanized antibody would be required. This would enable more effective and long-term prophylaxis with stronger hemostatic effect for hemophilia A patients by a subcutaneous formulation with a longer dosing interval.

In this paper, we report molecular identification and multidimensional optimization of a FVIII-mimetic bispecific antibody which can be used for clinical application. At the start of this study, we examined bispecific combinations of large number of monoclonal anti-FIXa and anti-FX antibodies, and identified the lead anti-FIXa/FX bispecific IgG antibody having FVIII-mimetic activity. Then, this lead bispecific antibody was subjected to multidimensional optimization processes [27] to improve both its therapeutic potential and manufacturability. We successfully generated a humanized bispecific IgG antibody having sufficient FVIII-mimetic activity for prophylactic use even in the presence of FVIII inhibitors, high subcutaneous bioavailability with an approximately 3-week plasma half-life in cynomolgus monkeys, and minimal immunogenicity risk. In addition, molecular engineering enabled purification on a large manufacturing scale and formulation into a liquid formulation of 150 mg/mL for subcutaneous delivery. We expect that this anti-FIXa/FX bispecific antibody mimicking FVIII cofactor activity will provide significant benefit for managing bleeding events in severe hemophilia A patients.

## Results

### Research Flow of Identification and Multidimensional Optimization of Lead FVIII-mimetic anti-FIXa/FX Bispecific Antibody

Figure 2 shows the flow of the screening process to identify the lead bispecific antibody with a common light chain (BS15). BS15 was generated by combinatorial screening of bispecific antibodies composed of anti-FIXa and anti-FX antibodies derived from immunization, followed by screening of common light chains and then framework/complementarity determining region (FR/CDR) shuffling.

Figure 3 represents the multidimensional optimization flow to generate the bispecific antibody with the most appropriate properties for clinical application (hBS910) from the lead bispecific antibody (BS15). BS15 was firstly humanized to generate hBS1, followed by engineering to improve FVIII-mimetic activity (hBS106), improve pharmacokinetics (hBS128 and hBS228), enable purification of target bispecific antibody (hBS366 and hBS376), improve solubility (hBS560), remove deamidation site (hBS660), and reduce immunogenicity risk (deimmunization) to generate a multidimensionally optimized bispecific antibody (hBS910). Through this multidimensional optimization process, the numbers of variable region variants that we have generated for anti-FIXa heavy chain, anti-FX heavy chain and common light chain were approximately 500, 300 and 400, respectively, and the number of bispecific IgG antibodies that we have prepared and evaluated is approximately 2,400. Supplementary table S1 represents the number of mutations which were introduced into hBS1 to generate bispecific antibodies described in this report.

### Identification of Lead Anti-FIXa/FX Bispecific Antibody with FVIII-mimetic Activity

Approximately 200 monoclonal antibodies against FIXa or FX were obtained from animals immunized with either human FIXa or FX. Approximately 40,000 bispecific IgG antibodies that comprised different combinations of anti-FIXa and anti-FX antibodies in each arm were expressed. Although the expression product of two different heavy chain and two different light chain genes consists of a mixture of ten different species with different heavy and light chain combinations, including the one desired bispecific antibody and the nine miss-paired antibodies (miss-paired antibody includes two homodimeric antibodies with a correct heavy and light chain pair), Fc heterodimerization mutations would theoretically enable expression of an antibody mixture containing at least approximately 20% of the target bispecific antibody (see methods for detail) [25]. A total number of 94 bispecific antibodies, or combinations of anti-FIXa heavy chain and anti-FX heavy chain, which had FVIII-mimetic activity were successfully identified by an enzymatic assay. Heavy chain combinations were selected from the point of high FVIII-mimetic activity, not from the point of the similarity of the cognate light chain.

Next, in order to identify a common light chain for the different heavy chains to FIXa and FX, the selected heavy chain combinations were expressed with either one of cognate light chains. Out of 188 light chain commonized bispecific antibodies, the most potent one, termed c1, which consisted of rat anti-FIXa $V_H$ and mouse anti-FX $V_H$ chimerized with human $IgG_4$ and rat anti-FIXa $V_L$ chimerized with human κ (c1L), was selected for the next step.

Finally, in order to design a more potent common light chain, we performed FR/CDR shuffling. Since the cognate light chain for the selected anti-FX $V_H$ (c2L) was not effective as a common light chain at all (data not shown), we tried to seek another effective common light chain with a high homology to c2L. From our antibody source, we identified a light chain (c3L) whose CDR sequence had >85% homology to that of c2L, and found out that it was effective as a common light chain for the selected two heavy chains. Therefore, we decided to use c3L for FR/CDR shuffling, too. Then, CDRs of c1L, c2L and c3L were shuffled among each other and grafted onto the FRs of c1L and c3L to generate twenty four light chain variants (supplementary Fig. S2A). Twenty four light chain variants were expressed with the two heavy chains, and the most potent common light chain, BS15L, a mouse–rat hybrid $V_L$ chimerized with human κ, was identified (supplementary Fig. S2B). Thus, the light chain communized bispecific antibody with BS15L was selected as the lead chimeric bispecific IgG antibody (termed BS15).

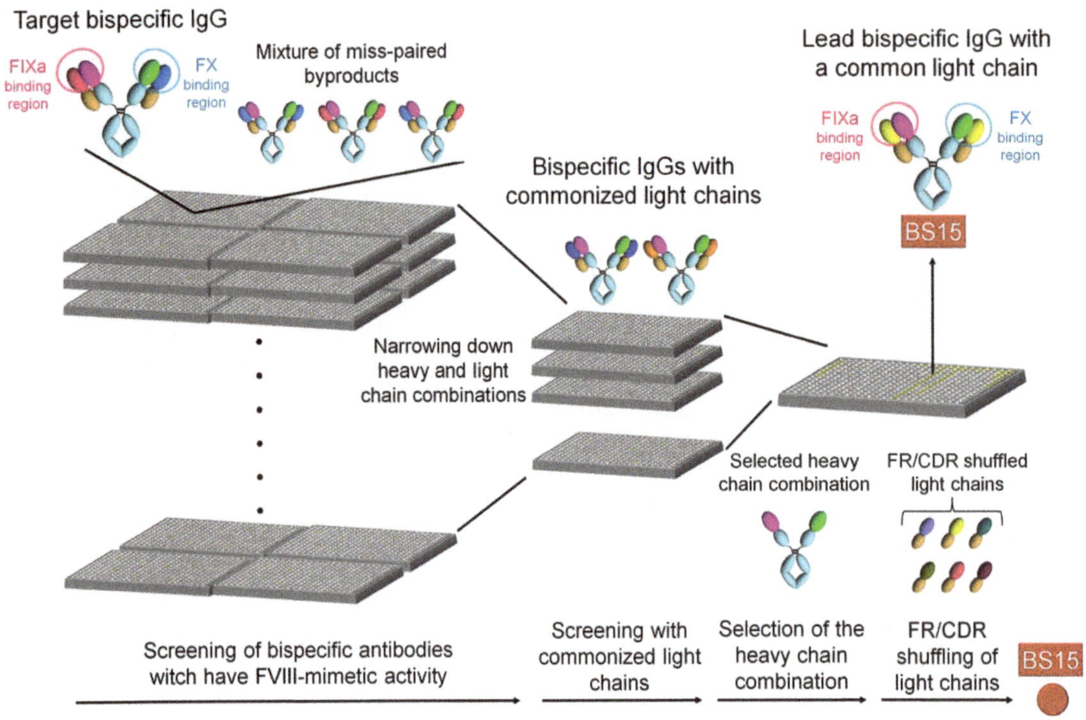


**Figure 2. Flow of process to identify the lead bispecific antibody (BS15).** BS15 was identified by combinatorial screening of bispecific antibodies, followed by screening of common light chains and then FR/CDR shuffling.

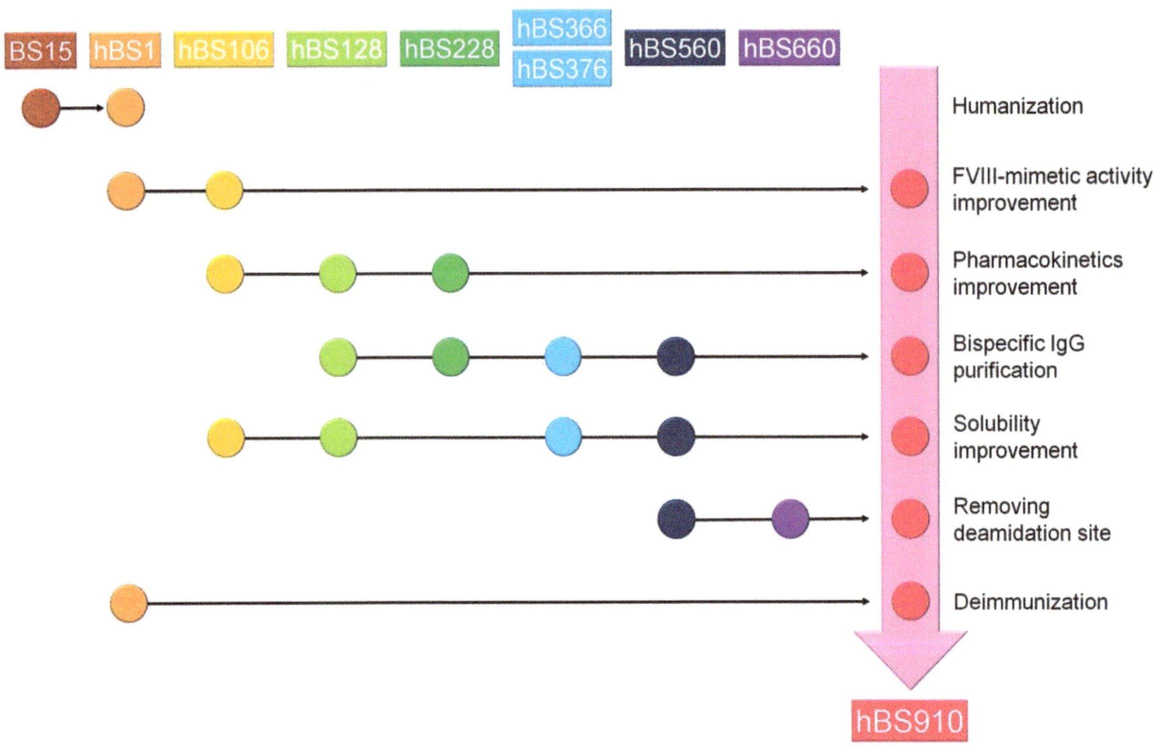


**Figure 3. Multidimensional optimization flow to generate the bispecific antibody with most appropriate properties (hBS910).** hBS910 was generated through multidimensional optimization with various antibody engineering technologies.

## Humanization of Lead Chimeric Bispecific Antibody

The lead chimeric bispecific IgG antibody, BS15, was subjected to humanization. CDRs of the anti-FIXa heavy chain, the anti-FX heavy chain and the common light chain were grafted onto homologous human antibody FRs, which were the FRs of $V_H3$, $V_H1$ and $V_\kappa1$ subfamilies respectively, by using a conventional CDR grafting approach [28]. A humanized bispecific $IgG_4$ antibody, termed hBS1, was successfully generated while maintaining FVIII-mimetic activity (supplementary Fig. S3).

## Improving FVIII-mimetic Activity of the Bispecific Antibody

Although the humanized bispecific antibody hBS1 enhanced FX activation dose-dependently, demonstrating FVIII-mimetic activity, its therapeutic potential was marginal and its FVIII-mimetic activity needed to be improved for therapeutic application. Therefore, we explored mutations in the CDRs of hBS1 to improve the FVIII-mimetic activity, and we identified several effective mutations in the CDRs of the three chains. Following extensive studies to identify effective combinations of mutations that additively or synergistically improved FVIII-mimetic activity, we successfully generated hBS106. hBS106 demonstrated marked improvement of FVIII-mimetic activity over hBS1, including maximum activity, in the enzymatic assay (Fig. 4A).

During the course of subsequent optimization of hBS106 from the point of other aspects, FVIII-mimetic activity was monitored for each mutation so that the mutation would not compromise the activity. Moreover, further screening for mutations to further improve the activity was performed in parallel with other optimizations. Finally, we successfully generated hBS910, whose activity was even higher than that of hBS106 (Fig. 4A).

## Improving Pharmacokinetics of the Bispecific Antibody

To assess the pharmacokinetics of hBS106, the variant with improved FVIII-mimetic activity, the time course of the plasma concentration of this antibody (Fig. 4B) and its pharmacokinetic parameters (supplementary Table S2) were determined in mice. Clearance of hBS106 (67 mL/day/kg) was unexpectedly larger than the clearance of human $IgG_4$ antibody reported in mice after subcutaneous injection (3–20 mL/day/kg) [29].

A homology model of hBS106 was used to explore the molecular features responsible for the poor pharmacokinetics, and a positive charge cluster was identified on the surface of Fv of the anti-FIXa arm (supplementary Fig. S4). To remove this cluster, we initially attempted subjecting lysine and arginine residues in the cluster to mutagenesis. However, these residues were found to be indispensable for the FVIII-mimetic activity (data not shown). Therefore, we explored the introduction of negatively charged residues near the cluster to neutralize the positive charge, and we identified a Tyr30Glu mutation in the common light chain that achieved this without any reduction in FVIII-mimetic activity. With hBS128, a variant of hBS106 with the single Tyr30Glu mutation, we observed improved plasma concentration and an approximately 4-fold improvement in the clearance compared to hBS106. Furthermore, the isoelectric point (pI) of hBS128 was lowered by introducing multiple mutations in the variable regions. hBS228, a variant of hBS128 with lowered pI, demonstrated further improved plasma concentration and approximately 2-fold improvement in clearance compared to hBS128 (Fig. 4B, supplementary Table S2). During the course of subsequent optimization of hBS228, we constantly made an effort to further improve the pharmacokinetics of the bispecific antibody.

## Isoelectric Point Engineering to Facilitate Purification of the Target Bispecific Antibody

Having a common light chain reduces the number of pairs of heavy and light chains to three, and engineering the $C_H3$ domain enables preferential secretion of heterodimerized heavy chains. However, it is still difficult to completely prevent miss-paired homodimerization in large-scale production. Therefore, a downstream purification process to remove homodimeric byproducts is essential for pharmaceutical development. Ion exchange chromatography (IEC) is the major purification process by which to remove impurities after Protein A purification. The retention of IgG antibodies by IEC is determined by the electrostatic charge of the antibody molecule, which can be measured as its pI. Therefore, the pIs of hBS128 and hBS228, variants with improved pharmacokinetics, and pIs of their homodimeric byproducts were determined by cIEF (Fig. 5A). For both hBS128 and hBS228, the pIs of the bispecific antibody and the homodimeric byproducts

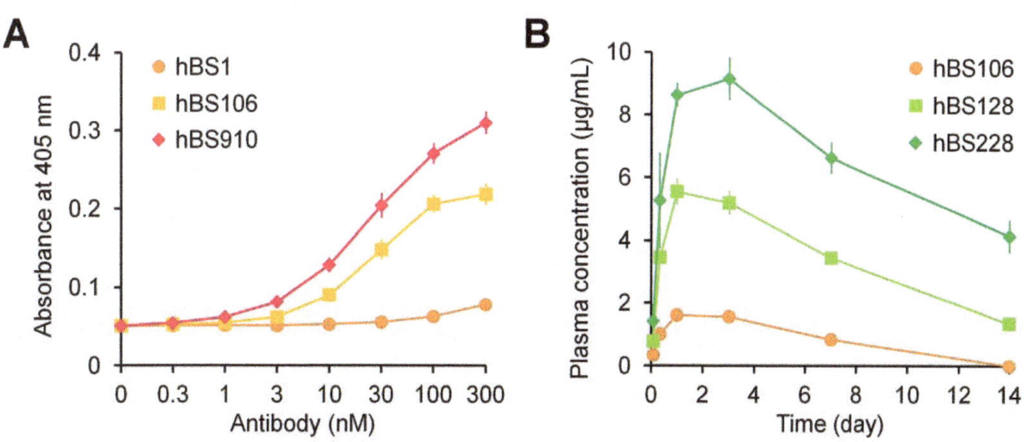


**Figure 4. Improvement of therapeutic potential of the bispecific antibody.** (A) Improving FVIII-mimetic activity of the bispecific antibody. Effect of hBS1 (circles), hBS106 (squares), and hBS910 (diamonds) on FX activation in the presence of FIXa, FX, and synthetic phospholipid is shown. The Y-axis indicates the absorbance at 405 nm of the chromogenic substrate assay (in many cases, the bars depicting s.d. are shorter than the height of the symbols). (B) Improving pharmacokinetics of the bispecific antibody. Time profiles of plasma concentration of hBS106 (circles), hBS128 (squares), and hBS228 (diamonds) in mice after subcutaneous injection at a dose of 1 mg/kg are shown. All the data were collected in triplicate and are expressed as mean ± s.d.

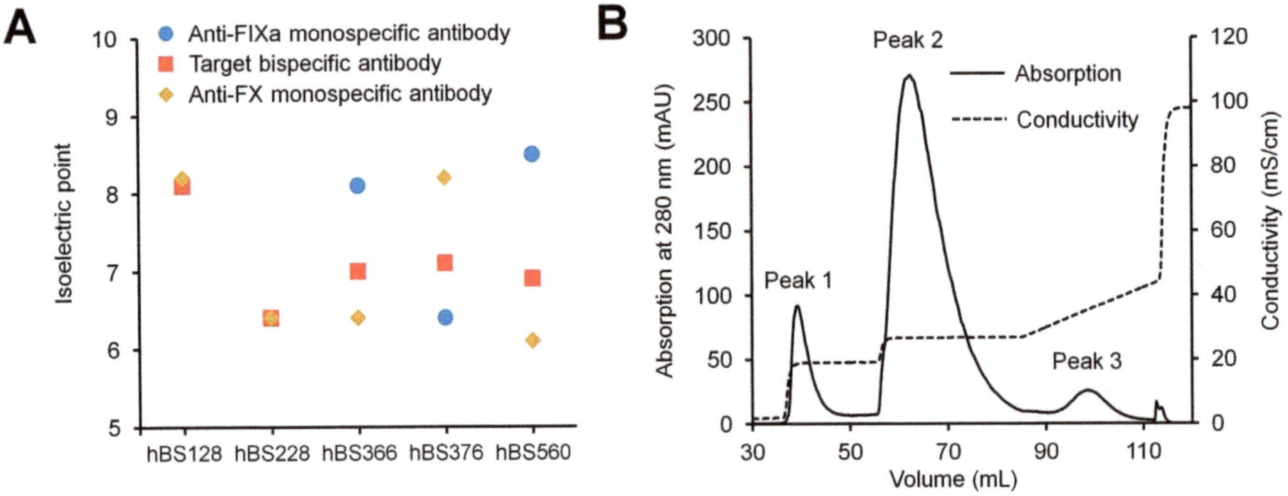


**Figure 5. Isoelectric point engineering to facilitate purification of the target bispecific antibody.** (A) Isoelectric points of target bispecific (squares) antibodies and homodimeric byproducts (anti-FIXa monospecific antibodies (circles) and anti-FX monospecific antibodies (diamonds)) determined by cIEF. (B) Cation exchange purification chromatogram of the target bispecific antibody of hBS560 from its homodimeric byproducts with step-wise elution with different NaCl concentrations. Peak 1, anti-FX homodimeric antibody; Peak 2, target bispecific antibody; Peak 3, anti-FIXa homodimeric antibody. Each peak area of peak 1, peak 2 and peak 3 was 9.9%, 85.7% and 4.4%, respectively.

were very close to each other, indicating that purification of the bispecific antibody hBS128 or hBS228 from the mixture of homodimeric byproducts is not feasible.

To facilitate the purification of the bispecific antibody, we implemented pI engineering into either one of the heavy chain variable regions to increase the pI difference between the bispecific antibody and the homodimeric byproducts. We successfully generated two variants, hBS366 (pI of the anti-FX heavy chain is lowered) and hBS376 (pI of the anti-FIXa heavy chain is lowered), in which FVIII-mimetic activity was maintained. The difference in the pI between the bispecific antibody and homodimeric byproducts markedly increased (Fig. 5A).

hBS366 was further optimized from the point of solubility (detail in the next paragraph), to generate hBS560. The target molecular form of hBS560 (heterodimeric bispecific antibody) was well separated from the two homodimeric byproducts by using cation exchange chromatography with step-wise elution (Fig. 5B). During the course of subsequent optimization of hBS560, the pI difference between the two heavy chains was carefully maintained. The multidimensionally optimized variant hBS910 was capable of being purified on a large production scale (2500-liter fermentation).

## Improving Solubility Properties of the Bispecific Antibody

hBS106, a variant with improved FVIII-mimetic activity, had unexpectedly low solubility, exhibiting either precipitation or liquid–liquid phase separation [30] (supplementary Fig. S5). This lack of solubility was partially due to the positive charge cluster, since the Tyr30Glu mutation described above markedly improved the solubility. However, for hBS376, the variant whose anti-FIXa heavy chain pI was lowered, at concentrations of 4 and 40 mg/mL, precipitation and phase separation still occurred in phosphate buffer of pH 5.5 to 7.0 and NaCl of 100 mM or less (Fig. 6A).

To improve the solubility of hBS376, mutations in the variable regions of hBS376 were explored. Several effective mutations including substitutions of hydrophobic residues into hydrophilic residues were identified, and their combinations successfully generated hBS560. Precipitation and phase separation of hBS560, the variant with improved solubility, was markedly suppressed and occurred only below 40 mM NaCl (Fig. 6A). The solubility of hBS560 was more than 100 mg/mL. During the course of subsequent optimization of hBS560, the effect of mutations on the solubility was constantly monitored and we made further efforts on improving the solubility. We successfully generated hBS910, a multidimensionally optimized variant, which did not exhibit precipitation or phase separation under the conditions tested (Fig. 6A). Furthermore, hBS910 could be concentrated up to at least 200 mg/mL.

## Removing Deamidation Site in the CDR of the Bispecific Antibody

An accelerated stability study revealed that hBS560 exhibited asparagine deamidation in the third complementarity-determining region of the heavy chain (HCDR3) (Asn99) of the anti-FIXa arm after incubation at 40°C for 2 weeks, as shown by the increase in the acidic peak in the cation exchange chromatography analysis (Fig. 6B) [31], and reduction of FVIII-mimetic activity was observed (data not shown). A single mutation of Asn99 to another amino acid to remove this deamidation site was not feasible due to the loss of FVIII-mimetic activity and solubility. Subsequently, a double mutation was explored, and hBS660, in which a His98Arg and Asn99Glu double mutation was introduced to hBS560, was identified to maintain the FVIII-mimetic activity. hBS660 showed no increase in the acidic peak after incubation, demonstrating that the deamidation site was removed (Fig. 6B). During the course of subsequent optimization of hBS660, deamidation was carefully monitored, and successfully generated hBS910, a multidimensionally optimized variant, which did not exhibit deamidation.

## Deimmunization of Humanized Bispecific Antibody by Removing T-cell Epitopes

During the course of multidimensional optimization of hBS1, the effects of each mutation on immunogenicity were evaluated by Epibase (Lonza), an *in silico* T-cell epitope prediction system [32]. Any mutation that was predicted to increase the potential

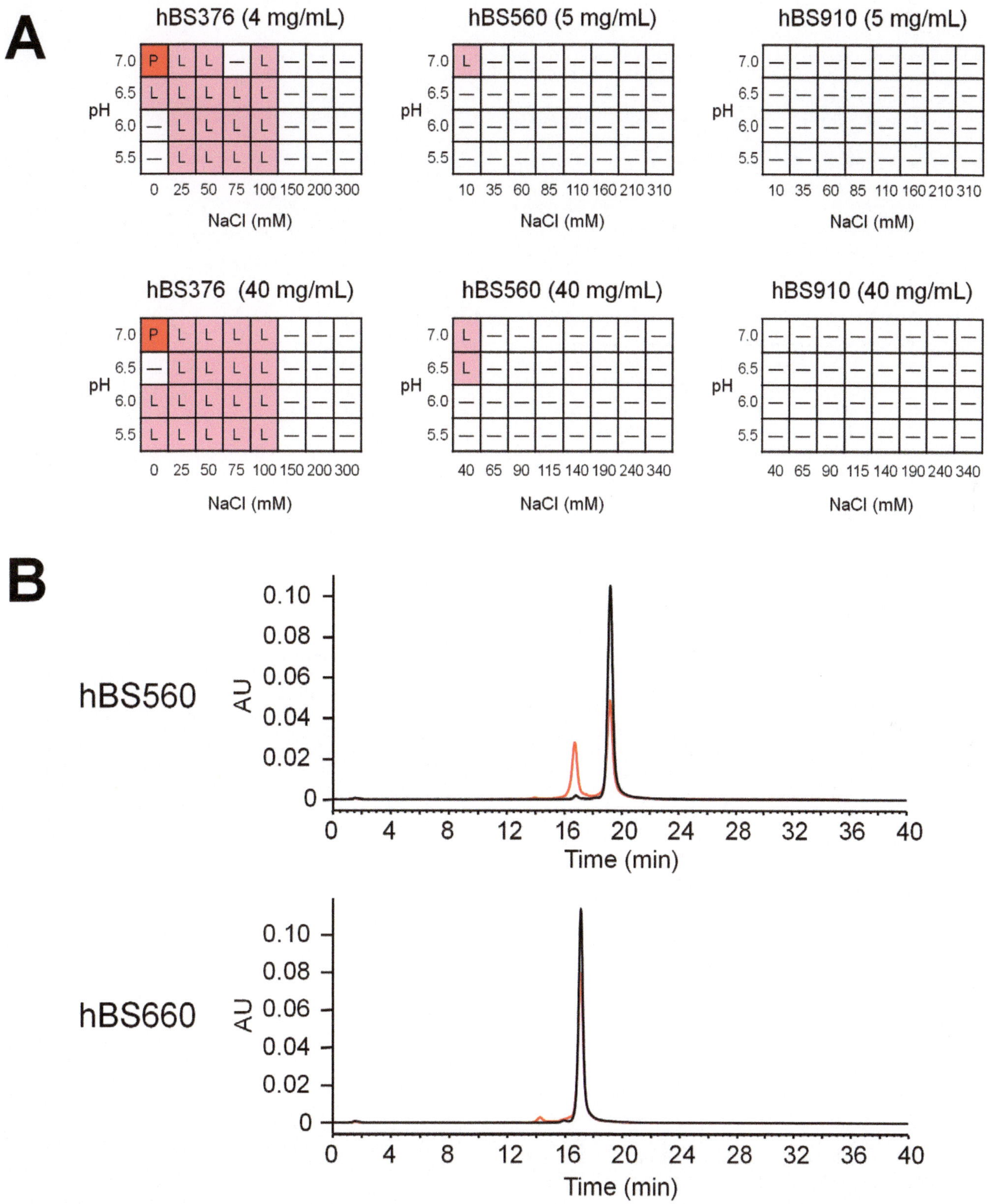


**Figure 6. Improvement of pharmaceutical properties of bispecific antibodies.** (A) Antibody solution profiles of hBS376 and hBS560 at different antibody concentrations, pH, and NaCl concentrations. The antibody solution under each condition was photographed and the state determined (P, precipitation; L, liquid–liquid phase separation; –, clear liquid). (B) Cation exchange chromatography of hBS560 and hBS660 before (black) and after incubation at 40°C for 2 weeks (red). Acidic peak indicating deamidation at HCDR3 increased after incubation at 40°C for 2 weeks for hBS560. No marked increase of acidic peak was observed for hBS660.

immunogenicity risk was avoided as much as possible. Simultaneously, to generate a bispecific antibody with minimum immunogenicity risk, any mutation that was predicted to decrease the potential immunogenicity risk by reducing the number of T-cell epitopes was screened.

The immunogenicity risk score of hBS910, the multidimensionally optimized variant, was markedly decreased compared to the lead chimeric antibody (BS15) and its humanized version (hBS1), and was comparable to trastuzumab and palivizumab which are non-immunogenic in clinical (supplementary Fig. S6A). Moreover, EpiMatrix (EpiVax), another *in silico* immunogenicity prediction system [33,34], also predicted that the sequence of hBS910 was minimally immunogenic (supplementary Fig. S6B).

### Therapeutic Potential of the Multidimensionally Optimized Bispecific Antibody, hBS910

To examine the therapeutic potential of hBS910, its FVIII-mimetic activity was compared with human FVIII by using thrombin generation assay (TGA) [23,35,36] in commercially available human FVIII-deficient plasma which was derived from a single donor with severe hemophilia A without FVIII inhibitors (Fig. 7A). hBS910 dose-dependently increased peak height (defined as the peak concentration of free thrombin) in the same manner as recombinant human FVIII (rhFVIII). The thrombin generation activity of hBS910 was also observed even in plasma of a hemophilia A donor who has FVIII inhibitors, whereas 1 IU/mL of rhFVIII did not exhibit any effects (data not shown). On the other hand, while polyclonal anti-idiotype antibodies to the anti-FIXa Fab or anti-FX Fab of hBS910 completely inhibited the activity of hBS910, they did not interfere with rhFVIII activity at all (data not shown).

To assess the potential for subcutaneous delivery with a long dosing interval, hBS910 was intravenously and subcutaneously administrated to cynomolgus monkeys, and the time course of the antibody plasma concentration and pharmacokinetic parameters were obtained (Fig. 7B, supplementary Table S3). The subcutaneous bioavailability was sufficiently high (86%) and the plasma half-life was approximately 3 weeks. Moreover, hBS910 could be formulated into a 150 mg/mL liquid formulation for subcutaneous delivery in the clinical setting without any significant aggregation or degradation during storage.

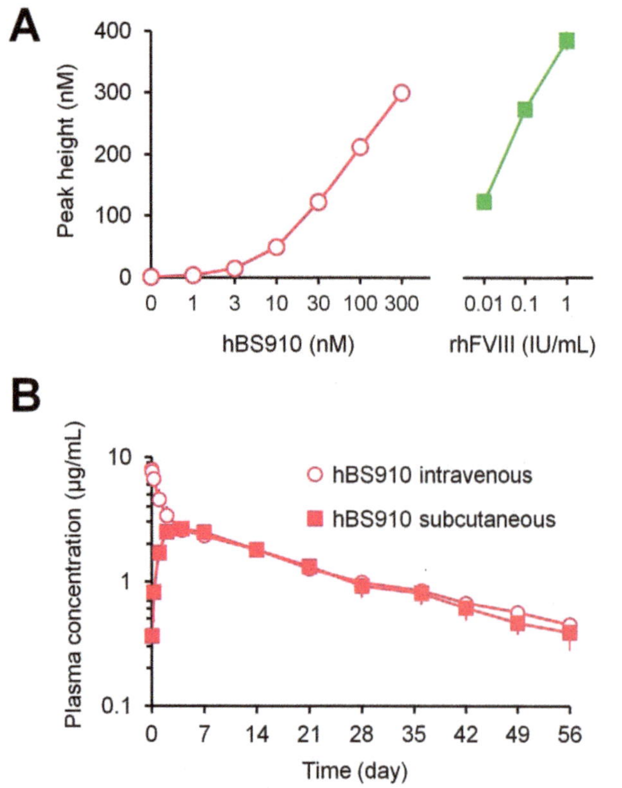


**Figure 7. Therapeutic potential of multidimensionally optimized bispecific antibody, hBS910.** (A) FVIII-mimetic activity of hBS910 in thrombin generation assay (TGA). Effect of hBS910 (circles) or recombinant human FVIII (squares) on thrombin generation in FVIII-deficient plasma is shown. The reaction was triggered by FXIa, synthetic phospholipid, and $Ca^{2+}$. The Y-axis indicates the peak height, a thrombin generation parameter (in many cases, the bars depicting s.d. are shorter than the height of the symbols). Data were collected in triplicate for each plasma lot and are expressed as mean ± s.d. (B) Pharmacokinetics of hBS910 in cynomolgus monkeys. Time profiles of plasma concentration of hBS910 after intravenous (circles) or subcutaneous (squares) injection are shown.

## Discussion

The lead bispecific antibody was identified from approximately 40,000 different bispecific antibodies. Bispecific antibodies meeting the criteria for FVIII cofactor activity were extremely rare (<0.3%). This seems reasonable since such a bispecific antibody requires simultaneous binding to the appropriate epitope of both FIXa and FX in order to place these two factors into a spatially appropriate position and precisely bring the catalytic site of FIXa close to the cleavage site of FX. Requirement of simultaneous binding to FIXa and FX by a single bispecific antibody was supported by the fact that only a bispecific antibody, and neither monospecific antibodies nor a mixture of them, exhibited FVIII-mimetic activity (supplementary Fig. S7).

Generally, the biological activity of antagonistic antibodies can be improved by increasing the binding affinity to the target antigen [37,38]. The biological activity of agonistic antibodies, on the other hand, is reported to be inversely correlated with the affinity to the antigen, presumably due to the necessity to dissociate from the antigen to repeatedly induce agonistic signals to the target [39]. In the case of our bispecific antibody, the antibody needs to bind to both FIXa and FX with sufficient affinity to promote the interaction between the factors, while after FX activation by FIXa, FXa needs to be rapidly dissociated from the antibody to proceed to the subsequent coagulation reaction and to enable the antibody to turn over. Therefore, we assumed that a simple increase in the binding affinity to both FIXa and FX would not necessarily improve the FVIII-mimetic activity. Consequently, generation of variants that had improved activity required screening of a large number of variants in which mutations had been introduced via structure-based or random mutagenesis. These mutations were mainly introduced to the residues that were predicted to directly or indirectly contact the antigen by the homology modeling of the lead antibody or experimentally identified to affect FVIII-mimetic activity.

To gain insights into the mechanisms underlying the improvement in activity, kinetic analyses of hBS1, hBS106, and hBS910 (bispecific antibodies each with different FVIII-mimetic activity) binding to FIXa and FX were performed by SPR analysis (supplementary Fig. S8A, B). However, because the kinetic parameters of these antibodies were differently affected by the fitting conditions of the sensorgrams, we were not able to obtain meaningful kinetic parameters with which to compare these antibodies. Nevertheless, the binding properties of these antibodies were obviously different, and these changes might have contributed to the improvement in activity. It appears that the binding affinity of hBS910 to FIXa is weaker than hBS1 and hBS106, and hBS910 has faster association and dissociation rate for FX. Although this tendency might result in rapid turnover of FX activation by the bispecific antibody and explain the highest FVIII-mimetic activity of hBS910, it does not explain the large difference of FVIII-mimetic activity between hBS1 and hBS106. Since bispecific antibody needs to strictly place catalytic site of FIXa to the cleavage site of FX and may require allosteric effect on FIXa in order to mimic the function of FVIII, it can be postulated that the improved FVIII-mimetic activity of hBS106 compared to hBS1 may be derived not only from the changes in the binding kinetics but also from the subtle changes in the binding epitope, binding angle or allosteric effect caused by the mutations introduced.

The FVIII-mimetic activity of 30 nM hBS910 was equivalent to rhFVIII activity at 0.01 IU/mL (1% of normal level), and the activity of 300 nM hBS910 was greater than rhFVIII activity at 0.1 IU/mL (10% of normal level) in the TGA using human FVIII-deficient plasma. The activity was also observed even in the presence of FVIII inhibitors, which is reasonable considering the lack of homology between the sequences of hBS910 and FVIII. Recently, we have demonstrated that FVIII-mimetic bispecific antibody, termed hBS23, exerts hemostatic activity *in vivo* in acquired hemophilia A model using cynomolgus monkey, which was considered sufficient for routine prophylaxis [23]. hBS23 is one of the FVIII-mimetic activity improved variants obtained during the multidimensional optimization to generate hBS910, and has similar FVIII-mimetic activity as hBS106. Kinetic analysis [23] of hBS23 and hBS910 demonstrated that hBS910 showed two times the effect on increasing $k_{cat}/K_m$ compared to hBS23 (unpublished data). In human FVIII-deficient plasma, hBS910 required only two third of plasma concentration to exhibit the equivalent activity to hBS23 in the TGA, suggesting that hBS910 has 1.5 times the activity to promote thrombin burst (unpublished data). When compared with FVIII, this *in vivo* hemostatic effect was consistent with its *in vitro* activity in TGA. Thus, the above results clearly demonstrate that hBS910, which is more potent than hBS23, could exert FVIII-mimetic activity in hemophilia A patients sufficient to achieve routine prophylaxis regardless of the presence of FVIII inhibitors. The amino acid sequence of hBS23 is described in our patent as an antibody name Q153-G4k/J142-G4h/L180-k, and that of hBS910 is described in the patent [Igawa T, Sampei Z, Kojima T, Soeda T, Muto A, et al. Multi-specific antigen-binding molecule having alternative function to function of blood coagulation factor VIII. WO/2012/067176].

It is highly desirable that our bispecific antibody be able to be administered subcutaneously with a long interval between doses. The poor pharmacokinetics of hBS106 in mice was partially attributed to the large positive charge cluster, and the single Tyr30Glu mutation to neutralize this charge cluster markedly improved the pharmacokinetics. Such a charge cluster may increase non-specific binding to the extracellular matrix which would increase the clearance of the molecule. Recently, it was reported that antibodies with lower pI have better pharmacokinetics, and engineering antibodies to lower the pI improved their pharmacokinetics [29]. This approach was successfully applied to our bispecific antibody. hBS910 exhibited a plasma half-life of approximately 3 weeks in cynomolgus monkeys, which is longer than the half-lives of hBS23 (approximately 2 weeks) [23] and other humanized or fully human IgG antibodies [40,41], presumably due to the benefit of pI engineering. Since the half-lives of IgG antibodies are generally longer in humans than in cynomolgus monkeys [42], we expect that hBS910 would have a half-life of at least 3 weeks in humans, which is overwhelmingly longer than that of exogenous FVIII (0.5 days) [4]. Furthermore, hBS910 exhibited 86% subcutaneous bioavailability, consistent with that of other IgG antibodies [24,43,44], providing a huge advantage over exogenous FVIII which requires intravenous administration [5]. This high subcutaneous bioavailability and long half-life of hBS910 strongly supports the feasibility of routine prophylaxis by subcutaneous administration with a long dosing interval.

We utilized an *in silico* T-cell epitope prediction system, Epibase, to minimize the number of T-cell epitopes present in the bispecific antibody. The immunogenicity risk of hBS910 was predicted not only by Epibase but also by EpiMatrix to be comparable with that of non-immunogenic antibodies. Considering that up to 30% of patients develop inhibitors against exogenous FVIII [45], immunogenicity is an important issue in the routine prophylaxis of hemophilia A. Although the true immunogenicity of hBS910 needs to be evaluated clinically, it is noteworthy that two different *in silico* systems predicted hBS910 to be non-immunogenic. However, the possibility that a small number of patients could develop anti-hBS910 antibodies cannot be ruled out. In case hBS910 becomes ineffective owing to development of anti-hBS910 antibodies, it is important that they do not cross-react with FVIII so that exogenous FVIII treatment remains as an alternative. In addition to the fact that there is no homology between the amino acid sequences of hBS910 and FVIII, the risk of such cross-reactivity was found to be negligible because polyclonal anti-idiotype antibodies against hBS910 did not inhibit the thrombin generation activity of FVIII. This demonstrates that the development of anti-hBS910 antibodies would not compromise the use of exogenous human FVIII therapy.

No recombinant bispecific IgG antibody has yet reached the market. One of the reasons is the difficulty in large-scale manufacturing at clinical grade. Identification of a common light chain is an important step for manufacturing asymmetric bispecific IgG antibodies. A previous study utilized a phage display library to identify a common light chain for the two heavy chains against different antigens [26]. However, since high-affinity binding to FIXa or FX would not necessarily result in high FVIII-mimetic activity, selection of a common light chain with potent activity based on the binding affinity by phage display was not feasible. We successfully identified a potent common light chain by a novel FR/CDR shuffling approach. Since we were able to obtain the common light chain (BS15L) that had much better potency than the parental light chain (c1L) from the initial twenty four light chain variants, we did not perform further shuffling of FRs. Because residues in the FR often affects antigen binding of the antibody, we propose that shuffling of both CDRs and FRs would be an efficient and general approach to identify potent common light chains for bispecific antibodies. This novel approach enables incorporation of the beneficial residues from each CDR and FR into a common light chain. Although it might be not suitable for screening common light chain against large panels of heavy chains, this approach can be generally applicable for identification of a common light chain for the selected pair of heavy chains without using a library display system. In our other asymmetric bispecific antibodies with the same molecular format, common light chains for two different heavy chains could be successfully identified by this approach. In some cases, it was possible to directly identify humanized common light chain by performing CDR shuffling on the human FRs. This could be a more efficient approach since it abbreviates the following humanization process. However, since FR residues are often important for antigen binding property, we took more cautious approach in this FVIII-mimetic bispecific antibody, and performed CDR shuffling on its cognate FRs to identify a potent common light chain, followed by humanization of the identified common light chain.

Although the heavy chain heterodimerization efficiency depended on the expression balance of each heavy chain, our Fc heterodimerization mutations achieved approximately 85% efficiency for hBS560 in the system of unoptimized expression balance (Fig. 5B). Nevertheless, expression of a small amount of monospecific homodimeric antibodies is inevitable even with Fc heterodimerization mutations. In the case of our bispecific antibody, byproducts are anti-FIXa and anti-FX bivalent monospecific antibodies. These byproducts are not simply impurities with no activity, but they have the potential to competitively inhibit the activity of the bispecific antibody by bivalent binding to the factors. Therefore, homodimeric byproducts need to be removed as much as possible by a downstream process. However, the only molecular difference between a homodimeric byproduct and the bispecific antibody is the heavy chain variable region of one arm. Since antibody variable regions generally have similar sequences except for the CDRs, it was assumed that separation of such byproducts from the bispecific antibody by IEC would be difficult. Therefore, we took the advantage of pI engineering, and engineered the heavy chain variable region to increase the pI difference between the bispecific antibody and the byproducts, thereby improving the separation by IEC. pI engineered hBS910 could be actually purified from 2500-liter fermentation by Protein A and IEC using conventional antibody purification processes. Such a novel pI engineering approach would be generally applicable to facilitate the purification of bispecific IgG antibodies.

To realize subcutaneous delivery in a clinical setting, bispecific antibodies need to be formulated into a high concentration (i.e. >100 mg/mL) since the volume that can be subcutaneously injected is generally limited to less than 1.5 mL [46]. However, hBS376 exhibited phase separation even at 4 mg/mL. Phase separation of antibody solutions into an upper phase with low antibody concentration and a lower phase with high antibody concentration not only precludes high concentration formulation but also makes downstream purification processes difficult [30]. It has been recently reported that low solubility of antibodies has been overcome by introducing specific mutations into the molecular surface [47], but mutations to prevent phase separation have not been reported. We demonstrated that phase separation of hBS376 could also be eliminated by introducing multiple mutations into the molecular surface. hBS910 exhibited no phase separation under the conditions tested, and could be concentrated up to at least 200 mg/mL without any issue.

Since FVIII is unstable under liquid formulation, all the marketed FVIII agents are distributed as lyophilized formulation, and therefore require reconstitution before injection. Liquid formulation allows injection without this process, and thus would be much more convenient for the patients. However, monoclonal antibodies stored in aqueous solution often undergo deamidation of the asparagine residues in the CDRs resulting in reduction of the biological activity of the antibody, which was the case for hBS560 [31]. Although the general strategy is to remove the deamidation site by mutating the asparagine residue itself to another amino acid, this strategy was not feasible in the case of hBS560. A simultaneous double mutation approach enabled storage of hBS910 at 40°C for 2 weeks without reduction of activity. Consequently, hBS910 could be stably stored in a patient-friendly 150 mg/mL liquid formulation, which would enable approximately 3 mg/kg subcutaneous delivery in humans (1.5 mL injection for a 75 kg patient). Considering the FVIII-mimetic activity determined by TGA and the pharmacokinetics in cynomolgus monkeys, such a formulation would provide effective prophylaxis by subcutaneous delivery with a long dosing interval.

In conclusion, we have generated a novel humanized anti-FIXa/FX bispecific IgG antibody, hBS910, through a process of identifying the lead candidate from approximately 40,000 bispecific combinations, followed by a multidimensional optimization process to improve both the therapeutic potential and the manufacturability. hBS910 overcomes the two major drawbacks of routine prophylaxis by exogenous FVIII. First, while exogenous FVIII requires frequent intravenous administration, hBS910 can be subcutaneously administered with a long dosing interval. Second, while the development of FVIII inhibitors is a critical issue for exogenous FVIII, our study suggests that hBS910 can be used without fear of developing FVIII inhibitors and can be used in patients who have already developed FVIII inhibitors. We believe that hBS910, with its multidimensionally optimized profile, will provide significant improvement in the quality of life of hemophilia A patients by reducing not only bleeding but also the burden on the patients themselves, their parents, and all medical staff. Potential of hBS910 (investigational drug name; ACE910) in hemophilia A patient is currently being evaluated in clinical study.

## Materials and Methods

### Ethics Statement

Animal studies were performed in accordance with the Guidelines for the Care and Use of Laboratory Animals at Chugai Pharmaceutical Co., Ltd. under the approval of the company's Institutional Animal Care and Use Committee. The company is fully accredited by the Association for Assessment and Accreditation of Laboratory Animal Care International (http://www.aaalac.org). Details of cynomolgus monkey care and maintenance, including shelter and availability of food, water, and environmental enrichment are as follows. Identification of individuals; individual animals were identified by microchip number. Each cage was identified by a cage number indicated on the cage rack, cage type; a stainless steel cage, housing density; 1 animal/cage, temperature; 18°C to 28°C, relative humidity; 35% to 75%, air change frequency; at least 10 times per h, illumination timing; 12 h per day, from 7:00 am to 7:00 pm, feed; cynomolgus monkeys were daily provided approximately 100 g of solid chow (Certified Primate Diet 5048, Japan SLC) and supplementary foods (1/2 peeled banana or 50 g of sweet potato) and drinking water; animal room tap water, provided ad libitum using the automatic water supply system. The studies involved injection of the agents and collection of blood samples which did not require procedures that would cause more than slight or momentary pain or distress to the animals. All injections and blood collection were conducted by trained and qualified primate therapeutic staff in Chugai Pharmaceutical. Provisions were made in the approved protocol for veterinary intervention in the case of any distress or morbidity from the injected agent; however no adverse events occurred during the studies or during the recovery period. Therefore, no scarification of the animals was conducted.

### Generation of Anti-FIXa/FX Bispecific Antibodies

Approximately 200 monoclonal antibodies to FIXa or FX were obtained from FIXa or FX immunized mice, rats, and rabbits. The $V_H$ and $V_L$ of those antibodies were then combined with engineered human $IgG_2$ or $IgG_4$ that included mutations to facilitate Fc heterodimerization. These engineered human $IgG_2$ or $IgG_4$, which included the knobs-into-holes mutations [26], were generated by introducing the same substitutions as those of $IgG_1$ knobs-into-holes to the corresponding positions of $IgG_2$ and $IgG_4$

heavy chains, and were used through the screening for the lead identification. Anti-FIXa/FX bispecific were generated by HEK293 cells co-transfected with mixture of four plasmids encoding anti-FIXa heavy and light chain and anti-FX heavy and light chain (supplementary Fig. S1). At the screening step for lead identification, we used knobs-into-holes mutations which could achieve at least 90% efficiency of the two heavy chain heterodimerization. If the assembly of a heavy chain with two light chains occurs with equal probability, 25% of heavy chain heterodimeric antibodies will have the correct heavy and light chain pair [25]. Theoretically, in the supernatant of transfected cells, this would result in at least approximately 20% (90% ×25% = 22.5%) of antibodies to be the target bispecific antibody, and less than 3% (10% ×25% = 2.5%) of antibodies to be homodimeric antibodies with a correct heavy and light chain pair. Transfected cells were cultured in 96-well culture plates, and either filtrated culture supernatants or Protein A purified antibodies were used for evaluation. We also generated bispecific antibodies with a common light chain, anti-FIXa monospecific IgG consisting of anti-FIXa heavy chain and a common light chain, and anti-FX monospecific IgG consisting of anti-FX heavy chain and a common light chain by the method described above. The engineered human $IgG_4$ including the knobs-into-holes mutations was used during the early stage of the multidimensional optimization, and a different engineered $IgG_4$ including electrostatic steering mutations [25] was used during the late stage optimization. By using the engineered $IgG_4$, approximately 85–95% of the Protein A purified antibody would be the target bispecific antibody and the rest would be homodimeric antibodies. For the detailed characterization of hBS910, homodimeric antibodies were removed by ion exchange chromatography.

### Screening Bispecific Antibodies for FVIII-mimetic Activity (Enzymatic Assay)

The ability of each antibody to enhance FIXa-catalized FXa generation was evaluated in an enzymatic assay with purified human coagulation factors (Enzyme Research Laboratories). The FXa generation reaction was performed in the presence of 1 nM human FIXa, 140 nM human FX, 20 μM synthetic phospholipid (10% phosphatidylserine, 60% phosphatidylcholine and 30% phosphatidylethanolamine; Avanti Polar Lipids) prepared as previously described [48], and antibodies at room temperature for 2 min in TBS containing 5 mM $CaCl_2$, 1 mM $MgCl_2$ and 0.1% (wt/vol) BSA (pH 7.6). The reaction was stopped by the addition of EDTA at appropriate time points. The activity of the FXa generated was determined by absorbance at 405 nm after the addition of chromogenic substrate S-2222 (Chromogenix). Data were collected in triplicate.

### FR/CDR Shuffling of the Light Chains

CDRs of three light chains (c1L, c2L and c3L) were shuffled among each other then grafted onto the FRs of either c1L or c3L to generate light chain variants (supplementary Fig. S2A). Since c2L and c3L had identical CDR1 and CDR2 sequences, we generated twenty four light chains (twelve CDR combinations in two FRs) including parental c1L and c3L. Light chain variant genes were generated by assembly PCR, and the twenty four bispecific antibodies with these light chains were prepared as described above. FVIII-mimetic activity of each antibody was evaluated in APTT assay with standard techniques using APTT reagent (Sysmex) and FVIII deficient plasma (Sysmex).

### Pharmacokinetic Study of Bispecific Antibodies in Mice

1 mg/kg doses of each bispecific antibody were administered to C57BL/6J normal mice (Charles River) by single subcutaneous injection ($n = 3$ for each group). Blood samples were collected at an appropriate time after each administration. Plasma concentration of bispecific antibodies was determined by human IgG-specific ELISA. Pharmacokinetic parameters were calculated by WinNonlin Professional software (Pharsight).

### Determination of Isoelectric Point (pI) by Capillary Isoelectric Focusing (cIEF)

cIEF analyses of antibodies were performed with a PA800 plus Pharmaceutical Analysis System and 32 Karat software (Beckman Coulter) as described in the cIEF application guide (PN A78788AA). Briefly, antibody solutions (approximately 1 mg/mL in PBS) were diluted 1:25 with cIEF master mix solution containing cIEF gel, urea, Pharmalyte 3–10, arginine, iminodiacetic acid, and pI markers. A neutral capillary was preconditioned by rinsing with urea solution, and samples were injected and focused. The pI of each of the antibodies was determined from the pI markers.

### Separation of Bispecific Antibodies by Cation Exchange Chromatography

Using an AKTAexplorer 10S (GE Healthcare), two HiTrap SP FF 1 mL columns (GE Healthcare) were connected in tandem and equilibrated by 20 mM sodium phosphate, pH 6.0. The elution buffer contained an appropriate concentration of NaCl in this equilibration buffer. The load sample was first prepared from culture supernatant using MabSelect Sure (GE Healthcare), and was then dialyzed with equilibration buffer. The sample antibody solution was applied to the HiTrap SP FF column at 1.5 mg/mL resin. After washing with equilibration buffer, antibodies were eluted with 10 column volumes (CV) of equilibration buffer containing 150 mM NaCl and then with 15 CV of 220 mM NaCl buffer in a stepwise manner, and finally with 15 CV of NaCl buffer in a linear gradient to 450 mM.

### Solubility Analysis of Bispecific Antibodies

Stock solutions of bispecific antibodies were prepared by dialysis against water or 50 mM NaCl solutions, followed by ultrafiltration concentration. Samples were prepared by adding formulation stock solution to antibody stock solution using a Hydra II Plus One liquid-handling robot (Matrix). After centrifugation, 1 μL of each sample was stored in an Intelli-Plate 96-2 (Art Robbins) at 20°C for 1 day, and then images of each well were taken by Rock Imager 54 (Formulatrix). The state of each antibody solution was determined as either a clear solution, precipitation, or liquid–liquid phase separation.

### Accelerated Stability Study of Bispecific Antibodies

Solutions of bispecific antibodies were dialyzed against PBS, pH 7.4 (Sigma). 1 mg/mL of each antibody solution was stored at 40°C for 2 weeks and then analyzed by cation exchange chromatography (IEC) using BioPro SP columns (YMC) at room temperature. The mobile phase (A) was 20 mM sodium phosphate, pH 5.8, and the mobile phase (B) was 20 mM sodium phosphate and 500 mM NaCl, pH 5.8.

### *In silico* Evaluation of Immunogenicity of the Variable Region of Bispecific Antibodies

T-cell epitope prediction of the bispecific antibody variants and immunogenicity risk scores of antibodies were provided by

Epibase (Lonza). Immunogenicity scales of antibodies were provided by EpiMatrix (EpiVax).

## Thrombin Generation Assay (TGA)

Calibrated automated thrombography [35] was employed using a 96-well plate fluorometer (Thermo Fisher Scientific) equipped with a 390/460 filter set, a dispenser, and the analyzing software (Thrombinoscope software version 3.0.0.29; Thrombinoscope) to measure thrombograms. Briefly, each concentration of bispecific antibody or rhFVIII (Bayer Healthcare) was added to FVIII-deficient plasma (<1% FVIII activity) either without inhibitors or with inhibitors against FVIII (George King Bio-Medical). Each concentration of bispecific antibody or rhFVIII was also added to plasma containing polyclonal rabbit anti-idiotype antibodies against anti-FIXa Fab or anti-FX Fab (300 μg/mL each). Into each well was dispensed 80 μL of the plasma, to which was then added 20 μL of the triggering solution containing 0.47 nM human FXIa (Enzyme Research Laboratories) and 20 μM synthetic phospholipid but no $Ca^{2+}$. For calibration, 20 μL of Thrombin Calibrator (Thrombinoscope) was added instead of the triggering solution. 20 μL of FluCa-reagent prepared from FluCa-kit (Thrombinoscope) was dispensed to initiate the reaction. The thrombograms and peak height were analyzed by the software. Data were collected in triplicate.

## Pharmacokinetic Study of hBS910 in Cynomolgus Monkeys

A single dose of 0.3 mg/kg of hBS910 was intravenously or subcutaneously administered to male cynomolgus monkeys ($n = 3$ for each group). Blood samples were collected at an appropriate time after each administration. Plasma concentration of bispecific antibodies was determined by human IgG-specific ELISA. Pharmacokinetic parameters were calculated by WinNonlin Professional software (Pharsight).

## Kinetic Analysis of Bispecific Antibodies Binding to FIXa and FX Using Surface Plasmon Resonance

Kinetic analysis of bispecific antibodies was performed by surface plasmon resonance (SPR) using a Biacore T200 system (GE Healthcare). MabSelect SuRe Ligand (Recombinant Protein A; GE Healthcare) was immobilized onto a CM4 sensor chip (GE Healthcare). Then, anti-FIXa or anti-FX monospecific antibodies were injected into flow cell 2 to be captured. Natalizumab (Biogen-Idec Inc.) as a control human $IgG_4$ antibody was also injected into flow cell 1 to be captured. Then, 0, 80, 160, 320, 640, or 1,280 nM human FIXa or FX dissolved in running buffer (10 mM HEPES [pH 7.4], 150 mM NaCl, 0.05% surfactant P20, 2.5 mM $CaCl_2$) was injected at a flow rate of 30 μL/min to monitor the association phase for 120 s and the dissociation phase for 30 s.

## Supporting Information

**Figure S1 Generation of anti-FIXa/FX bispecific antibodies.** Heavy chain variable regions ($V_H$) of anti-FIXa or anti-FX antibodies were fused with engineered human $IgG_2$ or $IgG_4$ constant region having mutations to facilitate Fc heterodimerization. Light chain variable regions ($V_L$) were fused with human κ or λ constant region. Bispecific antibodies were generated by expression with two pairs of genes, anti-FIXa and anti-FX heavy chain and light chain genes (or a common light chain gene).

**Figure S2 FR/CDR shuffling of the light chain.** (A) CDRs of three light chains (c1L, c2L and c3L) were shuffled among each other and grafted onto the FRs of c1L and c3L. Each light chain variant was expressed with the selected anti-FIXa and anti-FX heavy chains. (B) Effects of bispecific antibodies (67 nM) with light chain variants on APTT assay in FVIII-deficient plasma are shown. The Y-axis indicates the APTT (s). All the data were collected in duplicate and are expressed as mean.

**Figure S3 Effect of humanization of the lead chimeric antibody (BS15) on FVIII-mimetic activity.** Effect of chimeric antibody BS15 (circles) or humanized antibody hBS1 (squares) on FX activation in the presence of FIXa, FX, and synthetic phospholipid. The Y-axis indicates the 405 nm absorbance at 120 min of chromogenic development in the chromogenic substrate assay. All the data were collected in triplicate and are expressed as mean ± s.d (in many cases, the bars depicting s.d. are shorter than the height of the symbols).

**Figure S4 Positive charge cluster and Tyr30Glu mutation on anti-FIXa Fv of hBS106.** The positive charge cluster consists of arginine or lysine residues at Kabat position 60, 61, and 95 in the heavy chain and Kabat positions 24, 27, 31, 53, 54, 61, and 66 in the light chain of hBS106. Tyrosine located at Kabat position 30 in the light chain was mutated to glutamic acid to neutralize the positive charge cluster. Blue, red and gray colored surface indicates positively charged, negatively charged and neutral protein surface, respectively. Red and green line indicates heavy and light chain, respectively.

**Figure S5 Precipitation and liquid–liquid phase separation of bispecific antibody solution.** Micro CCD camera images of states of bispecific antibody solution showing a clear solution, precipitation, and liquid–liquid phase separation.

**Figure S6 *In silico* prediction of immunogenicity of bispecific antibodies.** (A) Immunogenicity risk score of BS15, hBS1, hBS910, trastuzumab, and palivizumab predicted by Epibase. (B) Immunogenicity scale of hBS910 and other marketed monoclonal antibodies by EpiMatrix. In both prediction systems, higher score indicates higher risk of immunogenicity in human.

**Figure S7 Necessity of bispecific binding to FIXa and FX for FVIII-mimetic activity.** Effect of the bispecific antibody (hBS910) (circles), monospecific anti-FIXa antibody (squares), monospecific anti-FX antibody (triangles), or a mixture of the two monospecific antibodies (diamonds) on FX activation in the presence of FIXa, FX, and synthetic phospholipid. The Y-axis indicates the 405 nm absorbance at 30 min of chromogenic development in the chromogenic substrate assay. All the data were collected in triplicate and are expressed as mean ± s.d (in many cases, the bars depicting s.d. are shorter than the height of the symbols). Monospecific antibodies against FIXa or FX antibodies or the mixture of them did not exhibit any detectable activity even at 120 min of chromogenic development.

**Figure S8 Surface plasmon resonance analysis of bispecific antibodies binding to FIXa and FX.** Sensorgrams of hBS1, hBS106, and hBS910 binding to FIXa (A) and FX (B) at a concentration of 80 nM, 160 nM, 320 nM, 640 nM, and 1280 nM.

## Acknowledgments

We thank our colleagues at Chugai Research Institute for Medical Science, Inc. and Chugai Pharmaceutical Co. Ltd., K. Nagano, H. Sano, T. Sakamoto, T. Matsuura, R. Takemoto, M. Hiranuma and H. Kitamura for carrying out *in vivo* studies; M. Fujii, Y. Nakata, H. Ishida and F. Isomura for antibody generation; S. Ohtsu for carrying out the physicochemical study; and M. Yoshida for carrying out the cIEF analysis.


## Author Contributions

Provided direction and guidance: YN. Provided the hypothesis of the bispecific antibody, directed and organized the program: K. Hattori. Conceived and designed the experiments: TI T. Kojima T. Kitazawa. Performed the experiments: ZS TI TS YON CM TW ET AM KY AH MF K. Haraya TT SS KE. Wrote the paper: ZS TI.

## References


1. Geraghty S, Dunkley T, Harrington C, Lindvall K, Maahs J, et al. (2006) Practice patterns in haemophilia A therapy - global progress towards optimal care. Haemophilia 12: 75–81.
2. Manco-Johnson MJ, Abshire TC, Shapiro AD, Riske B, Hacker MR, et al. (2007) Prophylaxis versus episodic treatment to prevent joint disease in boys with severe hemophilia. N Engl J Med 357: 535–544.
3. Ragni MV, Fogarty PJ, Josephson NC, Neff AT, Raffini LJ, et al. (2012) Survey of current prophylaxis practices and bleeding characteristics of children with severe haemophilia A in US haemophilia treatment centres. Haemophilia 18: 63–68.
4. Bjorkman S, Oh M, Spotts G, Schroth P, Fritsch S, et al. (2012) Population pharmacokinetics of recombinant factor VIII: the relationships of pharmacokinetics to age and body weight. Blood 119: 612–618.
5. Shi Q, Kuether EL, Schroeder JA, Fahs SA, Montgomery RR (2012) Intravascular recovery of VWF and FVIII following intraperitoneal injection and differences from intravenous and subcutaneous injection in mice. Haemophilia 18: 639–646.
6. Berntorp E, Shapiro AD (2012) Modern haemophilia care. Lancet 379: 1447–1456.
7. Astermark J, Donfield SM, DiMichele DM, Gringeri A, Gilbert SA, et al. (2007) A randomized comparison of bypassing agents in hemophilia complicated by an inhibitor: the FEIBA NovoSeven Comparative (FENOC) Study. Blood 109: 546–551.
8. Leissinger C, Gringeri A, Antmen B, Berntorp E, Biasoli C, et al. (2011) Anti-inhibitor coagulant complex prophylaxis in hemophilia with inhibitors. N Engl J Med 365: 1684–1692.
9. Chan AC, Carter PJ (2010) Therapeutic antibodies for autoimmunity and inflammation. Nat Rev Immunol 10: 301–316.
10. Weiner LM, Surana R, Wang S (2010) Monoclonal antibodies: versatile platforms for cancer immunotherapy. Nat Rev Immunol 10: 317–327.
11. Beck A, Wurch T, Bailly C, Corvaia N (2010) Strategies and challenges for the next generation of therapeutic antibodies. Nat Rev Immunol 10: 345–352.
12. Liu Z, Stoll VS, Devries PJ, Jakob CG, Xie N, et al. (2007) A potent erythropoietin-mimicking human antibody interacts through a novel binding site. Blood 110: 2408–2413.
13. Mayorov AV, Amara N, Chang JY, Moss JA, Hixon MS, et al. (2008) Catalytic antibody degradation of ghrelin increases whole-body metabolic rate and reduces refeeding in fasting mice. Proc Natl Acad Sci U S A 105: 17487–17492.
14. Bhaskar V, Goldfine ID, Bedinger DH, Lau A, Kuan HF, et al. (2012) A fully human, allosteric monoclonal antibody that activates the insulin receptor and improves glycemic control. Diabetes 61: 1263–1271.
15. Songsivilai S, Lachmann PJ (1990) Bispecific antibody: a tool for diagnosis and treatment of disease. Clin Exp Immunol 79: 315–321.
16. Baeuerle PA, Kufer P, Bargou R (2009) BiTE: Teaching antibodies to engage T-cells for cancer therapy. Curr Opin Mol Ther 11: 22–30.
17. Jackman J, Chen Y, Huang A, Moffat B, Scheer JM, et al. (2010) Development of a two-part strategy to identify a therapeutic human bispecific antibody that inhibits IgE receptor signaling. J Biol Chem 285: 20850–20859.
18. Fay PJ, Haidaris PJ, Smudzin TM (1991) Human factor VIIIa subunit structure. Reconstruction of factor VIIIa from the isolated A1/A3-C1-C2 dimer and A2 subunit. J Biol Chem 266: 8957–8962.
19. Fay PJ, Koshibu K (1998) The A2 subunit of factor VIIIa modulates the active site of factor IXa. J Biol Chem 273: 19049–19054.
20. Lapan KA, Fay PJ (1997) Localization of a factor X interactive site in the A1 subunit of factor VIIIa. J Biol Chem 272: 2082–2088.
21. Lapan KA, Fay PJ (1998) Interaction of the A1 subunit of factor VIIIa and the serine protease domain of factor X identified by zero-length cross-linking. Thromb Haemost 80: 418–422.
22. Lenting PJ, Donath MJ, van Mourik JA, Mertens K (1994) Identification of a binding site for blood coagulation factor IXa on the light chain of human factor VIII. J Biol Chem 269: 7150–7155.
23. Kitazawa T, Igawa T, Sampei Z, Muto A, Kojima T, et al. (2012) A bispecific antibody to factors IXa and X restores factor VIII hemostatic activity in a hemophilia A model. Nat Med 18: 1570–1574.
24. Kontermann R (2012) Dual targeting strategies with bispecific antibodies. MAbs 4: 182–197.
25. Klein C, Sustmann C, Thomas M, Stubenrauch K, Croasdale R, et al. (2012) Progress in overcoming the chain association issue in bispecific heterodimeric IgG antibodies. MAbs 4: 653–663.
26. Merchant AM, Zhu Z, Yuan JQ, Goddard A, Adams CW, et al. (1998) An efficient route to human bispecific IgG. Nat Biotechnol 16: 677–681.
27. Igawa T, Tsunoda H, Kuramochi T, Sampei Z, Ishii S, et al. (2011) Engineering the variable region of therapeutic IgG antibodies. MAbs 3: 243–252.
28. Jones PT, Dear PH, Foote J, Neuberger MS, Winter G (1986) Replacing the complementarity-determining regions in a human antibody with those from a mouse. Nature 321: 522–525.
29. Igawa T, Tsunoda H, Tachibana T, Maeda A, Mimoto F, et al. (2010) Reduced elimination of IgG antibodies by engineering the variable region. Protein Eng Des Sel 23: 385–392.
30. Mason BD, Zhang L, Remmele RL, Jr., Zhang J (2011) Opalescence of an IgG2 monoclonal antibody solution as it relates to liquid-liquid phase separation. J Pharm Sci 100: 4587–4596.
31. Vlasak J, Bussat MC, Wang S, Wagner-Rousset E, Schaefer M, et al. (2009) Identification and characterization of asparagine deamidation in the light chain CDR1 of a humanized IgG1 antibody. Anal Biochem 392: 145–154.
32. Van Walle I, Gansemans Y, Parren PW, Stas P, Lasters I (2007) Immunogenicity screening in protein drug development. Expert Opin Biol Ther 7: 405–418.
33. De Groot AS, McMurry J, Moise L (2008) Prediction of immunogenicity: in silico paradigms, ex vivo and in vivo correlates. Curr Opin Pharmacol 8: 620–626.
34. Weber CA, Mehta PJ, Ardito M, Moise L, Martin B, et al. (2009) T cell epitope: friend or foe? Immunogenicity of biologics in context. Adv Drug Deliv Rev 61: 965–976.
35. Hemker HC, Giesen P, Al Dieri R, Regnault V, de Smedt E, et al. (2003) Calibrated automated thrombin generation measurement in clotting plasma. Pathophysiol Haemost Thromb 33: 4–15.
36. Shima M, Matsumoto T, Ogiwara K (2008) New assays for monitoring haemophilia treatment. Haemophilia 14 Suppl 3: 83–92.
37. Maynard JA, Maassen CB, Leppla SH, Brasky K, Patterson JL, et al. (2002) Protection against anthrax toxin by recombinant antibody fragments correlates with antigen affinity. Nat Biotechnol 20: 597–601.
38. Johnson S, Griego SD, Pfarr DS, Doyle ML, Woods R, et al. (1999) A direct comparison of the activities of two humanized respiratory syncytial virus monoclonal antibodies: MEDI-493 and RSHZl9. J Infect Dis 180: 35–40.
39. Lacy SE, DeVries PJ, Xie N, Fung E, Lesniewski RR, et al. (2008) The potency of erythropoietin-mimic antibodies correlates inversely with affinity. J Immunol 181: 1282–1287.
40. Lin YS, Nguyen C, Mendoza JL, Escandon E, Fei D, et al. (1999) Preclinical pharmacokinetics, interspecies scaling, and tissue distribution of a humanized monoclonal antibody against vascular endothelial growth factor. J Pharmacol Exp Ther 288: 371–378.
41. Benincosa LJ, Chow FS, Tobia LP, Kwok DC, Davis CB, et al. (2000) Pharmacokinetics and pharmacodynamics of a humanized monoclonal antibody to factor IX in cynomolgus monkeys. J Pharmacol Exp Ther 292: 810–816.
42. Deng R, Iyer S, Theil FP, Mortensen DL, Fielder PJ, et al. (2011) Projecting human pharmacokinetics of therapeutic antibodies from nonclinical data: what have we learned? MAbs 3: 61–66.
43. Vugmeyster Y, Xu X, Theil FP, Khawli LA, Leach MW (2012) Pharmacokinetics and toxicology of therapeutic proteins: Advances and challenges. World J Biol Chem 3: 73–92.
44. Richter WF, Bhansali SG, Morris ME (2012) Mechanistic determinants of biotherapeutics absorption following SC administration. AAPS J 14: 559–570.
45. Fischer K, Lewandowski D, Marijke van den Berg H, Janssen MP (2012) Validity of assessing inhibitor development in haemophilia PUPs using registry data: the EUHASS project. Haemophilia 18: e241–246.
46. Shire SJ, Shahrokh Z, Liu J (2004) Challenges in the development of high protein concentration formulations. J Pharm Sci 93: 1390–1402.
47. Pepinsky RB, Silvian L, Berkowitz SA, Farrington G, Lugovskoy A, et al. (2010) Improving the solubility of anti-LINGO-1 monoclonal antibody Li33 by isotype switching and targeted mutagenesis. Protein Sci 19: 954–966.
48. Okuda M, Yamamoto Y (2004) Usefulness of synthetic phospholipid in measurement of activated partial thromboplastin time: a new preparation procedure to reduce batch difference. Clin Lab Haematol 26: 215–223.

# 16

# Engineering of Three-Finger Fold Toxins Creates Ligands with Original Pharmacological Profiles for Muscarinic and Adrenergic Receptors

**Carole Fruchart-Gaillard**[1❾], **Gilles Mourier**[1❾], **Guillaume Blanchet**[1,2], **Laura Vera**[1], **Nicolas Gilles**[1], **Renée Ménez**[1], **Elodie Marcon**[1], **Enrico A. Stura**[1], **Denis Servent**[1*]

**1** CEA, iBiTecS, Service d'Ingénierie Moléculaire des Protéines, Laboratoire de Toxinologie Moléculaire et Biotechnologies, Gif-sur-Yvette, France, **2** Université Pierre et Marie Curie, Paris, France

## Abstract

Protein engineering approaches are often a combination of rational design and directed evolution using display technologies. Here, we test "loop grafting," a rational design method, on three-finger fold proteins. These small reticulated proteins have exceptional affinity and specificity for their diverse molecular targets, display protease-resistance, and are highly stable and poorly immunogenic. The wealth of structural knowledge makes them good candidates for protein engineering of new functionality. Our goal is to enhance the efficacy of these mini-proteins by modifying their pharmacological properties in order to extend their use in imaging, diagnostics and therapeutic applications. Using the interaction of three-finger fold toxins with muscarinic and adrenergic receptors as a model, chimeric toxins have been engineered by substituting loops on toxin MT7 by those from toxin MT1. The pharmacological impact of these grafts was examined using binding experiments on muscarinic receptors M1 and M4 and on the $\alpha_{1A}$-adrenoceptor. Some of the designed chimeric proteins have impressive gain of function on certain receptor subtypes achieving an original selectivity profile with high affinity for muscarinic receptor M1 and $\alpha_{1A}$-adrenoceptor. Structure-function analysis supported by crystallographic data for MT1 and two chimeras permits a molecular based interpretation of these gains and details the merits of this protein engineering technique. The results obtained shed light on how loop permutation can be used to design new three-finger proteins with original pharmacological profiles.

**Editor:** Alejandra Bravo, Universidad Nacional Autonoma de Mexico, Instituto de Biotecnologia, Mexico

**Funding:** The authors have no support or funding to report.

**Competing Interests:** The authors have declared that no competing interests exist.

* E-mail: denis.servent@cea.fr

❾ These authors contributed equally to this work.

## Introduction

Protein engineering can draw inspiration from architectural motifs that have been selected and used in nature to support a large diversity of biological functions. Peptide toxins from venomous animals could be part of the answer. Indeed, although toxins exert their neurotoxic, cardiotoxic or cytotoxic effect by interacting with a large diversity of molecular targets, the number of protein templates selected during the course of evolution to accomplish such vast range of physiological effects appears to be rather limited [1,2,3]. The three-finger toxin (3FT) family is widely represented as structural motif, mainly in the *Elapidae* snake venoms, supporting various biological functions [4,5]. This fold is characterized by three adjacent loops rich in β-pleated sheets emerging from a small globular core containing four invariant disulfide bridges [6]. Despite the small number of residues in the 3FT motif, their targets are numerous and include: L-type calcium channels, integrin receptors, cell membrane phospholipids, acetylcholinesterase, nicotinic acetylcholine or muscarinic and adrenergic receptors [5], making this template attractive in protein engineering. Functional studies by site-directed mutagenesis have identified the sites by which some of these toxins interact with their respective targets. Examples are neurotoxins that interact with muscular and/or neuronal subtypes of nicotinic acetylcholine receptors [7,8,9] and with acetylcholinesterase [10]. Transfer of small portions of sequence between an acetylcholinesterase inhibitory toxin to a curaremimetic toxin has resulted in a committal transfer of function [11]. The structure of the modified toxin shows a replication of the loop conformation of the parent [12]. Similarly, grafting the extremity of the central loop of a neuronal nicotinic toxin on the scaffold of a muscular-type toxin confers to it an improved neuronal activity [13]. More recently, using a cDNA display strategy to produce a large peptide library, Naimuddin and coworkers engineered a 3FT by randomizing residues in the three loops in order to obtain modulators of interleukin-6 receptor [14]. All these results emphasize the structural and functional adaptability of the three-finger fold and the ease with which it can support protein engineering.

Muscarinic toxins (MTs) were isolated and purified from the venom of African mambas more than 20-years ago [15,16] while the identification and purification of adrenergic toxins from mamba or cobra venom occurred more recently [17,18,19]. Despite high primary structure identity, often greater than 55%, muscarinic and adrenergic toxins exhibit distinct pharmacological profiles and clear functional differences [17,18,20,21,22,23]. For example, MT7 acts as a highly potent and selective antagonist of the M1 receptor subtype [24,25,26,27,28], through a very stable interaction with the allosteric binding site on the M1 receptor [29]. On the other hand, the mode of interaction of MT1 with muscarinic receptors is less precisely defined, but its selectivity profile for M1 and M4 receptors has been clearly established (review in [20,21]). Based on previous data reporting adrenergic effects of MT1 in tissue preparations [30], the ability of various MTs to interact with adrenoceptors was investigated, showing that MT1 interacts efficiently with mAChRs and also with $\alpha_{1A}$- and $\alpha_{2B}$-adrenoceptors [23].

The 23 sequence variations between MT7 and MT1 correspond to residues equally distributed on the three loops of these toxins (Fig. 1). "Loop grafting" has been used to construct seven chimeric toxins to dissect how the affinity, selectivity and functionality for the muscarinic and adrenergic receptors depend on shuffling loops between various toxins. Three chimeras were obtained by substituting loops of the MT7 toxin with the corresponding loop from the MT1. These chimeras were named MT7-1/1, MT7-1/2 and MT7-1/3, numbered for the respective grafted MT1 loop. Given that grafts of the entire loop 2 had enormous effects, this loop was separated into two distinct regions. Two other chimeras were synthesized containing three modifications at the tip of loop 2 (MT7-1/2 tip) or four modifications at the top (MT7-1/2 top), respectively. For sake of completion, two further chimeras were synthesized. These were designed by combining MT7-1/2 tip and MT7-1/3 (MT7-1/2 tip+3) or MT7-1/1 and MT7-1/3 with two additional point mutations found in the C-terminal portion of the MT1 sequence (MT7-1/1+3; Fig. 1). The new designed toxins exhibit original selectivity profiles.

The crystallographic structures of wild-type MT1 and of the engineered chimeric toxins obtained in this study show that despite local structural deviations, the overall integrity of the three-finger fold structure is conserved and the intra-loop hydrogen-bonding maintained. Binding experiments of the natural and engineered toxins on different GPCRs reveal a significant gain in function on $\alpha_{1A}$-adrenoceptor and hM4 muscarinic receptor. It cannot be excluded that these results may have a broader general relevance on a wider spectrum of GPCRs. This engineering study has created novel ligands by combining pre-existing loops. Given the lack of GPCR selective ligands, this approach offers an additional strategy to fill this gap.

## Results

### Synthesis and Folding of Wild-Type and Modified Toxins

The muscarinic toxins, MT7 and MT1, have been synthesised by the one step Fmoc solid phase peptide synthesis approach previously described [25]. We have solved the problems previously reported during the synthesis of the MT1 toxin consisting of failures in the deprotection monitoring and lower couplings because of local chain aggregation. These occurred in two regions; the central part of the sequence (Val-32 - Tyr-35) and the N-terminal end (Leu-1 - Ser-8). The solution was to replace Thr-13 and Thr-39, two residues introduced during the synthetic process before the two critical regions, by a pseudo-proline residue. We therefore incorporated the dimethyloxazolidine of the di-peptide Ile-Thr at positions 12–13 and 38–39 during the synthesis. This chemical modification disrupts peptide chain aggregation in the

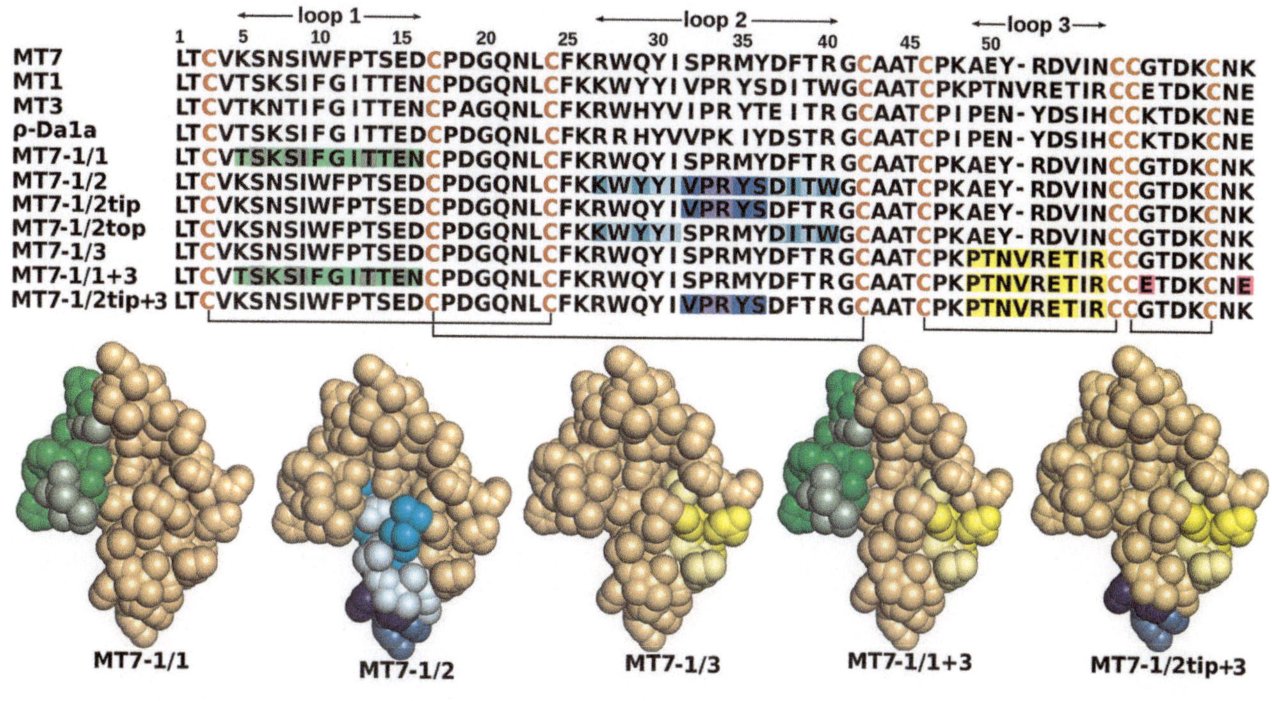


**Figure 1. Sequence alignment of the toxins MT7, MT1, MT3 and ρ-Da1a compared to the various chimeras.** The grafted residues of MT1 loop 1 are shown in green with conserved residues in a lighter shade generating the chimera of MT7-1/1. For MT7-1/2, MT7-1/2tip and MT7-1/2top the grafted residues of MT1 loop 2 are shown in cyan (top) and blue (tip) and the conserved residues in a lighter shade. For MT7-1/3, MT7-1/1+3 and MT7-1/2tip+3 the grafted residues of MT1 loop 3 are shown in yellow with the conserved residues in a lighter shade. The two extra mutations on the C-terminal section of MT7-1/1+3 that correspond to residues from MT1 are shown in magenta. Below the space-filling representation on the MT7 structure show the positioning of the grafted loops from MT1.

same manner as a proline does [31]. We took advantage of this improvement during the synthesis of the different chimeric toxins. The synthesis of MT7-1/2 top and MT7-1/3 were undertaken in a similar manner as that described for MT7 using Fmoc amino acids and the efficient coupling reagents HOAT or CL-HOBT/DCCI. However, for the other chimeric toxins, the dipeptide Ile-Thr was introduced using Fmoc-Ileu-Thr ($\Psi^{MeMe}$pro)-OH at positions 12–13 in the MT7-1/1 or MT7-1/1+3 and 38–39 in MT7-1/2, MT7-1/2 tip and MT7-1/2 tip+3. A longer coupling time was used for this reaction. During the synthesis of the different toxins no failure in the deprotection monitoring was observed. At the end of the synthesis, regeneration of the Thr from the oxazolidine occurs during the course of the normal TFA-mediated cleavage reaction and each toxin yielded a crude mixture in which the main component corresponds to the reduced form of each chimera toxin.

Folding of the different chimera toxins was achieved using the experimental conditions that were successfully applied to the MT7 and MT1 toxins. An efficient and rapid folding process was followed for MT7-1/1 and MT7-1/3 but not for MT7-1/2, MT7-1/2tip and MT7-1/2top. To find out a common optimal refolding condition, we screened different additives and various concentrations of reduced and oxidized glutathione. The final optimized refolding buffer containing 30% glycerol, 0.1 M Tris-HCl, 1 mM EDTA, 1 mM GSH, 1 mM GSSG, pH 8 was used to prepare large amounts of the different chimeras (refolding yield; 20–30%). Amino acid composition and electro-spray mass analysis confirmed the purity and identity of the different proteins. The masses were 7359.2 for MT7-1/1 (theoretical 7359.4), 7538.5 for MT7-1/1+3 (theoretical 7538.6), 7443.3 for MT7-1/2 (theoretical 7443.5), 7440.2 for MT7-1/2tip (theoretical 7440.5), 7546.5 for MT7-1/2tip+3 (theoretical 7546.6) 7475.2 for MT7-1/2top (theoretical 7475.6) and, 7578.7 for MT7-1/3 (theoretical 7578.8).

The CD spectrum analysis of the MT7, MT1 and chimeric toxins reveals a typical β-sheet signature in these toxins with the maximum and minimum signals observed at 196 and 210 nm, respectively (see Figure S1, Supporting Information). An additional distinct signature at 230 nm, previously reported for MT1 [25] and characteristic of the two substitutions Q29Y and R40W, is also observed for MT7-1/2 and MT7-1/2 top. A similar effect was also observed for erabutoxin, due to an aromatic residue at position 29 [32].

## Crystallization and Crystallographic Structure Determination

Crystals of MT1 and MT7-1/1 toxins were obtained by sitting drop vapor diffusion and MT7-1/3 by slow evaporation close to 48 mg/ml protein concentration. The crystallographic structures of the MT7-1/1 and MT7-1/3 toxins were solved by molecular replacement with the MT7 diiodo-Tyr51 (PDB code = 2VLW) [28] as a starting model while MT1 was solved using toxin ρ-Da1a (unpublished) as starting model. The structures were refined to 1.3 Å, 1.39 Å and 1.8 Å resolution, respectively (Table S1). The asymmetric unit of MT1 (PDB code = 4DO8) is composed of two toxins, an acetate and a thiocyanate ions, the latter ion, positioned between the two molecules in the asymmetric unit, might be important for the crystallization. The two molecules are virtually identical and superimpose with a mean deviation of 0.32 Å (rmsd = 0.74 Å) on all atoms and an rmsd 0.15 Å on 256 main chain atoms. The asymmetric unit of chimera MT7-1/1 (PDB code = 3FEV) consists of three toxins that superimpose well on each other but with larger deviations compared to MT1 (rmsd between chains A/B = 0.88 Å; A/C = 0.73 Å; B/C = 1.10 Å on all main chain atoms). The major variations are at the tips of loops 2 and 3, the region that was not manipulated. MT7-1/3 (PDB code = 3NEQ) counts two molecules in the asymmetric unit with an rmsd of 1.47 Å between the two molecules on all 207 main chain atoms. The superimposition of the different copies of each toxin infers that these structures are uniquely defined while retaining a certain degree of flexibility. The canonical three-finger fold structure comprising five β-strands forming a twisted β-sheet is strictly abided to (Fig. 2a).

## Structure Comparisons

The structures of the MT7 diiodo-Tyr51 (PDB code = 2VLW), MT1, MT7-1/1 and MT7-1/3 described here, were superimposed with COOT [33] and pyMOL to understand the structural effect of loop grafting on the individual loops (Fig. 2) and on the global superimposition of the various structures (see Figure S2, Supporting Information). The loops with the MT1 sequence grafted onto the MT7 scaffold superimpose well on those found on the parent toxin (Fig. 2). Loop 1 of each of the three molecules in the asymmetric unit of MT7-1/1 superimposes precisely on that of loop 1 of MT1 (Fig. 2b). The loop separates from loops 2 and 3 as it does in MT1 but to a greater extent. The comparison of loop 3 from each of the two molecules in the asymmetric unit of MT7-1/3 compared to MT1 shows again an excellent match (Fig. 2c). The comparisons of loop 2 from MT1 and MT7 shows that the backbone of this loop is conserved between the two natural toxins, but the side chains adopt a different orientation (Fig. 2d). The comparison of loop 2 from chain A of MT7-1/3 with loop 2 of MT1 shows that the replacement of loop 3 of MT7 for that of MT1 imposes a reorientation of the side chains to match that of MT1 although there is poor sequence identity (Fig. 2e).

MT1 superimposition onto MT7 by least-squares fit shows large deviations in both the loops and in the sequence-conserved disulphide-bridged scaffold leading to important differences when the main chain atoms are compared (rmsd MT1 vs MT7 = 3.22±0.01 Å, depending on the molecule in the asymmetric unit that are compared). A better match can be obtained by not matching consecutive residues but only the secondary-structure elements. Secondary-structure matching (SSM) [34] for the MT1/MT7 comparison leads to a value of around 1.34 Å where the matched residues have a reduced sequence identity of only 66%. Given that loop 3 of MT1 is longer than that of MT7, the least-squares fit method is also inappropriate. In supplementary figure S2 we use the relatively good conservation of the conformation of loop 2 and its interaction with the second strand of loop 3 to guide the overall structural superposition. This allows for a visual rather than mathematical scoring of the alignment. To better understand the comparison by the two methods we can consider the superposition of MT1 with chimera MT7-1/3 that counts the same number of residues (Fig. 1). By least-squares fit we have an rmsd for MT1 vs MT7-1/3 of 2.87±0.08 Å while by SSM the value falls to an rmsd of 0.89 Å on 61 aligned residues with a 79% sequence identity. The superimposition obtained by matching only loop 2 and the second strand of loop 3 agrees better with the SSM value (see Figure S2, Supporting Information). Comparing by least-squares fit MT1 to chimera MT7-1/1 is less reliable because of a difference in sequence length which gives an rmsd between MT1 and MT7-1/1 of 2.38±0.03 Å. By SSM the rmsd remains high at 1.64 Å. This is mainly because the algorithm aligns 64 residues instead of only 61 in the previous comparison where 3 extra residues on loop 1 were omitted due to excessive deviation. If MT7 is compared to chimera MT7-1/1, least-squares gives an rmsd of 3.13±0.02 Å while by SSM gives 1.4 Å. In this comparison SSM uses only 59 residues. The SSM superimposition

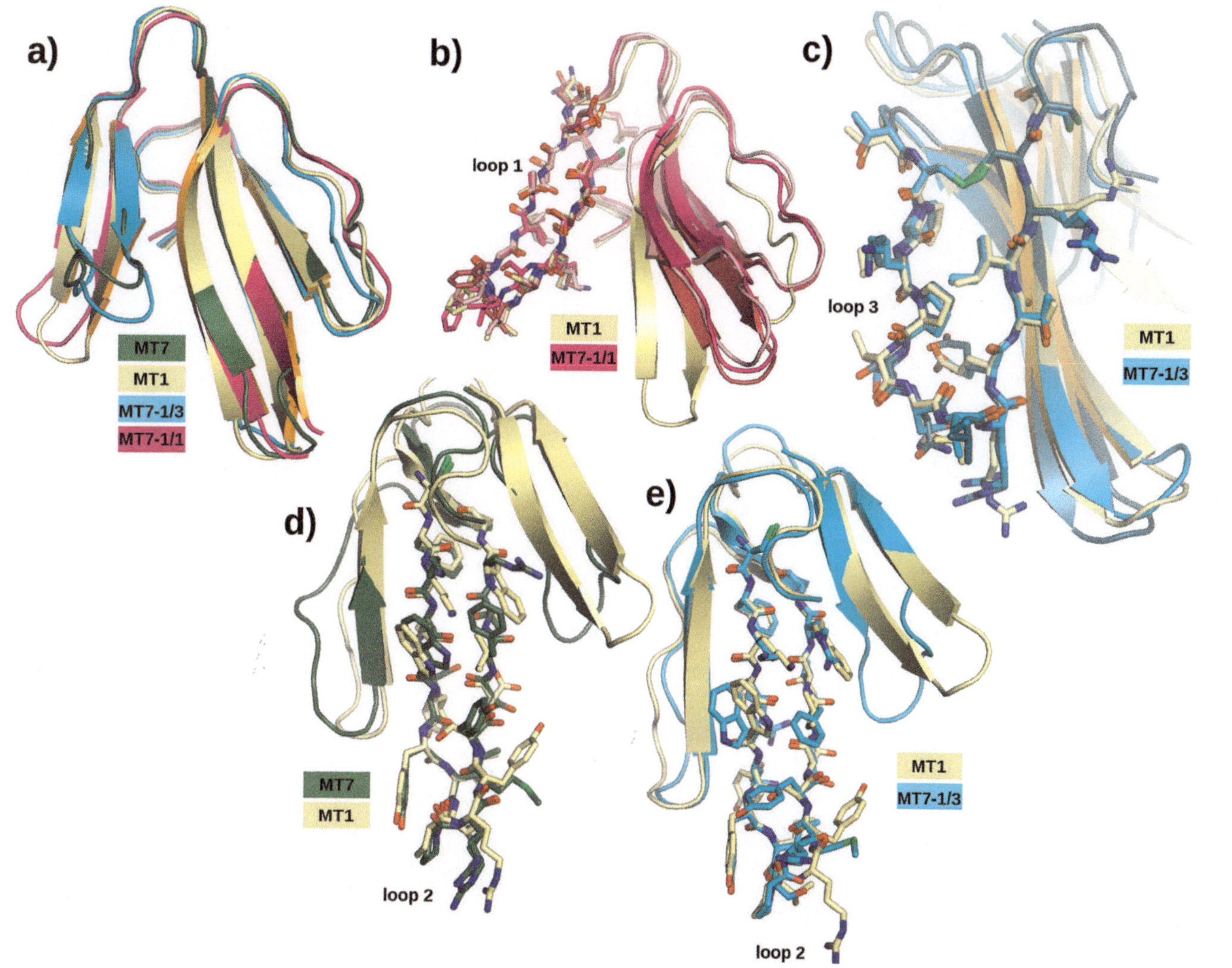


**Figure 2. Schematic representation of MT7 (green), MT1 (gold) and chimeras MT7-1/1 (magenta) and MT7-1/3 (cyan). a)** Toxins superimposed globally on each other. The largest deviations are observed on loop 1 and the tips of the various loops. **b)** Chimera MT7-1/1 superimposed on MT1 on loop 1 shown in stick representation. **c)** Chimera MT7-1/3 superimposed on MT1 on loop 3 showing that although loop 2 remains entirely that of MT7 and that there are strong interactions between loop 2 and loop 3, loop 3 maintains strictly its conformation as on MT1. **d)** As shown by the superimpositions of loop 2 from MT1 and MT7, although the backbone matches well, to maintain the interaction between loop 2 and loop 3 and the conformation of loop 3 as in MT1, the side chain orientations on loop 2 of MT7 require adaptive changes. **e)** The side chain orientations of loop 2 of chimera MT7-1/3 resemble those of loop 2 of MT1 although the sequence is that of MT7.

for MT7/MT7-1/3 gives an rsmd value of 2.8 Å with only 47 residues matched with a 91.5% identity, in good agreement with the SSM values (see Figure S2, Supporting Information).

## Equilibrium Binding Studies of Wild-Type and Chimeric Toxins to the hM1, hM4 Muscarinic Receptors and $\alpha_{1A}$-Adrenoceptor

MT7 and MT1 toxins interact with muscarinic and adrenergic receptors with an affinity of 34 pM and 24 nM for M1, 13 μM and 310 nM for M4 and greater than 20 μM and 62 nM for $\alpha_{1A}$, respectively. To identify the 3FT regions responsible for these large differences in affinity which could be submitted to engineering, we analyzed the impact of substitutions in each loop of MT7 on its binding affinity on the different receptor subtypes. The capacity of wild-type and chimeric toxins to inhibit the binding of [$^3$H]-NMS to hM1 or hM4 membrane preparations was measured (Fig. 3a, 3b), allowing the calculation of their $pK_i$ (Table 1). Substitutions within loops 1 and 3 weakly affect the toxin's interaction with the hM1 receptor, corresponding to a 2-fold decrease in affinity. In contrast, grafting the entire loop 2 of MT1 into the MT7 scaffold (MT71/2) induces a drastic 400-fold drop in affinity for the M1 receptor (Table 1). Interestingly, a similar effect was observed with the MT7-1/2 top chimera with only four modifications: R27K, Q29Y, F38I, and R40W. Changing the tip of the central loop (MT71/2 tip chimera) affects moderately toxin binding with a relatively low 7-fold decrease in affinity. Moreover, combining these later substitutions (S32V, M35Y and Y36S) with the grafting of loop 3, leading to the MT7-1/2 tip+3 chimera, confirms the low impact of the modifications introduced in these two regions of the toxin for the hM1 interaction. Finally, combining loops 1 and 3 substitutions with two additional modifications in the C-terminal part (G59E, K65E) (MT7-1/1+3) leads to a severe 360-fold decrease in affinity for the hM1 receptor (Table 1).

On the hM4 receptor, only MT1 was able to displace completely the radiotracer in the range of concentrations used

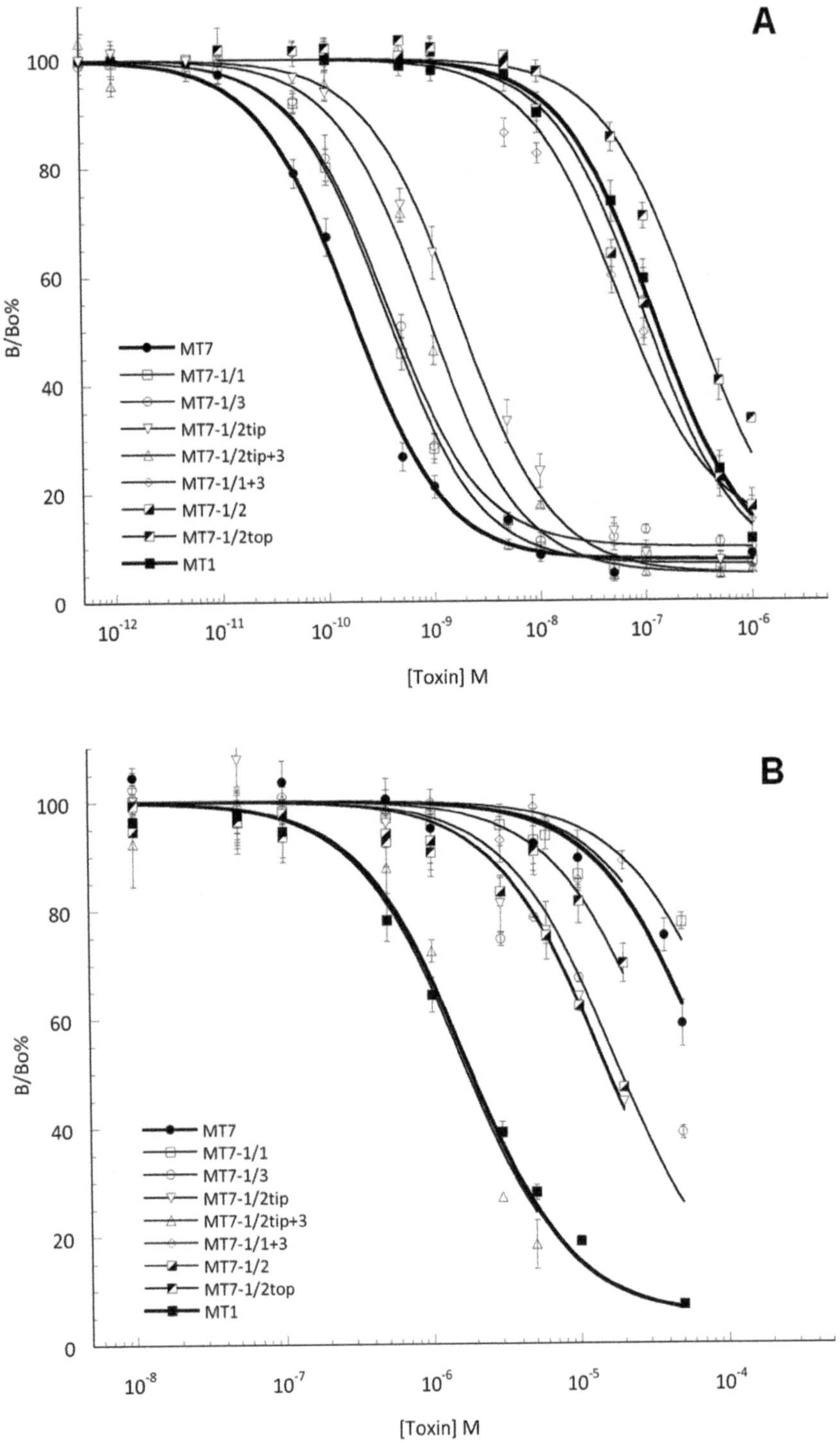


**Figure 3. Inhibition of [$^3$H]-NMS binding to hM1 and hM4 receptors by wild-type MT7, MT1 and chimera toxins.** Binding experiments were performed by incubating hM1 (A) or hM4 (B) membrane fractions of receptor with [$^3$H]-NMS (0.5 nM) and varying concentrations of toxin at room temperature and overnight. The total specific binding in each experiment was 1500±300 cpm. The results are expressed as the ratio of the specific [$^3$H]-NMS binding measured with (B) or without toxin ($B_o$). All experiments were performed at least three times in duplicate.

(Fig. 3b). Nevertheless, the incomplete inhibition curves obtained for all the other toxins suffice to give a good approximation of their affinity constants (Table 1). The affinity of MT7 for hM4 is very low (pKi: 4.87±0.06), approximately 40-fold less than that of MT1 (pKi: 6.50±0.04). Modifications introduced in the MT7-1/1, MT7-1/1+3 and MT7-1/2 top chimeras affected only weakly the affinity for the hM4 receptor (less than 2-fold decrease), whereas the MT7-1/2, MT7-1/2 tip and MT7-1/3 increase the affinity by 4- to 5-fold. It is worth noting that the gain in affinity obtained by combining MT7-1/2 tip and MT7-1/3, in the MT7-

**Table 1.** Affinity constants of wild-type MT7, MT1 and chimeric toxins for M1, M4 muscarinic receptors and $\alpha_{1A}$-adrenoceptor.

| | M1 | | M4 | | $\alpha_{1A}$ | |
|---|---|---|---|---|---|---|
| | pKi | Ki tox/Ki MT7 loss of function | pKi | Ki MT7/Ki tox gain of function | pKi | Ki MT7/Ki tox.gain of function |
| **MT7** | 10.47±0.04 | 1 | 4.87±0.06 | 1 | >4.7 | 1 |
| **MT7-1/1** | 10.16±0.05 | 2 | >4.7 | <0.67 | 7.10±0.11 | >252 |
| **MT7-1/2** | 7.86±0.02 | 406 | 5.59±0.04 | 5.3 | >4.7 | 1 |
| **MT7-1/3** | 10.04±0.11 | 2.6 | 5.51±0.08 | 4.4 | 6.97±0.09 | >187 |
| **MT7-1/2tip** | 9.57±0.13 | 7.5 | 5.60±0.08 | 5.4 | >4.7 | 1 |
| **MT7-1/2top** | 7.62±0.07 | 706 | >4.7 | <0.67 | >4.7 | 1 |
| **MT7-1/2tip+3** | 9.83±0.02 | 4.4 | 6.58±0.01 | 52 | 6.36±0.01 | >46 |
| **MT7-1/1+3** | 7.91±0.02 | 362 | >4.7 | <0.67 | 8.47±0.08 | >5900 |
| **MT1** | 7.62±0.04 | 706 | 6.50±0.04 | 43 | 7.21±0.16 | >325 |

1/2 tip+3 chimera, is additive achieving an overall 50-fold increase for the hM4 receptor (Table 1).

Equilibrium binding experiments with [3H]-Prazosin show that MT1 and MT7 toxins interact with $\alpha_{1A}$-adrenoceptor with relatively high (pKi: 7.21±0.16) and very low affinity (pKi >4.7), respectively. A 200-fold increase in affinity was obtained by the independent grafting of either loop 1 or 3, enough to elevate the affinity of MT7 to the level of MT1 (Table 1; Fig. 4). The partial or total insertion of the loop 2 was ineffective. Remarkably, a 6000-fold increase in affinity compared to MT7 was achieved on the $\alpha_{1A}$-adrenoceptor by grafting a combination of loops 1 and 3 (MT7-1/1+3; pKi: 8.47±0.08), corresponding to a 20-times affinity increase relative to MT1 (Table 1).

## Discussion

Because they are resistant to degradation, permissive to mutations and tolerant to large insertions or deletions, three-finger fold toxins (3FT) have been largely exploited by snakes to support various toxic functions in their venoms [5]. These characteristics make the three-finger fold template suitable for protein engineering [11,13,14]. Structural knowledge to drive the 3FT engineering is still lacking in the context of 3FT-GPCRs complexes although molecular dynamics

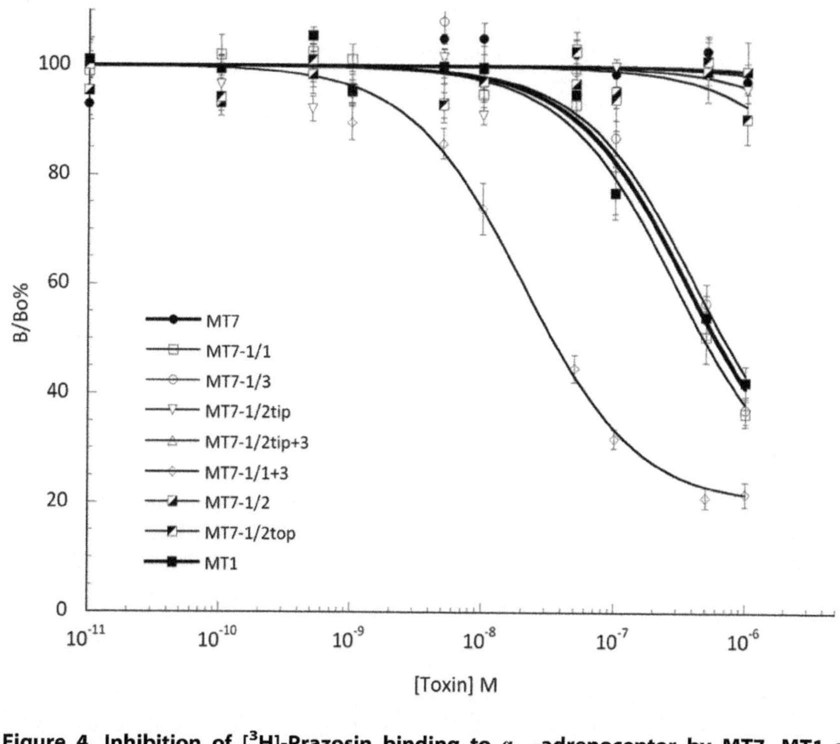


**Figure 4. Inhibition of [$^3$H]-Prazosin binding to $\alpha_{1A}$-adrenoceptor by MT7, MT1 and chimeric toxins.** Binding experiments were performed by incubating $\alpha_{1A}$-AR membrane fractions of receptor with [$^3$H]-Prazosin (1.5 nM) and varying concentrations of toxin at room temperature and overnight. The results are expressed as the ratio of the specific [$^3$H]-Prazosin binding measured with (B) or without toxin ($B_o$). All experiments were performed at least three times in duplicate.

simulations have identified regions susceptible to large-scale motions that could be permissive to modification [35]. Using the interaction of 3FT toxins with muscarinic or adrenergic receptors as a model system, we have addressed this question by a loop grafting strategy producing seven chimeric toxins by chemical synthesis. These chimeras are the results of the transfer of MT1 loops into the MT7 template (Fig. 1).

The structural impact of loop grafting was analyzed by X-ray crystallography. The three-dimensional structures of chimeras MT7-1/1 and MT7/1-3 were determined and compared to those of the natural toxins MT1 and MT7, previously reported [28].?The comparison shows which features are transferred from parent to chimera and which features are more variable and could give rise to the new functionality. Given the high resolution of the structures and the presence of multiple copies of each toxin in the crystal asymmetric unit, it leaves no doubt that the amino acid sequence determines the conformation of the individual loops and that the loop conformation is transferred together with the loop. The structures provide irrefutable evidence that for loops 1 and 3 grafted from MT1 onto the scaffold of MT7 retain their original conformation (Fig. 2a, 2b). Another invariable is that the backbone conformation of loop 2 is well conserved between MT7 and MT1, although the sequence is not. Loop grafting preserves the overall three-finger structure in all chimeric toxins. The spatial relationship of the grafted loops with the remaining part of the structure is variable. The grafted loop 1 of chimera MT7-1/1 preserves its structure as in MT1 but does not adopt the same relationship with loop 2 as in MT1 nor keep the relationship found between loop 1 and loop 2 of MT7. Thus a new toxin with novel properties is created by the grafting. The situation with loop 3 is different because of the conserved interaction between loop 2 and the second strand of loop 3. To preserve the interaction we have observed that the side chain orientations on the top part of loop 2 tend to conform to those from the parent toxin. Indeed, in chimera MT7-1/3 the side chains on loop 2 execute a concerted rotation and adopt conformers that match closely those on MT1 loop 2. This observation is linked to the change from Arg-40 at the top of loop 2 of MT7 to Trp in MT1. The flexible Arg side chain can adopt the same side chain orientation as the Trp (Fig. 2e) and because such orientation is not compatible with that of the other side chains on the loop, the side chains are forced to change. The effect that this could have on receptor binding is unpredictable. These movements are important in explaining the effect of grafting of loop 2, in particular the top of the loop. The tip of this loop is relatively independent of the rest of the toxin in MT7 but in MT1 with the longer loop 3 this is no longer the case (see Figure S3, Supporting Information). This is important when comparing toxins with different lengths of loop 3 (Fig. 1).

Since GPCRs are among the targeted receptors for the three-finger fold class of toxins, it can be envisaged that the scaffold forming the basis of their diversity could be engineered to yield novel ligands, some of which would be selective for GPCRs not yet targeted in nature. To evaluate this hypothesis, we have approached the process of diversification by the "loop grafting" strategy and we have produced 3FT chimeric toxins with new pharmacological profiles towards muscarinic and adrenergic receptors, with affinities that sometimes surpass those of the two parent toxins.

Grafting the entire MT1 loop 2 (chimera MT7-1/2) onto the MT7 template provokes a large decrease in affinity of the toxin for hM1 receptor (400-fold), a weak increase for hM4 (5-fold) and no effect on $\alpha_{1A}$-adrenoceptor (Table 1). Comparison of the affinities of the MT7-1/2top and MT7-1/2tip chimeras on the three receptors with those of the MT7-1/2 toxin confirms the lack of effect of the modifications introduced in this region on the $\alpha_{1A}$-adrenoceptor binding and suggests that the drastic loss in function on hM1 and the moderate gain in function on hM4 are mainly associated with the top and the tip of loop 2, respectively. Indeed, with only four substitutions compared to MT7 (R27K, Q29Y, F38I and R40W), MT7-1/2top reaches an affinity for hM1 equal to that of MT1 while all chimeras have a higher affinity. Of these four substitutions, the F38I mutation involves a 2-fold loss in affinity [36] and the R27K substitution is relatively conservative, suggesting that the MT7-specific residues, Gln-29 and Arg-40 might play an important role in the hM1 receptor interaction. The R40W is important for the relationship between loop 2 and loop 1 (Fig. 2e, see Figure S3, Supporting Information). Nevertheless, previous results using punctual alanine modification have shown the crucial role of residues at the tip of loop 2, such as Arg-34 and Tyr 30, in the MT7-hM1 interaction [25,29]. Conservation of these residues in MT7-1/2tip chimera may explain the relative high affinity of this toxin and suggests that the three substitutions included in this chimera (S32V, M35Y, Y36S) induced the moderate affinity decrease for hM1 receptor (7-fold). Similarly, these substitutions play a positive role in the hM4 recognition, as shown by the 5-fold increase in affinity observed on this receptor (Table 1). Thus, loop 2 is highly critical in the picomolar affinity of the MT7-hM1 interaction, weakly involved in the hM4 recognition and plays no role in the $\alpha_{1A}$-adrenoceptor binding.

On hM1 receptor, substitutions of MT7 loops 1 and 3 by those of MT1 (chimeras MT7-1/1 and MT7-1/3) have no significant impact on the toxin affinity (Table 1), suggesting that the seven modifications associated with each of these substitutions were not involved in the high affinity of the MT7-hM1 interaction. Even if these results do not exclude a complementary role of the toxin loops 1 and 3 in the MT7-hM1 receptor interaction, as proposed on the structural model of this complex [29], they are consistent with previous results obtained with punctual alanine modifications, reporting that some modifications in loops 1 and 3 of MT7 have weak or no effect on the toxin's affinity for the hM1 receptor [28]. While loop 1 transfer has no significant effect on the toxin affinity for hM4, a moderate gain of function (5-fold) is observed with the grafting of loop 3 (Table 1). Notably, a two orders of magnitude increase in affinity for the $\alpha_{1A}$-adrenoceptor is achieved for chimeras MT7-1/1 and MT7-1/3 by the grafting of loop 1 or 3. These toxins are characterized by an affinity close to that of MT1 for the $\alpha_{1A}$-adrenoceptor.

Combining some of the previous grafting in an attempt to optimize our engineering, leads to new chimeric toxins with affinities which surpasses that both of natural parents. For example, the weak gain in function on hM4 receptor consecutive to the loop 3 or loop 2tip substitutions is largely enhanced on the MT7-1/2tip+3 which combines both modifications. Thus, a 50-fold gain in function is observed with this chimera (Table 1). In nature, the MT3 toxin has evolved to reach an affinity of 2 nM for the hM4 receptor [23,37]. Comparing its sequence with those of MT7, MT1, MT7-1/2tip+3 (Fig. 1) reveals that the chimera comes close to matching the MT3 sequence. The most interesting changes are localized at the tip of loops 2 and 3. These are $Val_{MT1}/Ile_{MT3}/Ser_{MT7}$-32, $Tyr_{MT1\&MT3}/Met_{MT7}$-35, $Ser_{MT1}/Thr_{MT3}/Tyr_{MT7}$-36 at the tip of the loop 2 and $Pro_{MT1\&MT3}/Ala_{MT7}$-49, $Asn_{MT1\&MT3}/Tyr_{MT7}$-51, $Thr_{MT1}$-55/$Ser_{MT3}$-54/$Val_{MT7}$-54 and $Arg_{MT1}$-57/$His_{MT3}$-56/$Asn_{MT7}$-56 at the tip of the loop 3. The MT7 specific residues, Ser-32, Met-35, Tyr-36, Ala-49, Tyr-51, Val-54 and Asn-56 are likely to be responsible for the exceptional selectivity of this toxin for hM1 relative to the hM4. Loop 3 would appear to be responsible for the receptor subtype selectivity. More strikingly, the simultaneous grafting of the loops 1 and 3 of MT1 on the MT7 scaffold provokes on $\alpha_{1A}$-

adrenoceptor a cumulative effect as compared to those obtained with separate permutations. Thus, the MT7-1/1+3 chimera reaches an affinity which is 6000 times greater that MT7 and even 20 times more potent than MT1 on this receptor subtype (3.4 nM; Table 1). The major role of the loop 1 in this interaction is coherent with the large sequence identity (80–90%) in this region between MT1 and the two most potent toxins that target this receptor subtype: MT3 (1.6 nM) and ρ-Da1a (0.3 nM) (Fig. 1) [17,23]. Since loop 3 of MT3 and ρ-Da1a are identical, the differences in their affinity for $\alpha_{1A}$-adrenoceptor are probably due to sequence variations on loop 1. It is interesting to note that the chimera MT7-1/1+3 is built by loops 1 and 3 of MT1 grafted simultaneously on the MT7 skeleton with the additional G59E-K65E modifications present on MT1 (Fig. 1, see Figure S3, Supporting Information). This chimera is MT1-like with only the central loop of MT7 grafted onto the MT1 template and displays an affinity for hM1 receptor twice that of MT1. The transposition of loop 2 from MT1 to MT7 (chimera MT7-1/2) or from MT7 to MT1 (chimera MT7-1/1+3) leads to the same affinity on hM1. This corresponds to a 2-fold gain compared to MT1 but a 400-fold loss compared to MT7. This suggests first that the two Glu insertions and more particularly the C-terminal charge inversion would be involved in the decrease in affinity of the MT7-hM1 interaction and secondly that the relationship between the loops and not just loop 2 is important in the interaction with the hM1 receptor. This result is structurally very important as it highlights how selectivity can be modulated through loop grafting. The MT7-1/1+3 toxin possesses a unique pharmacological profile among all the muscarinic and adrenergic toxins with a high affinity for hM1 and $\alpha_{1A}$ receptors and no interaction with hM4.

In conclusion, the loop transfer engineering methodology described here is an effective means of creating new functionality onto the 3FT scaffold by connecting together a relatively small number of LEGO-like blocks. By this approach, we were able to produce chimeric toxins with affinity towards receptors that surpass that of the two parent toxins, thus creating ligands with unique pharmacological profiles. Since only seven chimeric toxins have been chemically synthesized starting from loops from just two toxins (MT1 and MT7; Fig. 1), the gain in function is very encouraging, considering that the engineering database from which loops could be selected is vast. Furthermore, functional regions that have been identified can later be improved by display methodology. Comparison of the chimeric toxins sequences developed in this study with those of natural toxins like MT3 and ρ-Da1a reveals certain similarities, suggesting that a "loop permutation-like" strategy might happen in nature to diversify the toxins functions to respond to modifications of the snake's environment. This hypothesis can be correlated with the recently proposed ASSET strategy, suggesting that an accelerated segment switch in exons in 3FT may alter their molecular targets selection by inducing change or gain of functions [38,39]. Since the three-finger toxin fold was known to interact with diverse targets including various biogenic amine GPCRs, our data are likely to be of broad general relevance.

## Materials and Methods

### Materials

[$^3$H]-N-Methylscopolamine ([$^3$H]-NMS; 78 Ci/mmol) and 7-Methoxy-[$^3$H]-prazosin ([$^3$H]-prazosin; 85 Ci/mmol) were obtained from PerkinElmer (Courtaboeuf, France). Atropine, N-methylscopolamine and prazosin were purchased from Sigma-Aldrich (St Quentin-Fallavier, France).

### Peptide Synthesis

Assembly of the different chimeric proteins were performed on a standard Applied Biosystems 433 peptide synthesizer (Applied Biosystems, France) and carried out using the stepwise solid-phase method with dicyclohexyl-carbodiimide/HOAT (1-hydroxy-aza-benzotriazole) or 6-Cl-HOBT (6-chloro-1-hydroxybenzotriazole) as coupling reagents and N-methyl pyrrolidone as solvent. Fmoc-protected amino acids were used with the following side-chain protections: t-butyl ester (Glu,Asp), t-butyl ether (Ser, Thr, Tyr), trityl (Cys, His, Asn, Gln), 2,2,5,7,8-pentamethyl-chromane-6-sulfonyl (Arg), t-butyloxy-carbonyl (Trp). The Fmoc-pseudoproline dipeptide, introduced at different positions in the chimera toxins was Fmoc-Ileu-Thr ($\Psi^{MeMe}$pro)-OH. The MT7-MT1- chimera toxins were assembled on a Fmoc-Glu(OtBu)-Wang resin (loading: 0.5 mMol/g) or Fmoc-Lys(Boc)-Wang resin (loading: 0.55 mMol/g). The different syntheses were run on a modified version of the Applied Biosystems standard 0.1 mmole small-scale program using 0.05 mmole of each resin. This program achieves UV monitoring of the deprotection step. When the deprotection is too slow after two and/or three successive deprotections of 3 min, it automatically extends the deprotection time by 20 min and the coupling time (normal coupling: 30 min or 60min in case of pseudoproline dipeptide) by an extra 30 min. After each coupling, the resin was acetylated by a mixture of 5% acetic anhydride, 6% 2,4,6-collidine in DMF. At the end of the synthesis the peptide-resins were treated with trifluoroacetic acid (9 ml), triisopropylsilan (0.5 ml), and 0.5 ml distilled water. The peptides were then cleaved from the resin and the protecting groups were removed from amino acid side-chains. After two hours of incubation, the mixture was filtered in cold tert-butyl methyloxide and centrifuged three times. The precipitates were dissolved in a solution of 10% acetic acid and lyophilized. The toxins were purified by reverse phase HPLC using a Discovery BioWidepore C5 column (Supelco,PA, USA) (250×10 mm) with a gradient of 40 to 60% of solvent B in 40 min, (A: 0.1% TFA in $H_2O$, B: 60% acetonitrile and 0.1% TFA in $H_2O$) The flow rate was 4 ml/min and the detection was followed at 280 and 214 nm.

### Optimization of Disulfide Bond Formation, Purification and Characterization

The reduced synthetic and chimera toxins were subjected to an oxidative reaction in 0.1 M Tris, 1mM EDTA buffer (pH 7.8) containing 0.5 M guanidine hydrochloride in the presence of reduced (GSH) and oxidized (GSSG) glutathione in a molar ratio of 1/10/100 peptide/GSSG/GSH at a concentration of 0.05 mg/ml as described elsewhere (12) or submitted to a refolding screening procedure. In this case, a reduced-denatured toxin, was screened against 40 different refolding conditions, a combination of 8 buffers (100 mM Tris-Cl, 1 mM EDTA pH 8 with or without 100 mM NaCl, or 30% glycerol or 0,25% triton X100 or 100 mM NDSB-256 or 0.05% PEG3500 (w/v) or 0.5 M Gnd-Cl or 0.5 M Arg) and 5 concentrations of the redox couple of reduced glutathione (GSH) and oxidized glutathione (GSSG): 10/1 mM, 5/1 mM, 1/1 mM, 5/0.5 mM, 1/0.5 mM. To minimize the adsorption of protein on to the vessel's surface the refolding was carried out in a microsorb eppendorf tube of 1 ml. Each sample was reduced for 2 h at 37°C in 6 M guanidine-HCl, 10 mM dithiothreitol (DTT), 50 mM Tris, pH 8 denaturing buffer. Refolding was initiated by a 1/200 fold rapid dilution in ice cold refolding buffer to a final concentration of 20 µg/ml and a final volume of 1 ml followed by incubation at 4°C for 72 h. Refolding efficiency was evaluated by integrating the surface of the peak corresponding to the refolded form of the protein after analytical HPLC separation. Typically, the pH of each sample was lowered to pH 3 by addition of 30% TFA and 200 µl loaded in an autosampler (Waters 717plus) connected to an HPLC chain (Waters 600) and a photodiode array detector (Waters 996). Fast separation was performed on a Chromolith SpeedRod, RP18e, 50–4.6 mm column (Merck, Darmstadt,Germany) by means of a 10 min linear gradient of acetonitrile in 0,1% (by vol.) trifluoroacetic acid from 25 to 35% at a flow rate of 4 ml/min, detection 214 nm.

Large scale refolding was performed by diluting the reduced protein 200 fold in the optimized refolding buffer namely 100 mM Tris-HCl, 1 mM EDTA, 30% glycerol, 1 mM GSH, 1 mM GSSG pH 8 to a final concentration of 0.1 mg/ml and then incubated at 4°C for 72 h. After acidification the purification of the refolded toxins were performed on a Discovery BioWidepore C5 column (Supelco, PA, USA) (250×10 mm) with a gradient of 40 to 60% of solvent B in 40 min, (A: 0.1% TFA in $H_2O$, B: 60% acetonitrile and 0.1% TFA in $H_2O$) The flow rate was 4 ml/min and the detection was followed at 280 and 214 nm. Finally, each protein was analytically characterized on analytical reversed-phase HPLC a Discovery BioWidepore C5 column (Supelco,PA, USA) (100×4.5 mm) with a gradient of 40 to 60% of solvent B in 40 min, (A: 0.1% TFA in $H_2O$, B: 60% acetonitrile and 0.1% TFA in $H_2O$). The flow rate was 1 ml/min and the detection was followed 214 nm. For the amino acid analysis, the hydrolysates obtained after acid hydrolysis in a sealed vial were heated at 120°C in the presence of 6 N HCl for 16 h and analysed using an Applied Biosystems model 130A automatic analyzer equipped with an online 420A derivatizer for the conversion of the free amino acid into phenyl thiocarbamoyl. Mass determination was performed on a Nermag spectrometer coupled to an analytical (Brandford) electrospray source. The concentrations of the different toxins were evaluated spectrometrically.

## Circular Dichroism Analysis

CD spectra were recorded on a Jasco J-815 CD spectrometer. Measurements were routinely performed at 20°C in 0.1 cm path-length quartz cells (Hellma, Paris, France) with a peptide concentration of $10^{-5}$ M in water. Spectra were recorded in the 186 to 260 nm wavelength range. Each spectrum represents the average of at least three spectra.

## Crystallization

The crystallization of the MT7 wild-type toxin in its diiodo-Tyr51 derivative had been reported previously [28]. Briefly, lyophilized synthetic toxin was re-suspended at 5 mg/ml in 0.02% azide, 50 mM sodium acetate, pH 4.6. The reservoir solution used was 1.25 M ammonium sulfate, 90 mM sodium citrate, and 10% methyl pentanediol (MPD), pH 5.5. Similarly, lyophilized synthetic toxin MT1 was re-suspended at 8.3 mg/ml in 0.02% azide, 50 mM sodium acetate, pH 5.5 and crystallized by sitting drop vapor diffusion over a reservoir consisting of 36% mono-methyl polyethylene glycol 2,000 (MPEG 2K), 450 mM NaCl, 90 mM KSCN, 100 mM imidazole-HCl, pH 7.5. Crystals appeared after 1 month. No cryoprotectant was needed for flash-cooling given the high MPEG concentration used to crystallize the toxin. The chimera MT7-1/1 was prepared from 1.3 mg of lyophilized synthetic toxin dissolved in 50 mM Na acetate, pH 5.5, with 0.02% azide to yield a 5 mg/ml solution. Screening for crystallization was carried out by sitting drop vapor diffusion at 20°C in a cooled incubator using 4 selected conditions from the "Stura" screens [40] (Molecular Dimensions) and the conditions that had yielded crystals for the wild type MT7. Conditions that yielded crystals were further refined. The crystals used for data collection were grown from 1.2 M ammonium sulfate, 90 mM sodium citrate, with 4% MPD, 6% propanol at pH 5.5. The crystals were transferred to a cryo-solution consisting of 1.5 M ammonium sulfate, 75 mM sodium citrate, pH 5.5, 25% glycerol and 5% MPD. After a short solvent exchange step the crystals were cryo-cooled in liquid ethane. Chimera MT7-1/2 was found to be less soluble than MT7-1/1, tended towards filamentous aggregation. A similar problem had been encountered with the first batch of toxin MT1 from which crystals could not be obtained given the limited quantities available insufficient to carry out a more complete screen. Chimeric toxin MT7-1/3 was prepared from 0.48 mg resuspended in 0.02% azide, 50 mM sodium acetate at pH 5.5 to yield a 5 mg/ml solution. Needle crystals of this chimera were obtained from 1.02 M ammonium sulfate, 50 mM sodium citrate, 1% MPD at pH 5.5 using the same conditions as for MT7-1/1 without any need for screening. Better crystals were obtained later by the slow evaporation method over a period of 3 months at 4°C in a laboratory refrigerator. The method consists in allowing moisture to escape slowly over a long period of time. This can be done by loosening the cap of an eppendorf tube containing the protein. However, we used a more reproducible method that consists in replacing the screw cap by one from a different manufacturer and screwing in the cap tightly. Because the specifications of the two manufacturers differ, this still allows water vapor to leak out at a rate that is compatible with crystallization while it is difficult to evaluate the rate of vapor escape from a loosened cap. When the crystals were harvested, the effective concentration of the protein was estimated from the decrease in sample volume to be 48 mg/ml. The crystals were stabilized in two steps: first in a solution consisting of 0.96–1.26 M ammonium sulfate, 65 mM sodium citrate with 5% MPD, pH 5.5 over a period of at least one day and then in 80% saturated lithium sulfate before cryo-cooling in liquid ethane.

## Data Collection and Crystal Structure Determination

Data for the all the crystals were collected at the European Synchrotron Radiation Facility (ESRF). As previously reported [28], data to 1.38 Å resolution were collected on beamline ID14-2 from a crystal cryo-cooled in 80% saturated Li2SO4 as a cryoprotectant. Data for MT7-1/1 were collected on beamline ID14-4 at 100°K from single crystal to 1.3 Å resolution. The crystal was found to belong to the space group P212121 with cell parameters 27.1 Å, 72 Å, 89.4 Å and three molecules in the asymmetric unit (Table S1). For MT7-1/3 data extending to 1.39 Å resolution were collected on beamline ID29. The crystals also belong to the space group P212121 but with different cell parameters: 25.8 Å, 56.6 Å, 80.7 Å; with two molecules in the asymmetric unit. The data were processed using MOSFLM [41] and reduced using programs from the CCP4 suite of programs. Molecular replacement was carried out with MOLREP [42] using the diiodo-MT7 (PDB code = 2VLW) as the starting model. After restrained refinement using REFMAC [43] the amino acid differences between the two loops were corrected according to the sequence. Density fitting and refinement were carried out with the aid of electron density maps (omit σA-weighted 2Fo- Fc and Fo-Fc) calculated and displayed using the XtalView [44] suite of programs and COOT [33]. Stereochemical analysis of the final refined model was checked with the validation tools in COOT. Waters were checked with phenix.refine [45]. Refinement statistics are detailed in Table S1 (MT1 R = 19.3% Rfree = 23.4%); (MT7-1/1 R = 21.4% Rfree = 23.3%); (MT7-1/3 R = 21.2% Rfree = 24.9%). The figures were made with PYMOL from DeLano Scientific LLC (The pyMOL Molecular Graphics System. DeLano Scientific, San Carlos, CA, USA).

## Protein Data Bank Accession Numbers

The coordinates and structure factors for the chimeric toxins have been deposited in the RCSB Protein Data Bank with the following ID codes: MT1:4DO8; MT7-1/1:3FEV; MT7-1/3:3NEQ.

## CHO Cells and Membrane Preparation

Profs. P. O. Couraud & A. D. Strosberg (ICGM, Paris, France) kindly provided CHO cells stably expressing the cloned human muscarinic $M_1$ and $M_4$ receptors. The cells were grown in plastic Petri dishes (Falcon) which were incubated at 37°C in an atmosphere of 5% $CO_2$ and 95% humidified air in Ham F12 medium pre-complemented with L-glutamine and bicarbonate (Sigma) supplemented with 10% fœtal calf serum and 1% penicillin/streptomycin (Sigma). At 100% confluence, the medium was removed and the cells were harvested using Versen buffer (PBS +5 mM EDTA). They were washed with ice-cold phosphate buffer and centrifuged at 1700 g for 10 min (4°C). The pellet was suspended in ice-cold buffer (1 mM EDTA, 25 mM Na phosphate, 5 mM $MgCl_2$, pH 7.4) and homogenized using an Elvehjem-Potter homogenizer (Fisher Scientific Labosi, Elancourt, France). The homogenate was centrifuged at 1700 g for 15 min (4°C). The sediment was resuspended in buffer, homogenized and centrifuged at 1700 g for 15 min (4°C). The combined supernatants were centrifuged at 35000 g for 30 min (4°C) and the pellet was suspended in the same buffer (0.1 ml/dish). Protein concentrations were determined according to the Lowry method using bovine serum albumin as standard. The membrane preparations were aliquoted and stored at –80°C. $\alpha_{1A}$-adrenoceptor cDNA inserted into the prK5 vector was kindly provided by Michael Brownstein (Craig Venter Institute, Rockville, MD). COS cells grown at 37°C under 5% $CO_2$ in Dulbecco's modified Eagle's medium containing 10% fetal calf serum, 1% penicillin and 1% glutamine (Sigma-Aldrich, St Quentin-Fallavier, France). At 80% confluence, cells were transfected using calcium phosphate precipitation to transiently express the $\alpha_{1A}$-adrenoceptor. After 48h incubation at 37°C, cells were harvested and the membranes were prepared as previously described [27].

## Equilibrium [$^3$H]-NMS and [$^3$H]-Prazosin Binding Assays

The effect of MT7, MT1 toxins and various chimeric toxins on the equilibrium binding of [$^3$H]-NMS for muscarinic $M_1$ and M4 receptors and [$^3$H]-Prazosin for α1a-adrenoceptor were determined with equilibrium inhibition binding experiments. For [$^3$H]-NMS binding, membrane protein concentrations, adjusted so no more than 10% of added radio-ligand was specifically bound (around 1500–2000 cpm), were incubated in PBS-BSA at 25°C for 18 to 22 h, with varying concentrations of toxin and [$^3$H]-NMS (0.5 nM), in a final assay volume of 300 µl. Non-specific binding was determined in the presence of 50 µM atropine. The reaction was stopped by addition of 3 ml of ice-cold buffer (PBS) immediately followed by filtration through Whatman GF/C glass fiber filters pre-soaked in 0.5% polyethylenimine. The filters were washed once again with 3 ml ice-cold buffer (PBS), dried and the bound radioactivity was counted by liquid scintillation spectrometry. Binding experiments with [$^3$H]-Prazosin (1.5 nM) were performed in a 100 µL reaction mix at room temperature in buffer composed of 50 mM Tris-HCl, pH 7.4, 10 mM $MgCl_2$, 1 g/L BSA. Incubation was stopped by filtration through 96 GF/C filter plates pre-incubated with 0.5% polyethylenimine. 25 µL of Microscint 0 were added onto each dry filter and the radioactivity was quantified on a TopCount beta counter (PerkinElmer, Courtaboeuf, France). Non specific binding were measured in presence of prazosin (10 µM). Each experiment was done in duplicate at least three times.

## Data Analysis

The binding data from individual experiments (n≥3) were analyzed by nonlinear regression analysis using Kaleidagraph 4.0 (Synergy Software, Reading, PA). After subtraction of the non-specific binding and normalization, data obtained with [$^3$H]-NMS and [$^3$H]-prazosin were analyzed using the Hill equation to estimate the $IC_{50}$ and the slope factor, $n_H$, of the inhibition curve. The affinities of toxins in inhibiting the binding of radiotracers, expressed as p$K_i$, were calculated from the $IC_{50}$ values by applying the Cheng-Prusoff correction. Following the results obtained on the[$^3$H]-NMS dissociation experiments on M1 receptor, highlighting the allosteric behaviour of MT7, MT7-1/1and MT7-1/3 toxins, the allosteric ternary complex model [46] was used to analyze the data on the effect of these toxins on the specific binding of the orthosteric radioligand [3H]-NMS.

## Supporting Information

**Figure S1 Overall far-UV CD spectra pattern of the different toxins and chimera.** The CD spectra were monitored in water, at 20°C with a peptide concentration of 10 µM.

**Figure S2 Schematic representation of MT7 (green), MT1 (gold) and chimeras MT7-1/1 (magenta) and MT7-1/3 (cyan) superimposed on each other on loop 2. a)** The superimposition of MT1 onto MT7 shows important deviation in both loops and in the sequence conserved disulphide-bridged scaffold but good conservation of the conformation of loop 2 and its interaction with the second strand of loop 3. This interaction is conserved in the chimeras. **b)** Chimera MT7-1/1 superimposes well on MT7 except for the grafted loop 1 which adopts the conformation and orientation seen in MT1. **c)** Chimera MT7-1/3 superimposes well on MT1 except for loop 1 which maintains the conformation and orientation seen in MT7. **d)** Chimera MT7-1/3 superimposes poorly on MT7. It maintains the interaction between loops 1 and 2 and but this implies that the disulphide-bridged scaffold adopts an orientation rotated with respect to the first two loops.

**Figure S3 Space-filling representation of MT7 and MT1.** The residues from loop 1 colored in green (conserved in lighter shade), residues from loop 2 in blue (tip) and cyan (top) with the conserved residues in lighter shades, loop 3 in yellow (conserved in lighter shade) and the C-terminal residue in magenta. **a)** Front view of MT7 showing the interacting residues from the three loops. **b)** Back view of MT7 showing the interacting residues from the three loops and the C-terminal residue. **c)** Front view of MT7 with the residues from loop 2 top removed to show the lack of interactions between the other elements of the toxin assembly apart from the interaction of the C-terminal residue with loop 1. **d)** Front view of MT1 showing that the overall shape of the toxin differs substantially from that of MT7. **e)** Back view of MT1 showing variations in most elements, except the tip that maintains its orientation relative to the rest of the toxin as in MT7. f) With the top of loop 2 removed, MT1 shows that its C-terminal residue mediates an interaction between loops 1 and 3 and that the tip of loop 2 is stabilized by the longer loop 3.

## Author Contributions

Conceived and designed the experiments: CFG GM EAS DS. Performed the experiments: CFG GM GB EAS LV EM RM NG. Analyzed the data: CFG GM GB EAS NG DS. Wrote the paper: DS EAS CFG GM NG.

## References


1. Ménez A (1998) Functional architectures of animal toxins: a clue to drug design? Toxicon 36: 1557–1572.
2. Mouhat S, Jouirou B, Mosbah A, De Waard M, Sabatier JM (2004) Diversity of folds in animal toxins acting on ion channels. Biochem J 378: 717–726.
3. King GF (2011) Venoms as a platform for human drugs: translating toxins into therapeutics. Expert Opin Biol Ther 11: 1469–1484.
4. Kini RM (2002) Molecular moulds with multiple missions: functional sites in three-finger toxins. Clin Exp Pharmacol Physiol 29: 815–822.
5. Kini RM, Doley R (2010) Structure, function and evolution of three-finger toxins: mini proteins with multiple targets. Toxicon 56: 855–867.
6. Galat A (2008) The three-fingered protein domain of the human genome. Cell Mol Life Sci 65: 3481–3493.
7. Trémeau O, Lemaire C, Drevet P, Pinkasfeld S, Ducancel F, et al. (1995) Genetic engineering of snake toxins. The functional site of Erabutoxin a, as delineated by site-directed mutagenesis, includes variant residues. J Biol Chem 270: 9362–9369.
8. Antil S, Servent D, Ménez A (1999) Variability among the sites by which curaremimetic toxins bind to *Torpedo* acetylcholine receptor, as revealed by identification of the functional residues of α-Cobratoxin. J Biol Chem 274: 34851–34858.
9. Antil-Delbeke S, Gaillard C, Tamiya T, Corringer PJ, Changeux JP, et al. (2000) Molecular determinants by which a long chain toxin from snake venom interacts with the neuronal {alpha}7 nicotinic acetylcholine receptor. J Biol Chem 275: 29594–29601.
10. Marchot P, Prowse CN, Kanter J, Camp S, Ackermann EJ, et al. (1997) Expression and activity of mutants of fasciculin, a peptidic acetylcholinesterase inhibitor from mamba venom. J Biol Chem 272: 3502–3510.
11. Ricciardi A, le Du MH, Khayati M, Dajas F, Boulain JC, et al. (2000) Do structural deviations between toxins adopting the same fold reflect functional differences? J Biol Chem 275: 18302–18310.
12. Le Du MH, Ricciardi A, Khayati M, Menez R, Boulain JC, et al. (2000) Stability of a structural scaffold upon activity transfer: X-ray structure of a three fingers chimeric protein. J Mol Biol 296: 1017–1026.
13. Mourier G, Servent D, Zinn-Justin S, Ménez A (2000) Chemical engineering of a three-fingered toxin with anti-α7 neuronal acetylcholine receptor activity. Protein Engineering 13: 217–225.
14. Naimuddin M, Kobayashi S, Tsutsui C, Machida M, Nemoto N, et al. (2011) Directed evolution of a three-finger neurotoxin by using cDNA display yields antagonists as well as agonists of interleukin-6 receptor signaling. Mol Brain 4: 2–16.
15. Adem A, Asblom A, Johansson G, Mbugua PM, Karlsson E (1988) Toxins from the venom of the green mamba Dendroaspis angusticeps that inhibit the binding of quinuclidinyl benzilate to muscarinic acetylcholine receptors. Biochim Biophys Acta 968: 340–345.
16. Karlsson E, Jolkkonen M, Satyapan N, Adem A, Kumlin E, et al. (1994) Protein toxins that bind to muscarinic acetylcholine receptors. Ann N Y Acad Sci 710: 153–161.
17. Quinton L, Girard E, Maiga A, Rekik M, Lluel P, et al. (2010) Isolation and pharmacological characterization of AdTx1, a natural peptide displaying specific insurmountable antagonism of the alpha(1A)-adrenoceptor. Br J Pharmacol 159: 316–325.
18. Rouget C, Quinton L, Maiga A, Gales C, Masuyer G, et al. (2010) Specific inhibition of the alpha2-adrenoceptors by a natural antagonist peptide toxin. Br J Pharmacol 161: 1361–1374.
19. Rajagopalan N, Pung YF, Zhu YZ, Wong PT, Kumar PP, et al. (2007) Beta-cardiotoxin: a new three-finger toxin from Ophiophagus hannah (king cobra) venom with beta-blocker activity. Faseb J 21: 3685–3695.
20. Servent D, Blanchet G, Mourier G, Marquer C, Marcon E, et al. (2011) Muscarinic toxins. Toxicon 58: 455–463.
21. Servent D, Fruchart-Gaillard C (2009) Muscarinic toxins: tools for the study of the pharmacological and functional properties of muscarinic receptors. J Neurochem 109: 1193–1202.
22. Koivula K, Rondinelli S, Nasman J (2010) The three-finger toxin MTalpha is a selective alpha(2B)-adrenoceptor antagonist. Toxicon 56: 440–447.
23. Nareoja K, Kukkonen J, Rondinelli S, Toivola D, Meriluoto J, et al. (2011) Adrenoceptor Activity of Muscarinic Toxins identified from Mamba Venoms. Br J Pharmacol 2b: 538–550.
24. Max SI, Liang JS, Potter LT (1993) Purification and properties of m1-toxin, a specific antagonist of m1 muscarinic receptors. J Neurosci 13: 4293–4300.
25. Mourier G, Dutertre S, Fruchart-Gaillard C, Ménez A, Servent D (2003) Chemical Synthesis of MT1 and MT7 Muscarinic Toxins: Critical Role of Arg-34 in Their Interaction with M1 Muscarinic Receptor. Mol Pharmacol 63: 26–35.
26. Olianas MC, Maullu C, Adem A, Mulugeta E, Karlsson E, et al. (2000) Inhibition of acetylcholine muscarinic M(1) receptor function by the M(1)-selective ligand muscarinic toxin 7 (MT-7). Br J Pharmacol 131: 447–452.
27. Fruchart-Gaillard C, Mourier G, Marquer C, Menez A, Servent D (2006) Identification of various allosteric interaction sites on M1 muscarinic receptor using 125I-Met35-oxidized muscarinic toxin 7. Mol Pharmacol 69: 1641–1651.
28. Fruchart-Gaillard C, Mourier G, Marquer C, Stura E, Birdsall NJ, et al. (2008) Different interactions between MT7 toxin and the human muscarinic M1 receptor in its free and NMS-occupied states. Mol Pharmacol 74: 1554–1563.
29. Marquer C, Fruchart-Gaillard C, Letellier G, Marcon E, Mourier G, et al. (2011) A structural model of a ligand/GPCR complex based on experimental double-mutant cycle data: the MT7 snake toxin bound to a dimeric hM1 muscarinic receptor. J Biol Chem 286: 31661–31675.
30. Harvey AL, Kornisiuk E, Bradley KN, Cervenansky C, Duran R, et al. (2002) Effects of muscarinic toxins MT1 and MT2 from green mamba on different muscarinic cholinoceptors. Neurochem Res 27: 1543–1554.
31. Mutter M, Nefzi A, Sato T, Sun X, Wahl F, et al. (1995) Pseudo-prolines (psi Pro) for accessing "inaccessible" peptides. Pept Res 8: 145–153.
32. Pillet L, Tremeau O, Ducancel F, Drevet P, Zinn-Justin S, et al. (1993) Genetic engineering of snake toxins. Role of invariant residues in the structural and functional properties of a curaremimetic toxin, as probed by site-directed mutagenesis. J Biol Chem 268: 909–916.
33. Emsley P, Lohkamp B, Scott WG, Cowtan K (2010) Features and development of Coot. Acta Crystallographica Section D-Biological Crystallography 66: 486–501.
34. Krissinel E, Henrick K (2004) Secondary-structure matching (SSM), a new tool for fast protein structure alignment in three dimensions. Acta Crystallographica Section D-Biological Crystallography 60: 2256–2268.
35. Gilquin B, Bourgoin M, Menez R, Le Du MH, Servent D, et al. (2003) Motions and structural variability within toxins: implication for their use as scaffolds for protein engineering. Protein Sci 12: 266–277.
36. Krajewski JL, Dickerson IM, Potter LT (2001) Site-directed mutagenesis of m1-toxin1: two amino acids responsible for stable toxin binding to M(1) muscarinic receptors. Mol Pharmacol 60: 725–731.
37. Jolkkonen M, van Giersbergen PL, Hellman U, Wernstedt C, Karlsson E (1994) A toxin from the green mamba Dendroaspis angusticeps: amino acid sequence and selectivity for muscarinic m4 receptors. FEBS Lett 352: 91–94.
38. Doley R, Pahari S, Mackessy SP, Kini RM (2008) Accelerated exchange of exon segments in Viperid three-finger toxin genes (Sistrurus catenatus edwardsii; Desert Massasauga). BMC Evol Biol 8: 196–207.
39. Doley R, Mackessy SP, Kini RM (2009) Role of accelerated segment switch in exons to alter targeting (ASSET) in the molecular evolution of snake venom proteins. BMC Evol Biol 9: 146–161.
40. Stura EA, Nemerow GR, Wilson IA (1992) Strategies in the Crystallization of Glycoproteins and Protein Complexes. Journal of Crystal Growth 122: 273–285.
41. Leslie AGW (2006) The integration of macromolecular diffraction data. Acta Crystallographica Section D-Biological Crystallography 62: 48–57.
42. Vagin A, Teplyakov A (2010) Molecular replacement with MOLREP. Acta Crystallographica Section D-Biological Crystallography 66: 22–25.
43. Murshudov GN, Vagin AA, Dodson EJ (1997) Refinement of Macromolecular Structures by the Maximum-Likelihood Method. Acta Cryst D53: 240–255.
44. McRee DE (1999) XtalView/Xfit–A versatile program for manipulating atomic coordinates and electron density. J Struct Biol 125: 156–165.
45. Adams PD, Afonine PV, Bunkoczi G, Chen VB, Davis IW, et al. (2010) PHENIX: a comprehensive Python-based system for macromolecular structure solution. Acta Crystallographica Section D-Biological Crystallography 66: 213–221.
46. Ehlert FJ (1988) Estimation of the affinities of allosteric ligands using radioligand binding and pharmacological null methods. Mol Pharmacol 33: 187–194.

# 17

# Structure of an Engineered β-Lactamase Maltose Binding Protein Fusion Protein: Insights into Heterotropic Allosteric Regulation

**Wei Ke[1], Abigail H. Laurent[2], Morgan D. Armstrong[2], Yuchao Chen[2], William E. Smith[2], Jing Liang[2], Chapman M. Wright[2], Marc Ostermeier[2], Focco van den Akker[1]***

**1** Departments of Biochemistry, Case Western Reserve University, Cleveland, Ohio, United States of America, **2** Department of Chemical and Biomolecular Engineering, Johns Hopkins University, Baltimore, Maryland, United States of America

## Abstract

Engineering novel allostery into existing proteins is a challenging endeavor to obtain novel sensors, therapeutic proteins, or modulate metabolic and cellular processes. The RG13 protein achieves such allostery by inserting a circularly permuted TEM-1 β-lactamase gene into the maltose binding protein (MBP). RG13 is positively regulated by maltose yet is, serendipitously, inhibited by $Zn^{2+}$ at low μM concentration. To probe the structure and allostery of RG13, we crystallized RG13 in the presence of mM $Zn^{2+}$ concentration and determined its structure. The structure reveals that the MBP and TEM-1 domains are in close proximity connected via two linkers and a zinc ion bridging both domains. By bridging both TEM-1 and MBP, $Zn^{2+}$ acts to "twist tie" the linkers thereby partially dislodging a linker between the two domains from its original catalytically productive position in TEM-1. This linker 1 contains residues normally part of the TEM-1 active site including the critical β3 and β4 strands important for activity. Mutagenesis of residues comprising the crystallographically observed $Zn^{2+}$ site only slightly affected $Zn^{2+}$ inhibition 2- to 4-fold. Combined with previous mutagenesis results we therefore hypothesize the presence of two or more inter-domain mutually exclusive inhibitory $Zn^{2+}$ sites. Mutagenesis and molecular modeling of an intact TEM-1 domain near MBP within the RG13 framework indicated a close surface proximity of the two domains with maltose switching being critically dependent on MBP linker anchoring residues and linker length. Structural analysis indicated that the linker attachment sites on MBP are at a site that, upon maltose binding, harbors both the largest local Cα distance changes and displays surface curvature changes, from concave to relatively flat becoming thus less sterically intrusive. Maltose activation and zinc inhibition of RG13 are hypothesized to have opposite effects on productive relaxation of the TEM-1 β3 linker region via steric and/or linker juxtapositioning mechanisms.

**Editor:** Bostjan Kobe, University of Queensland, Australia

**Funding:** FvdA is supported by National Institutes of Health (R01 AI062968). MO was supported by National Institutes of Health (R01 GM066972) and National Science Foundation (MCB 0919377). The funders had no role in study design, data collection and analysis, decision to publish, or preparation of the manuscript.

**Competing Interests:** The authors have declared that no competing interests exist.

* E-mail: focco.vandenakker@case.edu

## Introduction

Allosteric regulation is a common mechanism cells utilize to regulate protein activity. Heterotropic allosteric proteins have a ligand binding site distant from the substrate binding site to allow conformational changes to be transmitted from the ligand regulatory binding site. As such, engineered proteins with allosteric properties have attractive applications including molecular biosensors, therapeutic proteins, and gene regulation [1,2]. One approach to engineering allosteric proteins is to mimic non-homologous recombination via domain/module combinations and rearrangements. This approach can develop complex new functions involving molecular recognition and regulation [3] and can lead to development of analytical molecular sensors [4]. A successful example of this approach is the allosteric protein RG13 which is a molecular switch created by recombining the non-homologous genes encoding MBP and TEM-1 β-lactamase and subjecting the resulting library to evolutionary pressure [5]. In brief, a circular permutated TEM-1 gene was randomly inserted into the gene encoding MBP. The constructs were subjected to an *in vivo* selection for high ampicillin hydrolysis activity in the presence of maltose, followed by an *in vitro* screen for variants with maltose-dependent rate of β-lactam hydrolysis. The resulting RG13 construct was demonstrated, in the presence of maltose, to have β-lactamase activity comparable to the wild-type TEM-1 β-lactamase but RG13's activity ($k_{cat}/K_m$) decreased 25-fold in the absence of maltose [5]. As a consequence, the *RG13* gene confers a maltose-dependent ampicillin resistance to *E. coli* cells [5]. Intriguingly, RG13 activity was serendipitously found to be negatively regulated by $Zn^{2+}$ at low μM concentrations in a non-competitive and reversible mode, a characteristic that neither parent protein MBP nor TEM-1 possesses [6].

The primary structure of RG13 is unusual in that it is a 72kD fusion protein of TEM-1 inserted into MBP [5]. Due to the circularly permutated TEM-1 gene, the new TEM-1 N-terminal residue 229 is inserted after MBP residue 316 whereas the new C-terminal residue of TEM-1, residue 228, is connected to MBP residue 319 via an extra serine residue [5]. The original N- and

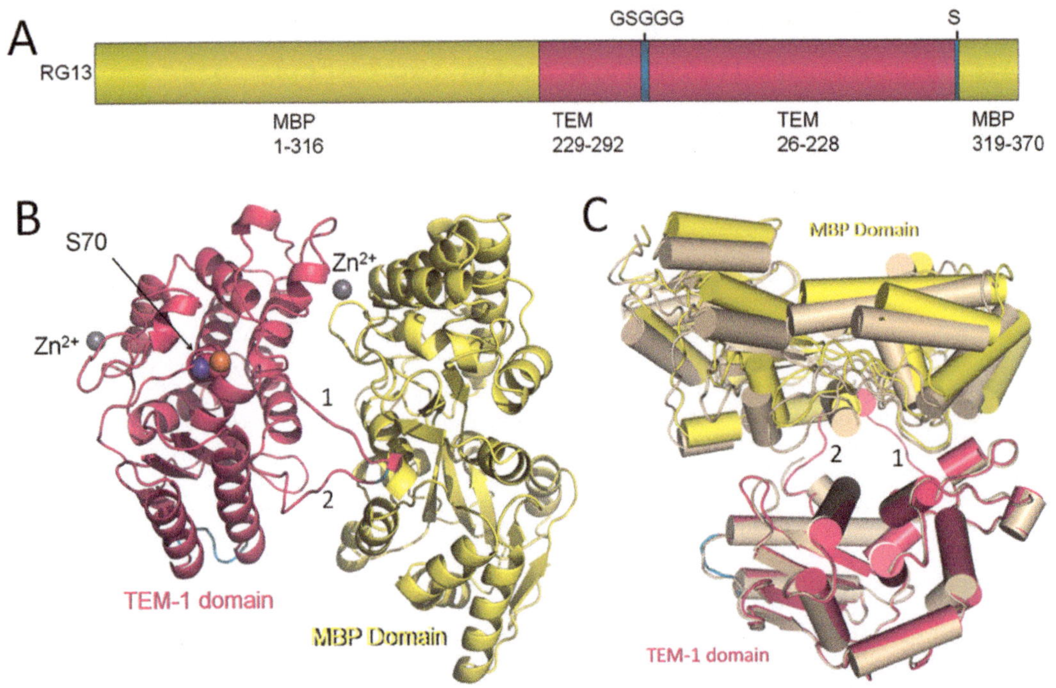


**Figure 1. Overall structure of RG13.** (A) RG13 primary structure. (B) Schematic diagram of RG13 crystal structure with zinc ions bound. TEM-1 and MBP domains are colored magenta and yellow, respectively. Zinc ions are shown as grey spheres and the engineered GSGGG linker is colored cyan. To indicate the position of the TEM-1 active, its catalytic serine residue is highlighted in spheres. The linker 1 and linker 2 regions between the MBP and TEM-1 subdomains are labeled as '1' and '2', respectively; (C) Superposition of the two RG13 crystal forms of the TEM-1 domain reveals a slightly different orientation of the MBP domain after TEM-1 domain superpositioning.

C-terminal residues of TEM-1 are connected via a linker GSGGG. In total, there are two connections between the TEM-1 and MBP domains in RG13 (Figure 1A).

To our knowledge, the structural basis for how allostery is achieved by non-homologous domain insertion has not been described. Initial insights into the allosteric mechanism of RG13 were obtained by the Ostermeier lab with mutational studies indicating that the extent of the domain closure angle in the MBP domain, which is maximally achieved by maltose binding, is responsible for affecting TEM-1 activity in RG13 [5,7]. A recent NMR study compared the NMR data of RG13 in the absence/ presence of maltose with data from previous NMR studies of MBP and TEM-1 and provided evidence that: (1) the individual MBP and TEM-1 domain structures in RG13 are substantially conserved; (2) the TEM-1 active site of RG13 is relatively unperturbed in the presence of maltose; and (3) in the absence of maltose the active site of RG13 is slightly perturbed with possible displacement of several residues [8]. Less is known about the negative allosteric regulation of RG13 by $Zn^{2+}$. Previous mutagenesis studies involving single and double mutations have potentially ruled out all Cys and His residues for $Zn^{2+}$ binding [6] but did suggest a possible role for TEM-1 residues H26 and H289 in the zinc inhibition mechanism. Upon mutating both of these residues to alanine, the $Zn^{2+}$ $K_i$ increased 30-fold although a corresponding loss of $Zn^{2+}$ affinity was not observed [6].

We are interested in further delineating the mechanism of switching of RG13 as to how the TEM-1 activity is positively regulated by the allosteric effector maltose and negatively regulated by $Zn^{2+}$. The RG13 crystal structure with mM $Zn^{2+}$ concentration in the crystal presented here shows RG13 with a distorted TEM-1 domain with a zinc ion bound at an inter-domain site. The structure shows that this zinc ion bridges the TEM-1 and MBP at an additional contact point away from the linker region between the two domains thereby negatively affecting the conformation of this linker region. An analogous distortion of this linker may be key for RG13's compromised catalytic activity in the absence of maltose. To gain additional insights, we carried out mutagenesis studies to probe both the linker region and the zinc site as well as carried out molecular modeling to investigate RG13 with an intact TEM-1 domain.

## Results

### Overall Structure

Two types of crystals of RG13 were grown in the presence of high zinc concentration (2.5 mM). These crystals belonged to space groups P1 and C2 and were refined to 2.30 and 2.29Å resolution, respectively (Table 1). The structure reveals that despite the domain shuffling, circular permutation, and different connections, RG13 contains two folded domains comprised of the regulatory MBP and the catalytic TEM-1 (Figure 1B). The overall dimensions of each RG13 molecule are roughly 50×70×70Å with the TEM-1 and MBP domains lying alongside each other connected by two linkers. The two RG13 molecules in the P1 space group have a similar conformation as the root-mean-square-deviation (r.m.s.d.) for Cα atoms is 0.39Å (for 636 residues). Superpositioning of the RG13 molecule from the C2 space group onto each of two molecules in the P1 space group yields r.m.s.d.'s of 1.7 Å (619 residues) and 1.6 Å (615 residues) for molecules A and B, respectively. The electron density for most of RG13 is well resolved including most, but not all, of the TEM-1 active site (Figure S1). The two connecting loops between TEM-1 and MBP could only be resolved in the P1 space group (in both monomers) (Figure S1) although these loops, not unexpectedly, have higher temperature factors than most of the rest of the protein (Figure S1); part of these linker regions comprising residues 317–327 and 583–584 are not included in the C2 space group structure due to lack of electron density. Similarly, the GSGGG linker introduced to circularly permute TEM-1 is resolved in the electron density maps but also has somewhat higher temperature factors compared to the cores of the two domains (Figure S1).

The orientation between RG13's TEM-1 and MBP domains in the C2 space group is somewhat different compared to that in the P1 space group (Figure 1C). When superimposing the TEM-1 domains of the RG13 structures, the MBP domain orientation between the two different space group structures is different by about 8.5°: the MBP domain appears to pivot across the TEM-1 surface with MBP residues furthest from the TEM-1 domain shifted more compared to the MBP residues closest to the TEM-1 domain likely a consequence of the bridging $Zn^{2+}$ and linkers restrictions. These observed inter-domain orientational differences indicate a degree of flexibility in orientation between the two domains, not unexpectedly due to the presence of the somewhat more mobile linker regions and the limited contacts between the two domains. The additional $Zn^{2+}$ ion contact between MBP and TEM-1 will be discussed next.

## $Zn^{2+}$ Binding

In addition to the two RG13 monomers in the P1 space group, there are four zinc ions present with two zinc ions bound per monomer. Similarly, there are two zinc ions present in the C2 space group structure of RG13. The zinc ions are readily identifiable being the strongest difference Fourier density features in difference density maps (Figure S2). The $Zn^{2+}$ ions bound to the three different RG13 monomers can be grouped in two different classes: the main $Zn^{2+}$ ion, which bridges the TEM-1 and MBP domains, and the non-interdomain bridging $Zn^{2+}$ ion. The main $Zn^{2+}$ ion in the C2 RG13 structure bridges the TEM-1 and MBP domains via residues H468 (*H112*, TEM-1 residue numbering shown in italics and brackets from here on forward), residue E477 (*E121*), and both oxygens of residue D164 (MBP residue) (Figure S2). In the P1 structure, residue E477 (*E121*) has slightly shifted away and therefore the main $Zn^{2+}$ ion only interacts with H468 (*H112*) and both oxygens of D164 (Figure S2). The main $Zn^{2+}$ ion binding site in RG13 is distant from the active site being 20Å from the Oγ atom of the catalytic S426 (*S70*) residue. The (other) non inter-domain bridging $Zn^{2+}$ ion makes liganding interactions with residues H509 (*H153*) and H514 (*H158*) of the TEM-1 domain (in C2 space group, Figure S2). A similar zinc binding site is also present in the P1 space group involving the same H509 and H514

**Table 1.** Data collection and refinement statistics.

| | | |
|---|---|---|
| **Data collection** | | |
| space group | *P1* | *C2* |
| cell dimensions a, b, c (Å), α, β, γ (deg) | 48.32, 74.18, 103.13, 83.54, 77.64, 89.98 | 124.12, 47.78, 107.87, 90.00, 114.17, 90.00 |
| wavelength (Å) | 1.0810 | 1.0810 |
| resolution (Å) [a] | 50.00–2.30 (2.38–2.30) | 50.00–2.30 (2.38–2.30) |
| Rsym | 5.4 (22.3) | 5.1 (32.8) |
| I/σI | 12.6 (3.8) | 15.4 (2.7) |
| Completeness (%) | 95.9 (95.5) | 88.0 (88.2) |
| Redundancy | 1.9 (1.9) | 2.3 (2.3) |
| **Refinement** | | |
| Resolution range (Å) | 31.45–2.29 (2.294–2.354) | 42.32–2.30 (2.360–2.301) |
| no. of reflections | 56130 | 22012 |
| Rwork/Rfree | 23.2/29.2 | 23.3/29.1 |
| no. of atoms: protein, zinc ion, water | 9851, 4, 490 | 4830, 2, 134 |
| rmsd [b] | | |
| bond length (Å) | 0.007 | 0.006 |
| bond angles (deg) | 0.960 | 0.942 |
| average B-factors (Å2) | | |
| Protein | 34.385 | 41.047 |
| zinc ion | 31.573 | 40.825 |
| Water | 34.750 | 36.184 |
| Ramanchandran plot statistics (%) | | |
| core regions | 91.7 | 92.1 |
| additional allowed regions | 8.1 | 7.5 |
| generously allowed regions | 0.2 | 0.2 |
| disallowed regions | 0.0 | 0.2 |

[a]Numbers in parentheses refer to the highest resolution shell.
[b]rmsd, root-mean-square deviation.

residues but this zinc binding site is expanded and also includes ligands from a crystallographically related RG13 MBP domain: H39′ and Y17′ (Figure S2).

## MBP Domain

MBP adopts a periplasmic binding fold and is capable of undergoing a ligand-induced subdomain closure in which the ligand maltose binds to both subdomain halves of MBP leading to a domain closure of ~35° [9]. Superpositioning shows that the MBP domain of RG13 conformation is more similar to the uncomplexed MBP structure (PDB ID: 1OMP). The latter superpositioning yielded relative low r.m.s.d. values of 1.25Å monomer A in P1, 1.39Å for monomer B in P1, and 1.81Å in space group C2 compared to superpositioning with the maltotriose complexed MBP structure (PDB ID: 3MBP) which yielded larger r.m.s.d.s. of 2.76 (P1 monA), 2.80 (P1 monB), and 2.0 Å (C2) for about 360 Cα atoms (Figure 2). This result is in agreement with a subdomain rotation analysis indicating that RG13 is in a more unliganded open conformation in both P1 and C2 spacegroups since its domain closure angle is about 10° and 17°, respectively, compared to that of the uncomplexed MBP structure (Figure 2). The linker regions, where TEM-1 and MBP are fused, extend from the backside of the MBP domain opposite of where maltose binds. Therefore, the TEM-1 domain in RG13 will see a changing surface upon maltose binding going from concave to more flat (Figure 2B). This will have likely steric consequences for the linker tension between the MBP and TEM-1 domains as will be discussed later. The residues of the linkers closest to the MBP domain that appear to be anchoring the linker to MBP are residues S585 and R316 as the mainchain and sidechain residues of these residues make numerous stabilizing interactions in the P1 space group (Figure 2C). Furthermore, S585 and R316 are also the last residues visible in the C2 RG13 structure whereas the rest of the linker regions are too disordered to be modeled in this C2 spacegroup.

## TEM-1 Domain

TEM-1 is a β-lactamase that hydrolyses bicyclic β-lactam compounds that contain a carboxyl moiety and a carbonyl moiety. Key active site elements for TEM-1 are the catalytic S70, the carboxyl binding pocket, the oxy-anion hole for carbonyl oxygen, and the deacylation-water priming E166 residue. The TEM-1 domain of RG13 has an overall fold similar to the wild-type TEM-1 structure (PDB ID: 1ZG4) with r.m.s.d. values of 1.02Å (P1 molA), 1.02 Å (P1 molB), and 1.1 Å (C2) for ~240 Cα as calculated using COOT [10]. The most significant structural change in the TEM-1 domain of RG13, compared to the wt TEM-1 structure, is the displacement of TEM-1 residues 229–244 which comprises the critical active site β3 strand, part of the β4 strand, and the connecting loop between these strands. This TEM-1 active site section in RG13 is now part of the linker 1 connecting the MBP and TEM domains (Figures 1B and 3). In addition, the preceding residues 214–228 are in a somewhat different conformation as the TEM-1 α10 helix extends in RG13 forming a longer helix (Figure 3). Despite these substantial active site differences of missing active site β-strands, the positions of most other active site TEM-1 residues including S70, Y105, S130, N132, E166, and N170 (in TEM-1 residue numbering) are relatively unchanged in RG13. The β3-strand plays an important role for TEM-1 activity because it forms one of the active site walls providing part of both the carboxyl binding pocket (via S235) and oxy-anion hole for the carbonyl oxygen (via the backbone nitrogen of residue A237) which are thus no longer in the correct position in the $Zn^{2+}$ bound structure of RG13. β3-strand residue K234 is also critical for β-lactamase functioning [11] and is no longer in the native TEM-1 position (Figure 3). Furthermore, β4-strand residue R244, which is also involved in electrostatic interactions with the β-lactam carboxyl moiety [12], has also shifted. This β3-β4-strand section, comprising TEM-1 residues 229–244 is in close proximity to one of the fusion sites, i.e. residue 229, between TEM-1 and MBP (Figures 1 and 3) [5]. In addition to lacking some key active site features, the entry to the TEM-1 active site in RG13 is partially blocked by this displaced region that now forms a short helix (Figure 3). Taken together, loss of β3-β4 strand section adopting a new conformation, now sterically blocking the active site, likely explains the lack of activity of the $Zn^{2+}$-bound RG13 structure at mM concentration of $Zn^{2+}$. Further away from the active site, residues V216 and A217 have also shifted somewhat and these shifts could also affect enzyme activity as those residues have been shown to not tolerate changes [13]. However, this disturbance is likely not a main contributor to the loss of activity.

## Mutagenesis

To probe the importance of the crystallographically observed main $Zn^{2+}$ binding site (Figure S2) for $Zn^{2+}$ inhibition of RG13, we mutated the liganding residues D164, H468, and E477 to alanine. In addition to the single mutants D164A and E477A, we also constructed the D164A/H468A double mutant as well as the D164A/H468A/E477A triple mutant. All proteins were purified and confirmed to maintain their ability to be activated by maltose. Both single mutants D164A and E477A were inhibited by $Zn^{2+}$ to the same extent as wild-type RG13. The $K_i$ for the D164A/H468A double mutant was about 4-fold higher; however, the triple mutant exhibited a $K_i$ that was only 2-fold higher compared to RG13's $K_i$ being (Table 2). The inhibition of all mutants fit well to a model of a single $Zn^{2+}$-binding site. This data indicate that the D164A/H468A/E477A site is not singly responsible for the μM $Zn^{2+}$ inhibition although some modest contribution to $Zn^{2+}$ inhibition was observed.

To probe the importance of the linker 1 and linker 2 regions at the site of fusion between TEM-1 and MBP, a set of point mutations and single amino acid additions/deletions were constructed (Figure 4A). Switching was measured as a ratio of initial rates +/− maltose at a particular concentration of substrate that is around the $K_m$ (hence RG13 switching is not 25-fold). Mutations to the linker 1 region either greatly diminished activity (e.g. ΔW317) or resulted in a loss of switching. The effects of mutations to the linker 2 varied from a loss of switching to a 2-fold increase in switching (e.g. ΔG584). As a general rule, mutations that caused an increased enzyme activity in the absence of maltose also decreased switching (Figure 4B). Similarly, the ΔG584 mutation increased switching by decreasing the activity in the absence of maltose. To test specifically potential MBP anchor residues S585 and R316 (Figure 2B–2C) we mutated residues S585 to an Ala (S585A) and separately deleted residue R316 (ΔR316); these mutants were found to have abrogated maltose-switching. Overall, the linker 1 region could not tolerate any of the three changes made to the linker without losing switching or overall activity whereas 3 out of 5 linker 2 modifications still yielded substantial switching activity. These results point to a critical role for linker 1 in maltose switching likely due to being adjacent to the key TEM-1 active site β3 strand section. In accord with linker 1 and its anchoring residue(s) being critical in maltose switching, loss of switching and restoration of maximal wt activity in the absence of maltose can be achieved by increasing slack immediately adjacent to the linker 1 anchor residue R316 via insertion of a Gly residue (+G228) (Figure 4). Linker 2 is important

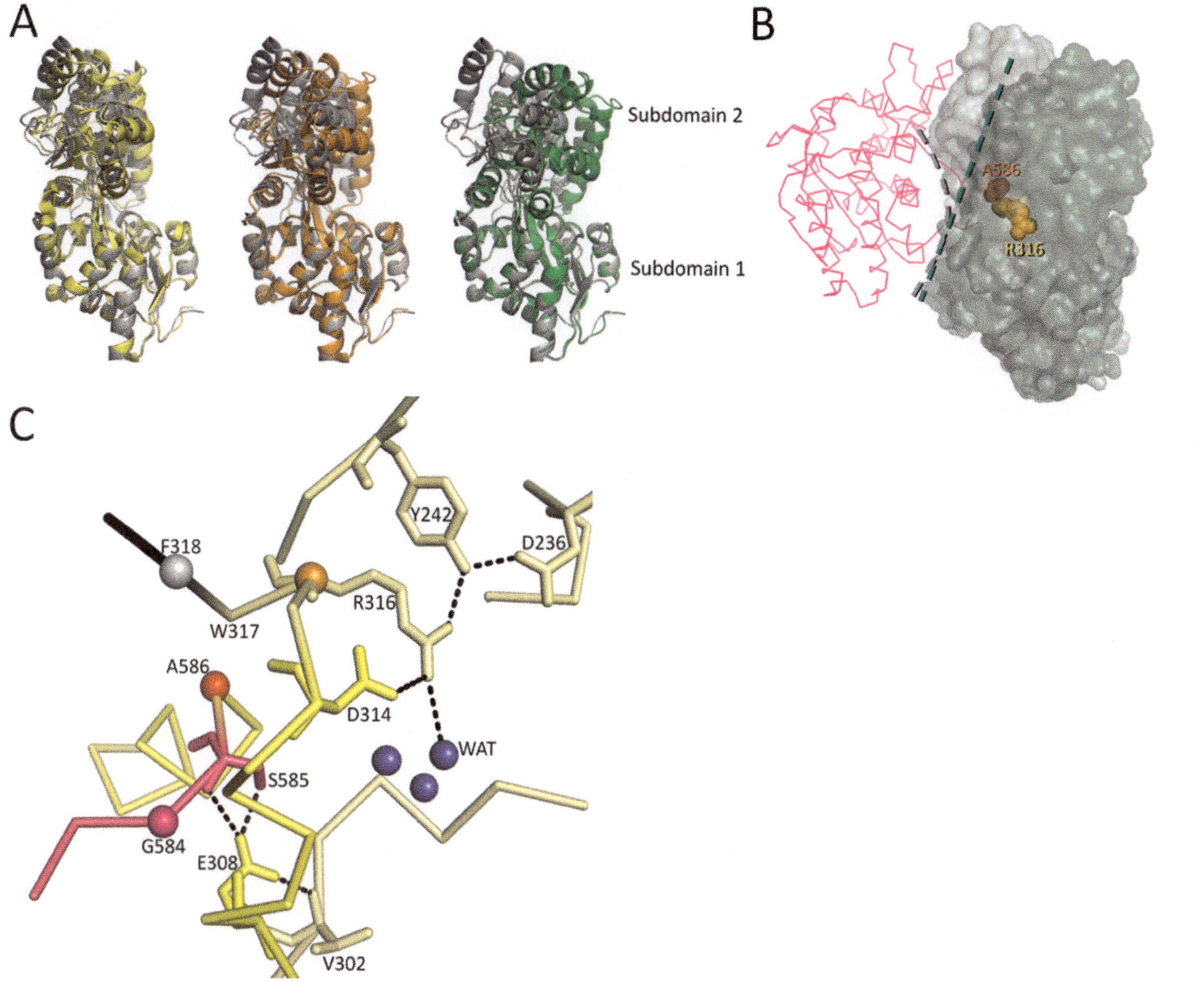


**Figure 2. Structural analysis of the MBP subdomain angle variations.** (A) Superposition of subdomain 1 of MBP reveals different domain closure angles of subdomain 2. The superposed structures include uncomplexed MBP structure (PDB ID: 1OMP, grey), RG13 structure of P1 space group (yellow), RG13 structure of C2 space group (wheat), and maltotriose complexed MBP structure (PDB ID: 3MBP, green). Subdomain 1 of MBP with RG13 is defined by residues 1–109 and 261–316 and subdomain 2 is defined as 110–260 & 586–637. (B) Hypothesized surface change of MBP upon maltose binding in the vicinity the fusion sites R316 and A586 (A319). The surface of the maltose-bound MBP (green) and uncomplexed MBP domains (grey) are shown within the RG13 framework with TEM-1 depicted in a magenta Cα trace. The surface change from concave (grey curved interrupted line) to flat (green straight interrupted line) going from uncomplexed to maltose-bound is indicated. (C) Close-up view of the fusion site between MBP and TEM-1 showing the RG13 linker anchor residues R316 and S585 and their interactions with the rest of the MBP domain.

as well yet positive and negative effects on switching can be observed upon altering the linker. As mentioned above, mutating the anchor residue S585 to an Ala caused loss of switching. Deleting either residue adjacent to S585 (i.e. G584 or A586) had completely opposite effects on switching and points to an important role for A586. Keeping the entire α15 anchoring region intact (i.e. residues S585-A586) yet deleting G584 yielded a hyper maltose switching variant (ΔG584). In contrast, deleting the A586 residue, which makes key hydrophobic interactions and is part of the α15 helix of MBP, eliminates the switching. This is likely due to disruption of this S585-A586 α15 helix anchor region as deletion of A586 will shift S585 to the position of this deleted residue but this residue is likely unable to be accommodated due to the serine's larger size and more hydrophilic nature compared to alanine. The significant switching activity still present for the variant with the deletion of residue S585 is likely a consequence of two opposite effects negating each other. The negative effect of removing the anchoring abilities of residue S585, as observed for the S585A mutation, is compensated for by the hyper-switching effect of shortening the linker 2 region by the deletion as observed for the ΔG584 mutant (while maintaining at least the A586 part of the α15 helix anchor region). The insertion of a tryptophan in the linker 2 region (+W229) still maintained some switching activity but the structural consequences of increasing slack via insertion of the largest hydrophobic amino acid are not easily explained since this residue might generate additional (hydrophobic) anchoring interactions.

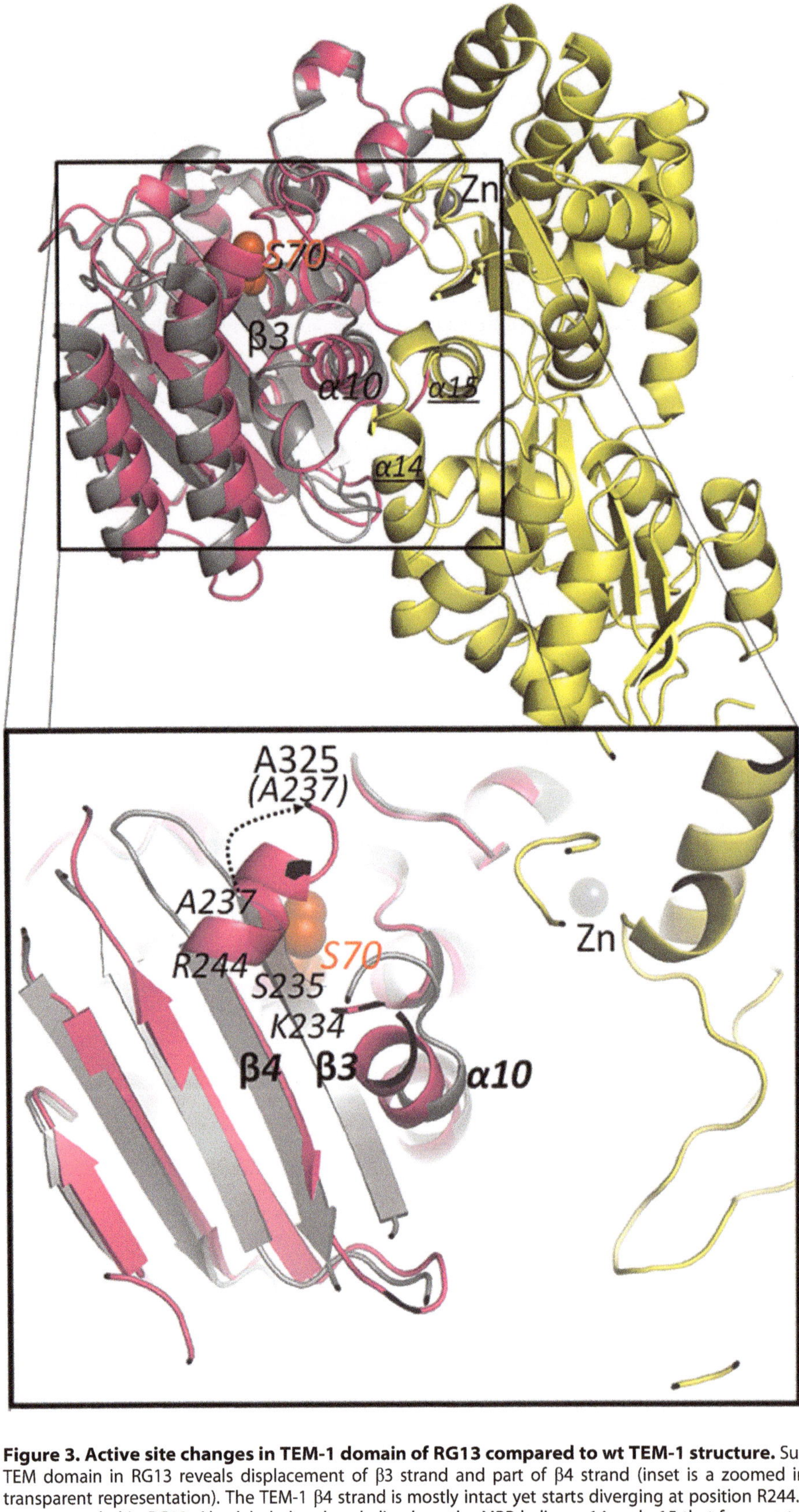


**Figure 3. Active site changes in TEM-1 domain of RG13 compared to wt TEM-1 structure.** Superpositioning of wt TEM-1 structure onto the TEM domain in RG13 reveals displacement of β3 strand and part of β4 strand (inset is a zoomed in view with secondary structure elements in transparent representation). The TEM-1 β4 strand is mostly intact yet starts diverging at position R244. The α10 TEM-1 helix which is observed to be more extended in RG13. Also labeled and underlined are the MBP helices α14 and α15 that form part of the linker anchor points in the RG13 fusion protein. TEM-1 β3 strand residue positions K234, S235, A237, and β4 strand residue K244 are indicated as well as the new position of A237 with the RG13 sequence A325. The zinc ion is indicated via a grey sphere.

**Table 2.** Kinetic constants for nitrocefin hydrolysis and $Zn^{2+}$ inhibition of TEM-1, RG13, and mutants thereof.

| Protein | $k_{cat}$ ($s^{-1}$) | $K_m$ (μM) | $k_{cat}/K_m$ ($s^{-1}μM^{-1}$) | $K_i$ for $Zn^{2+}$ (μM) | Ref. |
|---|---|---|---|---|---|
| TEM-1 | 900 | 110 | 8.18 | 340–1100 | [5,6] |
| RG13 | 200±40 | 550±120 | 0.36 | 2.7±0.1, 3.2±0.5 | [5,6], this study |
| RG13+maltose | 620±60 | 68±4 | 9.12 | 2.1±0.2 | [5,6] |
| RG13 D164A/H468A | nd* | Nd | nd | 8.7±1.4 | this study |
| RG13 D164A/H468A/E477A | nd | Nd | nd | 4.1±1.0 | this study |
| RG13 H375A/H382A | nd | Nd | nd | 73±5 | [6] |

*nd, not determined; however, it was confirmed that all mutants were activated by maltose like RG13 using nitrocefin assays (100 μM nitrocefin) in the presence and absence of maltose.

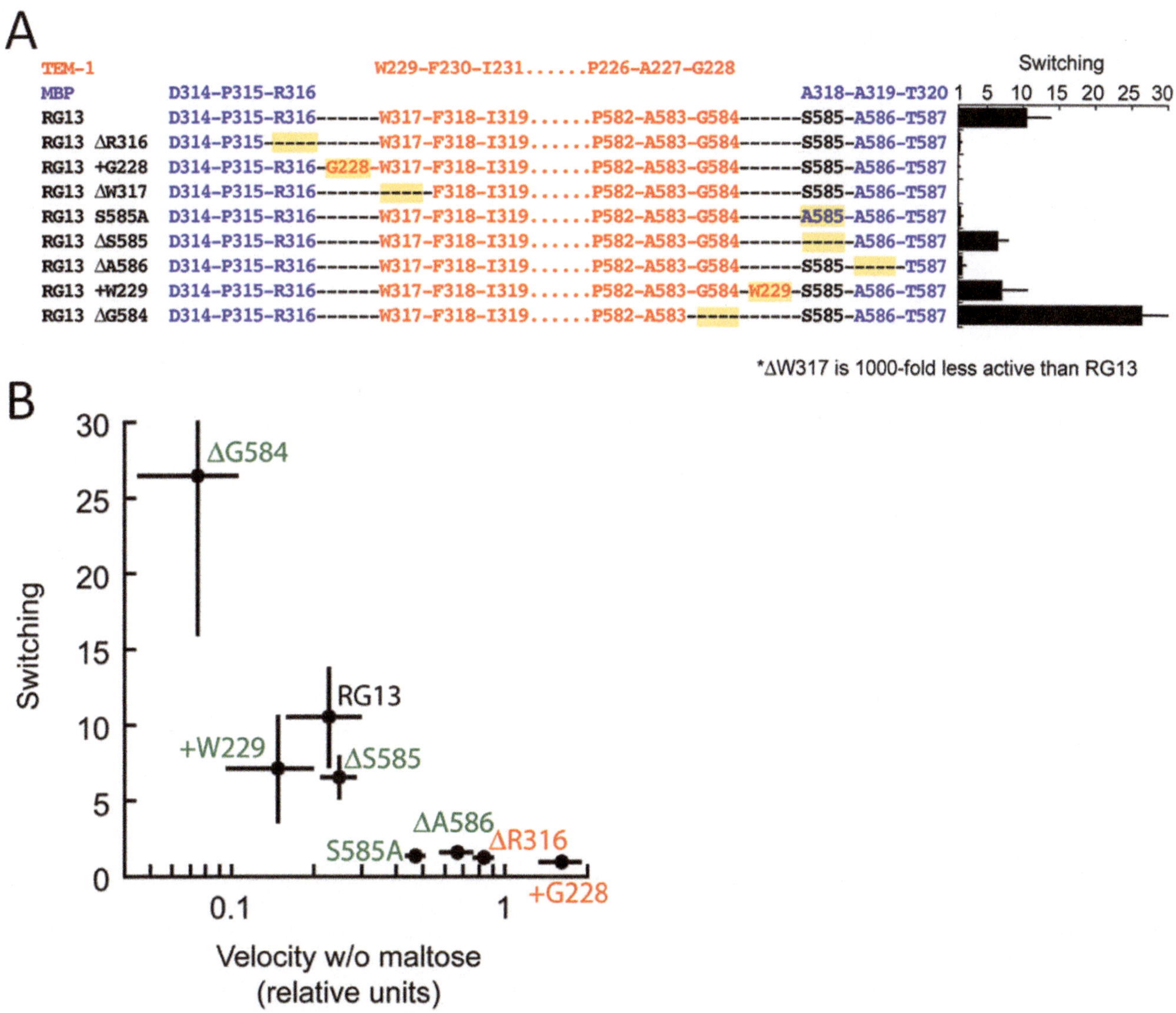


**Figure 4. The effect of mutations in linker 1 and linker 2 on maltose-induced switching.** (A) Mutation locations and their effect on switching. Residues derived from MBP are colored blue and those from TEM-1 are colored red. The corresponding numbers in the wt proteins are indicated above. Mutations relative to RG13 are highlighted in yellow. The switching activity (fold increase in catalytic activity in the presence of a saturating concentration of maltose) of each mutant is indicated on the right. Error bars are standard deviations (n≥3). (B) Switching activity shown as a function of enzyme activity in the absence of maltose. Mutations to linker 1 are shown in green and mutations to linker 2 are shown in red.

### Molecular Modeling of RG13

To gain insights into how an intact TEM-1 structure would fit within the RG13 framework, we employed a rigid body modeling approach. In short, this was done by modeling an intact TEM-1 molecule (with the β3-β4-strand section in the wild-type position) in close proximity to the maltose-bound MBP within the RG13 framework (Figure 5). Firstly, maltotriose bound (PDB id: 3MBP) and unbound (PDB id: 1OMP) MBP structures were superposed onto the RG13-$Zn^{2+}$ crystal structure by superpositioning of residues 1–109 and 261–316 which is one of the sub-domains of MBP. The superposition indicated that the MBP anchor point helices α14 and α15 move relative to each other upon maltose/maltotriose binding (Figure 5A). Subsequently, G584 (*G228*) and F318 (*F230*) were chosen as the modeling target points to attach TEM-1 to the MBP in RG13 thereby keeping the R316/W317 and S585 MBP anchor regions intact (we kept W229 in its RG13 position)(Figure 5B). In the modeled TEM-MBP RG13-like complex, the TEM-1 domain has a three dimensional location fitting reasonably well only some van der Waals clashes as the two domains are in very close proximity (Figure 5C). Nevertheless, in the modeled complex, the TEM-1 G228 modeling target point is in closer proximity to F318 compared to where it is in the RG13-Zn structure, being RG13 residue G584 (Figure 5B). As noted above, this target attachment point G584 is however anticipated to shift in closer proximity upon maltose binding based upon superpositioning of the maltose-free and maltose/maltotriose-bound MBP (Figure 5A). Thus, the movement extrapolated to occur upon maltose binding (Figure 5A) and the movement needed to occur to have residues G584 (*228*) and F318 (*230*) in wt TEM-1 proximity (Figure 5B) are in roughly the same direction suggesting that maltose binding could juxtaposition TEM-1 residues G228 and F230 to a shorter more native like distance. This shift is likely key as residue G228 is the start of the important active site β3 strand. Therefore, we propose that maltose binding likely tweaks this RG13 region such that the β3-β4 strand region involving TEM-1 residues 228–244 is relaxed enough to fit correctly and productively in the active site such that RG13 yields wild-type TEM-1 activity in the presence of maltose. Concomitantly, the local surface change from concave to flat near the linker attachments points on MBP upon maltose binding (Figure 2B) will likely result in less steric clashes between the two domains and also less pull on the TEM-1 β3 strand.

## Discussion

The crystal structure of RG13 bound to the non-competitive $Zn^{2+}$ inhibitor provides insights into the heterotropic allosteric signaling of RG13 although the relative importance of the inter-domain $Zn^{2+}$ site is not immediately evident. The structure reveals that the TEM-1 and MBP domains are in close proximity to each other connected via two linkers and a bridging $Zn^{2+}$ ion (Figure 1B). Whereas the MBP domain is quite intact adopting a relatively open/unliganded conformation, the TEM-1 domain active site is compromised with the active site β3-β4 strand section mostly displaced from its wild-type TEM-1 position with this displaced section now partially blocking substrate entry while also forming a linker between the two domains (Figure 3). Comparison of two different space group RG13 structures reveals that there is still some flexibility in the orientation between the MBP and TEM-1 domains as well as within the two lobes of MBP (Figure 1C and 2A). The zinc ion at high mM concentration therefore seems to inhibit RG13 by generating an additional anchor point between the two domains and therefore acts to 'twist-tie' the original anchor points and linkers such that these linkers are contorted and no longer have enough slack to be inserted productively into the active site (in particular β3-β4 strand residues 226–244). The β-strand expulsion and re-insertion is reminiscent of how the bloodclotting serpin anti-thrombin is regulated [14,15]. The zinc ion observed at the inter-domain site acts thus as a non-competitive inhibitor binding distant from the active site yet clearly affecting β-lactamase activity. Before discussing the relative importance of the $Zn^{2+}$ site, we will first describe the structural consistencies of the RG13 structure with the currently available data.

### Structural Agreement with Non-Crystallographic Data

Many crystallographically observed details of the RG13 structure correlate well with previously published data on RG13 (in the presence or absence of maltose) or are in agreement with mutagenesis data presented herein: (1) the position of the small TEM-1 α11 helix comprising residues 220–225 is perturbed compared to wt TEM-1 (Figure 6) in agreement with NMR data that pointed to a different position for this region in both the absence and presence of maltose [8]. This is not too surprising since MBP is fused to TEM-1 at nearby position 228. (2) The original C-terminal helix of TEM-1, α12, is somewhat shifted (Figure 6) as was also observed by NMR [8]. A possible explanation for this α12 shift is the shift of the flanking α11 helix and possibly the presence of the new GSGGG linker. (3) The RG13 structure contains an inhibitory inter-domain $Zn^{2+}$ site that forces RG13 to adopt a conformation that provides a β3-strand involving mechanism regarding explaining the non-competitive inhibitory properties of $Zn^{2+}$ [6], although the observed $Zn^{2+}$ site is likely not singly responsible for the entire inhibitory effect (as will be discussed below). (4) MBP is observed in a mostly open/unliganded conformation since no maltose was included during the crystallization. (5) Most, but not all, of the TEM-1 active site residues are in unperturbed positions as also observed using NMR [8]. (6) The structure pointed to important structural roles for the linker anchor residues S585 and R164 as also determined by mutagenesis. (7) The N-terminal section of the TEM-1 β3 strand region adjacent to the 229 fusion site was previously pinpointed by NMR as potentially mechanistically important; this section was observed by NMR to be displaced in the absence of maltose [8]. The RG13 structure shows also a displacement of this section, albeit more drastic as the entire β3 strand is displaced. Nevertheless, both independent observations are consistent and point to a crucial role for part or the whole β3 strand which is adjacent to the 229 fusion site.

### Insights into a Maltose-Activated RG13 Structure

From the above analyses it is evident that RG13 if it has $Zn^{2+}$ bound as observed in the crystal is in an inhibited state due to the displacement of the critical β3-β4-strand region now partially blocking the active site. There are several lines of data that indicate that in the absence of $Zn^{2+}$, RG13 adopts a much more catalytically conducive conformation compared to the observed $Zn^{2+}$ bound structure, even before maltose binding. First, previous kinetic studies showed that in the presence of maltose, RG13 has wt TEM-1 activity. Even when maltose is absent, there is still significant activity present since the $k_{cat}/K_m$ is ~25-fold lower activity compared to wild-type TEM-1 with $k_{cat}$ and $K_m$ each only improving roughly 3- and 8-fold, respectively, upon maltose binding (Table 2). Note that if the β3-strand would be displaced entirely in the presence (or absence) of maltose, the effect on activity would be drastic as there would be no oxyanion hole, no K234 and R244, and such an active site configuration would likely therefore also have little or no catalytic activity. For example, the

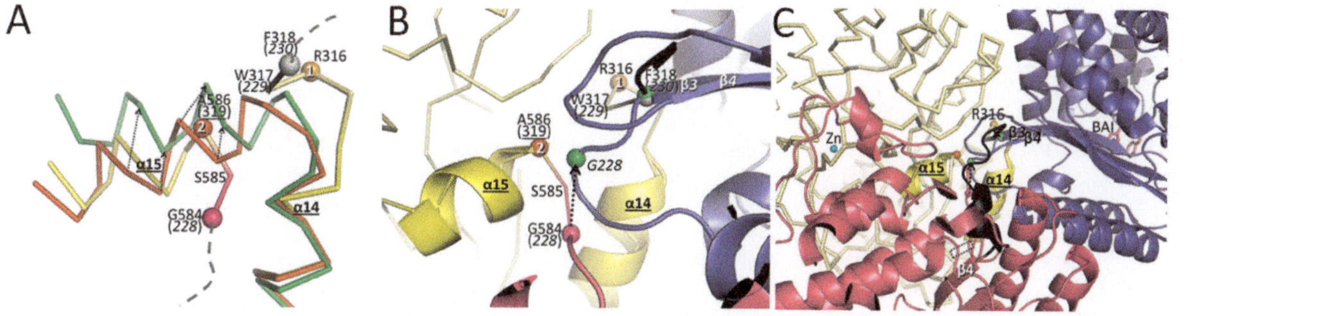


**Figure 5. Modeling of RG13 fusion protein with a catalytically competent TEM-1 domain near the MBP fusion site to relay maltose-induced conformational changes.** (A) Superpositioning of the subdomain 1 of MBP (residues 1–109 and 261–316) from maltose-free MBP (red, PDBid 1OMP), maltotriose-bound MBP (green, PDBid 3MBP), and RG13 P1 space group structure (yellow). MBP anchor/fusion points labeled '1' (R316) and '2' (A586(*319*)) to which TEM-1 is fused are labeled and colored as orange and red spheres, respectively. The respective MBP helices to which these two MBP anchor points are adjacent to are helices α14 and α15, respectively. The positions of S585, introduced as part of constructing RG13 (Figure 1A), and the TEM-1 residues 228, 229, and 230 in the RG13 fusion protein are labeled; the other connecting TEM-1 residues are shown as dashed grey lines. Shifts in the Cα positions of the α15 helix going from unbound to maltotriose bound MBP conformation are indicated by dashed arrows. (B) Close-up view of RG13 fusion site showing the MBP (yellow) and TEM-1 domains (magenta) of RG13, and the modeled position of an intact TEM-1 domain (blue, PDBid 1ERO which includes a boronic acid inhibitor to indicate the position of the active site). Modeling target points F318 (*230*) and G584 (*228*) in RG13 are shown as a grey and magenta sphere, respectively, and the equivalent positions F230 and G228 in wt TEM-1 are depicted as green spheres. The movement that is needed for G584 (*228*) in RG13 to reach the G228 position as found in wt TEM-1 is depicted by a dashed arrow. The MBP anchor point helices α14 and α15, are shown in yellow helix cartoon representation. TEM-1 β3 and β4 strands are labeled; (C) zoomed out view as in (B). TEM-1 bound boronic acid inhibitor (BAI) is shown in magenta stick representation to pinpoint the position of the TEM-1 active site. The black line and small black helix depict the RG13 residues that normally form the TEM-1 β3 and part of β4 strand yet now form the linker between the TEM-1 and MBP domains in the zinc inhibited RG13 structure. The rectangular dashed box in RG13 highlights the position of where the β3 strand used to be in an intact TEM-1 conformation.

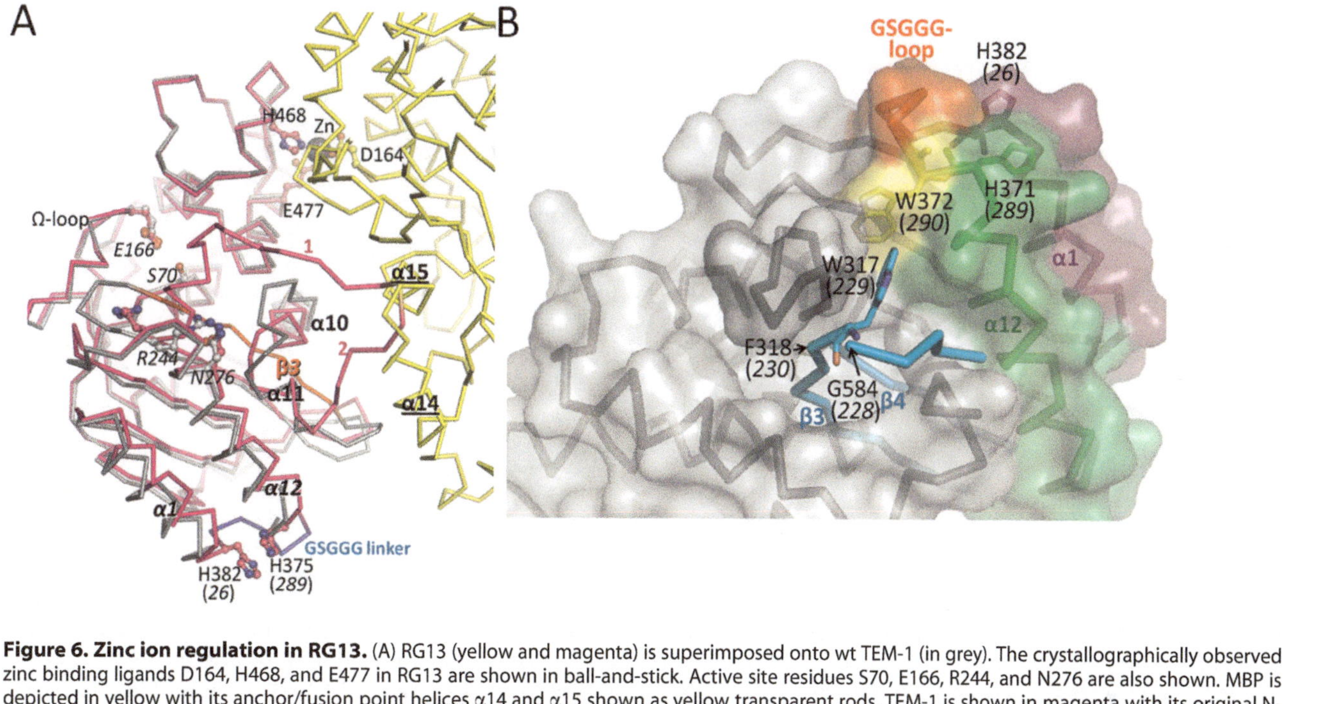


**Figure 6. Zinc ion regulation in RG13.** (A) RG13 (yellow and magenta) is superimposed onto wt TEM-1 (in grey). The crystallographically observed zinc binding ligands D164, H468, and E477 in RG13 are shown in ball-and-stick. Active site residues S70, E166, R244, and N276 are also shown. MBP is depicted in yellow with its anchor/fusion point helices α14 and α15 shown as yellow transparent rods. TEM-1 is shown in magenta with its original N- and C-terminal helices *α1* and *α12* shown in magenta transparent rods. The engineered GSGGG linker between these helices is shown in blue. The position of the original TEM-1 β3 strand position is indicated by the transparent red strand. The linkers 1 and 2 between MBP and TEM-1 are labeled 1 and 2, respectively. The positions of the RG13 residues previously shown to be involved in the zinc-mediated inhibition of RG13, H375 and H382 (i.e. TEM-1 residues 26 and 289), are shown in ball and stick and labeled. TEM-1 helix α10 is also labeled due to its close proximity to the active site, close proximity to TEM-1 α12, and observed structural differences between RG13 and TEM-1. (B) Surface representation depicting the TEM-1 fusion site and GSGGG-loop region of a modeled TEM-1 with an intact β3-β4 strand section and the GSGGG-loop within the RG13 framework. Depicted are residues H26 and H289 from the original N- and C-terminal helices of TEM-1, respectively. Also shown are residues W290 and W229 as well as the residues 230 and 228. This surface is facing the MBP domain in RG13 revealing the potential steric and structural consequences of mutations in the histidine residues.

mutation K234T in TEM-1 affects substrate affinity 50-fold [11], and the K234A mutation in a related β-lactamase decreased $k_{cat}$/ $K_m$ multiple orders of magnitude [16]. Second, in agreement that a $Zn^{2+}$-free and maltose-bound RG13 structure likely adopts a more catalytically proficient conformation is that a previous NMR study found that, in the presence of maltose, the entire β3-strand can be assigned to a wt position (even in the absence of maltose, roughly two-thirds of the C-terminal part of the TEM-1 β3 strand in RG13 can be assigned to a relatively wt conformation including K234) [8]. In conclusion, it seems therefore likely that in maltose-bound RG13, the entire β3-strand is in the wt TEM-1 position. With the RG13 fusion sites between TEM-1 and MBP being at the beginning of this β3 strand (i.e. residues 228 and 229), and that this entire β3 strands adopts a wt-like position in RG13 in the presence of maltose, according to the NMR study, yielded a starting direction to model RG13 in a more activated state. This was done via rigid body modeling an active wt TEM-1 structure within an RG13 framework dictated by the fusion points (Figure 5). The rigid body modeling indicated that juxtapositioning an intact TEM-1 molecule adjacent to where it is fused to MBP in RG13 results in a very close distance between the two domains. Furthermore, the fusion point distance between TEM-1 is only moderately too large when MBP is in the unliganded conformation (Figure 5). However, this potential fusion distance mismatch is likely shortened upon the extrapolated maltose-induced MBP conformational changes (Figure 5A). This suggests that maltose brings about a conformational change such that RG13 residues G584 (*G228*) and F318(*F230*) can adopt a distance more closely observed in wt TEM-1 for these corresponding residues G228 and F230 thereby likely allowing a catalytic proficient conformation for this entire region encompassing strands β3 and β4. In addition to the fusion points juxtapositioning upon maltose binding, the MBP surface near the fusion sites becomes less sterically repulsive as it changes from concave to relatively flat upon maltose binding (Figure 2B). As such, there will likely be less inter-domain steric pull on the TEM-1 β3 fusion section by MBP when maltose is bound.

## Insights into the (Maltose Absent) apo RG13 Structure

There are three points of evidence that indicate that the N-terminal half of the β3 strand of TEM-1 in RG13 in the absence of maltose is in a more tensed, non-wt position. Firstly, this stretch of residues could not be assigned to the wt TEM-1 position using NMR measurements on RG13 in the absence of maltose but, upon maltose binding, changes conformation and could be assigned [8]. Second, the ease of dislodging the β3 strand, and adjacent β4 strand region, by binding zinc at mM concentrations at an interdomain site as evidenced in the RG13 crystal structures suggests that some strain or partial displacement is already present in this β3 region. Thirdly, the molecular modeling of an intact TEM-1 within the RG13 framework indicated a very close proximity between the TEM-1 and MBP domains since the fusion points on either domain are not on protruding, flexible loop regions. We therefore hypothesize that, in the absence of maltose and zinc, that by MBP sterically pulling slightly on the TEM-1 β3-strand, which starts at TEM-1 residue F230, this strand is modestly shifted in maltose-free RG13 affecting both $k_{cat}$ (oxyanion needed) and $K_m$ (carboxyl moiety interaction needed). Since the N-terminal part of the β3 strand is very close to both K234 and A237 of TEM-1 located in the middle of the β3 strand (even though these two residues were assigned in the NMR spectra), subtle changes in their dynamics and position could account for the 25-fold maltose-dependent activity switching of RG13. A slight pulling on the N-terminal end of the β3 by a maltose-free MBP domain in RG13 could affect the position of the important oxyanion hole, involving A237, by perhaps 0.1–0.4Å which will both have an effect on $K_m$ and $k_{cat}$ (as the substrate needs to place its carbonyl oxygen in this site and also it needs to be primed for ~2 steps during catalysis (each only 5-fold so changes need to be small). Shifts of that magnitude in the oxy-anion hole residues of a closely related β-lactamase [17] have been observed to affect $k_{cat}$ in the 5-fold range [18].

We note that TEM-1 residue E166 was previously suggested to be functionally perturbed prior to maltose binding due to (a) a decreased deacylation of RG13 in the absence of maltose and (b) disappearance of the E166 peak in the NMR spectra of apo-RG13 (although this peak in the HSQC-TROSY of RG13 in the presence of maltose was however already quite weak) [7,8]. Residue E166 has a key role in deacylation [19] and therefore seem to be a candidate residue to be involved in maltose switching in RG13 yet the structure found this residue to be ordered and quite distant from the fusion site. We do however not rule out an indirect effect on the conformation of E166 via β3 strand changes as this strand interacts with the E166-containing Ω loop. Such β3-strand mediated Ω-loop changes were speculated for the G238S mutant of TEM-1 [20] and we also recently observed a large Ω loop displacement in variants of a closely related β-lactamase [21] (although such a large loop shift was not observed in RG13 by NMR [8]).

## MBP Fusion Site

Our mutagenesis studies indicate that proper anchoring of the linkers to MBP, via R316, S585 and even A586, is critical to transmit the maltose signal as altering these residues led to loss of maltose-switching (Figure 4). The mutations involving these residues likely caused these anchor regions to be no longer tightly interacting with MBP such that MBP no longer can control the activity of TEM-1 thereby explaining their abrogated maltose switching. This maltose switching is likely done by a positional change, as illustrated by the shift in anchor points (Figure 5A), and/or by relieving steric clashes between the concave surface of the MBP domain upon maltose binding (Figure 2B). Previously, the MBP mutation A319T (RG13 residue A586) was also found to cause a loss of maltose switching [7]. This mutation is adjacent to the S585 linker anchor residue and, since there is no space to accommodate the threonine, this mutation will likely lead to a shift of these few residues thereby likely displacing S585 from its anchoring role. In addition to stable anchor residues, linker length was also critical for switching as introducing slack into linker 1 via a glycine insertion led to loss of switching as the maltose-free activity increased to wt TEM-1 activity. Furthermore, by decreasing slack in the linker 2 region, via deletion of G584, the maltose-free activity could be further dampened and maltose switching increased. Proper linker anchoring and linker length are thus critical attributes for maltose switching by RG13.

In addition to RG13, another TEM-1 MBP fusion protein has previously been obtained that uses the same MBP fusion site of residue 316 to also obtain maltose-dependent signaling [2]. But instead of TEM-1 residue 229 being fused to MBP residue 316, this other MBP-TEM-1 construct fused TEM-1 residue N170 to MBP residue 316. This suggests that this MBP region, near residue 316, has unique and robust features that can transmit maltose-dependent conformational changes to an unrelated protein in different ways. This other MBP-TEM-1 maltose-regulated fusion construct was shown to have its maltose-dependent abilities to be negatively affected by increasing the linker length by one and two additional residues [2]. This agrees with our RG13 structure-based allostery hypothesis as the linker length and amount of slack are hypothesized to be key for proper maltose signaling in RG13. Another example indicating that short linkers tend to have better switching properties in engineered allosteric fusion proteins is a TEM-1:cytochrome $b_{562}$ construct [22,23].

It is interesting as to why MBP residue 316 is selected for during the generation of two different maltose-regulated TEM-1 fusion constructs as residue 316 is not near where the largest conformational differences are observed upon maltose binding that include a 35° domain closure conformational change. Residue 316 is actually situated on the backside of the hinge region of the bi-lobal MBP structure. However, one has to consider as to how these allosteric fusion proteins are generated as the TEM-1 domain is inserted into a single site of MBP (with possibly some residues deleted as in RG13, Figure 1). Although the linker anchor points of MBP are not where there are the largest differences in distance between an uncomplexed and complexed MBP are found, the anchor points are near where the largest differences are when one only considers Cα positions of a few residues downstream from each other in the continuous polypeptide chain to allow for insertion of the non-homologous TEM-1 gene at a single site. By calculating Cα-Cα Δdistance measurement of residues i and i+x via substracting the distances from the maltose-bound from the maltose-free MBP structure, the MBP region at residue 312 lights up as to having the largest Cα-Cα difference for i and i+3 and i and i+6 (Figure 7). These calculations suggest that the MBP region from 312 to 312+i (i.e. 312–318) has the largest shifts thereby likely explaining why insertions near residue 316, in particular since R316 has adopted the linker anchor role in RG13 (Figure 2C), are capable of transmitting maltose signals to TEM-1 (as is also evident mechanistically in Figure 5A). An additional potential benefit of this region, being on the backside of MBP, is that this region could exert a lever/torque effect as the fusion points are close to the hinge region. As such, this region is thus perhaps able to exert a larger force on the movements near its rotation point, which might be needed as the β3 TEM-1 strand is normally well lodged into the active site. Furthermore, the backside of MBP, near the hinge region where TEM-1 is attached, changes from concave to more flat upon maltose binding (Figure 2B) and could therefore decrease steric contacts between TEM-1 and MBP and thus provide additional slack for the β3 linker regions such that they can adopt the *wt* catalytically competent conformation. It is remarkable that the forced evolutionary pressure resulted in this successful RG13 of having indeed found the MBP 312–318 region and the 230 β3-strand region of TEM-1 indicating that this approach works to find this narrow signaling window region being almost a 'needle in a hay stack'.

## $Zn^{2+}$ Binding and Regulation in RG13

Biochemical and mutagenesis studies indicated that $Zn^{2+}$ inhibits RG13 in a non-competitive, yet reversible manner and suggested the presence of a single low μM $Zn^{2+}$ site with an inhibitory role for residues H382 (*H26*) and H375 (*H289*) [6]. These residues likely have an indirect role as mutating these residues did not affect $Zn^{2+}$ binding, only $Zn^{2+}$ inhibition. In agreement, the structure of RG13 revealed also no direct involvement of these histidine residues in $Zn^{2+}$ binding even with mM $Zn^{2+}$ concentration. Instead, the RG13 structure, with 2.5 mM $Zn^{2+}$ present, revealed a $Zn^{2+}$ binding role and potentially regulatory role for role for residues D164, H468 and potentially E477. However, mutating 2 or 3 of these residues only caused a minor effect (2- to 4-fold; Table 2) on the $K_i$ for $Zn^{2+}$ indicating no exclusive role for D164/H468/E477 in the $Zn^{2+}$ inhibition mechanism. In agreement, previous mutagenesis studies found that the H452A/H468L mutation pair did not alter $Zn^{2+}$'s inhibitory effect [6]. The observed second $Zn^{2+}$ site, which also involves residues from two neighboring crystallographically related RG13 molecules (in the P1 space group), is likely also not functionally relevant for RG13 as mutating H509 (*H153*) or H514 (*H158*) did not affect $Zn^{2+}$ inhibition [6].

There are three possible explanations for the apparent discrepancy between the mutagenesis results and the crystallographically observed $Zn^{2+}$ binding site in RG13. First, $Zn^{2+}$ binding to RG13 could be kinetically controlled such that at early time points, the lowest energy barrier binding site will be occupied first whereas the energetically most favorable binding site might take longer to reach due to a higher initial energy barrier. The latter state could perhaps be reached during the several days it took for RG13 crystals to form. This could be compounded by the difference in "induced-fit" and "conformational selection" as modes of binding [24]. At low concentrations, conformational selection takes place, whereas at higher concentrations, induced fit takes place (i.e. ligand binding first to low affinity conformation thereby inducing conformational change). A second possibility is that only the observed zinc-bound conformations could be crystallized such that crystallization selected for a particular conformation that might perhaps not be the dominant one in solution. A third possible explanation is that there is more than one low μM inhibitory zinc binding site that are all mutually exclusive since they involve juxtapositioning of 1–2 liganding residues of MBP and 1–2 liganding residues of TEM-1. This last possibility is especially intriguing as previous mutagenesis experiments, indirectly, indicated that the presumed single $Zn^{2+}$ binding site in RG13 would fall into a rare category of $Zn^{2+}$ binding sites: single and double mutant studies had ruled out all histidine and cysteine residues in RG13 being involved in zinc binding [6] if the hypothesis were correct regarding a single unique inhibitory $Zn^{2+}$ site. However, a $Zn^{2+}$ binding site without histidine and without cysteine is quite rare as it has only been observed in ~1% out of the over 230 different $Zn^{2+}$ sites [25]. Thus, if there is indeed a single unique non-histidine/non-cysteine $Zn^{2+}$ inhibitory site in RG13, it would fall into this rare category that could not even be observed crystallographically at mM zinc concentrations in the RG13 structures whereas two different, histidine-containing, $Zn^{2+}$ sites were observed instead. Therefore, an alternative interpretation is that RG13 has multiple, yet mutually exclusive low μM inhibitory zinc binding sites that likely will have one or more histidine liganding residues (cysteine residues are occupied in disulfide bonds).

That there are multiple mutually exclusive μM $Zn^{2+}$ inhibitory sites is an attractive hypothesis for two additional reasons. First, RG13 has a potential mechanism to form multiple mutually exclusive inhibitory $Zn^{2+}$ binding sites via mere rotation and repositioning of the MBP and TEM-1 domains to juxtaposition 1–2 $Zn^{2+}$ liganding residues from each domain. Such juxtapositioning could perhaps entail $Zn^{2+}$ first binding to one of the (histidine-containing) multi-residue "half-sites" situated on one domain (i.e. H468/E477, H509/H514, or Y17/H39; see Figure S2) followed by 1–2 additional ligands from the other domain such as D164 combining with H468/E477 (Figure S2). Note that μM Zn binding can be readily be engineered into a variety of proteins via juxtapositioning of at least 2 His residues [26–29]. Due to avidity, all RG13 would need to do is to rotate, and perhaps shift the two domains somewhat such that His/Asp/Glu and/or other $Zn^{2+}$ liganding residues from the different domains are juxtaposed locking the RG13 in a more fixed conformation thus affecting the critical linker regions. As RG13 has a low pI of 5.53, it has ample opportunity to do this via 40 Asp, 47 Glu, and 9 His residues such

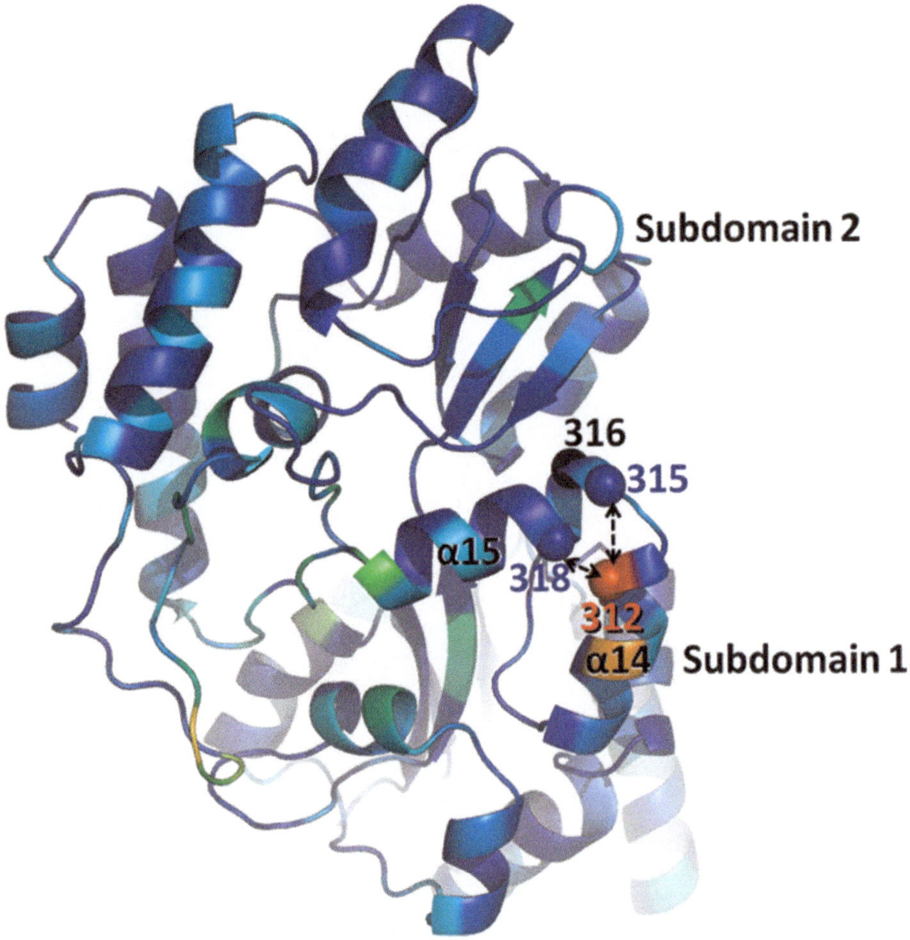


**Figure 7. Structure of uncomplexed MBP (1OMP) showing maltose-induced local Cα-Cα distance shifts.** Maltose-dependent Cα-Cα distance differences for residues i = i+3 and i = i+6 are averaged and mapped onto the structure with residues color ramped from blue to red for averaged shifts ranging from 0 to 1.4Å, respectively. Residue 312 yielded the largest maltose-dependent local shifts (1.4Å). Labeled also are residues 315 (i+3) and 318 (i+6) with arrows to residue 312 (i) indicating the local distances that were maximally shifted by maltose. Helices α14 and α15 are labeled as well as residue 316 (black), the site of the RG13's TEM-1 insertion in MBP, to show that the insertion occurred within the maltose-dependent locally shifting region of 312–318 situated between the two helices. Labeled residues are also shown as spheres. The orientation of MBP is similar to that of Figure 5.

that 15% of its residues can be involved in liganding $Zn^{2+}$ (96 out of 637 residues). Since most of these residues are at the surface, the relative amount of potential $Zn^{2+}$ liganding residues is thus even higher. We hypothesize that the multiple yet mutually exclusive inhibitory $Zn^{2+}$ sites of RG13 share common features such as bridging both TEM-1 and MBP and also affecting the critical inter-domain linker regions. A second additional reason for the multiple mutually exclusive $Zn^{2+}$ sites hypothesis is that it provides an explanation for the mutagenesis results as the double and triple mutant neither fully nor negligibly affected the $K_i$ (2- to 4-fold change was observed); these mutants likely thus eliminated one of the inhibitory $Zn^{2+}$ sites from the population causing a partial $Zn^{2+}$ binding site redistribution. Note that inter-domain $Zn^{2+}$ binding is the most likely possibility as MBP and TEM-1 themselves as individual proteins are not regulated by $Zn^{2+}$ [6]. Also, if the μM $Zn^{2+}$ binding site were located within just one of the domains, some $Zn^{2+}$ electron density would have likely become apparent due to the mM $Zn^{2+}$ concentration used during crystallization. It is also not uncommon to have two mutually exclusive zinc binding sites as this was observed previously in an engineered zinc biosensor [30]. This third possibility to explain the discrepancy between the mutagenesis data and the RG13 structure would thus render the RG13:$Zn^{2+}$ crystal structure as one of the two or more modes of $Zn^{2+}$ inhibition in RG13 with individual $Zn^{2+}$ affinities such that the apparent $K_i$ shifts by about 3-fold upon mutating the crystallographically observed one (from 2μM to 4–8μM; Table 2). Future experiments are needed to further probe these three possible explanations.

Regarding the indirect role of H382 (*H26*) and H375 (*H289*) in $Zn^{2+}$ inhibition [6], we hypothesize that the above noted close steric proximity between the MBP and TEM-1 domains could explain why mutating TEM-1 residues H26 and/or H289 had an effect on $Zn^{2+}$ inhibition but not on $Zn^{2+}$ binding [6]. Each of these histidines is at the end of a helix either after or before the GSGGG loop. Their side chains both protrude to the solvent forming an aromatic stacking interaction with each other (Figure 6). We postulate that mutating either or both of these histidines to alanine results not only in removal of a sterically protruding histidine sidechain, but could also lead to a shift of the other histidine concomitant with shift or disorder of this entire engineered loop region comprising residues $\mathbf{H_{375}}W_{376}GSGGG\mathbf{H_{382}}$ (H375 is TEM-1 residue *H289* and

H382 is TEM-1 residue *H26*). This loop region is adjacent to the fusion site (Figure 6) and being potentially less protruding and less ordered upon a H→A mutation could lessen the steric repulsion between the MBP and TEM-1 domains in that region and thus allow $Zn^{2+}$ to bind to the interdomain site(s) without having to dislodge the β3-β4 region when the MBP and TEM-1 domains pivot to juxtaposition $Zn^{2+}$ liganding residues. An alternative explanation, but sharing similar features, is that the histidine mutation(s) cause destabilization of the circularly-permutated loop region that could alter the W376(*W290*)-W317(*W229*) interaction (Figure 6B). These tryptophan residues in TEM-1 are not tolerant to substitution [13]; W229 is near the fusion site (Figure 1) and is displaced 25Å in RG13 compared to its position in the *wt* TEM-1 structure. A H26A and/or H289A mutation could thus indirectly disrupt the W229:W290 interaction thereby affecting whether the critical β3-β4 region will be dislodged when inter-domain $Zn^{2+}$ is bound.

Here we propose a mechanistic model for RG13 from the above findings and discussion: The N-terminal section of the β3 strand adjacent to the TEM-1 228 fusion point is likely strained or partially displaced in RG13. This strain in the absence of maltose is likely due to a combination of the close proximity of the TEM-1 and MBP domains when fused in RG13 that is exacerbated by the concave character of that MBP domain surface and the increased distance between the anchor points. This tensed state of the TEM-1 domain can undergo two different allosteric pathways. First, activation via relieving the strain by maltose binding will alleviate the steric repulsion between the MBP and TEM-1 domains due to flattening the surface near the MBP fusion site from its concave state in the absence of maltose. Furthermore, maltose binding will change the position of MBP helices α14 and α15 such as to shorten the distance between the anchor points about 2Å. A second pathway the strained apo-RG13 molecule could undertake is inhibition via $Zn^{2+}$ mediated reorientation of the MBP and TEM-1 domains by forming a 3rd ($Zn^{2+}$-mediated) contact point between these domains (in addition to the two linkers already present in RG13). This would allow the strained β3 strand region to completely pop out of its strained wt-type-like position as the inter-domain bridging zinc ion 'twist-ties' the linkers to gain the needed additional linker slack for $Zn^{2+}$ to bind to both domains (at multiple yet mutually exclusive sites). The described findings regarding maltose and $Zn^{2+}$ regulation in an engineered fusion protein could be useful for future allosteric protein engineering efforts.

## Materials and Methods

### Expression and Purification

RG13 was expressed and purified as previously described [6]. Briefly, the expression plasmid containing RG13 was transformed into *Escherichia coli* strain PM9F'. The culture were inducted by 1 mM IPTG and grew overnight at 20°C. Cells were lysed by French press and RG13 was subsequently purified using an amylase affinity resin (New England Biolabs) and eluted using maltose. The eluted RG13 protein was dialyzed at 4°C, to remove maltose, and subsequently concentrated to ~11 mg/mL in 20 mM Tris-Cl pH7.0 before mixing with glycerol for storage at −80°C. Aliquots of RG13 were thawed and subjected to gel filtration using a Superdex200 column (GE Biosciences) that was equilibrated with 20 mM Tris/HCl pH 7.2 and 500 mM NaCl. RG13 was subsequently buffer exchanged to 10 mM Tris/HCl pH 7.2, and concentrated to 10–15 mg/ml, as assessed by the Bradford assay, prior to crystallization. RG13 mutants were expressed as above with the following exceptions. The strain used was BL21, the overnight growth temperature was 25°C, and purification was performed using an MBPTrap column (GE Biosciences).

### Crystallization and Soaking

Prior to crystallization, $ZnCl_2$ was added to RG13 to a final concentration of 2.5 mM. Initial crystallization conditions were found by using Hampton screen kits. After optimization, single crystals were obtained by vapor diffusion sitting drop method at room temperature: a 1.0 µl drop was prepared using 0.5 µl protein mixture (13.8 mg/ml RG13 and 2.5 mM $ZnCl_2$) and 0.5 µl reservoir solution (containing 0.2M Ammonium Acetate, 0.1M Tris pH 8.5–9.5, 15–30% PEG3350) and equilibrated over a 1 ml reservoir solution. RG13-Zn co-crystals grew to full size in one week and were subsequently cryo-protected in the reservoir solution with 33%PEG3350 and 2.5 mM $ZnCl_2$.

### Data Collection and Structure Determination

Crystals of RG13 grew in two different space groups. Diffraction data for the RG13-Zn complex P1 and C2 spacegroup crystals were collected at the National Synchrotron Light Source X-29A and processed using HKL2000 [31,32]. The data from the P1 space group was double checked but could not be processed in a higher symmetry space group. The structures for each of the space groups was solved by Molecular Replacement using the program Phaser [33] with two RG13 molecules in the P1 space group and one RG13 molecule in C2 space group. Based on the folding pattern of MBP of having two subdomains and the circular permutation of RG13 [5], molecular replacement searches were carried out using search models encompassing each of the MBP subdomains and most of the TEM-1 structure to search for the two copies of RG13 molecules in the asymmetric unit of P1 space group. The search models were obtained from the MBP structure (PDBid 1OMP) and the TEM-1 structure (PDBid 1ZG4). The Molecular Replacement program Phaser [33] successfully placed the subdomains in the asymmetric unit of P1 and C2 space groups to yield two and one RG13 molecules, respectively, in the asymmetric unit. After several rounds of REFMAC (CCP4 suite [34]) refinement and adjusted using COOT [10], most residues of the full length RG13 protein in both space groups could be built. During the refinement, inspection of Fourier difference maps indicated the presence of very strong non-protein electron density peaks which were identified, and refined subsequently, as zinc ions. Water molecules were included in the final stages of refinement. The simulated annealing protocol in CNS [35] was also used in refinement. Iterative rebuilding and refinement allowed the final model to converge with an $R/R_{free}$ 23.2/29.2% for the P1 space group and 23.3/29.1% for the C2 space group (Table 1 for more refinement stats). The C2 space group contains RG13 residues 1–316, 328–582, 585–637, 2 zinc ions, and 134 water molecules whereas the P1 space group structure contains two RG13 copies of residues 1–637, 4 zinc ions, and 490 water molecules. PROCHECK [36] and DDQ [37] were used during the model building process and to validate the final structures. The atomic coordinates and structure factors of the RG13 with zinc ions for the P1 and C2 space groups have been deposited in the Protein Data Bank under accession code 4DXB and 4DXC, respectively. For the rigid body positioning of an intact TEM-1 domain within a RG13 framework, we used the program COOT and the coordinate files as described in the results section.

### Biochemical Characterization of RG13 and RG13 Mutants

The $K_i$ for $Zn^{2+}$ for was determined as previously described by nitrocefin hydrolysis assay [6]. Maltose-induced switching was determined at 25°C by nitrocefin hydrolysis assay with and without maltose in which the initial rate of nitrocefin hydrolysis was monitored at 486 nm as previously described [5]. Solution conditions were 100 mM sodium phosphate buffer (pH 7.4) with 26 nM protein, 100 μM nitrocefin, and 0 or 10 mM maltose.

## Supporting Information

**Figure S1** Electron density for active site and linker regions of RG13 (A) $2F_o$-$F_c$ density map depicting the TEM-1 active site region; (B) $2F_o$-$F_c$ density maps for residues 315–330 (the linker 1 region) and (C) residues 571–587 (the linker 2 region); (D) $2F_o$-$F_c$ density map for residues 377–381 (the engineered GSGGG linker); (E) Temperature representation of the RG13 structure in the P1 space group. The thicker and the more red the backbone is, the more flexible it is due to its higher refined temperature factors. All density maps shown are contoured at 1.0σ.

**Figure S2** Zinc ion binding sites in RG13. (A) main zinc ion RG13 binding site in space group C2 structure bridges MPB domain and TEM domains of RG13 involving residues D164, H468, and E477. (B) The main zinc ion binding site in space group P1 involves residues D164 and H468; (C) Second zinc ion binding site in the C2 space group structure involving residues of H509 and H514 in the TEM domain of RG13. (D) The second zinc binding site in the P1 space group involving H509, H514, Y17′, and H39′ (the latter two belong to a crystallographically related RG13-MBP domain). The $2F_o$-$F_c$ density maps were contoured at 1.5σ and colored blue for the amino acid residues of RG13 and the water molecules. Omit $F_o$-$F_c$ maps were contoured at 10σ and colored green for the Zn ions. Domain coloring is same as in Figure 1 with the crystallographic related neighboring molecules depicted in grey. Water molecules and zinc ions are shown as red and grey spheres, respectively.

## Acknowledgments

We thank the beamline personnel at NSLS X-29A for help with data collection.

## Author Contributions

Conceived and designed the experiments: WK AHL MDA YC WES JL CMW MO FvdA. Performed the experiments: WK AHL MDA YC WES JL CMW. Analyzed the data: WK AHL MDA YC WES JL CMW MO FvdA. Wrote the paper: WK MO FvdA.

## References

1. Swain JF, Gierasch LM (2006) The changing landscape of protein allostery. Curr Opin Struct Biol 16: 102–108.
2. Guntas G, Mansell TJ, Kim JR, Ostermeier M (2005) Directed evolution of protein switches and their application to the creation of ligand-binding proteins. Proc Natl Acad Sci U S A 102: 11224–11229.
3. Koide S (2009) Generation of new protein functions by nonhomologous combinations and rearrangements of domains and modules. Curr Opin Biotechnol 20: 398–404.
4. Ferraz RM, Vera A, Aris A, Villaverde A (2006) Insertional protein engineering for analytical molecular sensing. Microb Cell Fact 5: 15.
5. Guntas G, Mitchell SF, Ostermeier M (2004) A molecular switch created by in vitro recombination of nonhomologous genes. Chem Biol 11: 1483–1487.
6. Liang J, Kim JR, Boock JT, Mansell TJ, Ostermeier M (2007) Ligand binding and allostery can emerge simultaneously. Protein Sci 16: 929–937.
7. Kim JR, Ostermeier M (2006) Modulation of effector affinity by hinge region mutations also modulates switching activity in an engineered allosteric TEM1 beta-lactamase switch. Arch Biochem Biophys 446: 44–51.
8. Wright CM, Majumdar A, Tolman JR, Ostermeier M (2010) NMR characterization of an engineered domain fusion between maltose binding protein and TEM1 beta-lactamase provides insight into its structure and allosteric mechanism. Proteins 78: 1423–1430.
9. Quiocho FA, Spurlino JC, Rodseth LE (1997) Extensive features of tight oligosaccharide binding revealed in high-resolution structures of the maltodextrin transport/chemosensory receptor. Structure 5: 997–1015.
10. Emsley P, Cowtan K (2004) Coot: model-building tools for molecular graphics. Acta Crystallogr D Biol Crystallogr 60: 2126–2132.
11. Lenfant F, Labia R, Masson JM (1991) Replacement of lysine 234 affects transition state stabilization in the active site of beta-lactamase TEM1. J Biol Chem 266: 17187–17194.
12. Thomson JM, Distler AM, Prati F, Bonomo RA (2006) Probing active site chemistry in SHV beta-lactamase variants at Ambler position 244. Understanding unique properties of inhibitor resistance. J Biol Chem 281: 26734–26744.
13. Huang W, Petrosino J, Hirsch M, Shenkin PS, Palzkill T (1996) Amino acid sequence determinants of beta-lactamase structure and activity. J Mol Biol 258: 688–703.
14. Jin L, Abrahams JP, Skinner R, Petitou M, Pike RN, et al. (1997) The anticoagulant activation of antithrombin by heparin. Proc Natl Acad Sci U S A 94: 14683–14688.
15. Schreuder HA, de Boer B, Dijkema R, Mulders J, Theunissen HJ, et al. (1994) The intact and cleaved human antithrombin III complex as a model for serpin-proteinase interactions. Nat Struct Biol 1: 48–54.
16. Ellerby LM, Escobar WA, Fink AL, Mitchinson C, Wells JA (1990) The role of lysine-234 in beta-lactamase catalysis probed by site-directed mutagenesis. Biochemistry 29: 5797–5806.
17. Totir MA, Padayatti PS, Helfand MS, Carey MP, Bonomo RA, et al. (2006) Effect of the Inhibitor-Resistant M69V Substitution on the Structures and Populations of trans-Enamine beta-Lactamase Intermediates. Biochemistry 45: 11895–11904.
18. Helfand MS, Hujer AM, Sonnichsen FD, Bonomo RA (2002) Unexpected advanced generation cephalosporinase activity of the M69F variant of SHV beta-lactamase. J Biol Chem 277: 47719–47723.
19. Adachi H, Ohta T, Matsuzawa H (1991) Site-directed mutants, at position 166, of RTEM-1 beta-lactamase that form a stable acyl-enzyme intermediate with penicillin. J Biol Chem 266: 3186–3191.
20. Saves I, Burlet-Schiltz O, Maveyraud L, Samama JP, Prome JC, et al. (1995) Mass spectral kinetic study of acylation and deacylation during the hydrolysis of penicillins and cefotaxime by beta-lactamase TEM-1 and the G238S mutant. Biochemistry 34: 11660–11667.
21. Sampson JM, Ke W, Bethel CR, Pagadala SR, Nottingham MD, et al. (2011) Ligand-dependent disorder of {Omega} loop observed in extended-spectrum SHV-type {beta}-lactamase. Antimicrob Agents Chemother.
22. Edwards WR, Busse K, Allemann RK, Jones DD (2008) Linking the functions of unrelated proteins using a novel directed evolution domain insertion method. Nucleic Acids Res 36: e78.
23. Edwards WR, Williams AJ, Morris JL, Baldwin AJ, Allemann RK, et al. (2010) Regulation of beta-lactamase activity by remote binding of heme: functional coupling of unrelated proteins through domain insertion. Biochemistry 49: 6541–6549.
24. Hammes GG, Chang YC, Oas TG (2009) Conformational selection or induced fit: a flux description of reaction mechanism. Proc Natl Acad Sci U S A 106: 13737–13741.
25. Auld DS (2001) Zinc coordination sphere in biochemical zinc sites. Biometals 14: 271–313.
26. Higaki JN, Haymore BL, Chen S, Fletterick RJ, Craik CS (1990) Regulation of serine protease activity by an engineered metal switch. Biochemistry 29: 8582–8586.
27. Greene AC, Trent AM, Bachand GD (2008) Controlling kinesin motor proteins in nanoengineered systems through a metal-binding on/off switch. Biotechnol Bioeng 101: 478–486.
28. Suetomi K, Rojo D, Navarro J (2002) Identification of a signal transduction switch in the chemokine receptor CXCR1. J Biol Chem 277: 31563–31566.
29. Schmidt AM, Muller HN, Skerra A (1996) A Zn(II)-binding site engineered into retinol-binding protein exhibits metal-ion specificity and allows highly efficient affinity purification with a newly designed metal ligand. Chem Biol 3: 645–653.
30. Telmer PG, Shilton BH (2005) Structural studies of an engineered zinc biosensor reveal an unanticipated mode of zinc binding. J Mol Biol 354: 829–840.
31. Minor W, Tomchick D, Otwinowski Z (2000) Strategies for macromolecular synchrotron crystallography. Structure 8: R105-R110.
32. Otwinowski Z, Minor W (1997) Processing of X-ray diffraction data collected in oscillation mode. Methods Enzymol 276: 307–326.
33. McCoy AJ, Grosse-Kunstleve RW, Adams PD, Winn MD, Storoni LC, et al. (2007) Phaser crystallographic software. J Appl Crystallogr 40: 658–674.
34. Murshudov GN, Vagin AA, Dodson EJ (1997) Refinement of macromolecular structures by the maximum-likelihood method. Acta Crystallogr D Biol Crystallogr 53: 240–255.
35. Brunger AT, Adams PD, Clore GM, DeLano WL, Gros P, et al. (1998) Crystallography & NMR system: A new software suite for macromolecular structure determination. Acta Crystallogr D 54: 905–921.

36. Laskowski RA, MacArthur MW, Moss DS, Thornton JM (2001) PROCHECK - a program to check the stereochemical quality of protein structures. J Appl Cryst 26: 283–291.

37. van den Akker F, Hol WG (1999) Difference density quality (DDQ): a method to assess the global and local correctness of macromolecular crystal structures. Acta Crystallogr D Biol Crystallogr 55 (Pt 1): 206–218.

# 18

# An Automated Flow for Directed Evolution Based on Detection of Promiscuous Scaffolds Using Spatial and Electrostatic Properties of Catalytic Residues

**Sandeep Chakraborty***

Department of Biological Sciences, Tata Institute of Fundamental Research, Mumbai, India

**Abstract**

The aspiration to mimic and accelerate natural evolution has fueled interest in directed evolution experiments, which endow or enhance functionality in enzymes. Barring a few de novo approaches, most methods take a template protein having the desired activity, known active site residues and structure, and proceed to select a target protein which has a pre-existing scaffold congruent to the template motif. Previously, we have established a computational method (CLASP) based on spatial and electrostatic properties to detect active sites, and a method to quantify promiscuity in proteins. We exploit the prospect of promiscuous active sites to serve as the starting point for directed evolution and present a method to select a target protein which possesses a significant partial match with the template scaffold (DECAAF). A library of partial motifs, constructed from the active site residues of the template protein, is used to rank a set of target proteins based on maximal significant matches with the partial motifs, and cull out the best candidate from the reduced set as the target protein. Considering the scenario where this 'incubator' protein lacks activity, we identify mutations in the target protein that will mirror the template motif by superimposing the target and template protein based on the partial match. Using this superimposition technique, we analyzed the less than expected gain of activity achieved by an attempt to induce β-lactamase activity in a penicillin binding protein (PBP) (PBP-A from *T. elongatus*), and attributed this to steric hindrance from neighboring residues. We also propose mutations in PBP-5 from *E. coli*, which does not have similar steric constraints. The flow details have been worked out in an example which aims to select a substitute protein for human neutrophil elastase, preferably related to grapevines, in a chimeric anti-microbial enzyme which bolsters the innate immune defense system of grapevines.

**Editor:** Vladimir N. Uversky, University of South Florida College of Medicine, United States of America

**Funding:** This work was funded by the Tata Institute of Fundamental Research (Department of Atomic Energy). The funders had no role in study design, data collection and analysis, decision to publish, or preparation of the manuscript.

**Competing Interests:** The author has declared that no competing interests exist.

* E-mail: sandeepchak@tifr.res.in

## Introduction

Directed evolution experiments aspire to mimic and accelerate natural evolution. The methodology typically consists of random mutations [1], in vitro recombination [2–6], continuous evolution [7,8], intramolecular relocation of the N and C termini of a protein [9,10] and high-throughput screening [11,12]. While a few exceptional de novo approaches that endow function in proteins from scratch have been successful [13–17], most techniques target residues in the catalytic site or in its close vicinity for mutational purposes [18–20]. Directed evolution techniques have also sought out the remnants of secondary activities [21–23], with the rationale that the presence of a pre-existing catalytic scaffold increases the odds of success in endowing the activity to the protein [24–28].

Previous work by our group has established a computational method (CLASP) for the detection of active sites [29]. CLASP analysis of the catalytic residues in Class A β-lactamases identified the dichotomy in the proton abstraction mechanism [30], Based on this idea, we proposed a method that enumerates the possible pathways for proton abstraction [31]. We also outlined a computational methodology to quantify promiscuous activities in a wide range of proteins [32]. Jensen hypothesized that ancient enzymes were sparse and promiscuous [33], and this promiscuity formed the basis of the evolution of complex organisms through gene duplication and specialization [34,35]. This hypothesis germinated the idea of selecting promiscuous active sites to serve as a pre-existing scaffold for directed evolution.

In the current work, we present a computational method that selects a protein which has a significant match with a desired catalytic scaffold - *D*irected *e*volution using *CLASP*: *a*n *a*utomated *f*low (DECAAF).

We create a library of partial motifs from the template scaffold and rank a set of proteins based on maximal significant matches with this library. Manual screening, considering further substantiating evidence, is used at this stage to cull the best candidate from the reduced set of proteins with significant congruence. For example, a particular protein would be more suitable if related proteins in the same superfamily are known to have the function under consideration.

The second part of the flow addresses the problem of predicting mutations that mirrors catalytic properties (efficiency, substrate specificity, etc.) of the template protein. In order to replicate the

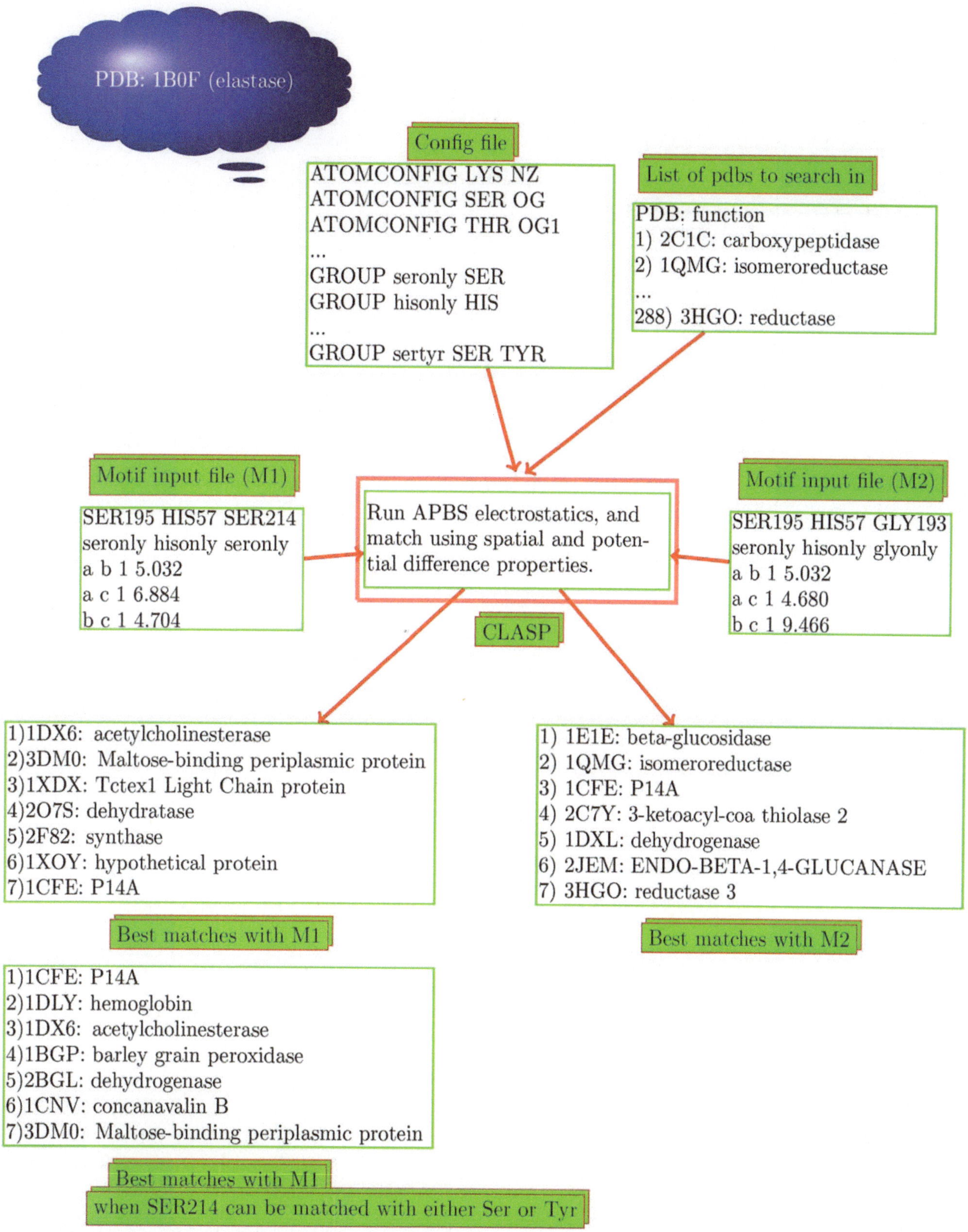


**Figure 1. DECAAF flow to select a plant protein with a significant presence of an elastase-like scaffold.** A pathogenesis related protein (PDB id: 1CFE) was selected as the best possible candidate. The ability to specify a set of residues to match a particular position allows this protein (PDB id: 1CFE) to score much better when a Tyr (Tyr36) is matched instead of a Ser (Ser120).

template scaffold in the target protein, we first superimpose the template and target proteins based on the partial match. The residues in the target protein that are spatially close to the residues in the template motif, if they exist and are not part of the partial motif, are enumerated as possible candidates for site directed mutagenesis techniques to build on [18–20]. In contrast to the structural homology used by DECAAF, a recent method uses phylogenetic and sequence analyses to predict such amino acid substitutions [36].

The evolution of serine β-lactamases (Blase) (divided into Classes A, C and D) and the related penicillin binding proteins (PBP) has fascinated and confounded researchers for decades [37–39]. The physiological relevance of these proteins remains enigmatic [40,41]. Surprisingly, few directed evolution experiments have reported success in enhancing deacylation in PBPs [42,43], the catalytic step that Blases use to hydrolyze β-lactams [44]. Even when successful, the activity in these PBP mutants remains much lower than the Blases (110-fold in [42] and 90-fold in [43]). Sequence alignment identified L158E as a potential mutation that would modify the active site in PBP-A from *T. elongatus* to replicate the active site of Class A Blases, but reported only a 90-fold enhancement in the activity [43]. Firstly, CLASP analysis revealed that the L158E mutant indeed gains potential congruence with the Class A Blase as compared to the wildtype PBP. Subsequently, we demonstrated that Pro159 and Asp160 hinders the approach of substrate, using our superimposition technique. This probably explains the 'puzzling' lack of significant improvement in spite of the fact that 'in the L158E mutant, all the catalytic residues and the interactions characteristic of β-lactamases seem to be present' [43]. Finally, we chose another PBP-5 from *E. Coli* [45], which is also known to have a similar loop present in Class A Blases and PBP-A from *T. elongatus* [43,46], and propose that a L153E mutant of this protein can gain significant β-lactamase activity, as the side chain resembles Class A Blase to a greater extent.

There have been several methods developed for identifying enzymes with a partial catalytic structure [47–49]. The use of 'exact' electrostatic properties in the search process at an early stage is a key innovation of DECAAF when compared to these methods. Other related methods have used binding energy and energy minimization considerations at a later stage of the search [24,25,27]. Although, the CLASP signature for any function comprises of a few residues, it implicitly encodes the surroundings. Potential congruence implies a favorable milieu for the particular function (hydrophilicity, polarity, presence in a cleft, etc.). Consequently, DECAAF filters out unfeasible configurations at a much lower computational cost.

**Table 1.** DECAAF SCORES (DScore) - best matches for motifs of length 3: The best matching protein (PDBid:1DX6) is *from Torpedo californica* (Pacific electric ray), an organism not related to plants.

| PDB | Description | Dscore |
|---|---|---|
| 1DX6 | acetylcholinesterase | 0.048 |
| 1CFE | pathogenesis-related protein P14A | 0.06 |
| 1QMG | acetohydroxy-acid isomeroreductase | 0.068 |
| 2EFJ | 3,7-dimethylxanthine methyltransferase | 0.069 |
| 1W1O | cytokinin dehydrogenase 1 | 0.076 |
| 3DMO | maltose-binding periplasmic protein | 0.077 |
| 1E1E | beta-glucosidase | 0.077 |
| 1AIR | pectate lyase c | 0.084 |
| 2V6G | progesterone 5-beta-reductase | 0.084 |
| 3HGO | 12-oxophytodienoate reductase 3 | 0.087 |

The acetylcholinesterase protein figures in the set of proteins being queried since it has a plant alkaloid galanthamine bound in active site, and we used a keyword search for 'plant' in http://www.pdb.org/to obtain the query set of 288 proteins. Hence, the best match was with the P14A protein from Solanum lycopersicum (tomato).

We have worked out the detailed steps in DECAAF for identifying an elastase-like protein in plants. Our motivation is to find an alternative for human neutrophil elastase (HNE) [50] in a chimeric enzyme that provides enhanced resistance to the Pierce's disease causing Gram-negative pathogen *Xylella fastidiosa* [51]. The substitute should be preferably from an organism related to grapevines. P14A [52] from *Solanum lycopersicum* (tomato), a member of the PR-1 group of pathogenesis-related proteins [53], is obtained as a suitable candidate. The structural homology shared by P14A and a snake venom protein, which was previously demonstrated to be an elastase [54], suggests with increasing certitude that P14A might have elastase activity. Furthermore, protease function has also been associated with the pathogenesis related proteins [55]. We have also identified residues in P14A that do not match the corresponding residue in HNE, and thus predicts mutations in the P14A which should replicate the HNE scaffold. In case we do not find detectable activity in the wild type P14A, we believe that the identified P14A mutants might gain elastase function, Finally, we propose that a L153E mutant of PBP-5 from *E. coli* (PDBid:1NZO) is likely to show better results than the L158E mutant of PBP-A from *T. elongatus* (PDBid:2J9O), which achieved a 90-fold increase in β-lactamase activity. Such experimental validations will be the ultimate litmus test for the DECAAF flow.

## Results

We demonstrate our method (DECAAF) by working out the steps in selecting a elastase-like protein from a set of plant proteins. Human neutrophil elastase (HNE) is a serine protease that is present in the primary granules of polymorphonuclear neutrophils, the other two being proteinase 3 and cathepsin G [56]. The bactericidal properties of HNE have been exploited to design a anti-microbial protein that targets the outer-membrane of the bacteria [57]. HNE is combined with cecropinB, which lyses gram-negative bacteria, in this chimeric protein using a flexible linker and provides enhanced resistance to grapevines against the Gram-negative pathogen Xylella fastidiosa. A protein from a plant related to grapevines that has elastase activity, if found, can substitute HNE in the anti-microbial enzyme.

### 1 Selecting the Best Candidate

We chose the HNE protein with PDBid:1B0F as the template protein. The active site residues comprised of the following residues - Ser195, His57, Asp102, Ser214 and Gly193. The keyword search for 'plants' in http://www.pdb.org/was pruned for redundancy based on a 40% sequence similarity, and yielded 288 proteins (Table S1). From the active site residues ($N = 5$), we generated all possible motifs of size $k < N$ (3, 4 and 5 in this case). An example motif le is shown in Fig. 1. By default each position in the match can be occupied by the same amino acid, i.e. we do not consider stereochemical equivalence at this stage.

Let us consider the case when the partial motifs have three residues - there are 10 such motifs ($^5C_3 = 10$). M1 = (Ser195, His57, Ser214), M2 = (Ser195, His57, Gly193) ... M10 = (Ser195, His57,

**Table 2.** Expanding the set of residues that can match a single position in the input motif.

| | Predicted residues | | | Pairwise Distances in Å | | | Pairwise Potential difference | | | |
|---|---|---|---|---|---|---|---|---|---|---|
| **PDB:** | **a** | **b** | **c** | **ab** | **ac** | **bc** | **ab** | **ac** | **bc** | **Score** |
| 1B0F | Ser195 | His57 | Ser214 | 5 | 6.8 | 4.7 | 53.7 | 105.7 | 51.9 | 0 |
| 1CFE (Ser214 can be Ser only) | Ser49 | His48 | Ser120 | 0.1 | −0.2 | 1.9 | 15.3 | 83.9 | 68.6 | 0.111 |
| 1CFE (Ser214 can be Ser or Tyr) | Ser49 | His48 | Tyr36 | 0.1 | 0.3 | −0.3 | 15.3 | 88.6 | 73.3 | 0.023 |

When the position Ser214 in the input motif (Ser195, His57, Ser214) can be matched by a Ser or Tyr, Tyr36 has a much better spatial orientation with respect to Ser49 and His48 as compared to Ser120 in the P14A protein (PDBid:1CFE). The distances are specified in the reference protein (PDBid:1B0F) in A. For the remaining, we show the deviation from the reference value. Potential differences are in units of kT/e (k is Boltzmann's constant, T is the temperature in K and e is the charge of an electron). They are absolute values and not shown as deviations.

Asp102). We ran each motif (M1, M2, ... M10 ) on each protein in the target list, resulting in a score that quantifies the degree to which a motif exists in a protein based on spatial and electrostatic congruence [29]. Each run of a single motif on the list of 288 proteins took less than 30 mins on a simple workstation (2 GB Ram). Fig. 1 shows the best matches with M1 and M2.

We then compute a cumulative score from the runs for different motifs for each protein that reflects the likelihood of endowing elastase to that protein (Table 1). Manual inspection now chose the best protein from the reduced set of the proteins. The small number of proteins to be manually evaluated made this feasible. The best matching protein (PDBid:1DX6) is from *Torpedo californica* (Pacific electric ray), an organism not related to plants. The acetylcholinesterase protein is included in the set of proteins being queried since it has a plant alkaloid galanthamine bound in active site, and we used a keyword search for 'plant' in http://www.pdb.org/to obtain the query set of 288 proteins. The P14A (PDB id: 1CFE), the next best ranked protein, from *Solanum lycopersicum* (tomato) was therefore the desired target protein [52]. Table 2 shows the congruence of the partial motif (Ser195, His57, Ser214) from the HNE protein with the match from P14A.

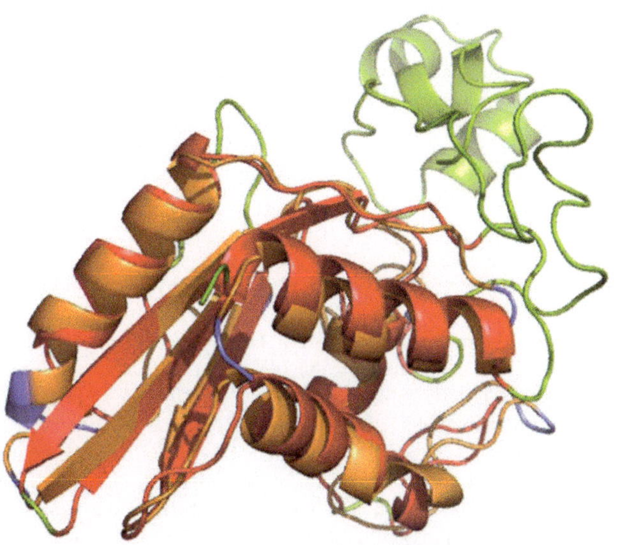

**Figure 2. Superimposition of pseudechetoxin (PDB id: 1CFE, in green) and P14a (PDB id: 1B0F, in blue) and using TopMatch.** The overlapping regions are colored red and orange for pseudechetoxin and P14a respectively. It is known that pseudechetoxin has elastase activity.

P14A is a member of the PR-1 group of pathogenesis-related proteins [53]. While P14A has not been specifically known to have elastase activity, it is structurally homologous to a snake venom protein (Fig. 2) which was previously demonstrated to be an elastase [54]. Pathogenesis related proteins are known to possess proteolytic function [55]. Furthermore, it has been noted that "His48, Ser49 and His93 are in close proximity, so that the conservation of this group of residues could be taken as an indication of it being important as an active site in P14a" [52]. His48 and Ser49 are the residues predicted by DECAAF to be a part of the catalytic scaffold (Table 2 and Fig. 3).

A similar analysis using partial motifs of size k = 4 and k = 5 was done. We have not been able to design a reasonable algorithm for comparing scores with different motif lengths. The P14A protein does not have a good match when the input scaffold includes Asp102. Hence, any motif figuring the Asp102 was not found in P14A with significant congruence. We discuss this aspect subsequently.

## 2 Improving Matches through Stereochemical Equivalence

We extended the matches by allowing stereochemically equivalent residues to match in a motif. While stereochemical equivalence can be hardcoded for amino acids with similar properties, there are instances where residues with different properties occupy the same sequence and spatial location, and perform the same functionality. A well-known example is the equivalence of Ser130 and Tyr150 in Class A and C - lactamases respectively [58]. In the motif (M1 = Ser195, His57, Ser214), we allowed the position Ser214 to be matched by either a Ser or a Tyr. Table 2 shows that Tyr36 has a much better spatial orientation than the Ser120 in the P14A protein, although the potential differences are almost similar. Similarly extending the motif length (n = 4) allowed us to match Gly193 in the HNE protein to Ala51 in the P14A protein (Table 3 and Fig. 3). Gly and Ala both have non-polar, aliphatic R groups.

## 3 Superimposing the Partial Scaffolds

As mentioned previously, the P14A protein does not have a good match when the input scaffold includes Asp102. Although, unconventional serine proteases with variation in the catalytic triad have been observed in many cases, it is possible that P14A is not one them [59]. In such a scenario, we would have to resort to directed evolution techniques to impart elastase function to the protein. Table 4 shows the coordinates of the atoms before and after the transformations when the partial match - (Ser195/OG, His57/ND1, Ser214/OG, Gly193/N - PDBid:1B0F) and (Ser49/OG, His48/ND1, Tyr36/OH, Ala51/N - PDBid:1CFE) - are

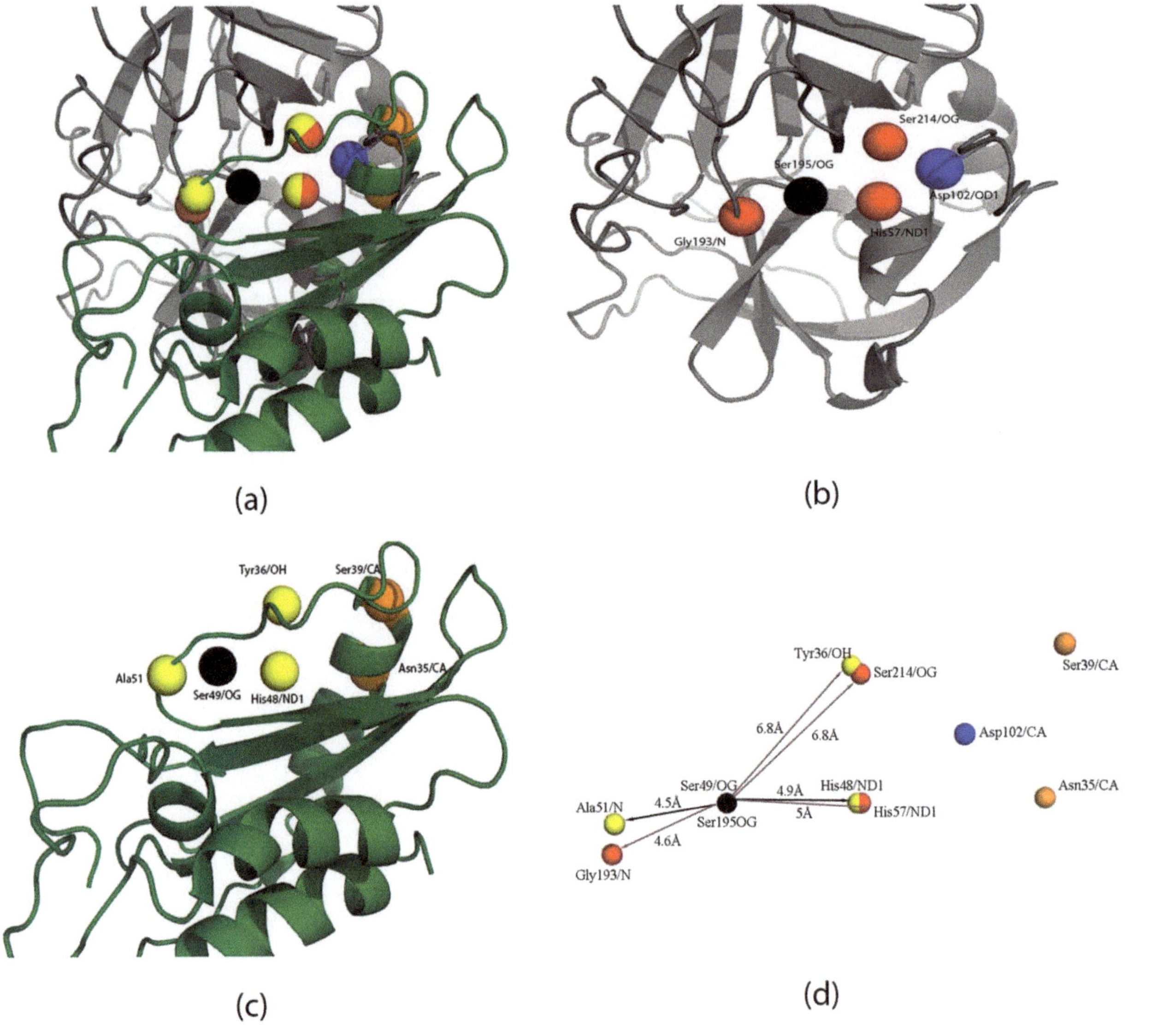


**Figure 3. Superimposing proteins based on partial matches (Ser195, His57, Ser214 from PDBid:1B0F) and (Ser49, His48, Tyr36 from PDBid:1CFE).** Ser195 from PDBid:1B0F and Ser49 from PDBid:1CFE coincide and is colored in black. (**a**) Superimposition of both proteins: Human neutrophilelastase (PDBid:1B0F) is in grey, and P14A (PDBid:1CFE) is in green. (**b**) Human neutrophil elastase (PDBid:1B0F) with active site atoms - Ser195/OG, His57/ND1, Ser214/OG, Gly193/N. (**c**) P14A protein (PDBid:1CFE) with predicted active site residues - Ser49/OG, His48/ND1, Tyr36/OH, Ala51/N. (**d**) Distance between pairs of residues in the partial matches. It can be seen that the Asp102 in PDBid:1B0F is close to Asn35 and Ser39 in PDBid:1CFE.

superimposed. Note, that the superimposition algorithm uses the first three residues in the match. Fig. 3 shows the superimposed proteins. It can be seen that the pair (Gly193/N, Ala51/N), although not used for superimposition, are spatially close. The spatial position of the Asp102 in the HNE is in the vicinity of Asn35 and Ser39 in the P14A protein. Hence, it is possible that the Asn35Asp or Ser39Asp mutant might gain elastase function.

## 4 Analyzing Results of Directed Evolution of Penicillin Binding Proteins to Enhance β-lactamase Activity

The technique for superimposing the active sites of proteins based on the partial matches has been applied to analyze the lower than expected gain in β-lactamase (Blase) activity of a penicillin binding protein (PBP) using site directed mutations [43]. Class A Blases are differentiated from other classes of Blases (C and D) by virtue of the presence of a loop, which houses the critical Glu166 [46]. PBP-A from *T. elongatus* possesses a similar loop, and sequence alignment suggested that Leu158 occupies the position corresponding to Glu166 (of the Blase) in this protein. However, the mutant L158E showed a meager 50-fold gain in activity as compared to the wildtype. This gain in activity is corroborated by CLASP analysis of the mutant active site, which gains potential congruence in cognate pairs as compared to the Class A Blase catalytic residues (Table 5). This electrostatic congruence was missing in the wildtype (Table 5).

The superimposition of the two proteins, Class A Blase (PB Did:1E25) and L158E PBP mutant (PDBid:2J9O) based on their partial matches - (Ser70/OG, Lys73NZ, Ser130/OG) and (Ser61/OG, Lys64/NZ, Ser122/OG) respectively, suggested that the residues adjacent to Glu158 in the mutant protein (Pro159 and Asp160) are responsible for hindering substrate access (Fig. 4).

**Table 3.** Extending the partial match.

| | a | b | c | d | | |
|---|---|---|---|---|---|---|
| 1B0F Motif | SER195 | HIS57 | SER214 | GLY193 | | |
| 1CFE motif | SER49 | HIS48 | TYR36 | ALA51 | | |
| | ab | ac | ad | bc | bd | cd |
| PDB | Pairwise Distances | | | | | |
| 1B0F | 5 | 6.8 | 4.6 | 4.7 | 9.4 | 11 |
| 1CFE | 4.9 | 6.8 | 4.5 | 5 | 9.1 | 10 |
| | Pairwise Potential difference | | | | | |
| 1B0F | 53.7 | 105.7 | −111.1 | 51.9 | −164.9 | −216.8 |
| 1CFE | 15.3 | 88.6 | −58.1 | 73.3 | −73.4 | −146.8 |

Spatial and electrostatic potential difference (PD) congruence in cognate pairs in the human neutrophil elastase (PDBid:1B0F) and the pathogenesis related P14A protein (PDBid:1CFE) for an extended set of four residues in the active site.

**Table 4.** Superimposing proteins based on partial matches (Ser195, His57, Ser214 from PDBid:1B0F) and (Ser49, His48, Tyr36 from PDBid:1CFE).

| | Atom | Before | | | After | | |
|---|---|---|---|---|---|---|---|
| | Coordinates | X | Y | Z | X | Y | Z |
| PDB: 1B0F (template) | Ser195/OG | 64.4 | 57 | 53.8 | 0 | 0 | 0 |
| | His57/ND1 | 63.3 | 54.8 | 58.2 | 5 | 0 | 0 |
| | Ser214/OG | 63.6 | 50.6 | 56.1 | 5 | 4.7 | 0 |
| PDB: 1CFE (target) | Ser49/OG | 8.8 | −6.3 | −4.8 | 0 | 0 | 0 |
| | His48/ND1 | 9.2 | −2.4 | −1.8 | 4.9 | 0 | 0 |
| | Tyr36/OH | 13.7 | −1.6 | −3.7 | 4.7 | 5 | 0 |

After applying linear and rotational transformations on both proteins, Ser195 and Ser49 lie at the center of the coordinate axis, His57 and His48 lie on the -X-Y axis (Y = 0) and Ser214 and Tyr36 lie on the X–Y plane (Z = 0).

Pro159 and Asp160 are spatially close to Leu158/Glu158 in the wildtype/mutant, whereas in the Class A Blase there are no residues within 3 Å of Glu166 (Table 6). Several combinations of mutations failed to improve the activity any more than 90-fold. Pro159 was conserved in all mutants, and hence possibly continued to deny access to substrate in the mutations.

Subsequently, we chose PBP-5 from *E. Coli*, which is also known to have a similar loop like the Class A Blases and PBP-A [45]. It has been noted that a His151 occupies the same spatial position as Glu166, and is possible a good target for mutation – "spatial equivalence of carbonyl group of His-151 in PBP 5 with the carboxyl group of Glu-166 in TEM-1" [60]. However, Leu153 showed up as the best candidate for mimicking Glu166 when we superimposed the Class A Blase and the PBP-5 (PDBid:1NZO) based on partial matches. Furthermore, the adjacent residues to Leu153 do not impede the substrate (Table 6 and Fig. 5). We hypothesize that a mutational study similar to the L158E of PBP-A, if applied to PBP-5 to generate L153E, might provide greater success in replicating Blase enzymatic efficiency in PBPs.

## Discussion

Mutations resulting from the innate inaccuracies of biological processes and environmental factors accrue over millions of years and are a key determinant of evolution. Rapid technological advances in the last few decades have allowed enzyme engineers to apply similar mutation generating methods, and shrunk the timescales for evolving new enzymes to a few months [1–5]. Directed evolution is the generic term applied to experiments that

**Table 5.** Gain of potential congruence in L158E mutant of PBP-A.

| | a | b | c | d | e |
|---|---|---|---|---|---|
| 1E25 | Ser70/OG | Lys73/NZ | Ser130/OG | Lys234/NZ | Glu166/OE1 |
| 2J7V | Ser61/OG | LYS64/NZ | Ser122/OG | LYS219/NZ | Leu158/CD2 |
| 2J9O | Ser61/OG | Lys64/NZ | Ser122/OG | Lys219/NZ | Glu158/OE1 |
| 1NZO | Ser44/OG | Lys47/NZ | Ser110/OG | Lys213/NZ | Leu153/CD2 |

| | | ab | ac | ad | ae | bc | bd | be | cd | ce | de |
|---|---|---|---|---|---|---|---|---|---|---|---|
| 1E25 | PD | −125.6 | 22.4 | −189.1 | 310.7 | 148.1 | −63.5 | 436.3 | −211.5 | 288.2 | 499.7 |
| 1E25 | D | 2.8 | 3.2 | 4.7 | 5.5 | 3.6 | 5.6 | 5 | 2.9 | 8 | 10 |
| 2J7V | PD | −134.5 | 38.7 | −135.6 | −112.4 | 173.3 | −1 | 22.1 | −174.3 | −151.1 | 23.2 |
| 2J7V | D | 3 | 3 | 4.4 | 7.1 | 3.1 | 5.4 | 5.8 | 2.9 | 8.7 | 10.1 |
| 2J9O | PD | −162.3 | −16.4 | −198.3 | 247.7 | 146 | −36 | 410 | −181.9 | 264 | 445.9 |
| 2J9O | D | 2.7 | 3 | 4.5 | 4.5 | 3 | 5.3 | 4 | 2.8 | 6.7 | 8.6 |
| 1NZO | PD | −241.6 | −68.8 | −277.9 | −188.2 | 172.8 | −36.2 | 53.5 | −209.1 | −119.4 | 89.7 |
| 1NZO | D | 3.1 | 4.2 | 6.3 | 6.5 | 5.1 | 6.7 | 5.5 | 2.7 | 10 | 12.1 |

Pairwise potential difference (PD) and distance (D) between cognate pairs in a Class A Blase (PDBid:1E25), a PBP-5 protein (PDBid:1NZO), the wildtype PBP-A (PDBid:2J7V) and the L158E PBP-A mutant (PDBid:2J9O). The L158E PBP-A mutant has gains potential congruence with the Class A Blase, as compared to the wildtype PBP-A. Distances are in A. See Methods section for units of potential.

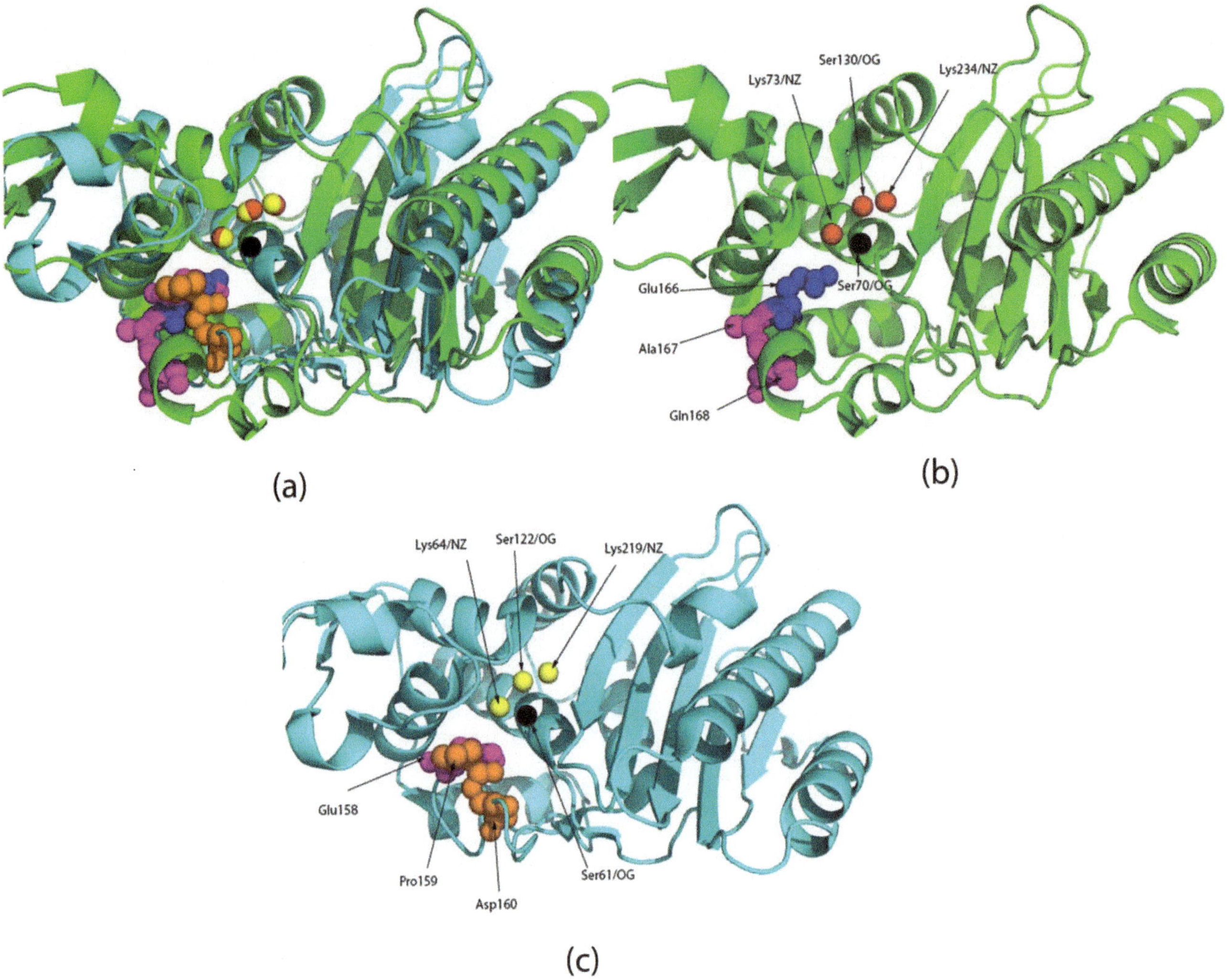


**Figure 4. Superimposition of Class A Blase (PBDid:1E25 in green) and L158D PBP-A mutant (PDBid:2J9O in light blue) based on their partial matches - (Ser70/OG, Lys73NZ, Ser130/OG) and (Ser61/OG, Lys64/NZ, Ser122/OG) respectively.** (**a**) It is seen that transformations aimed at superimposing the partial matches results in a good superimposition of the complete proteins. (**b**) Class A Blase (PBDid:1E25) - it can be seen that the Glu166/OE1 has no steric hindrance from the adjacent residues (Ala167 AND Gln168). (**c**) Mutant L158D PBP-A (PDBid:2J9O) - neighboring residues, Pro159 and Asp160, can be seen obstructing access to substrate.

mimic and accelerate natural evolution [13–20,61]. Several groups have reviewed the current developments in directed evolution in detail [62–65].

Promiscuity, the catalysis of reactions distinct from the one the protein has evolved to perform, is another characteristic feature of enzymes that has shaped the evolution of primitive cells [33,66,67]. Gene duplication unshackled such 'leaky'/'messy' domains from selection pressures, and subsequent specialization culminated into new enzymes [34,35,68]. Identifying pre-existing scaffolds in proteins is a critical step as it implies better odds of success when attempting to endow catalytic activity to proteins [24–28].

The current work has tried to exploit this prospect of a promiscuous active site to serve as a pre-existing scaffold for directed evolution. Previously, we have defined a quantitative measure of promiscuity in a wide range of proteins [32]. The spatial and electrostatic congruence of the active site residues in a heme cytochrome C peroxidase and a $Zn^{2+}$ carboxypeptidase was vindicated by the fact that carboxypeptidase is known to gain pyruvate oxidase catalysis abilities on the replacement of $Zn^{2+}$ with $Cu^{2+}$ [69]. This germinated the idea of defining a generic methodology that would seek an 'incubator' protein for any enzymatic activity with known 3D structure and active site residues.

We present a computational method that selects a protein which has a significant match with a desired catalytic scaffold - ***D****irected* ***e****volution using* ***C****LASP:* ***a****n* ***a****utomated* ***f****low* (DECAAF). Barring the rare chance of finding a protein matching the complete motif, the problem at hand is to find a protein with the best partial matches. This has been solved using heuristics that generates a library of partial motifs and results in a cumulative score reflecting the likelihood of endowing the desired function to that protein. The intractability of the problem is evident, and the many methods are inclined to remain in a local minima [70]. Hence the final step might involve manual screening of a few best matching proteins. Expert eyes can also figure out other supporting facts - for example, it might be known that a related protein from the same superfamily possesses the function, instilling further confidence before one embarks on laborious mutational work.

The possibility that the selected protein does not possess the activity, despite the presence of a subset of the catalytic motif in its

**Table 6.** Atoms close to the Glu166/OE1 in Class A β-lactamase after superimposition of the partial motifs.

| 1E25 | | | 2J7V | | | 2J9O | | | 1NZO | | |
|---|---|---|---|---|---|---|---|---|---|---|---|
| D | Residue | Atom | D | Residue | Atom | D | Residue | Atom | D | Residue | Atom |
| 0 | Glu166 | OE1 | 1.6 | Pro159 | C | 1.6 | Pro159 | C | 1.4 | LEU153 | CD2 |
| 1.3 | Glu166 | CD | 1.6 | Pro159 | CA | 1.6 | Pro159 | CA | 2.7 | LEU153 | CG |
| 2.2 | Glu166 | OE2 | 1.8 | Pro159 | O | 1.9 | Glu158 | OE1 | 2.8 | HOH475 | O |
| 2.4 | Glu166 | CG | 2.4 | Leu158 | CD1 | 1.9 | Pro159 | O | | | |
| 2.7 | Glu166 | CB | 2.7 | Asp160 | N | 2.6 | Asp160 | N | | | |
| 3 | Glu166 | CA | 2.7 | Pro159 | N | 2.7 | Glu158 | CA | | | |
| 3.1 | HOH2076 | O | 2.7 | Leu158 | CA | 2.7 | Pro159 | N | | | |
| | | | 2.8 | Pro159 | CB | 2.8 | Glu158 | CB | | | |
| | | | 2.9 | Leu158 | CB | 2.8 | Glu158 | CD | | | |
| | | | 3.1 | Leu158 | CG | 2.9 | Pro159 | CB | | | |
| | | | 3.2 | Leu158 | C | 3.1 | Glu158 | C | | | |
| | | | 3.4 | HOH2146 | O | 3.3 | HOH2234 | O | | | |

PDBid:1E25 is the Class A Blase, PDBid:2J7V is the wildtype PBP-A, PDBid:2J9O is the L158E PBP-A mutant and PDBid:1NZO is the PBP-5 protein. It can be seen that in the Class A Blase (PDBid:1E25) and PBP-5 there are no atoms within a radius of about 3 Å of Glu166/OE1. In the mutant L158E PBP, both Pro159 and Asp160 are spatially close to Glu158, and can be seen to be obstructing access to substrate (Fig. 4). Distances are in Å, and have a cut-off till the first water molecule is encountered.

structure, leads to the next logical question - can one predict mutations that would bestow function? We have addressed this requirement by superimposing the template and target proteins based on the partial matches, and identified target residues that are in the vicinity of 'unmatched' residues of the template protein. Such residues can be the target of site-directed methods [18–20].

This superimposition technique has been used to analyze the results of an experiment intended to transform a penicillin binding protein (PBP) into a serine β -lactamase (Blase). PBPs are involved in the synthesis and remodeling of bacterial cell wall [71]. Owing to the lack of a counterpart in mammalian organisms, PBPs are the target of the β lactam family of antibiotics (penicillin, ampicillin, etc.) which are structurally homologous to the D-ala-D-ala moiety present on peptidoglycan precursors. These drugs acylate the nucleophilic serine residue and, unlike the peptide substrates, form long-lived complexes. Blases, which share three highly conserved motifs with PBPs (the SXXK tetrad, the SXN triad, and KTG triad) are similarly acetylated, but have evolved to gain significantly higher rates of deacylation resulting in the regeneration of the original uncomplexed enzyme [37–39]. This sets up the scene for a perfect directed evolution experiment - what are the mutations needed to transform a PBP into a Blase? However, this task has proved to be surprisingly difficult [42,43].

The serine Blases are divided into three classes (A, C and D) based on sequence homology [72]. Of these, the Class A Blases have evolved to use a Glu-X-X-Asn motif present on a loop (the loop) for catalysis [46]. PBP-A from *T. elongatus*, a thermophilic unicellular rod shaped cyanobacterium, possesses a similar loop, and sequence alignment suggested that Leu158 occupies the position corresponding to Glu166 in this protein [73]. However, the L158E PBP-A mutant could only achieve a meager 90-fold gain of activity, even when aided with other adjacent mutations [43]. This was 'puzzling' since 'in the L158E mutant, all the catalytic residues and the interactions characteristic of β-lactamases seem to be present' [43].

We have corroborated that the electrostatic properties of the active site residues in the L158E mutant now mirrors the Class A Blase catalytic site properties, explaining the increase in activity. Moreover, superimposition of the reactive atoms of the catalytic residues in both the protein provides a possible explanation of the modest gains - the residues adjacent to Glu158 in the mutant PBP-A (Pro159 and Asp160) is seen to be restricting access to the catalytic cleft, in stark contrast to the Class A Blase. Since Pro159 was conserved in all of the mutants, it might be a key determinant for the less than expected gain of activity.

Next, we chose PBP-5 from *E. coli* since it possesses a loop similar to Class A Blases and PBP-A [45]. PBP-5 helps in maintaining the normal morphology of the cell and provides resistance to β-lactams antibiotics [74–76]. While it has been commented that His151 is the residue that corresponds to the Glu166 in Class A Blase [60], we noted that Leu153 is a better candidate after superimposition of the proteins using the technique detailed in this work. Also, unlike in PBP-A, the adjacent residues to Leu153 in PBP-5 have a conformation similar to that in Class A Blases, and seem to provide unrestricted access of substrate to the catalytic site. We hypothesize that the L153E PBP-5 mutant might achieve better gain of function as compared to the L158E mutant(s).

The use of 'exact' electrostatic properties in the search process at an early stage is a key innovation of DECAAF when compared to similar methods seeking partial catalytic structure [47–49]. Pruning based on electrostatic properties reduces considerable false positives compared to purely 3D matching methods. A few methods have used binding energy and energy minimization considerations, albeit at a later stage of the search [24,25,27]. Electrostatic congruence of a few active site residues implies a favorable milieu for the catalytic activity and thus encodes residues in the close vicinity as well. This allows DECAAF to prune unfeasible configurations at a much lower computational cost.

We describe in the details the steps (Fig. 1) involved in choosing a plant protein to substitute human neutrophil elastase (HNE) [50] in a chimeric anti-microbial enzyme designed to bolster the innate immune defense system in grapevines [51]. The flow identifies P14A [52] from *Solanum lycopersicum* (tomato) as a very significant match (Tables 1, 2, 3). P14A is a member of the PR-1 group of pathogenesis-related proteins [53]. While P14A has not been

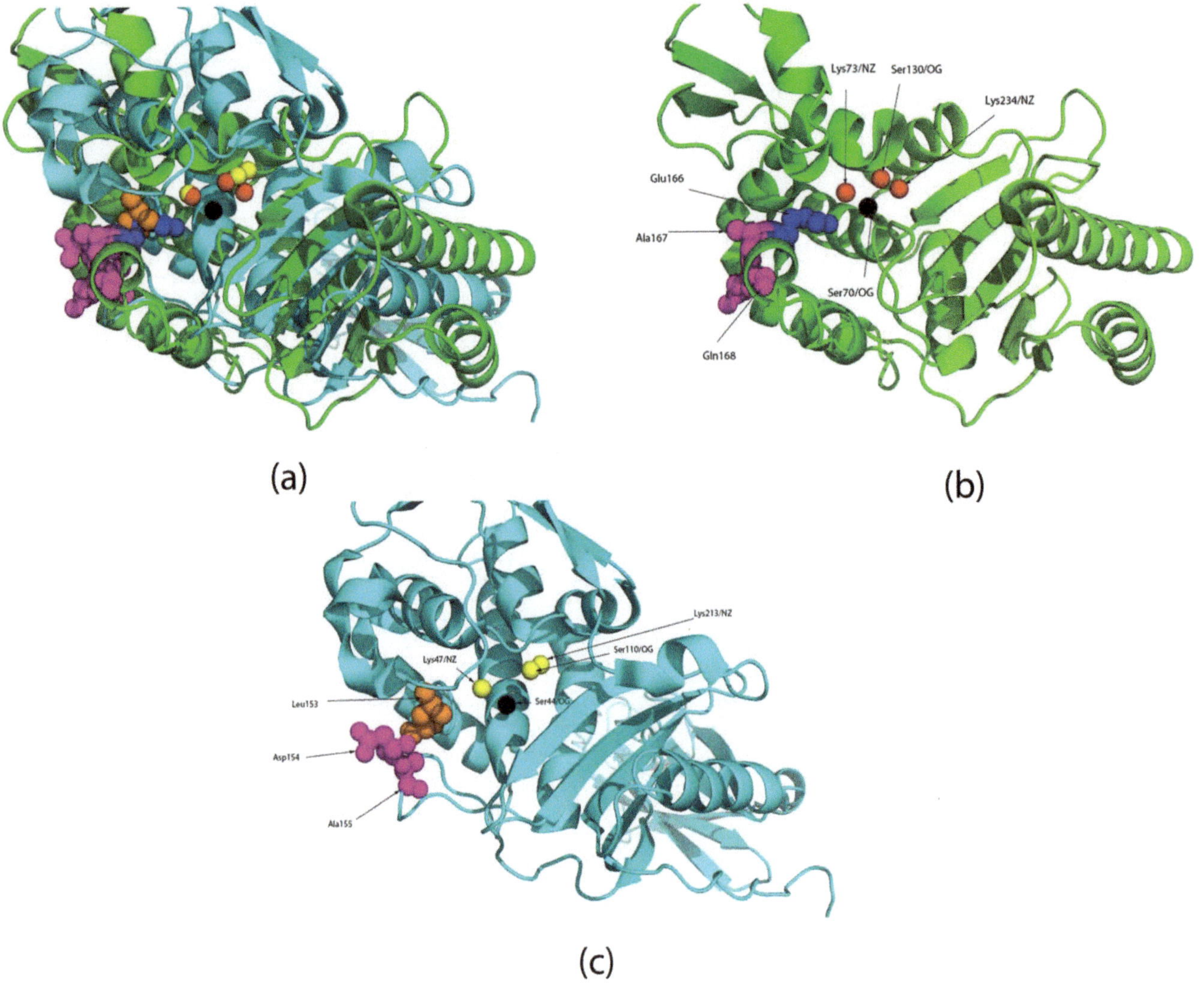


**Figure 5. Superimposition of Class A Blase (PBDid:1E25 in green) and PBP-5 (PDBid:1NZO in light blue) based on their partial matches - (Ser70/OG, Lys73NZ, Ser130/OG) and (Ser44/OG, Lys47/NZ, Ser110/OG) respectively.** (**a**) It is seen that the superimposition of PBP-5 and this Class A Blase is less significant than that with PBP-A. (**b**) Leu153 corresponds to the spatial location of Glu166, and it is seen that Asp154 and Ala155 do not restrict access to the catalytic cleft.

specifically known to have elastase activity, it is structurally homologous to a snake venom protein (Fig. 2), previously demonstrated to be an elastase [54]. Interestingly, the complete scaffold of the HNE does not exist in the P14A protein. While Ser195, His57, Ser214 and Gly193 from the input motif has an highly matching scaffold in P14A (Fig. 3), the spatial position of the Asp102 in the HNE is close to Asn35 and a Ser39 in the P14A protein when the proteins were superimposed. We intend to express this protein and test for its elastase activity. In case we do not find detectable activity, we believe that the Asn35Asp or the Ser39Asp P14A mutant might gain elastase function and thus validate the proposed rational design flow DECAAF.

## Methods

### 1 Choosing a Consensus Group of Proteins

We start with a template protein with known 3D structure and a set of active site residues

($\Phi_{residues}$, | $\Phi_{residues}$ | = N) (Equation 1).

Our goal is to select the best protein from a set of proteins ($\Phi_{proteins}$, | $\Phi_{proteins}$ | = M) with known 3D structures (Equation 2), that has a domain in its structure which is significantly congruous to $\Phi_{residues}$

$$\Phi_{residues} = \{R_1, R_2, \ldots R_N,\} \quad (1)$$

$$\Phi_{proteins} = \{PDB_1, PDB_2, \ldots PDB_M,\} \quad (2)$$

We first create partial motifs of size k from $\mathbf{\Phi_{residues}}$, for k = >3, 4…N (Equation 3). Let us denote the number of elements in $^{n}C_{k}$ by n.

$$_{motifs}^{k}\Phi = {}_{k}^{n}C\,\Phi_{residues} \text{ for, i} = 3, 4 .. \text{N} \quad (3)$$

CLASP analysis of each motif of size k results in n scores for each protein Pi (CLASPScorei). We combine this score into a cumulative score Score (PDBki ) (Equation 4).

$$Score({}^{k}_{i}PDB) = \sum CLASPScore_i \text{ for, i} = 1 \,..\, \text{n and } k = 3, 4 \,..\, N \quad (4)$$

Now, for each k, we obtained the sorted set of scores for each protein (Equation 5).

$$ {}_{sorted}^{k}\Phi = \{PDB_x, PDB_y, PDB_z, \ldots PDB_N\} $$
$$ \text{such that, } Score(PDB_x) < Score(PDB_y), \quad (5)$$
$$ Score(PDB_y) < Score(PDB_z) \ldots $$

At this stage, we manually screen the proteins which occur in the top matches (**$PDB_x$, $PDB_y$, $PDB_z$ ...**) in $\Phi^k_{sorted}$ for each k, and decide the best possible candidate.

## 2 Superimposing Proteins Based on Partial Matches and Predicting Mutations

We apply transformations on the template and target proteins such that the partial matches overlap. The superimposition algorithm takes the first three atoms of the partial match from both the template and target proteins - let us refer to them as (A,B,C) and (P,Q,R). We apply both linear and rotational transformations on both the proteins such that A and P are at the center of the coordinate axis (coordinates = [0,0,0]), B and Q lies on the X-Y axis (i.e. Y coordinate is 0) and C and R lie on the X-Y plane (i.e. Z coordinate is 0). In short, after the transformations, (A,B,C) and (P,Q,R) both lie on the same plane (Z = 0) such that A and P are at the center, and B and Q lie on the x-y axis. Once the two structures are superimposed, we proceed to find the residues in the target protein which are spatially close to the unmatched residues in the template protein. In case the wild type protein does not have the desired functions, mutating these residues might bestow function.

## 3 Tools

Adaptive Poisson-Boltzmann Solver (APBS) and PDB2PQR packages were used to calculate the potential difference between the reactive atoms of the corresponding proteins [77,78]. The APBS parameters were set as follows - solute dielectric: 2, solvent dielectric: 78, solvent probe radius: 1.4 A, temperature: 298 K and zero ionic strength. APBS writes out the electrostatic potential in dimensionless units of kT/e where k is Boltzmann's constant, T is the temperature in K and e is the charge of an electron. We extensively integrated and used the freely available BioPerl [79] modules and Emboss [80] tools. 3D structure was aligned by the TopMatch server [81]. All protein structures were rendered by PyMol (http://www.pymol.org/).

## Supporting Information

**Table S1 Target set of 288 proteins.** The keyword search for 'plants' in http://www.pdb.org/was pruned for redundancy based on a 40% sequence similarity, and yielded 288 proteins.

## Acknowledgments

I gratefully acknowledge B. J. Rao for technical discussions and infrastructural support. I am also indebted to Abhaya M. Dandekar from Plant Sciences Department, University of California, Davis, CA for technical inputs on the chimeric anti-microbial enzyme designed by their group [51]. I would like to thank Anindya S. Ghosh from Department of Biotechnology, IIT Kharagpur for discussions on PBP-5 from *E. coli*.

## Author Contributions

Conceived and designed the experiments: SC. Performed the experiments: SC. Analyzed the data: SC. Contributed reagents/materials/analysis tools: SC. Wrote the paper: SC.

## References

1. Cirino PC, Mayer KM, Umeno D (2003) Generating mutant libraries using error-prone PCR. Methods Mol Biol 231: 3–9.
2. Hall BG, Zuzel T (1980) Evolution of a new enzymatic function by recombination within a gene. Proc Natl Acad Sci USA 77: 3529–3533.
3. Stemmer WP (1994) Rapid evolution of a protein in vitro by DNA shuffling. Nature 370: 389–391.
4. Zhao H, Giver L, Shao Z, A holter JA, Arnold FH (1998) Molecular evolution by staggered extension process (StEP) in vitro recombination. Nat Biotechnol 16: 258–261.
5. Kolkman JA, Stemmer WP (2001) Directed evolution of proteins by exon shuffling. Nat Biotechnol 19: 423–428.
6. Engler C, Gruetzner R, Kandzia R, Marillonnet S (2009) Golden gate shuffling: a one-pot DNA shuffling method based on type IIs restriction enzymes. PLoS ONE 4: e5553.
7. Esvelt KM, Carlson JC, Liu DR (2011) A system for the continuous directed evolution of biomolecules. Nature 472: 499–503.
8. Johns GC, Joyce GF (2005) The promise and peril of continuous in vitro evolution. J Mol Evol 61: 253–263.
9. Graf R, Schachman HK (1996) Random circular permutation of genes and expressed polypeptide chains: application of the method to the catalytic chains of aspartate transcarbamoylase. Proc Natl Acad Sci USA 93: 11591–11596.
10. Guntas G, Kanwar M, Ostermeier M (2012) Circular Permutation in the Ω-Loop of TEM-1 β-lactamase Results in Improved Activity and Altered Substrate Specificity. PLoS ONE 7: e35998.
11. Goddard JP, Reymond JL (2004) Enzyme assays for high-throughput screening. Curr Opin Biotechnol 15: 314–322.
12. Stapleton JA, Swartz JR (2010) Development of an in vitro compartmentalization screen for high-throughput directed evolution of [FeFe] hydrogenases. PLoS ONE 5: e15275.
13. Bolon DN, Mayo SL (2001) Enzyme-like proteins by computational design. Proc Natl Acad Sci USA 98: 14274–14279.
14. Jiang L, Altho EA, Clemente FR, Doyle L, Rothlisberger D, et al. (2008) De novo computational design of retro-aldol enzymes. Science 319: 1387–1391.
15. Faiella M, Andreozzi C, de Rosales RT, Pavone V, Maglio O, et al. (2009) An artificial di-iron oxo-protein with phenol oxidase activity. Nat Chem Biol 5: 882–884.
16. Siegel JB, Zanghellini A, Lovick HM, Kiss G, Lambert AR, et al. (2010) Computational design of an enzyme catalyst for a stereoselective bimolecular Diels-Alder reaction. Science 329: 309–313.
17. Rothlisberger D, Khersonsky O, Wollacott AM, Jiang L, DeChancie J, et al. (2008) Kemp elimination catalysts by computational enzyme design. Nature 453: 190–195.
18. Reetz MT, Carballeira JD (2007) Iterative saturation mutagenesis (ISM) for rapid directed evolution of functional enzymes. Nat Protoc 2: 891–903.
19. Climie S, Ruiz-Perez L, Gonzalez-Pacanowska D, Prapunwattana P, Cho SW, et al. (1990) Saturation site-directed mutagenesis of thymidylate synthase. J Biol Chem 265: 18776–18779.
20. Reetz MT, Carballeira JD, Peyralans J, Hobenreich H, Maichele A, et al. (2006) Expanding the substrate scope of enzymes: combining mutations obtained by CASTing. Chemistry 12: 6031–6038.
21. Kimura M (1986) DNA and the neutral theory. Philos Trans R Soc Lond, B, Biol Sci 312: 343–354.
22. Amitai G, Gupta RD, Taw k DS (2007) Latent evolutionary potentials under the neutral mutational drift of an enzyme. HFSP J 1: 67–78.
23. Wroe R, Chan HS, Bornberg-Bauer E (2007) A structural model of latent evolutionary potentials underlying neutral networks in proteins. HFSP J 1: 79–87.
24. Zanghellini A, Jiang L, Wollacott AM, Cheng G, Meiler J, et al. (2006) New algorithms and an in silico benchmark for computational enzyme design. Protein Sci 15: 2785–2794.
25. Dahiyat BI, Mayo SL (1997) De novo protein design: fully automated sequence selection. Science 278: 82–87.
26. Malisi C, Kohlbacher O, Hocker B (2009) Automated scaffold selection for enzyme design. Proteins 77: 74–83.
27. Georgiev I, Lilien RH, Donald BR (2008) The minimized dead-end elimination criterion and its appli-cation to protein redesign in a hybrid scoring and search

algorithm for computing partition functions over molecular ensembles. J Comput Chem 29: 1527–1542.
28. Lovell SC, Word JM, Richardson JS, Richardson DC (2000) The penultimate rotamer library. Proteins 40: 389–408.
29. Chakraborty S, Minda R, Salaye L, Bhattacharjee SK, Rao BJ (2011) Active site detection by spatial conformity and electrostatic analysis-unravelling a proteolytic function in shrimp alkaline phosphatase. PLoS ONE 6: e28470.
30. Damblon C, Raquet X, Lian LY, Lamotte-Brasseur J, Fonze E, et al. (1996) The catalytic mechanism of beta-lactamases: NMR titration of an active-site lysine residue of the TEM-1 enzyme. Proc Natl Acad Sci USA 93: 1747–1752.
31. Chakraborty S (2012) Enumerating pathways of proton abstraction based on a spatial and electrostatic analysis of residues in the catalytic site. PLoS ONE, In Press.
32. Chakraborty S, Rao BJ (2012) A measure of the promiscuity of proteins and characteristics of residues in the vicinity of the catalytic site that regulate promiscuity. PLoS ONE 7: e32011.
33. Jensen RA (1976) Enzyme recruitment in evolution of new function. Annu Rev Microbiol 30: 409–425.
34. LEWIS EB (1951) Pseudoallelism and gene evolution. Cold Spring Harb Symp Quant Biol 16: 159–174.
35. Taw k DS (2010) Messy biology and the origins of evolutionary innovations. Nat Chem Biol 6: 692–696.
36. Cole MF, Gaucher EA (2011) Exploiting models of molecular evolution to efficiently direct protein engineering. J Mol Evol 72: 193–203.
37. Ghuysen JM (1991) Serine beta-lactamases and penicillin-binding proteins. Annu Rev Microbiol 45: 37–67.
38. Hall BG, Barlow M (2004) Evolution of the serine beta-lactamases: past, present and future. Drug Resist Updat 7: 111–123.
39. Matagne A, Dubus A, Galleni M, Frere JM (1999) The beta-lactamase cycle: a tale of selective pressure and bacterial ingenuity. Nat Prod Rep 16: 1–19.
40. Denome SA, Elf PK, Henderson TA, Nelson DE, Young KD (1999) Escherichia coli mutants lacking all possible combinations of eight penicillin binding proteins: viability, characteristics, and implications for peptidoglycan synthesis. J Bacteriol 181: 3981–3993.
41. Bhullar K, Waglechner N, Pawlowski A, Koteva K, Banks ED, et al. (2012) Antibiotic resistance is prevalent in an isolated cave microbiome. PLoS ONE 7: e34953.
42. Chesnel L, Zapun A, Mouz N, Dideberg O, Vernet T (2002) Increase of the deacylation rate of PBP2x from Streptococcus pneumoniae by single point mutations mimicking the class A beta-lactamases. Eur J Biochem 269: 1678–1683.
43. Urbach C, Evrard C, Pudzaitis V, Fastrez J, Soumillion P, et al. (2009) Structure of PBP-A from Ther-mosynechococcus elongatus, a penicillin-binding protein closely related to class A beta-lactamases. J Mol Biol 386: 109–120.
44. Matagne A, Lamotte-Brasseur J, Frere JM (1998) Catalytic properties of class A beta-lactamases: efficiency and diversity. Biochem J 330 (Pt 2): 581–598.
45. Nicholas RA, Krings S, Tomberg J, Nicola G, Davies C (2003) Crystal structure of wild-type penicillin-binding protein 5 from Escherichia coli: implications for deacylation of the acyl-enzyme complex. J Biol Chem 278: 52826–52833.
46. Banerjee S, Pieper U, Kapadia G, Pannell LK, Herzberg O (1998) Role of the omega-loop in the activity, substrate specificity, and structure of class A beta-lactamase. Biochemistry 37: 3286–3296.
47. Savile CK, Janey JM, Mundor EC, Moore JC, Tam S, et al. (2010) Biocatalytic asymmetric synthesis of chiral amines from ketones applied to sitagliptin manufacture. Science 329: 305–309.
48. Chen CY, Georgiev I, Anderson AC, Donald BR (2009) Computational structure-based redesign of enzyme activity. Proc Natl Acad Sci USA 106: 3764–3769.
49. Sandstrom AG, Wikmark Y, Engstrom K, Nyhlen J, Backvall JE (2012) Combinatorial reshaping of the Candida antarctica lipase A substrate pocket for enantioselectivity using an extremely condensed library. Proc Natl Acad Sci USA 109: 78–83.
50. Macdonald SJ, Dowle MD, Harrison LA, Clarke GD, Inglis GG, et al. (2002) Discovery of further pyrrolidine trans-lactams as inhibitors of human neutrophil elastase (HNE) with potential as development candidates and the crystal structure of HNE complexed with an inhibitor (GW475151). J Med Chem 45: 3878–3890.
51. Dandekar AM, Gouran H, Ibanez AM, Uratsu SL, Aguero CB, et al. (2012) An engineered innate immune defense protects grapevines from Pierce disease. Proc Natl Acad Sci USA 109: 3721–3725.
52. Fernandez C, Szyperski T, Bruyere T, Ramage P, Mosinger E, et al. (1997) NMR solution structure of the pathogenesis-related protein P14a. J Mol Biol 266: 576–593.
53. Stintzi A, Heitz T, Prasad V, Wiedemann-Merdinoglu S, Kau mann S, et al. (1993) Plant 'pathogenesis-related' proteins and their role in defense against pathogens. Biochimie 75: 687–706.
54. Bernick JJ, Simpson W (1976) Distribution of elastase-like enzyme activity among snake venoms. Comp Biochem Physiol, B 54: 51–54.
55. Milne TJ, Abbenante G, Tyndall JD, Halliday J, Lewis RJ (2003) Isolation and characterization of a cone snail protease with homology to CRISP proteins of the pathogenesis-related protein superfamily. J Biol Chem 278: 31105–31110.
56. Korkmaz B, Moreau T, Gauthier F (2008) Neutrophil elastase, proteinase 3 and cathepsin G: physic-ochemical properties, activity and physiopathological functions. Biochimie 90: 227–242.
57. Kunkel M, Vuyisich M, Gnanakaran G, Bruening GE, Dandekar AM, et al. (2007) Rapid clearance of bacteria and their toxins: development of therapeutic proteins. Crit Rev Immunol 27: 233–245.
58. Lobkovsky E, Moews PC, Liu H, Zhao H, Frere JM, et al. (1993) Evolution of an enzyme activity: crystallographic structure at 2-A resolution of cephalosporinase from the ampC gene of Enterobacter cloacae P99 and comparison with a class A penicillinase. Proc Natl Acad Sci USA 90: 11257–11261.
59. Ekici OD, Paetzel M, Dalbey RE (2008) Unconventional serine proteases: variations on the catalytic Ser/His/Asp triad configuration. Protein Sci 17: 2023–2037.
60. Davies C, White SW, Nicholas RA (2001) Crystal structure of a deacylation-defective mutant of penicillin-binding protein 5 at 2.3-A resolution. J Biol Chem 276: 616–623.
61. Kuhn I, Harden P, Bauzon M, Chartier C, Nye J, et al. (2008) Directed evolution generates a novel oncolytic virus for the treatment of colon cancer. PLoS ONE 3: e2409.
62. Nannemann DP, Birmingham WR, Scism RA, Bachmann BO (2011) Assessing directed evolution methods for the generation of biosynthetic enzymes with potential in drug biosynthesis. Future Med Chem 3: 809–819.
63. Dalby PA (2011) Strategy and success for the directed evolution of enzymes. Curr Opin Struct Biol 21: 473–480.
64. Lutz S (2010) Beyond directed evolution-semi-rational protein engineering and design. Curr Opin Biotechnol 21: 734–743.
65. Antikainen NM, Martin SF (2005) Altering protein specificity: techniques and applications. Bioorg Med Chem 13: 2701–2716.
66. James LC, Taw k DS (2003) Conformational diversity and protein evolution-a 60-year-old hypothesis revisited. Trends Biochem Sci 28: 361–368.
67. Depristo MA (2007) The subtle benefits of being promiscuous: adaptive evolution potentiated by enzyme promiscuity. HFSP J 1: 94–98.
68. Harada E, Nakagawa J, Asano T, Taoka M, Sorimachi H, et al. (2012) Functional evolution of dupli-cated odorant-binding protein genes, Obp57d and Obp57e, in Drosophila. PLoS ONE 7: e29710.
69. Yamamura K, Kaiser ET (1976) Studies on the oxidase activity of copper(ii) carboxypeptidase a. J Chem Soc, Chem Commun : 830–831.
70. Gumulya Y, Sanchis J, Reetz MT (2012) Many pathways in laboratory evolution can lead to improved enzymes: how to escape from local minima. Chembiochem 13: 1060–1066.
71. Macheboeuf P, Contreras-Martel C, Job V, Dideberg O, Dessen A (2006) Penicillin binding proteins: key players in bacterial cell cycle and drug resistance processes. FEMS Microbiol Rev 30: 673–691.
72. Bush K, Jacoby GA (2010) Updated functional classification of beta-lactamases. Antimicrob Agents Chemother 54: 969–976.
73. Urbach C, Fastrez J, Soumillion P (2008) A new family of cyanobacterial penicillin-binding proteins. A missing link in the evolution of class A beta-lactamases. J Biol Chem 283: 32516–32526.
74. Ghosh AS, Young KD (2003) Sequences near the active site in chimeric penicillin binding proteins 5 and 6 a ect uniform morphology of Escherichia coli. J Bacteriol 185: 2178–2186.
75. Sarkar SK, Dutta M, Chowdhury C, Kumar A, Ghosh AS (2011) PBP5, PBP6 and DacD play different roles in intrinsic β-lactam resistance of Escherichia coli. Microbiology (Reading, Engl) 157: 2702–2707.
76. Ghosh AS, Chowdhury C, Nelson DE (2008) Physiological functions of D-alanine carboxypeptidases in Escherichia coli. Trends Microbiol 16: 309{317.
77. Baker NA, Sept D, Joseph S, Holst MJ, McCammon JA (2001) Electrostatics of nanosystems: appli-cation to microtubules and the ribosome. Proc Natl Acad Sci USA 98: 10037–10041.
78. Dolinsky TJ, Nielsen JE, McCammon JA, Baker NA (2004) PDB2PQR: an automated pipeline for the setup of Poisson-Boltzmann electrostatics calculations. Nucleic Acids Res 32: W665–667.
79. Stajich JE, Block D, Boulez K, Brenner SE, Chervitz SA, et al. (2002) The bioperl toolkit: Perl modules for the life sciences. Genome research 12: 1611–1618.
80. Rice P, Longden I, Bleasby A (2000) EMBOSS: the European Molecular Biology Open Software Suite. Trends Genet 16: 276–277.
81. Sippl MJ, Wiederstein M (2008) A note on difficult structure alignment problems. Bioinformatics 24: 426–427.

# 19

# Engineered Monoclonal Antibody with Novel Antigen-Sweeping Activity *In Vivo*

**Tomoyuki Igawa*, Atsuhiko Maeda, Kenta Haraya, Tatsuhiko Tachibana, Yuki Iwayanagi, Futa Mimoto, Yoshinobu Higuchi, Shinya Ishii, Shigero Tamba, Naoka Hironiwa, Kozue Nagano, Tetsuya Wakabayashi, Hiroyuki Tsunoda, Kunihiro Hattori**

Research Division, Chugai Pharmaceutical Co., Ltd., Gotemba, Shizuoka, Japan

## Abstract

Monoclonal antibodies are widely used to target disease-related antigens. However, because conventional antibody binds to the antigen but cannot eliminate the antigen from plasma, and rather increases the plasma antigen concentration by reducing the clearance of the antigen, some clinically important antigens are still difficult to target with monoclonal antibodies because of the huge dosages required. While conventional antibody can only bind to the antigen, some natural endocytic receptors not only bind to the ligands but also continuously eliminate them from plasma by pH-dependent dissociation of the ligands within the acidic endosome and subsequent receptor recycling to the cell surface. Here, we demonstrate that an engineered antibody, named sweeping antibody, having both pH-dependent antigen binding (to mimic the receptor-ligand interaction) and increased binding to cell surface neonatal Fc receptor (FcRn) at neutral pH (to mimic the cell-bound form of the receptor), selectively eliminated the antigen from plasma. With this novel antigen-sweeping activity, antibody without *in vitro* neutralizing activity exerted *in vivo* efficacy by directly eliminating the antigen from plasma. Moreover, conversion of conventional antibody with *in vitro* neutralizing activity into sweeping antibody further potentiated the *in vivo* efficacy. Depending on the binding affinity to FcRn at neutral pH, sweeping antibody reduced antigen concentration 50- to 1000-fold compared to conventional antibody. Thereby, sweeping antibody antagonized excess amounts of antigen in plasma against which conventional antibody was completely ineffective, and could afford marked reduction of dosage to a level that conventional antibody can never achieve. Thus, the novel mode of action of sweeping antibody provides potential advantages over conventional antibody and may allow access to the target antigens which were previously undruggable by conventional antibody.

**Editor:** Maria Gasset, Consejo Superior de Investigaciones Cientificas, Spain

**Funding:** The authors have no support or funding to report.

**Competing Interests:** All authors are employees of Chugai Pharmaceutical Co., Ltd. T.I. and A.M. are inventors of the patents which claim pH-dependent binding antibody and sweeping antibody.

* E-mail: igawatmy@chugai-pharm.co.jp

## Introduction

Therapeutic monoclonal antibodies are now becoming an important option for treating various diseases [1,2]. Although high affinity antibodies with neutralizing activity against various antigens have been generated and shown to be therapeutically effective *in vivo*, some clinically important antigens have proved difficult to target by conventional antibody because of the huge antibody dosage required.

It has been reported that administering conventional antibodies to target soluble antigens, such as amyloid beta [3], MCP1 [4], hepcidin [5], IL6 [6], CD23 [7] and VEGF [8], results in more than 1000-fold increased antigen concentration over the baseline due to the accumulation of antibody-antigen complex in plasma. Since the half-life of the IgG antibody is very much longer than that of the antigen, the binding of antigen to antibody results in an increase in the plasma antigen concentration by reducing the clearance of the antigen [9]. The extent of increase in antigen concentration is determined by the difference in clearance between antigen and antibody-antigen complex [10]. As a striking example, administration of high affinity antibody against hepcidin, which has very rapid clearance, resulted in approximately 5,000-fold increase in plasma hepcidin concentration, requiring a huge antibody dosage of 300 mg/kg weekly to neutralize the hepcidin, which is an unrealistic dosage for therapeutic development [5]. In other cases, the baseline plasma concentration of the antigen may itself be extremely high, as in that of complement factor C5, the target antigen of eculizumab, which is in the range of μg/mL, in contrast to most therapeutic antibodies in the pg/mL or ng/mL range. Because of such high C5 concentration, eculizumab requires huge antibody dosage for efficient C5 neutralization [11], which makes eculizumab one of the highest annual dosages of the approved therapeutic antibodies. In theory, even an antibody with infinite affinity would need to be at a concentration higher than that of the total antigen to neutralize that antigen *in vivo* [12]. Therefore, when targeting soluble antigens with rapid clearance or high baseline plasma concentration, even conventional antibody with infinite affinity requires huge antibody dosage to achieve therapeutic efficacy. This impedes not only the development of subcutaneous formulations, which are important for chronic disease, but also the commercial development itself, because of increased manufacturing cost.

Since these issues stem from the fact that conventional antibody can only bind to the antigen and accumulates the antigen in plasma, engineered antibody that enables active and selective elimination of the antigen from plasma could overcome these issues. In the natural system, several cell surface endocytic receptors, such as asialoglycoprotein receptor [13], low-density lipoprotein receptor [14] and epidermal growth factor receptor [15], deliver ligands to the lysosome to eliminate the ligands from plasma. These receptors bind to the ligands at the cell surface and internalize the ligands into the cell. Since these receptors bind to the ligands pH dependently, they release the ligands in the acidic endosome, and while the released ligands are transferred to lysosome and degraded, the free receptors rapidly recycle back to the cell surface for another round of ligand elimination from plasma. Such properties make these receptors ideal for "sweeping" the ligands from the plasma. In this study, we engineered monoclonal antibody to mimic the endocytic receptor-like property so that it can exert antigen-sweeping activity and eliminate the antigen from the plasma.

We have investigated engineered monoclonal antibody to exert novel antigen-sweeping activity by simultaneously engineering both pH-dependent antigen binding, to mimic the pH-dependent receptor-ligand binding of such endocytic receptors, and increased FcRn binding affinity at neutral pH, to mimic the cell-bound form of the receptors. We demonstrate that the conversion of conventional antibody into sweeping antibody, which directly eliminates the antigen from plasma, could afford huge reduction of the antibody dosage to a level that conventional antibody even with infinite affinity cannot achieve and may allow access to target antigens which have previously been undruggable by conventional antibody.

## Materials and Methods

### Ethics Statement

Animal studies were performed in accordance with the Guidelines for the Care and Use of Laboratory Animals at Chugai Pharmaceutical Co., Ltd. under the approval of the company's Institutional Animal Care and Use Committee. The company is fully accredited by the Association for Assessment and Accreditation of Laboratory Animal Care International (http://www.aaalac.org).

### Generation of Anti-IL-6R Antibodies with Increased Binding Affinity to FcRn at Neutral pH

pH-dependent binding antibody against human soluble IL-6 receptor (hsIL-6R) with neutralizing activity (PH-IgG1) were generated from non-pH-dependent binding antibody (NPH-IgG1) as previously described [16]. pH-dependent binding antibodies against hsIL-6R without neutralizing activity (PHX-IgG1) in BaF/gp130 assay [17] were also generated. To increase the binding affinity to either mouse FcRn (mFcRn) or human FcRn (hFcRn) at neutral pH, various Fc-engineered variants were generated by site-directed mutagenesis of human IgG1. Amino acid substitutions were introduced at the positions 251–258, 286, 288, 307–316, 428 and 433–436 in the EU numbering system which were reported to affect FcRn binding [18–26]. Mutation was comprehensively introduced at each position, and effective mutations identified were combined to generate Fc variants with increased binding affinity to FcRn at neutral pH. More than 1,000 variants were generated and assessed for their binding affinity ($K_D$) to recombinant mFcRn or hFcRn [18] at pH 7.0 using Biacore T200 (GE Healthcare). Each variant was captured onto a Protein L (ACTIgen) immobilized CM4 sensor chip, then FcRn was injected over the flow cell. $K_D$ was determined using Biacore T200 Evaluation Software (GE Healthcare). Fc variants with the desired affinity to FcRn were identified. NPH-IgG1 (conventional antibody with neutralizing activity but without pH-dependent binding), PH-IgG1, PHX-IgG1 and their Fc variants were expressed transiently and purified. NPH-IgG1, PH-IgG1 and PHX-IgG1 were assessed for their $K_D$ to recombinant hsIL-6R at pH 7.4 and pH 6.0 as previously described [16].

### *In vivo* Study of Antibodies in Normal Mice and hFcRn Transgenic Mice Co-injection Model

All animal experiments in this study were performed in accordance with the Guidelines for the Care and Use of Laboratory Animals at Chugai Pharmaceutical Co., Ltd. In co-injection model, C57BL/6J normal mice (Charles River) or hFcRn transgenic mice (hFcRn-Tgm, B6.mFcRn−/−.hFcRn Tg line 276+/+ mouse, Jackson Laboratories) [27] were administered by single i.v. injection with hsIL-6R alone or with hsIL-6R pre-mixed with antibody. The first group received 50 μg/kg hsIL-6R but the other groups additionally received 1 mg/kg of anti-IL-6R antibodies. Total hsIL-6R plasma concentration was determined as previously described [16].

### *In vivo* Study of Antibodies in a Normal Mice hsIL-6R Trans-signaling Model

To evaluate the effect of antibodies on hsIL-6R trans-signaling inhibition *in vivo*, C57BL/6J normal mice were initially i.v. injected with hsIL-6R [16] (250 μg/kg). Then antibodies with designated doses and MR16-1 [28] (15 mg/kg, rat anti-mouse IL-6R antibody) were administered at 2 h after the initial injection. 8 μg/kg of human IL-6 (Toray) was injected at 24 h. Blood samples were collected at 30 h after the initial injection and total hsIL-6R and serum amyloid A (SAA) plasma concentrations were determined as previously described [16].

### *In vivo* Study of Single Doses of Antibodies in Normal Mice and hFcRn Transgenic Mice Steady-state Model

An infusion pump (alzet) filled with 92.8 μg/mL hsIL-6R was implanted under the skin on the back of C57BL/6J normal mice or hFcRn-Tgm (B6.mFcRn−/−.hFcRn Tg line 32+/+ mouse, Jackson Laboratories) [27] to prepare model mice with constant plasma concentration of hsIL-6R. Monoclonal anti-mouse CD4 antibody GK1.5 [29] was administered by i.v. injection to inhibit the production of mouse antibody against hsIL-6R by depleting CD4+ T-cells. Antibodies against hsIL-6R were administered at 1 mg/kg to normal mice or hFcRn-Tgm with or without a single i.v. injection of 1 g/kg of hIgG (Intravenous immunoglobulin, CSL Behring) to mimic endogenous human IgG.

Plasma anti-hsIL-6R antibody concentration in the presence of human IgG was determined using anti-idiotype antibody coated on ELISA 96-well plates, and detected by hsIL-6R, biotinylated anti-hIL-6R antibody (R&D Systems) and Streptavidin-PolyHRP80 (Stereospecific Detection Technologies) using peroxidase substrate. Plasma total hsIL-6R, antibody concentration in the absence of hIgG and pharmacokinetic parameters were determined as previously described [16]. The theoretical free hsIL-6R concentration was calculated from antibody concentration, total hsIL-6R concentration and the $K_D$ of the antibody by equilibrium reaction formula.

### *In vivo* Study of Multiple Doses of Antibodies in hFcRn Transgenic Mice Steady-state Model with High hsIL-6R Concentration

Study was performed as described in the single dose study but with 320 μg/mL hsIL-6R in the pump, and doses were administered to hFcRn-Tgm (B6.mFcRn−/−.hFcRn Tg line 32+/+ mouse, Jackson Laboratories) [27] every other day (except the first dose which was injected together with a single i.v. injection of 1 g/kg of human IgG). Total hsIL-6R plasma concentrations were determined as described above. To determine free hsIL-6R plasma concentration, samples were treated by rProtein A (GE healthcare) to remove antibody and antibody-antigen complex. Because rProtein A treatment requires 10 μL of plasma, samples of n = 3–5 were equally pooled before the treatment. Subsequently, the free hsIL-6R plasma concentrations were determined by the same method as for total hsIL-6R, and hsIL-6R neutralization percentages were obtained by calculating the percentage reduction of free hsIL-6R plasma concentration over control group.

### Pharmacokinetic Analysis and Simulation using Antibody-antigen Dynamic Model

The plasma concentration–time profiles of antibodies and total hsIL-6R obtained in the study of hFcRn-Tgm steady-state model were fitted to an antibody-antigen dynamic model [30] and parameters were optimized for conventional, pH-dependent binding, and v4-type sweeping antibodies. The $k_a$ and $k_d$ values were from surface plasmon resonance (SPR) data. Simulation study was carried out using the obtained pharmacokinetic parameters, and antibody dosages required to neutralize 95% of the antigen (baseline 250 ng/mL (6.6 nM)) at trough by dosing once a month were obtained for each type of antibody with antigen binding affinity ($K_D$) of 0.001, 0.01, 0.1, 1 and 10 nM.

## Results

### Antigen Sweeping by pH-dependent Binding Antibody with Increased FcRn Binding at Neutral pH

In order to evaluate the effect of pH-dependent antigen binding and increased FcRn binding affinity at neutral pH on antigen pharmacokinetics, we used a non-pH-dependent binding antibody against hsIL-6R with hIgG1 constant region (NPH-IgG1) and a pH-dependency-engineered variant (PH-IgG1) (Table S1). hIgG1 has almost no detectable binding to FcRn at pH 7.4, and very weak binding at pH 7.0. In order to increase the binding affinity of PH-IgG1 to either mouse FcRn (mFcRn) or human FcRn (hFcRn) at neutral pH, various hIgG1 Fc variants with mutation(s) in the FcRn binding region (positions 251–258, 286, 288, 307–316, 428 and 433–436 in the EU numbering system) were generated by site-directed mutagenesis. More than 1000 variants were screened by binding to either mFcRn or hFcRn at pH 7.0, and hIgG1 Fc variants for *in vivo* studies were selected (Table 1). The v1 variant was used to evaluate the *in vivo* effect of pH-dependent antigen binding antibody with increased binding affinity to mFcRn at pH 7.0 by administering hsIL-6R to normal mice either on its own or in complex with NPH-IgG1, PH-IgG1 or PH-v1 (Fig. 1A). In this co-injection model, NPH-IgG1 significantly reduced the clearance of hsIL-6R because antigen-antibody complex has lower clearance than the antigen [9,10]; PH-IgG1 increased the clearance of hsIL-6R to some extent, but was still slower than hsIL-6R alone; while PH-v1 accelerated the clearance of hsIL-6R faster than hsIL-6R alone.

Since this co-injection model may not have reflected the actual therapeutic situation where antibody is exposed to plasma in which steady-state baseline concentration of soluble antigen is present, we evaluated antigen sweeping in a mouse model which maintains steady-state plasma antigen concentration. We administered NPH-IgG1, PH-IgG1 and PH-v1 into normal mice steady-state model (Fig. 1B). Consistent with the co-injection model, NPH-IgG1 significantly increased hsIL-6R plasma concentration; PH-IgG1 reduced that increase but an increase over the baseline was still observed; and PH-v1 actively eliminated hsIL-6R from the plasma and reduced the plasma hsIL-6R concentration approximately 150-fold below the baseline, demonstrating that engineered antibody with pH-dependent binding antibody and increased binding affinity to FcRn at neutral pH can eliminate the antigen form plasma *in vivo*.

### Effect of Antigen Sweeping by Sweeping Antibody on Antigen Antagonism *in vivo*

To evaluate the effect of antigen sweeping by sweeping antibody on antigen antagonism *in vivo*, *in vivo* efficacy of the anti-hsIL-6R sweeping antibodies was tested in a normal mouse hsIL-6R trans-signaling model [31], which exhibits an increase in SAA dependent on hIL-6/hsIL-6R-mediated trans-signaling. We generated PHX-IgG1, a pH-dependent binding antibody against hsIL-6R without hsIL-6R neutralizing activity *in vitro* (Table S1), and its Fc variant PHX-v1 with increased binding affinity to mFcRn at neutral pH. In the first study, PHX-IgG1 and PHX-v1 were administered at antibody dosage of 30 mg/kg, and plasma concentration of hsIL-6R and SAA, as a pharmacodynamic marker of hsIL-6R antagonism, are shown (Fig. 2A,B). While PHX-IgG1 could not inhibit SAA production at all, PHX-v1 significantly inhibited SAA production *in vivo* by directly sweeping hsIL-6R from the plasma, despite having no neutralizing activity *in vitro*. In the next study, antibodies with hsIL-6R neutralizing activity, NPH-IgG1, PH-IgG1 and PH-v1, were administered at antibody dosage of 0.03 mg/kg. While NPH-IgG1 and PH-IgG1 with hsIL-6R neutralizing activity *in vitro* could not completely inhibit SAA production at this dosage, PH-v1, with both neutralizing and sweeping activity, completely inhibited SAA production (Fig. 2C,D).

### Antigen Sweeping Requires Both pH-Dependent Antigen Binding and Increased FcRn Binding Affinity at Neutral pH

For clinical application of sweeping antibody, further studies of antigen sweeping were conducted using an hFcRn system. The v2 variant with increased binding affinity to hFcRn at neutral pH was generated (Table 1). As a control for v2 variant, a YTE variant previously reported as improving the half-life [19] with increased binding affinity to hFcRn at acidic pH but not significantly at neutral pH, was used. In a co-injection model, hsIL-6R was administered to hFcRn-Tgm either on its own or in complex with NPH-IgG1, PH-IgG1, PH-YTE and PH-v2 (Fig. 3A). Consistent with the study using normal mice, PH-v2 markedly accelerated the clearance of hsIL-6R faster than hsIL-6R alone in hFcRn-Tgm. On the other hand, PH-YTE exerted slightly slower clearance of hsIL-6R than PH-IgG1.

To further clarify the molecular requirement to achieve antigen sweeping, we administered NPH-IgG1, PH-IgG1, NPH-v2, PH-v2 and PH-v0 to hFcRn-Tgm steady-state model (Fig. 3B). NPH-v2, a non-pH-dependent binding antibody with increased binding affinity to hFcRn at neutral pH, increased hsIL-6R plasma concentration above the baseline to a similar level to NPH-IgG1;

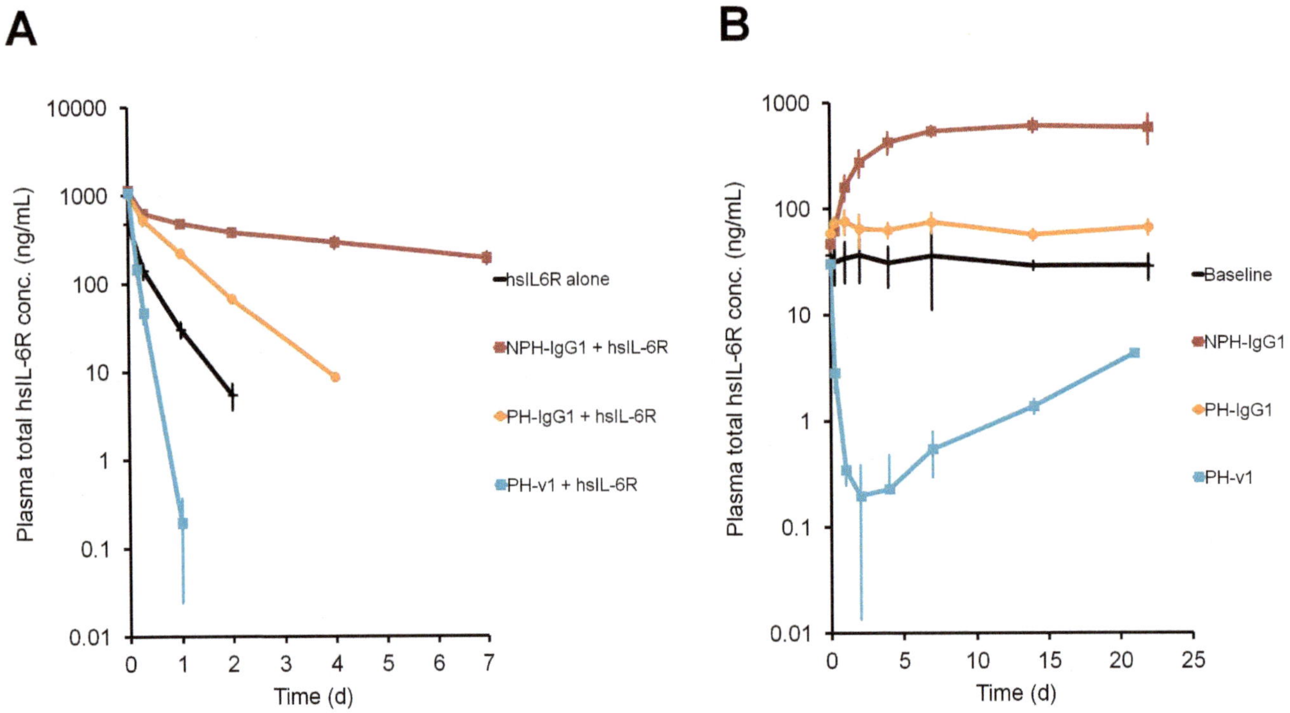


**Figure 1. Antigen sweeping by pH-dependent antigen binding antibody with increased FcRn binding at neutral pH.** *In vivo* study of NPH-IgG1, PH-IgG1 and PH-v1 in normal mice. Effect of antibodies on the total hsIL-6R plasma concentration was evaluated in a co-injection model and a steady-state model. In the co-injection model, hsIL-6R, hsIL-6R+NPH-IgG1, hsIL-6R+PH-IgG1 and hsIL-6R+PH-v1 were intravenously administered as single doses of 50 μg/kg for hsIL-6R and 1 mg/kg for antibody and a time profile of total hsIL-6R plasma concentration (A) is shown. Each data point represents the mean ± s.d. (n = 3 each). In the steady-state model, steady-state plasma concentration of approximately 20 ng/mL hsIL-6R was maintained using an infusion pump filled with hsIL-6R solution, and NPH-IgG1, PH-IgG1 and PH-v1 were intravenously administered as single doses of 1 mg/kg and a time profile of total hsIL-6R plasma concentration (B) is shown. Each data point represents the mean ± s.d. (n = 3 each).

PH-v0, a pH-dependent binding antibody with no hFcRn binding [32], also increased hsIL-6R plasma concentration to a similar level to PH-IgG1, but only transiently; and only PH-v2, a pH-dependent binding antibody with increased binding affinity to hFcRn at neutral pH, actively eliminated hsIL-6R from the plasma. This clearly demonstrates that both pH-dependent antigen binding and increased binding affinity to FcRn at neutral pH are required for antigen sweeping.

## Effect of Endogenous IgG Competition on Antigen Sweeping

Because mouse IgG does not bind to hFcRn [33], hFcRn-Tgm has substantially no endogenous IgG competing with sweeping antibody for hFcRn, which might not reflect the clinical situation in which there is high endogenous human IgG (hIgG) concentration [20]. To evaluate the effect of endogenous IgG on antigen sweeping, NPH-IgG1, PH-IgG1 and PH-v2 alone or together with

**Table 1.** Mutations and FcRn binding affinity of hIgG1 Fc variants.

| Fc variant | $K_D$ (nM) at pH7.0 | | $K_D$ (nM) at pH6.0 | | Mutations |
|---|---|---|---|---|---|
| | mouse FcRn | human FcRn | mouse FcRn | human FcRn | |
| IgG1 | 3918 | 88000 | 237 | 1377 | - |
| v1 | 52 | NT | 3 | NT | I332V/N434Y |
| v2 | NT | 155 | NT | 6 | M252W/N434W |
| v3 | NT | 288 | NT | 15 | M252Y/N434Y |
| v4 | NT | 120 | NT | 8 | M252Y/N286E/N434Y |
| v5 | NT | 77 | NT | 5 | M252Y/T307Q/Q311A/N434Y |
| v6 | NT | 35 | NT | 3 | M252Y/V308P/N434Y |
| v0 | no binding | no binding | no binding | no binding | I253A |

Binding affinity ($K_D$) of IgG1 and v1 to mFcRn at pH 7.0 and pH 6.0, binding affinity ($K_D$) of IgG1, v2-v6 and v0 to hFcRn at pH 7.0 and pH 6.0, and mutations introduced in the Fc region are shown. Mutation sites in the Fc region are described in EU numbering. NT, not tested.

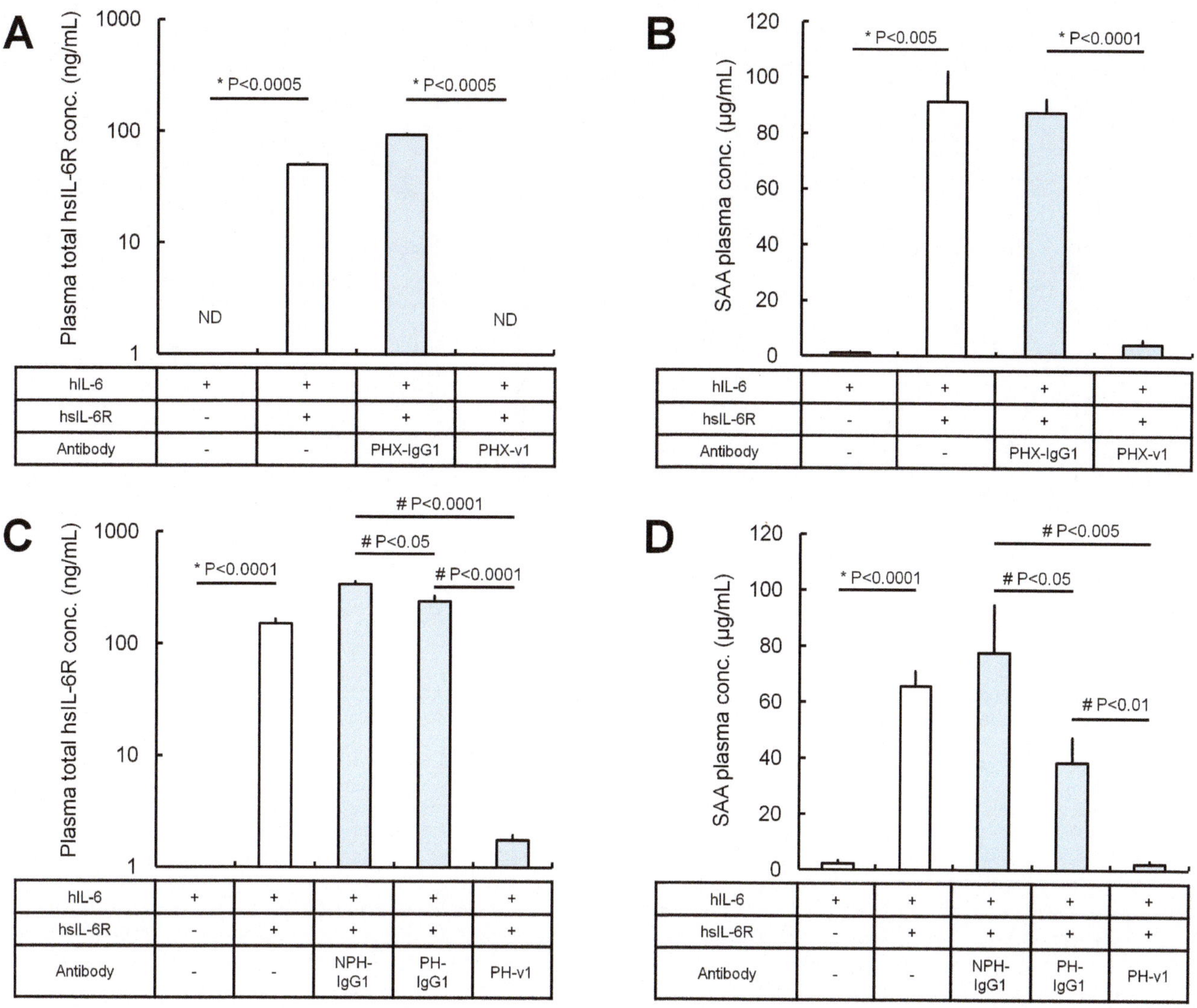


**Figure 2. *In vivo* study of sweeping antibodies in a normal mice hsIL-6R trans-signaling model.** Effect of antibodies on the total hsIL-6R plasma concentration and SAA plasma concentration (as a marker for hsIL-6R antagonism) were evaluated. hsIL-6R was intravenously administered as a single dose of 250 µg/kg. At 2 h, non-neutralizing antibodies PHX-IgG1 and PHX-v1 were intravenously administered as single doses of 30 mg/kg (A, B), and neutralizing antibodies NPH-IgG1, PH-IgG1 and PH-v1 were intravenously administered as single doses of 0.03 mg/kg (C, D). At 24 h, hIL-6 was intravenously administered as a single dose of 8 µg/kg. Total hsIL-6R plasma concentration (A, C) and SAA plasma concentration (B, D) at 30 h is shown. Each data represents the mean ± s.d. for total hsIL-6R plasma concentration and the mean ± s.e. for SAA plasma concentration (n = 3–7 each). ND, not detected (below 0.195 ng/mL). Statistical significance was determined by t-test (*) or Tukey's multiple comparison test (#) for total hsIL-6R and SAA plasma concentration.

1 g/kg of hIgG, which mimics endogenous IgG, were administered to hFcRn-Tgm steady-state model. hsIL-6R sweeping by PH-v2 was attenuated when hIgG as endogenous IgG was present (Fig. S1).

## Effect of hFcRn Binding Affinity at Neutral pH on Antigen Sweeping in Human FcRn Transgenic Mice

Since FcRn binding at neutral pH is required for antigen sweeping, it is assumed that binding affinity ($K_D$) to FcRn at neutral pH would affect the antigen sweeping profile. In addition, previous studies have shown that increasing FcRn binding affinity at neutral pH either increased or did not affect the antibody clearance [21–24]. In order to assess the effect of FcRn binding affinity at neutral pH on antigen sweeping and antibody pharmacokinetics, Fc variants (v3-v6) with various binding affinity to hFcRn at pH 7.0 were generated (Table 1).

Antigen sweeping and antibody pharmacokinetics of NPH-IgG1, PH-IgG1 and its Fc variants were evaluated in hFcRn-Tgm steady-state model in the presence of hIgG (Fig. 4A,B, Table S2). Compared to IgG1, the v3 variant, with $K_D$ 288 nM at pH 7.0, slightly prolonged the antibody pharmacokinetics; moreover, PH-v3 reduced total hsIL-6R plasma concentration to a similar level as baseline concentration. Notably, the v4 variant, with $K_D$ 120 nM at pH 7.0, reduced total hsIL-6R plasma concentration below the baseline level while the antibody pharmacokinetics was maintained. Total hsIL-6R plasma concentration of PH-v4 was 50-fold lower than NPH-IgG1 while the antibody plasma concentration was comparable. The study using the v5 and v6 variants with, respectively, $K_D$ 77 and 35 nM at pH 7.0 has demonstrated that variants with stronger hFcRn binding affinity exhibited more

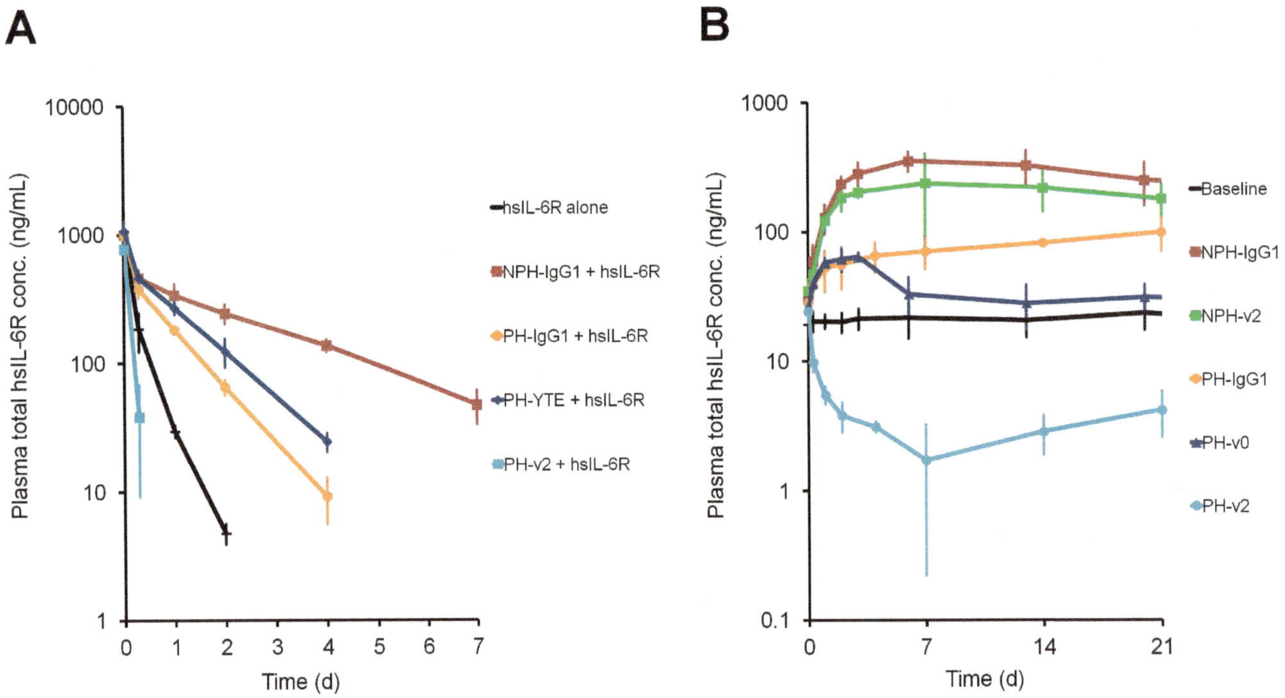


**Figure 3. Characterization of sweeping antibody in hFcRn-Tgm.** (A) *In vivo* study of NPH-IgG1, PH-IgG1, PH-YTE and PH-v2 in hFcRn-Tgm. Effect of antibodies on the total hsIL-6R plasma concentration was evaluated in a co-injection model. hsIL-6R, hsIL-6R+NPH-IgG1, hsIL-6R+PH-IgG1, hsIL-6R+PH-YTE and hsIL-6R+PH-v2 were intravenously administered as single doses of 50 µg/kg for hsIL-6R and 1 mg/kg for antibody and a time profile of total hsIL-6R plasma concentration is shown. Each data point represents the mean ± s.d. (n = 3 each). (B) Effect of pH-dependent antigen binding and increased binding affinity to FcRn at neutral pH on antigen sweeping in hFcRn-Tgm steady-state model with hsIL-6R plasma concentration of approximately 20 ng/mL. NPH-IgG1, NPH-v2, PH-IgG1, PH-v2 and PH-v0 were intravenously administered as single doses of 1 mg/kg. Time profile of total hsIL-6R plasma concentration is shown. Each data point represents the mean ± s.d. (n = 3 each).

extensive antigen sweeping and lower minimum total hsIL-6R plasma concentration, but had increased antibody clearance and a shorter duration of antigen sweeping (faster recovery to baseline). Specifically, the v6 variant reduced total hsIL-6R plasma concentration approximately 1000-fold compared to NPH-IgG1, while the antibody clearance was increased only by 4-fold.

## Sweeping Antibody Antagonizes High Concentration Antigen where Conventional Antibody is Ineffective

The v6-type sweeping antibody with hFcRn binding affinity of 35 nM provided short-lasting but extensive 1000-fold reduction of antigen plasma concentration compared to conventional antibody. To understand its therapeutic advantage, NPH-IgG1, PH-IgG1 and PH-v6 at doses of 0.01 mg/kg were administered to hFcRn-Tgm steady-state model with high plasma hsIL-6R concentration (250 ng/mL) every other day in the presence of hIgG (Fig. 5 A, B). Multiple dosing of NPH-IgG1 and PH-IgG1 achieved no hsIL-6R neutralization throughout the study because molar hsIL-6R concentration was higher than that of antibody (molar antibody concentration was approximately 5-fold lower than the total antigen concentration immediately after the first administration); however, PH-v6 gradually reduced the total hsIL-6R plasma concentration, enabling a neutralization of hsIL-6R at Day 8.

## Discussion

In this study, we have demonstrated that simultaneous engineering of pH-dependent antigen binding and increased FcRn binding affinity at neutral pH actively eliminated the antigen from the plasma, creating "sweeping antibody". Importantly, both pH-dependent antigen binding and increased FcRn binding affinity at neutral pH was required for antigen sweeping, mimicking the function of the ligand-sweeping endocytic receptors that we previously mentioned.

When targeting soluble antigen with monoclonal antibody, conventional antibody (NPH-IgG1) remains bound to the soluble antigen within the acidic endosome (Fig. S2A) and thereby inhibits the antigen degradation by lysosome, resulting in accumulation of the antigen in the plasma. We have recently reported that engineered antibody with pH-dependent antigen binding (PH-IgG1), named recycling antibody, dissociates the soluble antigen in the acidic endosome and the dissociated antigen is then transferred to lysosome and degraded (Fig. S2B) [16]. However, our current study demonstrates that pH-dependent antigen binding alone could not actively eliminate the antigen from plasma. This is because intact IgG1 does not bind to FcRn on the cell surface at neutral pH [25], and the antibody-antigen complex is only marginally taken up into the cell by pinocytosis, which limits the rate of antigen degradation.

Previous studies have demonstrated that Fc-engineering to increase the binding affinity to FcRn at acidic pH improved the endosomal recycling efficiency and prolonged the pharmacokinetics of the antibody [20,25,26]. However, a simultaneous increase of binding affinity at neutral pH did not prolong [21,22], or even shortened [23,24], the pharmacokinetics because of inefficient antibody release from FcRn back to plasma after transporting it back to the cell surface, providing no therapeutic merit. Increasing FcRn binding affinity at neutral pH would anchor the antibody to the cell surface, similarly to an endocytic receptor, and enhance cellular uptake of antibody-antigen complex by FcRn-mediated

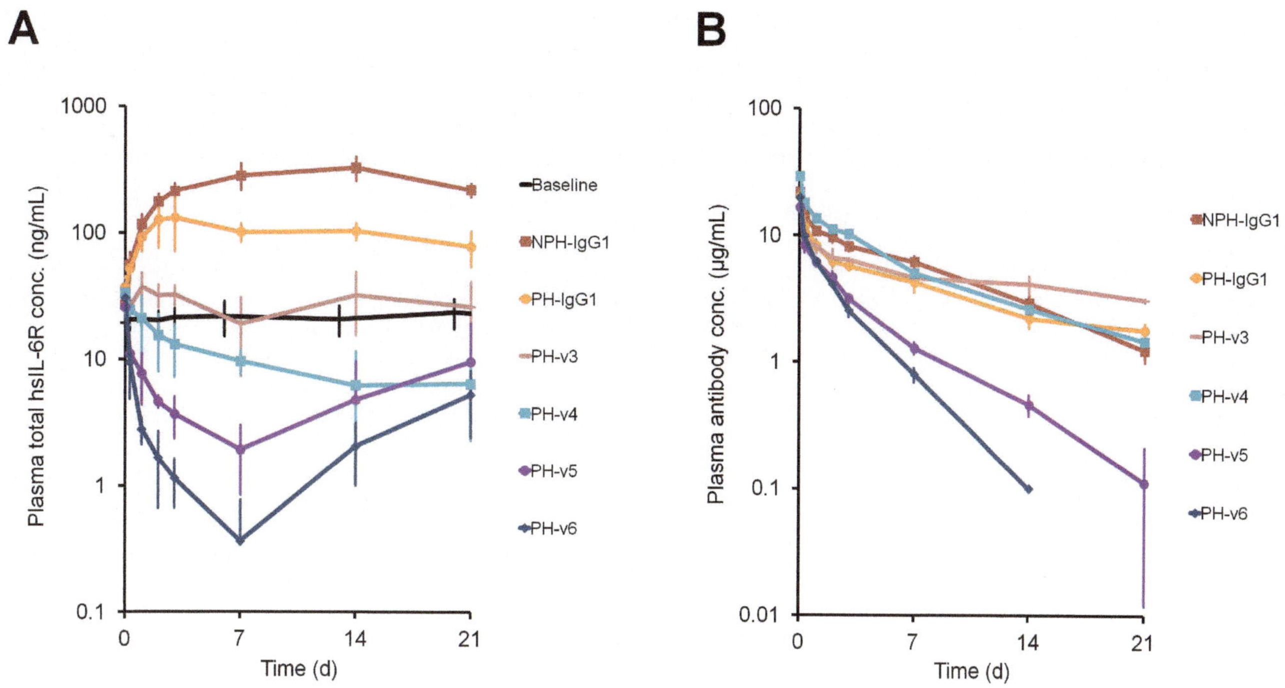


**Figure 4. Effect of hFcRn binding affinity at neutral pH on antigen sweeping profile in hFcRn-Tgm.** (A, B) Effect of FcRn binding affinity at pH 7.0 on antigen sweeping and antibody pharmacokinetics in hFcRn-Tgm steady-state model with hsIL-6R concentration of approximately 20 ng/mL in the presence of human IgG. NPH-IgG1, PH-IgG1, PH-v3, v4, v5 and v6 were intravenously administered as single doses of 1 mg/kg with 1 g/kg of hIgG. Time profiles of total hsIL-6R plasma concentration (A) and antibody plasma concentration (B) are shown. Each data point represents the mean ± s.d. (n = 3–6 each).

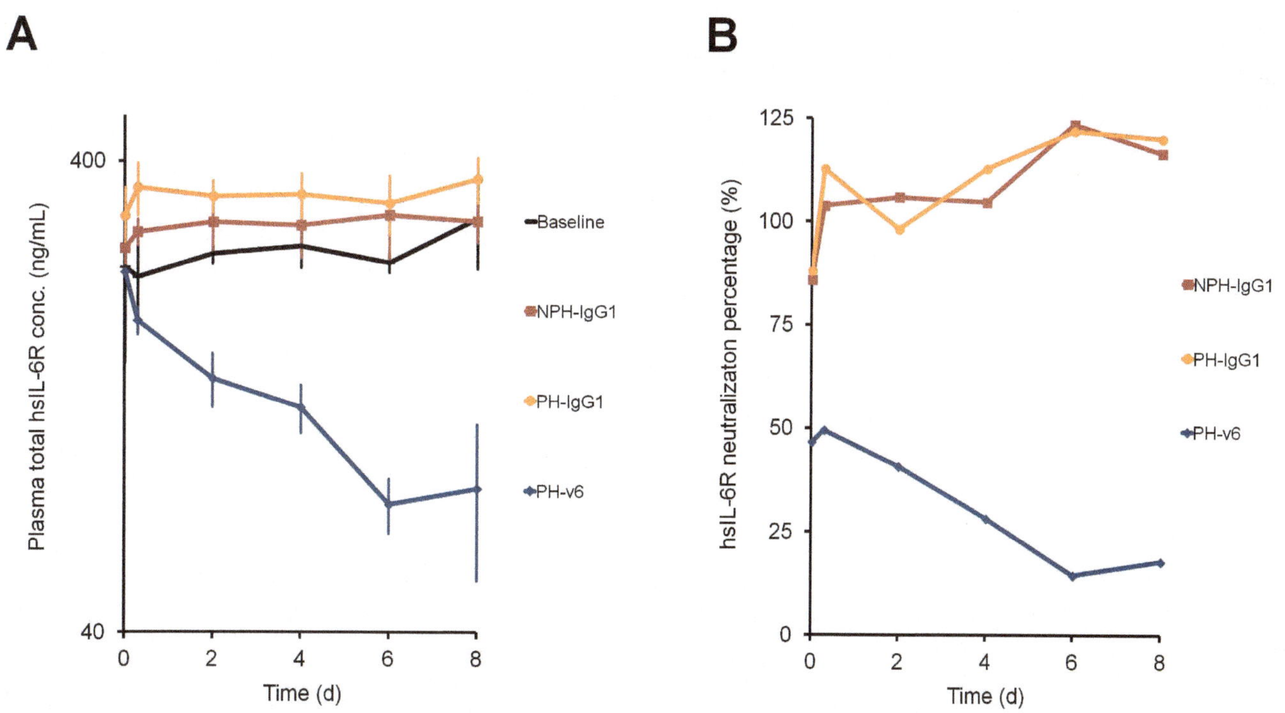


**Figure 5. Effect of sweeping antibody on high plasma concentration antigen.** Effect of NPH-IgG1, PH-IgG1 and PH-v6 on a hFcRn-Tgm steady-state model with high hsIL-6R concentration of approximately 250 ng/mL in the presence of human IgG. NPH-IgG1, PH-IgG1 and PH-v6 were intravenously administered as multiple doses of 0.01 mg/kg every other day. Molar baseline hsIL-6R concentration (6.6 nM) is 5-fold higher than antibody concentration at 15 min (1.3 nM). Time profiles of total hsIL-6R plasma concentration (A) and free hsIL-6R percentage over control (B) are shown. Each data point represents the mean ± s.d. for total hsIL-6R concentration (n = 3–5 each). Free hsIL-6R percentage over control is determined from the pooled plasma sample of n = 3–5 each.

endocytosis. However, our study demonstrated that antibody with increased FcRn binding affinity at neutral pH without pH-dependent antigen binding (NPH-v2) could not actively eliminate the antigen from plasma because the antigen is also recycled back to plasma by FcRn in the antibody-antigen complex.

Combining a pH-dependent antigen binding with increased FcRn binding at only acidic pH (PH-YTE) attenuated rather than accelerated antigen clearance compared to pH-dependent antigen-binding antibody with wild type IgG1 (PH-IgG1), probably because the improved endosomal recycling efficiency [19] also applied to the antigen that remained bound to the antibody. In addition, non-FcRn binding Fc (PH-NB) could not actively eliminate the antigen from plasma because, similar to PH-IgG1, the uptake of antibody-antigen complex into the cell was marginal.

Antigen sweeping was only achieved by the combination of pH-dependent antigen binding with increased FcRn binding affinity at neutral pH (PH-v2). These studies support the following mechanism of sweeping antibody, mimicking the process of rapid ligand sweeping by endocytic receptors [13–15] (Fig. S2C): (i) increasing binding affinity to FcRn at neutral pH anchors the antibody to the cell surface and provides FcRn-mediated cellular uptake of antibody-antigen complex, (ii) pH-dependent dissociation of antibody-antigen complex enables selective degradation of the antigen, (iii) FcRn-mediated recycling of the free antibody to the cell surface enables another round of the cycle. Because this cycle has a rapid turnover rate and FcRn is broadly expressed in the body, sweeping antibody can effectively eliminate the antigen from plasma.

As our results show, this antigen-sweeping activity can be successfully applied either to antibody which has no *in vitro* hsIL-6R neutralizing activity (PHX-v1) to create *in vivo* inhibition of hsIL-6R/hIL-6-mediated trans-signaling or to convert conventional antibody with *in vitro* activity (NPH-IgG1) into sweeping antibody (PH-v1) to further potentiate the *in vivo* signaling inhibition. This study demonstrated that *in vivo* efficacy of sweeping antibody required no *in vitro* biological activity, indicating that sweeping antibody could antagonize multi-epitope antigen (antigen with multiple functional epitope) or toxic antigen with no functional epitope, which cannot be antagonized with a conventional antibody, by directly eliminating the antigen from plasma. Moreover, this study also demonstrated that conventional antibody with biological activity *in vitro* can be further potentiated *in vivo* by engineering the antibody into sweeping antibody, indicating that engineering conventional antibody into sweeping antibody could be an alternative approach to enhancing the therapeutic efficacy of conventional antibody.

Since sweeping antibody would compete with endogenous IgG to bind to FcRn, it was assumed that endogenous IgG would affect the efficiency of antigen sweeping. A similar phenomenon has been reported for antibody-dependent cellular cytotoxicity mediated by Fc gamma receptor binding, which was significantly inhibited by the presence of endogenous hIgG [34]. As expected, antigen sweeping in hFcRn-Tgm was significantly attenuated in the presence of hIgG when hIgG concentration was maintained at an average of 10 mg/mL (reflecting the clinical situation where endogenous hIgG is approximately 10 mg/mL [20]). This demonstrates that endogenous IgG is an important factor in the efficacy of sweeping antibody when considering clinical applications. It has been reported that IgG with increased FcRn binding at neutral pH (Abdeg) accelerates the clearance of endogenous IgG by blocking FcRn [35]. However, we did not observe accelerated clearance of hIgG in the presence of v4-type sweeping antibody (data not shown). Since reported Abdeg (with FcRn binding affinity of 7.4 nM at pH 7.2) accelerated the clearance of endogenous IgG at a dose of approximately 8 mg/kg, it is expected that v4-type sweeping antibody (with FcRn binding affinity of 120 nM at pH 7.0 significantly lower than Abdeg) would not accelerate the clearance of endogenous IgG at a therapeutically relevant dosage, although a high dosage of v6-type sweeping antibody (with stronger FcRn binding affinity of 35 nM at pH 7.0) may have some effect on the clearance of endogenous IgG.

The effect of hFcRn binding affinity at neutral pH on the antigen sweeping profile and antibody pharmacokinetics was investigated by evaluating Fc variants v3 to v6 in hFcRn-Tgm in the presence of hIgG. The results clearly demonstrate that both antigen sweeping and antibody pharmacokinetics depend on hFcRn binding affinity at neutral pH. By increasing the binding affinity at neutral pH, the extent of antigen sweeping (reflected by minimum total hsIL-6R plasma concentration (Table S2)) was enhanced but the duration of antigen sweeping was shortened and antibody clearance was increased. Compared to conventional antibody (NPH-IgG1), all sweeping antibody exhibited stronger reduction of free antigen plasma concentration, which determines the *in vivo* efficacy as a therapeutic antibody, and the extent and the duration of free antigen reduction depended on hFcRn binding affinity (Fig. S3).

Sweeping antibody with moderate hFcRn binding affinity at neutral pH provides moderate but long-acting antigen sweeping. Specifically, compared to conventional antibody (NPH-IgG1), sweeping antibody with hFcRn binding affinity of 120 nM (PH-v4) maintains a similar antibody plasma concentration and provides long-lasting approximately 50-fold reduction of total antigen plasma concentration. Importantly, this demonstrates that the antigen, not the antibody, is selectively eliminated from the plasma. To systematically understand the therapeutic advantage of this v4-type sweeping antibody, modeling and simulation [30] was conducted based on the experimental result of the hFcRn-Tgm study (Fig. S4A, Table S3). The simulation was conducted to calculate the dosage required to neutralize 95% of hsIL-6R (baseline 250 ng/mL) by once-a-month dosing using conventional, pH-dependent antigen binding, and v4-type sweeping antibodies with different binding affinity to hsIL-6R (Fig. S4B). In the simulation study, the dosage of conventional antibody cannot be lowered below 45 mg/kg even with infinite affinity, whereas sweeping antibody with only 0.1 nM affinity can be effective at 1.4 mg/kg. This demonstrates that v4-type sweeping antibody provides more than 30-fold reduction of dosage over conventional antibody even with infinite affinity, a level which can never be achieved with conventional antibody.

On the other hand, sweeping antibody with hFcRn binding affinity below 80 nM provides short-lasting but extensive reduction of antigen plasma concentration compared to conventional antibody. Specifically, sweeping antibody with hFcRn binding affinity of 35 nM (PH-v6) reduces antigen concentration approximately 1000-fold compared to conventional antibody, while the antibody clearance is increased only 4-fold. To understand the therapeutic advantage of v6-type sweeping antibody, antibodies were tested under conditions in which an excess molar amount of antigen was present in plasma, mimicking the therapeutic situation where antigen is present at a high concentration. This excess amount of antigen, which, as expected, conventional antibody (NPH-IgG1) or pH-dependent binding antibody (PH-IgG1) could not antagonize, was antagonized by sweeping antibody (PH-v6) by reducing the plasma antigen concentration below the baseline. This study demonstrated that sweeping antibody could antagonize high concentration antigen against which conventional or pH-dependent antigen binding antibody, even with infinite affinity, would be completely ineffective.

We believe that sweeping antibody, an engineered monoclonal antibody with novel antigen-sweeping activity, provides potential advantages over conventional antibody that can only bind to the antigen and accumulates the antigen in plasma. First, sweeping antibody could be applied to high concentration antigens or antigens with rapid clearance which conventional antibodies, even with infinite affinity, have previously had difficulty in targeting. Second, by directly eliminating the antigen from plasma, sweeping antibody could be applied to antagonize multi-epitope antigen or toxic antigens without functional epitope, which cannot be simply antagonized by a conventional antibody. These two points suggest that sweeping antibody may expand the target antigen space of therapeutic monoclonal antibody to include target antigens which were previously undruggable by conventional monoclonal antibody. Third, sweeping antibody could provide an alternative approach to affinity maturation against the antigen by reducing the plasma antigen concentration to potentiate the efficacy of conventional antibody [36]. Fourth, sweeping antibody could provide a significant advantage over conventional antibody (even assuming infinite affinity) in dosing by enabling the convenience of subcutaneous and less frequent injections, or in manufacturing by reducing the cost. Since changing the binding affinity to hFcRn generates antibodies with different extent and duration of antigen sweeping, antigen-sweeping profiles can be readily customized. We have applied sweeping antibody technology to various antigens such as IL6, IgA, soluble plexin A1, soluble CD4 and other antigens. We have identified pH-dependent antibodies against each of these antigens and engineered them to bind to FcRn at neutral pH. All of these antibodies demonstrated similar antigen sweeping effect that is shown in this study using hsIL-6R (data not shown). These results suggest that sweeping antibody can be broadly applicable to various antigens.

## Supporting Information

**Figure S1 Effect of high concentration hIgG on antigen sweeping in hFcRn-Tgm.** NPH-IgG1, PH-IgG1 and PH-v2 were intravenously administered as single doses of 1 mg/kg either with or without 1 g/kg of hIgG to hFcRn-Tgm with steady-state hsIL-6R concentration of approximately 20 ng/mL. Time profile of total hsIL-6R plasma concentration is shown. Each data point represents the mean ± s.d. (n = 3 each).

**Figure S2 Proposed mode of action of sweeping antibody in comparison with conventional and pH-dependent binding antibody.** (A) Conventional antibody bound to soluble antigen is non-specifically taken up by pinocytosis, and binds to FcRn in acidic endosome. Antibody-antigen complex is recycled back to the cell surface and released from FcRn back to plasma. (B) pH-dependent binding antibody (recycling antibody) bound to soluble antigen is non-specifically taken up by pinocytosis, and binds to FcRn in acidic endosome, while antigen is dissociated from the antibody, transferred into lysosome and degraded. Antibody is recycled back to the cell surface by FcRn, released from FcRn back to plasma and binds to another antigen, allowing single antibody to bind to antigen multiple times. (C) Sweeping antibody bound to soluble antigen is rapidly taken up by FcRn-mediated endocytosis. In acidic endosome, antibody binds to FcRn, and antigen is dissociated from the antibody, transferred into lysosome and degraded. Antibody is recycled back to the cell surface and either released from FcRn back to plasma or stays bound to FcRn on the cell surface to bind to another antigen. Rapid FcRn-mediated uptake allows enhanced lysosomal antigen degradation rate.

**Figure S3 Antigen sweeping profile of antibodies with different hFcRn binding affinity at neutral pH in hFcRn-Tgm.** Effect of FcRn binding affinity at pH 7.0 on antigen sweeping and antibody pharmacokinetics in hFcRn-Tgm with steady-state hsIL-6R concentration of approximately 20 ng/mL in the presence of human IgG. NPH-IgG1, PH-IgG1, PH-v3, v4, v5 and v6 were intravenously administered as single doses of 1 mg/kg with 1 g/kg of hIgG. Theoretical free hsIL-6R plasma concentration was calculated from plasma antibody concentration, total hsIL-6R concentration and binding affinity to hsIL-6R. Time profile of theoretical free hsIL-6R plasma concentration is shown.

**Figure S4 Modeling and simulation of sweeping antibody.** (A) Antibody-antigen dynamic model of sweeping antibody. Antibody is injected intravenously to the central compartment and distributed to the peripheral compartment. Antibody binds to the antigen in the central compartment. Antibody, antigen and antibody-antigen complex are eliminated from the central compartment. Effect of pH-dependent binding and increased binding affinity to FcRn is reflected in the elimination rate of antibody-antigen complex. Parameters used in this model are $k_{synthe}$ (rate constant of antigen synthesis), $C_{antigen,baseline}$ (baseline concentration of antigen), $k_{el,antigen}$ (elimination rate constant of antigen), $Vd1_{antigen}$ (volume of distribution of antigen), $k_{12,antigen}$ (transfer rate constant of antigen from central to peripheral compartment), $k_{21,antigen}$ (transfer rate constant of antigen from peripheral to central compartment), $k_{el,mab}$ (elimination rate constant of antibody), $Vd1_{mab}$ (volume of distribution of antibody), $k_{12,mab}$ (transfer rate constant of antibody from central to peripheral compartment), $k_{21,mab}$ (transfer rate constant of antibody from peripheral to central compartment) and $k_{el,complex}$ (elimination rate constant of antigen in complex with antibody). Note that antibody in complex with antigen is eliminated at the rate of $k_{el,mab}$. (B) Simulation of required dosage to neutralize antigen (baseline concentration 250 ng/mL) by 95% at trough with dosing once a month using conventional antibody (non-pH dependent binding IgG1 antibody), pH-dependent binding IgG1 antibody and v4-type sweeping antibody with different binding affinity to the antigen. Relationship between the antigen binding affinity ($K_D$) and the antibody dosage required to achieve once monthly dosing is shown.

**Table S1 Binding affinity of the anti-IL-6R IgG1 antibodies.**

**Table S2 Antibody clearance and minimum or maximum total hsIL-6R plasma concentration for tested antibodies.**

**Table S3 Fitted pharmacokinetic parameters in antibody-antigen dynamic model.**

## Acknowledgements

We thank colleagues in Chugai Research Institute for Medical Science, Inc, and Chugai Pharmaceutical Co. Ltd., Y. Kawase and H. Tateishi for breeding hFcRn-Tgm; M. Hirabayashi, M. Kinoshita, Yuichiro Ochiai, K. Koguchi, H. Sano, Yuichi Ochiai, T. Sakamoto, M. Kawaharada, T. Yokoyama and T. Nishimura for carrying out *in vivo* studies; M. Fujii, S.

Masujima and A. Takara for antibody generation; and W. Hatakeyama, M. Saito and Y. Okura for carrying out SPR and *in vitro* analysis.

## Author Contributions

Provided direction and guidance: HT K. Hattori. Conceived and designed the experiments: TI. Performed the experiments: AM K. Haraya TT YI KN YH SI. Analyzed the data: K. Haraya TT YI FM. Contributed reagents/materials/analysis tools: NH ST TW. Wrote the paper: TI.

## References


1. Chan AC, Carter PJ (2010) Therapeutic antibodies for autoimmunity and inflammation. Nature Rev Immunol 10: 301–316.
2. Weiner LM, Surana R, Wang S (2010) Monoclonal antibodies, versatile platforms for cancer immunotherapy. Nature Rev Immunol 10: 317–327.
3. Davda JP, Hansen RJ (2010) Properties of a general PK/PD model of antibody-ligand interactions for therapeutic antibodies that bind to soluble endogenous targets. MAbs: 2, 576–588.
4. Haringman JJ, Gerlag DM, Smeets TJ, Baeten D, van den Bosch F, et al. (2006) A randomized controlled trial with an anti-CCL2 (anti-monocyte chemotactic protein 1) monoclonal antibody in patients with rheumatoid arthritis. ARTHRITIS and RHEUMATISM 54: 2387–2392.
5. Xiao JJ, Krzyzanski W, Wang YM, Li H, Rose MJ, et al. (2010) Pharmacokinetics of anti-hepcidin monoclonal antibody Ab 12B9m and hepcidin in cynomolgus monkeys. AAPS J 12: 646–657.
6. Martin PL, Cornacoff J, Prabhakar U, Lohr T, Treacy G, et al. (2005) Preclinical Safety and Immune-Modulating Effects of Therapeutic Monoclonal Antibodies to Interleukin-6 and Tumor Necrosis Factor-alpha in Cynomolgus Macaques. J Immunotoxicol 3: 131–139.
7. Byrd JC, O'Brien S, Flinn IW, Kipps TJ, Weiss M, et al. (2007) Phase 1 study of lumiliximab with detailed pharmacokinetic and pharmacodynamic measurements in patients with relapsed or refractory chronic lymphocytic leukemia. Clin Cancer Res 13: 4448–4455.
8. Jayson GC, Mulatero C, Ranson M, Zweit J, Jackson A, et al. (2005) Phase I investigation of recombinant anti-human vascular endothelial growth factor antibody in patients with advanced cancer. Eur J Cancer 41: 555–563.
9. Finkelman FD, Madden KB, Morris SC, Holmes JM, Boiani N, et al. (1993) Anti-cytokine antibodies as carrier proteins. Prolongation of in vivo effects of exogenous cytokines by injection of cytokine-anti-cytokine antibody complexes. J Immunol 151: 1235–1244.
10. Davis CB, Tobia LP, Kwok DC, Oishi CM, Kheterpal N, et al. (1999) Accumulation of Antibody-Target Complexes and the Pharmacodynamics of Clotting after Single Intravenous Administration of Humanized Anti-Factor IX Monoclonal Antibody to Rats. Drug Delivery 6: 171–179.
11. Zareba KM (2007) Eculizumab, A novel therapy for paroxysmal nocturnal hemoglobinuria. Drugs Today (Barc) 43: 539–46.
12. Rathanaswami P, Roalstad S, Roskos L, Su QJ, Lackie S, et al. (2005) Demonstration of an in vivo generated sub-picomolar affinity fully human monoclonal antibody to interleukin-8. Biochem Biophys Res Commun 334: 1004–1013.
13. Feinberg H, Torgersen D, Drickamer K, Weis WI (2000) Mechanism of pH-dependent N-acetylgalactosamine binding by a functional mimic of the hepatocyte asialoglycoprotein receptor. J Biol Chem 275: 35176–35184.
14. Yamamoto T, Chen HC, Guigard E, Kay CM, Ryan RO (2008) Molecular studies of pH-dependent ligand interactions with the low-density lipoprotein receptor. Biochemistry 47: 11647–11652.
15. French AR, Tadaki DK, Niyogi SK, Lauffenburger DA (1995) Intracellular trafficking of epidermal growth factor family ligands is directly influenced by the pH sensitivity of the receptor/ligand interaction. J Biol Chem 270: 4334–4340.
16. Igawa T, Ishii S, Tachibana T, Maeda A, Higuchi Y, et al. (2010) Antibody recycling by engineered pH-dependent antigen binding improves the duration of antigen neutralization. Nat Biotechnol 28: 1203–1207.
17. Jostock T, Müllberg J, Ozbek S, Atreya R, Blinn G, et al. (2001) Soluble gp130 is the natural inhibitor of soluble interleukin-6 receptor transsignaling responses. Eur J Biochem 268: 160–167.
18. Dall'Acqua WF, Woods RM, Ward ES, Palaszynski SR, Patel NK, et al. (2002) Increasing the affinity of a human IgG1 for the neonatal Fc receptor:biological consequences. J Immunol 169: 5171–5180.
19. Dall'Acqua WF, Kiener PA, Wu H (2006) Properties of human IgG1s engineered for enhanced binding to the neonatal Fc receptor (FcRn). J Biol Chem 281: 23514–23524.
20. Petkova SB, Akilesh S, Sproule TJ, Christianson GJ, Al Khabbaz H, et al. (2006) Enhanced half-life of genetically engineered human IgG1 antibodies in a humanized FcRn mouse model, potential application in humorally mediated autoimmune disease. Int Immunol 18: 1759–1769.
21. Yeung YA, Leabman MK, Marvin JS, Qiu J, Adams CW, et al. (2009) Engineering human IgG1 affinity to human neonatal Fc receptor, impact of affinity improvement on pharmacokinetics in primates. J Immunol 182: 7663–7671.
22. Datta-Mannan A, Witcher DR, Tang Y, Watkins J, Wroblewski VJ (2007) Monoclonal antibody clearance. Impact of modulating the interaction of IgG with the neonatal Fc receptor. J Biol Chem 282: 1709–1717.
23. Deng R, Loyet KM, Lien S, Iyer S, DeForge LE, et al. (2010) Pharmacokinetics of humanized monoclonal anti-tumor necrosis factor-{alpha} antibody and its neonatal Fc receptor variants in mice and cynomolgus monkeys. Drug Metab Dispos 38: 600–605.
24. Dall'Acqua WF, Kiener PA, Wu H (2006) Properties of human IgG1s engineered for enhanced binding to the neonatal Fc receptor (FcRn). J Biol Chem. 281, 23514–23524.
25. Hinton PR, Xiong JM, Johlfs MG, Tang MT, Keller S, et al. (2006) An engineered human IgG1 antibody with longer serum half-life. J Immunol 176: 346–356.
26. Zalevsky J, Chamberlain AK, Horton HM, Karki S, Leung IW, et al. (2010) Enhanced antibody half-life improves in vivo activity. Nat Biotechnol 28: 157–159.
27. Roopenian DC, Christianson GJ, Sproule TJ (2010) Human FcRn transgenic mice for pharmacokinetic evaluation of therapeutic antibodies. Methods Mol Biol 602: 93–104.
28. Okazaki M, Yamada Y, Nishimoto N, Yoshizaki K, Mihara M (2002) Characterization of anti-mouse interleukin-6 receptor antibody. Immunol Lett 84: 231–240.
29. Rashid A, Auchincloss H Jr, Sharon J (1992) Comparison of GK1.5 and chimeric rat/mouse GK1.5 anti-CD4 antibodies for prolongation of skin allograft survival and suppression of alloantibody production in mice. J Immunol 148: 1382–1388.
30. Betts AM, Clark TH, Yang J, Treadway JL, Li M, et al. (2010) The application of target information and preclinical pharmacokinetic/pharmacodynamic modeling in predicting clinical doses of a Dickkopf-1 antibody for osteoporosis. J Pharmacol Exp Ther 333: 2–13.
31. Ozbek S, Peters M, Breuhahn K, Mann A, Blessing M, et al. (2001) The designer cytokine hyper-IL-6 mediates growth inhibition and GM-CSF-dependent rejection of B16 melanoma cells. Oncogene 20: 972–979.
32. Deng R, Meng YG, Hoyte K, Lutman J, Lu Y, et al. (2012) Subcutaneous bioavailability of therapeutic antibodies as a function of FcRn binding affinity in mice. MAbs 4: 101–109.
33. Ober RJ, Radu CG, Ghetie V, Ward ES (2001) Differences in promiscuity for antibody-FcRn interactions across species, implications for therapeutic antibodies. Int Immunol 13: 1551–1559.
34. Preithner S, Elm S, Lippold S, Locher M, Wolf A, et al. (2006) High concentrations of therapeutic IgG1 antibodies are needed to compensate for inhibition of antibody-dependent cellular cytotoxicity by excess endogenous immunoglobulin G. Mol Immunol 43: 1183–1193.
35. Vaccaro C, Zhou J, Ober RJ, Ward ES (2005) Engineering the Fc region of immunoglobulin G to modulate in vivo antibody levels. Nat Biotechnol 23: 1283–1288.
36. Finch DK, Sleeman MA, Moisan J, Ferraro F, Botterell S, et al. (2011) Whole-molecule antibody engineering: generation of a high-affinity anti-IL-6 antibody with extended pharmacokinetics. J Mol Biol 411: 791–807.

# 20

# Fitness Loss and Library Size Determination in Saturation Mutagenesis

**Yuval Nov***

Department of Statistics, University of Haifa, Haifa, Israel

## Abstract

Saturation mutagenesis is a widely used directed evolution technique, in which a large number of protein variants, each having random amino acids in certain predetermined positions, are screened in order to discover high-fitness variants among them. Several metrics for determining the library size (the number of variants screened) have been suggested in the literature, but none of them incorporates the actual fitness of the variants discovered in the experiment. We present the results of an extensive simulation study, which is based on probabilistic models for protein fitness landscape, and which investigates how the result of a saturation mutagenesis experiment – the fitness of the best variant discovered – varies as a function of the library size. In particular, we study the loss of fitness in the experiment: the difference between the fitness of the best variant discovered, and the fitness of the best variant in variant space. Our results are that the existing criteria for determining the library size are conservative, so smaller libraries are often satisfactory. Reducing the library size can save labor, time, and expenses in the laboratory.

**Editor:** Alexandre G. de Brevern, UMR-S665, INSERM, Université Paris Diderot, INTS, France

**Funding:** The author has no support or funding to report.

**Competing Interests:** The author has declared that no competing interests exist.

* E-mail: yuval@stat.haifa.ac.il

## Introduction

Directed evolution has emerged as an indispensable vehicle for engineering biocatalysts, therapeutics, and other protein-based tools. Through iterative rounds of mutagenesis and selection that mimic natural evolution, directed evolution "breeds" proteins with improved or novel chemical properties [1]. A key concept in evolutionary biology is that of *fitness* – the capacity of an organism to survive and reproduce. This meaning of the term originates from the phrase "survival of the fittest," coined by Herbert Spencer and adopted by Charles Darwin. John Maynard Smith extended the notion of the so-called "fitness landscape" of entire organisms to the molecular world of proteins [2], and more recent writers have used the term "fitness" in the context of protein engineering, to denote any desirable protein property that is subject to optimization [3–5]; we follow this usage below.

Though the very essence of directed evolution protocols depends on screening large numbers of mutants, recent works increasingly advocate relatively small, yet carefully designed libraries. Such targeted libraries enjoy a significantly higher fraction of functional mutants, compared to completely random libraries, and thus achieve remarkable experimental results. Various principles are used in the design of targeted libraries, including reduced amino-acid alphabets [6], inference on ancestral sequences [7], and combinatorial spiking [8]; for a recent review, see Goldsmith and Tawfik [9]. In this work we provide another argument in favor of smaller libraries – one which is related to an experimental procedure called saturation mutagenesis.

Saturation mutagenesis is a simple *in vitro* directed evolution technique, whereby the amino acids at certain preselected positions along a protein sequence are replaced by random ones, in a manner called *randomization*. A large number of the resulting mutants are generated to form a library, and are then screened, in the hope of discovering among them a highly improved variant of the wildtype protein. Saturation mutagenesis has been used successfully to improve a host of desirable protein properties, including thermostability, specificity, catalytic activity, enantioselectivity, binding affinity, and photostability [10–14]. The method can also be used iteratively, to probe efficiently larger portions of protein space [15].

All randomization schemes for saturation mutagenesis originate at the DNA level. Most commonly, the three wild-type nucleotides at each randomized codon are replaced with random ones. If all four nucleotides are allowed at each of the codon's three positions – a randomization scheme known as NNN – then all $4^3=64$ possible codons may be formed. Other randomization schemes use reduced codon sets, and include NNB, NNS, and NNK (where N = A/C/G/T, B = C/G/T, S = C/G, K = G/T), which are all superior to NNN randomization as they induce a lower probability of introducing a premature stop codon, as well as a more uniform probability distribution across the twenty amino acids [16]. Still better schemes eliminate completely, or almost completely, both codon bias and the stop codon problem; this can be achieved either by replacing entire wild-type codons by random ones [17], or as shown recently, by using an appropriately weighted mixture of primers [18,19].

Regardless of the chosen randomization scheme, the larger the library, the higher the likelihood of exploring more distinct protein variants at the screening stage, and hence the higher the probability of discovering a novel, high-fitness variant. On the other hand, generating and screening large libraries is laborious, time consuming, and expensive. To balance judiciously the benefits and drawbacks associated with the library size, one needs a *metric* – a performance measure that can be calculated before the actual fitness of the library's variants is observed – and understand how this metric depends on the library size.

Two well known such metrics are: (i) the probability that all possible variants are represented in the library ("probability of full coverage"); and (ii) the expected percentage of all possible variants that are represented in the library ("expected coverage"). Exact and approximate mathematical expressions for these two metrics can be found in [20–22].

Recently, another metric has been proposed: the probability that the library contains at least one of the $k$ top-performing variants (in terms of their fitness), for some small integer $k$ [23]. The probability that the library contains the best variant – i.e., the case $k=1$ – is clearly a meaningful metric; under very mild assumptions on the distribution of fitness, this case can be shown to be the same as metric (i) above. The rationale behind this metric for $k \geq 2$ is that often, especially when the number of all possible distinct variants is very large, there are several variants whose fitness satisfies the design requirements, and discovering any of these top-performing variants will be considered a success. Using a metric of the latter type, even with $k$ as low as 2, allows the designer to reduce significantly the library size, and thus to save laboratory resources and labor.

However, none of the above metrics incorporates directly the actual fitness of the variants to be discovered in the process, which is arguably the most relevant metric. The goal of this work is to fill this gap, and to study *in silico* how the fitness of the best variant discovered in a saturation mutagenesis experiment varies as a function of the library size. We wish to emphasize that it is not the intention of this work that protein engineering experimentalists should repeat themselves any of the computational or mathematical procedures described below (running simulations, estimating parameters, etc.); rather, we aim to derive general guidelines that are applicable to a wide range of saturation mutagenesis experiments, and that can be used by those practitioners.

To attain this study's goal, one needs a model for "protein landscape" – the relationship between a protein's sequence and fitness. This relationship is notoriously complex, and indeed, several attempts have been made to model it mathematically. We use for this purpose the probabilistic approach, i.e., view the sequence–fitness relationship as a realization of a *random* phenomenon, so that the fitness of each variant, before its exact value is revealed in an experiment, is only known to have a certain distribution. Among the probabilistic models for the sequence–fitness relationship are Kauffman and Weinberger's *NK* model [24], Aita and Husimi's "rough Mount Fuji" model [25], the model of Nov and Wein [26], and machine-learning techniques such as Fox et al.'s ProSAR [27].

Our main modeling vehicle in this work is the aforementioned model of Nov and Wein; this model was devised specifically for protein design purposes, and was used successfully to improve the reduction activity of the *E. coli* enzyme ChrR [28], and the oxidizing activity of the toluene-4-monooxygenase (T4MO) enzyme [29]. To ground our simulation results in experimental data, we use below the parameters estimated in these studies, as well as those estimated from another empirical work [30]. For further support of our conclusions and as a control, we also use the *NK* model and the "rough Mount Fuji" model.

## Methods

### Probability

Let the randomized protein positions be labeled $i=1,\ldots, M$, and let $Q_{a,i}$ be the probability that the randomization will result in amino acid $a$ at position $i$. The probabilities $Q_{a,i}$ are induced by the randomization scheme and the genetic code: for example, under NNN randomization at position $i$, since 4 out of the 64 codons encode alanine, $Q_{\text{Ala},i}=4/64$. The probability that the randomization will result in a nonsense mutation (a premature stop codon) at the codon corresponding to amino acid $i$ is $Q_{\text{stp},i}=1-\sum_a Q_{a,i}$; depending on the randomization scheme, this probability may be zero or positive.

Denote by $A_i$ the number of amino acids with positive probability to appear at position $i$, i.e., $A_i=|\{a : Q_{a,i}>0\}|$. Note that $A_i<20$ in some randomization schemes [6,31]. We use the term *variant space* to denote the set of all possible distinct protein variants that may be formed in the experiment; the number of elements in this space is $n=A_1A_2\ldots A_M$. Assuming that the randomizations across the $M$ positions are independent of each other, the probability that a random variant will have amino acid $a_1$ in randomized position 1, amino acid $a_2$ in randomized position 2, etc., is $Q_{a_1,1}Q_{a_2,2}\cdots Q_{a_M,M}$. Let $p_1,\ldots, p_n$ be an enumeration of these product probabilities, corresponding to the $n$ distinct variants in variant space. The probability that a random variant will have at least one nonsense mutation, and will thus be completely dysfunctional, is $p_{\text{stp}}=1-\Pi_{i=1}^{M}(1-Q_{\text{stp},i})$.

Let $L$ be the library size, i.e., the number of protein variants whose fitness is measured, let $X_v$, $v=1,\ldots, n$, be the number of copies of variant $v$ that are included in the library, and let $X_{\text{stp}}$ be the number of variants with a nonsense mutation. Then the random vector $(X_1,\ldots, X_n, X_{\text{stp}})$ has a multinomial distribution with parameters $L, p_1,\ldots, p_n, p_{\text{stp}}$. Denote by $V$ the (random) subset of variant space that is represented in the library: $V=\{v : X_v\geq 1\}$. In this notation, the probability of full coverage is $P(|V|=n)$ and the expected coverage is $E(|V|/n)$.

Let $F_v$ be the fitness of variant $v$. We denote by $v^*$ the variant with the highest fitness discovered in the experiment (i.e., the best among those that happened to be represented in the library), and by $v^{**}$ the globally fittest variant in variant space (i.e., the best among all distinct variants that could possibly be generated in the experiment). Formally,

$$F_{v*}=\max_{v\in V} F_v,\ F_{v**}=\max\{F_1,\ldots,F_n\}, \qquad (1)$$

There are two sources of randomness contributing to the variability of $F_{v*}$. The first is the aforementioned randomness of the set $V$, which is *experimental* in origin, i.e., stems from the randomness of the set of variants chosen by the experimenter to constitute the library. The second is the randomness of the mapping $v\longmapsto F_v$ (the protein landscape), which is *natural* in origin: even when the sequence of a variant is known, its fitness is determined by the laws of chemistry in such an intricate way, that we resort to model it probabilistically. Only the second type of randomness plays a role in the randomness of $F_{v**}$. While the first type of randomness requires only elementary probability to model and analyze, the second, as stated earlier, is more intricate and challenging to study. Three approaches to modeling the relationship between sequence and fitness are outlined in the next subsection.

Assuming that fitness is a continuous characteristic, so that there are no fitness ties, and assuming that no variant may be considered *a priori* to be better than another, the distribution of the fitness ranking (from best to worst) across variant space is uniform over all $n!$ permutations of its elements. Let $T_k$ be the event "the library contains at least one of the top $k$ variants in variant space," and let $L_k^\alpha$ be the minimal library size required to ensure a probability $\alpha$ of discovering at least one of the top $k$ variants, i.e.,

$L_k^\alpha = \min\{L : P(T_k) \geq \alpha\}$. See [23] for methods to compute $P(T_k)$ and $L_k^\alpha$.

## Sequence–fitness Models

**The model of nov and wein.** The model of Nov and Wein [26] aims to capture the following three characteristics of the sequence–fitness relationship:

- **Wild-type dominance**. Mutations in the wild type are more likely to decrease fitness, rather than to increase it, and the more mutations a variant has, the lower is its resulting fitness, on average [32–35].
- **Clustering**. Given a favorable mutation at a certain position, other mutations at this position are more likely to be favorable as well. Indeed, the basic rationale behind the saturation mutagenesis approach is that certain "hotspots" are more likely than others to accommodate favorable mutations [36].
- **Additivity**. The change in fitness due to multiple mutations equals approximately to the sum of the corresponding individual fitness changes [37–39]. Importantly, additivity may be weak when measured using the raw fitness values, but can become stronger if the measurements are transformed to a different scale; see below.

In this model, the raw fitness values are transformed so that the wild type, denoted by $\hat{v}$, has zero fitness: $F_{\hat{v}} = 0$. The fitness of any other variant $v \neq \hat{v}$ is viewed as a realization of a random variable $F_v$, and the collection of the random variables $(F_1, \ldots, F_n)$ is a realization of a random vector whose distribution is governed, in a manner described below, by four parameters: the drift $m$, the position variance $\sigma_P^2$, the amino-acid variance $\sigma_A^2$, and the non-additivity variance $\sigma_\varepsilon^2$.

A mutation at position $i$, which replaces the wild-type amino acid $\hat{a}_i$ with another amino acid $a$, contributes a random quantity $f_{i,a} \sim N(\mu_i, \sigma_A^2)$ to the fitness of $v$. The $M$ means $\mu_1, \ldots, \mu_M$ are not constants, but rather, independent random variables themselves (in a way resembling the random effect model from statistics), having a $N(m, \sigma_P^2)$ distribution, where $m < 0$. The fitness of a variant $v$ that has amino acid $a_1$ in position 1, amino acid $a_2$ in position 2, etc., is then given by

$$F_v = \sum_{i : a_i \neq \hat{a}_i} f_{i,a_i} + \varepsilon_v, \quad v = 1, \ldots, n \tag{2}$$

where $\varepsilon_1, \ldots, \varepsilon_n$ are $N(0, \sigma_\varepsilon^2)$ random variables, independent of each other, of the $f_{i,a}$'s, and of the $\mu_i$'s.

By setting the drift $m$ to be negative, the model captures the "wild-type dominance" characteristic. By having the 19 random variables $\{f_{i,a}, a \neq \hat{a}_i\}$ at each position $i$ share a common random mean $\mu_i$, we establish a positive correlation among them, which captures the clustering characteristic. Finally, the additive structure of (2) captures the additivity characteristic, with $\sigma_\varepsilon^2$ determining its magnitude.

It can be shown that the resulting collection of random fitness variables $(F_1, \ldots, F_n)$ is a multivariate Gaussian process, and that for two variants $v$ and $v'$, having respectively the amino acids $a_1, a_2, \ldots, a_M$ and $a'_1, a'_2, \ldots, a'_M$ at the randomized positions, this process satisfies

$$E(F_v) = d(v, \hat{v})m$$

$$\mathrm{Var}(F_v) = d(v, \hat{v})(\sigma_P^2 + \sigma_A^2) + \sigma_\varepsilon^2 \tag{3}$$

$$\mathrm{Cov}(F_v, F_{v'}) = \sigma_P^2 U_1(v, v') + \sigma_A^2 U_2(v, v'),$$

where $d(v, \hat{v}) = |\{i : a_i \neq \hat{a}_i\}|$ is the Hamming distance between $v$ and $\hat{v}$ (the number of positions in which $v$ and the wild type differ); $U_1(v, \hat{v}) = |\{i : a_i, a'_i \neq \hat{a}_i\}|$ (the number of positions in which both $v$ and $v'$ differ from the wild type); and $U_2(v, \hat{v}) = |\{i : a_i = a'_i \neq \hat{a}_i\}|$ (the number of positions in which $v$ and $v'$ have the same mutation).

For transforming the raw fitness values into the values $F_v$, we use in this work mainly the logarithmic transformation (relative to the wild type), in which a raw fitness $x$ is transformed by $\log(x/\hat{x})$, where $\hat{x}$ is the wild-type's raw fitness. The scaling by $1/\hat{x}$ guarantees that $F_{\hat{v}} = 0$, as required by the model. This logarithmic transformation was used in the two empirical studies involving the model of Nov and Wein [28,29].

The four parameters of the model used below were estimated using the maximum likelihood approach: Given a vector of fitness observations $\mathbf{F} = (F_{v_1}, F_{v_2}, \ldots, F_{v_N})$ for $N$ variants $v_1, v_2, \ldots, v_N$, the estimated parameters are those that maximize the likelihood of $\mathbf{F}$, i.e., that solve the maximization problem

$$\max \frac{1}{(2\pi)^{N/2}|\boldsymbol{\Sigma}|^{1/2}} \exp\left\{ -\frac{1}{2}(\mathbf{F} - \boldsymbol{\mu})^T \boldsymbol{\Sigma}^{-1} (\mathbf{F} - \boldsymbol{\mu}) \right\},$$

where $\boldsymbol{\mu} = \boldsymbol{\mu}(m)$ and $\boldsymbol{\Sigma} = \boldsymbol{\Sigma}(\sigma_P^2, \sigma_A^2, \sigma_\varepsilon^2)$ are the mean vector and covariance matrix of $\mathbf{F}$, respectively, whose entries satisfy equation (3). See [26] for more details about this estimation techniques, and [28,29] for the method used to address sampling bias problems.

## The *NK* Model

The *NK* model, introduced by Kauffman and Weinberger [24], is a spin-glass-like model of random epistatic interactions, conceived to study evolutionary adaptive walks in protein space. According to this model (and using its original notation, which is different from that used in the previous subsection), the total fitness of a protein variant is the average $\frac{1}{N}\sum_{i=1}^{N} W_i$ of the fitness contributions $W_i$ from the protein's $N$ positions. The fitness contribution of each position is a $U(0, 1)$ random variable that depends not only on the amino acid at that position, but also on the amino acids at $K < N$ other positions.

The value of $K$ determines the ruggedness of the protein landscape: When $K = 0$, the fitness contributions from the $N$ positions are independent of each other, there is a strong positive correlation between the fitness of neighboring variants (i.e., variants whose sequences differ by a single amino acid), and the resulting smooth landscape contains a single peak. At the other extreme, when $K = N - 1$, the fitness values of all $20N$ variants in protein space are independent of each other, resulting in a highly rugged landscape with many local peaks.

The $K$ positions influencing the fitness contribution of each position $i$ may be either randomly chosen, or the nearest $K$ positions flanking position $i$ in the primary sequence. We follow the simulations of Kauffman and Weinberger, and use the latter option.

## The Rough Mount Fuji Model

According to the "rough Mount Fuji" model of Aita and Husimi [26], and using its original notation, the fitness of a variant having sequence P is the sum $W_P$+$\omega_P$, where $W_P$ is an additive term and $\omega_P$ is a non-additive term. The non-additive term is modeled as a $\mathcal{N}(0, \varrho^2)$ random variable, and the additive term is the sum of the fitness contributions from the protein's $v$ positions: $W_P = \sum_{j=1}^{v} w_j(\alpha_{P_j})$, where $\alpha_{P_j}$ is the $j$th amino acid in the sequence P, and the fitness contribution of amino acid $\alpha_i$ at position $j$ is $w_j(\alpha_i) = \varepsilon i/10$, $i = 0, 1,\ldots, 19$. The parameter $\varepsilon < 0$ is the mean fitness contribution resulting from placing an "incorrect" amino acid (i.e., $i \neq 0$) at each position.

By placing the "correct" amino acid ($i = 0$) at each of the $v$ positions, the additive term $W_P$ attains its maximal value, zero. However, the final fitness of a variant having only "correct" amino acids (i.e., with $\alpha_{P_j} = \alpha_0$ for all $j$) may not be the global optimum, because of the non-additive terms $\omega_P$. The ratio $|\varepsilon|/\varrho$ is a measure of smoothness of the fitness landscape, with low values corresponding to a rugged landscape and high values corresponding to a smooth one.

## Computation and Simulation

The $L_k^\alpha$ values and the "full coverage" probabilities were computed by a C computer program called TopLib; see stat.haifa.ac.il/~yuval/toplib for documentation. The simulations involving the sequence–fitness models were coded in the R programming language (The R Foundation for Statistical Computing, www.r-project.org); each simulation was based on 5000 replications. When using the model of Nov and Wein, the simulated landscapes were initially generated following equation (2) and then exponentially transformed, so that the resulting values need to be logarithmically transformed to fit the model.

# Results

The goal of this work is to study the relationship between the library size $L$ and the loss of fitness in a saturation mutagenesis experiment, i.e., the difference between the fitness of the globally fittest variant, $F_{v**}$, and the fitness of the best variant discovered in the experiment, $F_{v*}$ (see (1)).

Both $F_{v*}$ and $F_{v**}$ depend on the sequence–fitness model used, and in particular, on the model's parameters; $F_{v*}$ (but not $F_{v**}$) depends also on the randomization scheme (NNN, NNK, etc.). We now explore, via computer simulations, how the loss of fitness depends on $L$ under the various models and parameters.

## Studying Loss of Fitness

We first study how $E(F_{v*})$ converges to $E(F_{v**})$ as $L$ increases. Figure 1 depicts this convergence under NNK randomization of $M = 1$, 2 and 3 positions; the underlying probabilistic model is that of Nov and Wein, with the parameters estimated in the T4MO study [29]: the variance parameters are $\sigma_P^2 = 0.38$, $\sigma_A^2 = 0.19$, $\sigma_\varepsilon^2 = 0.04$, and the drift parameter was taken to be the median of the values used in the study, $m = -0.3$. Still in accordance with that study, it is assumed that the raw activity values need to be logarithmically transformed in order to meet the model's probabilistic assumptions.

In each of the three panels of Figure 1, the horizontal upper dashed line intersects the vertical axis at $E(F_{v**})$, and thus serves as an upper bound for the curve of $E(F_{v*})$. The three dotted lines correspond to $L_1^{0.95}$, $L_2^{0.95}$, and $L_3^{0.95}$, i.e., to the library sizes required to ensure a 0.95 probability of discovering the best variant (rightmost line), any of the top two variants (middle), and any of the top three (left). The solid line corresponds to a library size which we denote by $\tilde{L}_{0.95}$: the size required to reach 95% of the distance from the expected fitness of a single randomized variant (i.e., the lowest point of the curve, corresponding to $L = 1$) to $E(F_{v**})$. This $\tilde{L}_{0.95}$ serves as a rough indication for a reasonable library size, which nearly exhausts, on average, the fitness that may be extracted from the fitness landscape.

It is evident that as $L$ increases, the difference between $E(F_{v*})$ and $E(F_{v**})$ becomes negligible. For example, when randomizing $M = 2$ positions (middle panel of Figure 1), a library size of about 1,000 appears to be satisfactory: When $L = L_1^{0.95} = 2130$ variants, we have $E(F_{v*}) = 5.03$, a loss of about 0.8% relative to $E(F_{v**}) = 5.07$, so the requirement $P(T_1) = 0.95$ seems overly stringent. When $L = L_2^{0.95} = 875$ or $L = L_3^{0.95} = 533$, we have $E(F_{v*}) = 4.88$ (3.7% loss) or 4.72 (6.9% loss), respectively. When $L = \tilde{L}_{0.95} = 842$, we have $E(F_{v*}) = 4.86$ (4.1% loss), which indeed appears reasonable, though it is up to the experiment designer to decide what magnitude of loss is considered negligible. The library size required to achieve a 0.95 probability of "full coverage" (metric (i) mentioned in the Introduction) is 8,128 – almost an order of magnitude more than the ~1,000 that appear to suffice, and is therefore extremely wasteful.

Figure 1 also shows that as expected, $\tilde{L}_{0.95}$ increases sharply in absolute numbers as the number of randomized positions increases, from 50 to 842 to 15,562, corresponding to $M = 1$, 2, 3, respectively. However, $\tilde{L}_{0.95}$ *decreases* relative to the landmarks $L_k^{0.95}$: when $M = 1$, $\tilde{L}_{0.95}$ is about half way between $L_1^{0.95}$ and $L_2^{0.95}$; when $M = 2$, $\tilde{L}_{0.95}$ is close to $L_2^{0.95}$; and when $M = 3$, $\tilde{L}_{0.95}$ is about half way between $L_2^{0.95}$ and $L_3^{0.95}$.

For the remainder of this work we fix the number of randomized positions at $M = 2$, and study how the choice of model and its parameters influence the difference between $F_{v*}$ and $F_{v**}$, as a function of $L$.

The only other study in which all four parameters of the model of Nov and Wein were estimated is [26]. That study analyzed data which was used to improve the binding affinity of an $\alpha_v\beta_3$-specific humanized monoclonal antibody called Vitaxin [30]. The estimated parameters are $m = -1.35$, $\sigma_P^2 = 0.25$, $\sigma_A^2 = 0.18$, $\sigma_\varepsilon^2 = 0.14$, and the raw data were logarithmically transformed. The left panel of Figure 2 shows again the convergence of $E(F_{v*})$ to $E(F_{v**})$ when randomizing two NNK positions, this time under the Vitaxin parameters. In this case, $\tilde{L}_{0.95} =$ 932, slightly above $L_2^{0.95} =$ 875. The main factor contributing tothis moderate increase in $\tilde{L}_{0.95}$, relative to the corresponding value under the T4MO parameters (middlepanel of Figure 1), is the higher proportion of the non-additivity variance from the total fitness variance:when randomizing two positions, this proportion is $L_2^{0.95} =$ 875. The main factor contributing tothis moderate increase in $\sigma_\varepsilon^2/[2(\sigma_P^2 + \sigma_A^2) + \sigma_\varepsilon^2] = 0.14$ (see equation (3)) inthe Vitaxin parameters, whereas it is only 0.034 in the T4MO parameters. To demonstrate the influenceof high non-additivity variance, we repeated the T4MO simulation under the (unrealistic and extreme) parameterization in which this proportion is 1, without changing the marginal distribution of the fitnessvalues. This was done by keeping $m = -0.3$, setting $\sigma_P^2 = \sigma_A^2 = 0$, and letting $\sigma_\varepsilon^2 = 1.18$, which is thefitness variance of an arbitrary variant (with two randomized positions) under the T4MO parameters.The results are shown in the right panel of Figure 2; as expected, $\tilde{L}_{0.95}$ has significantly increased, to 1,277.It can be shown mathematically, using Slepian's inequality [40], that a high proportion of non-additivityvariance yields more conservative

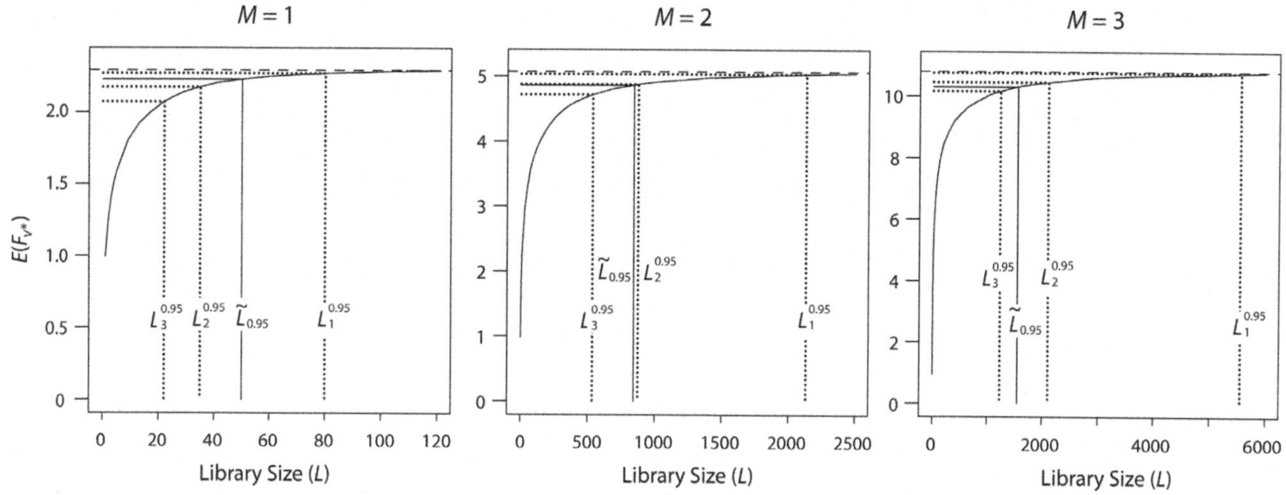


**Figure 1. Expected fitness as a function of library size.** Expected fitness of the best variant discovered in the experiment, $E(F_{\nu*})$, as a function of the library size $L$, when randomizing $M = 1, 2, 3$ NNK positions. The horizontal dashed lines denote the expected fitness of the best variant in variant space, $E(F_{\nu**})$. The dotted lines correspond to the library sizes required to ensure a 0.95 probability of discovering the top variant ($L_1^{0.95}$), any of the top two ($L_2^{0.95}$), and any of the top three ($L_3^{0.95}$). The solid line correspond to $\tilde{L}_{0.95}$, the library size required to reach 95% of the distance from the lowest point of the curve (at $L = 1$) to $E(F_{\nu**})$. Simulations are based on the model of Nov and Wein, with the T4MO parameter and a logarithmic transformation.

estimates for the library size, so that the right panel of Figure 2 is themost challenging scenario.

We now turn to modeling the fitness landscape by models other than the model of Nov and Wein. Figure 3 shows the results when randomizing two NNK positions using the $NK$ model, with parameters $N = 24$ (the median value of the $N$'s used in the simulations in [24]) and $K = 6$ (the median value of the $K$'s for $N = 24$, in that study). Given $K$, it still remains to determine the degree of overlap betweenthe $K$+ 1 positions influencing the first randomized position, and the $K$ + 1 influencing the second. Weconsider therefore three scenarios: maximal overlap of $K$ = 6 positions, moderate overlap of 3 positions,and no overlap. The resulting values of $\tilde{L}_{0.95}$ are 446, 491, and 358, respectively.

Figure 4 shows the results when modeling the fitness landscape through the rough Mount Fuji model. The parameters are $\nu = 60$, $\varepsilon = -1$, and $\varrho = 0.5$, 1, and 2 (the values used in [25]). The

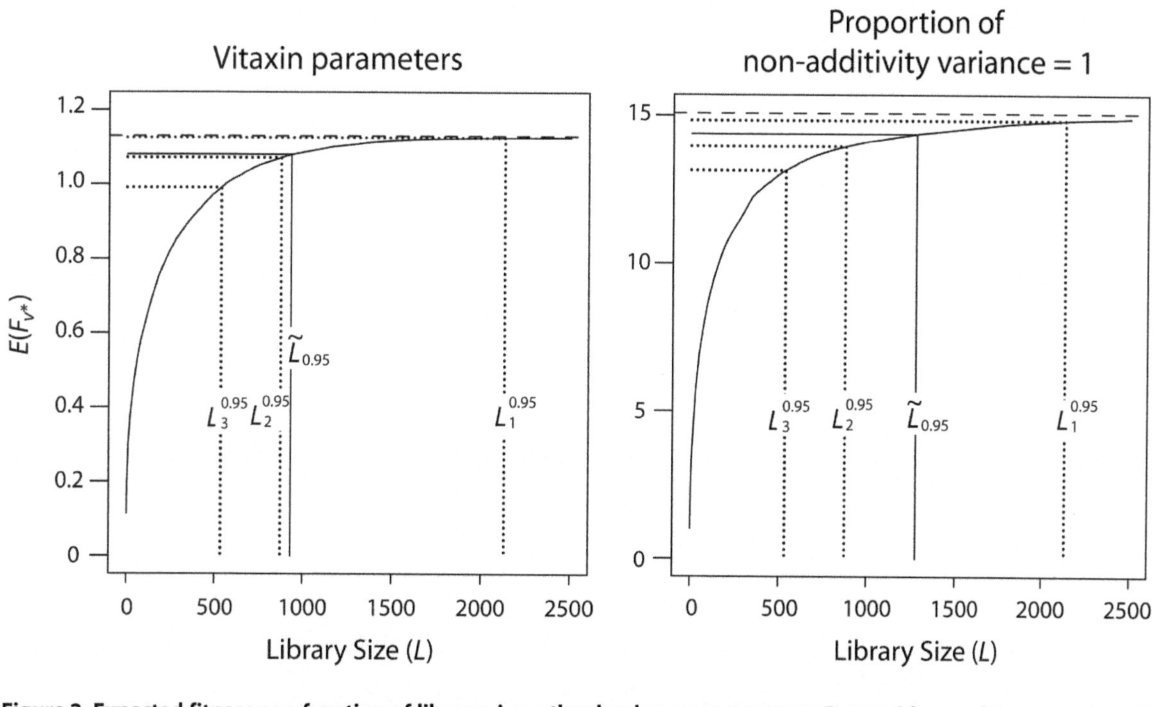


**Figure 2. Expected fitness as a function of library size, other landscape parameters.** Expected fitness of the best variant discovered in the experiment, $E(F_{\nu*})$, as a function of the library size $L$, when randomizing two NNK positions. Left: Vitaxin parameters. Right: same marginal fitness distribution as with the T4MO parameters, but the proportion of non-additivity variance equals 1.

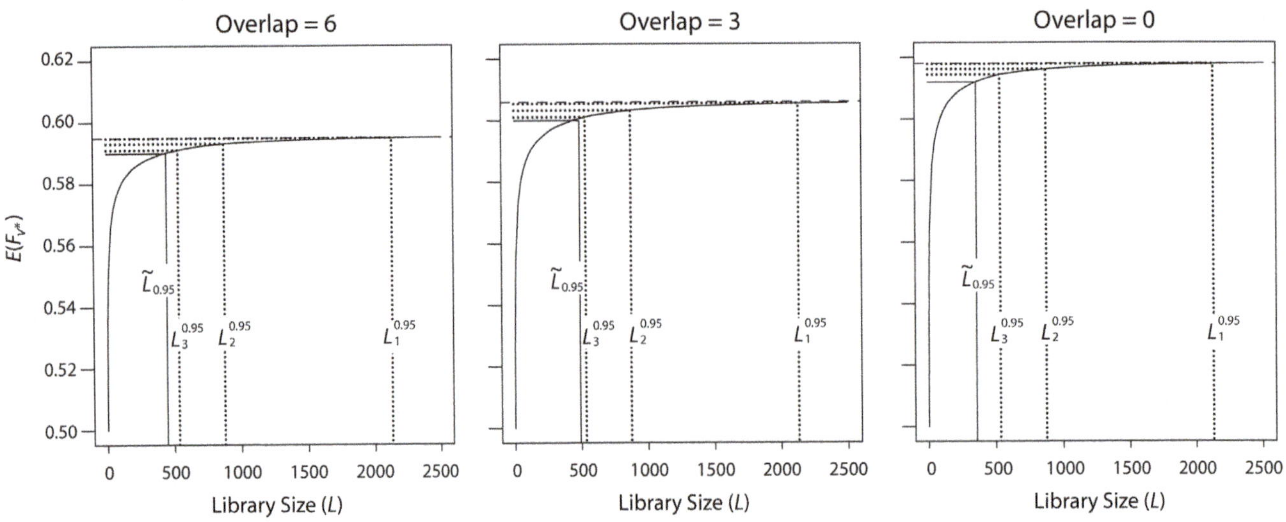


**Figure 3. Expected fitness as a function of library size, the *NK* model.** Expected fitness of the best variant discovered in the experiment, $E(F_{\nu *})$, as a function of the library size $L$, when randomizing two NNK positions, under the $NK$ model. Parameters are $N=24$ and $K=6$. The number of overlapping positions among those influencing the fitness contributions from the two randomized positions is 6, 3, and 0. Given $K$, it still remains to determine the degree of overlap between the $K+1$ positions influencing the first randomized position, and the $K+1$ influencing the second. We consider therefore three scenarios: maximal overlap of $K=6$ positions, moderate overlap of 3 positions, and no overlap. The resulting values of $\tilde{L}_{0.95}$ are 446, 491, and 358, respectively.

resulting valuesof $\tilde{L}_{0.95}$ are 543, 663, and 644, corresponding to the three values of $\varrho$.

Thus, under both the $NK$ model and the rough Mount Fuji model, $\tilde{L}_{0.95}$ is significantly smaller than the value 842, obtained under the model of Nov and Wein, so the latter model is more conservative. All the results presented so far concerned only mean values. We define the percent loss of fitness to be.

$$\Lambda = \frac{F_{\nu**} - F_{\nu*}}{F_{\nu**}} \cdot 100 \tag{4}$$

which is a random variable taking values in the set [0, 100], and whose distribution strongly depends on the library size $L$. Clearly, lower values of $\Lambda$ are desirable, with $\Lambda = 0$ corresponding to the (ideal) event $\{F_{\nu**} = F_{\nu*}\}$.

Figure 5 shows the loss probabilities $P(\Lambda > a)$ for $a = 5\%, 10\%$, and 20% (under the model of Nov and Wein, T4MO parameters, and a logarithmic transformation). For example, when the library size is $L = L_2^{0.95} = 875$, there is a 0.069 probability that the fitness of the best variant discovered will be lower than that of the best variant in variant space by more than 20% (the dotted lines).

Figure 6 shows histograms depicting the distribution of $\Lambda$ for three library sizes: $L = L_1^{0.95}$, $L_2^{0.95}$, and $L_3^{0.95}$. The spiking, leftmost column in each of the three histograms corresponds almost in its entirety to the case $\Lambda = 0$. As expected, the height of this column

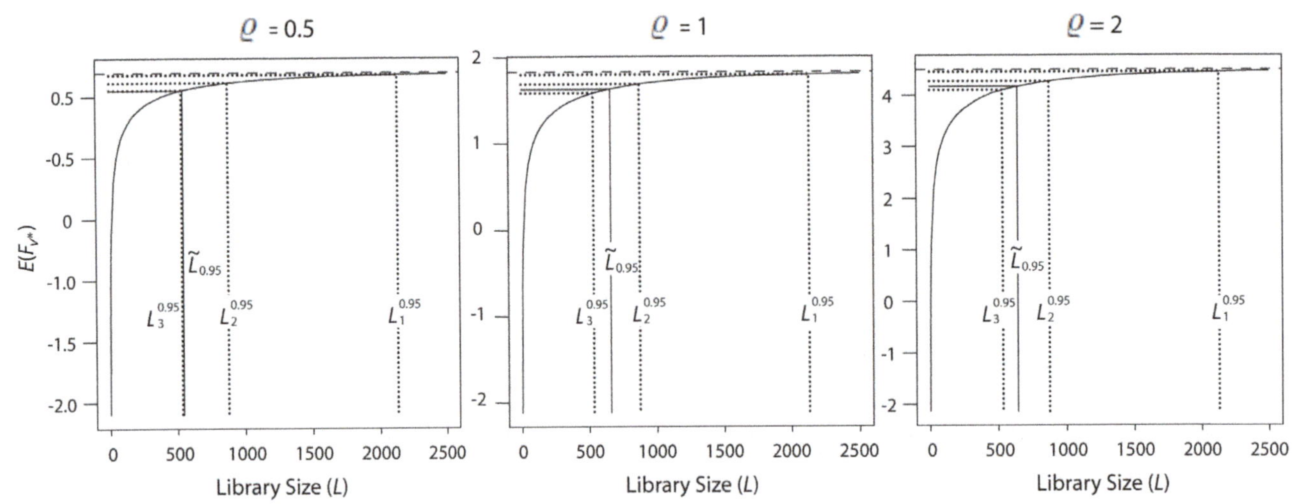


**Figure 4. Expected fitness as a function of library size, the rough Mount Fuji model.** Expected fitness of the best variant discovered in the experiment, $E(F_{\nu *})$, as a function of the library size $L$, when randomizing two NNK positions, under the rough Mount Fuji model. Parameters are $\nu=60$, $\varepsilon=-1$, and $\varrho=0.5, 1, 2$.

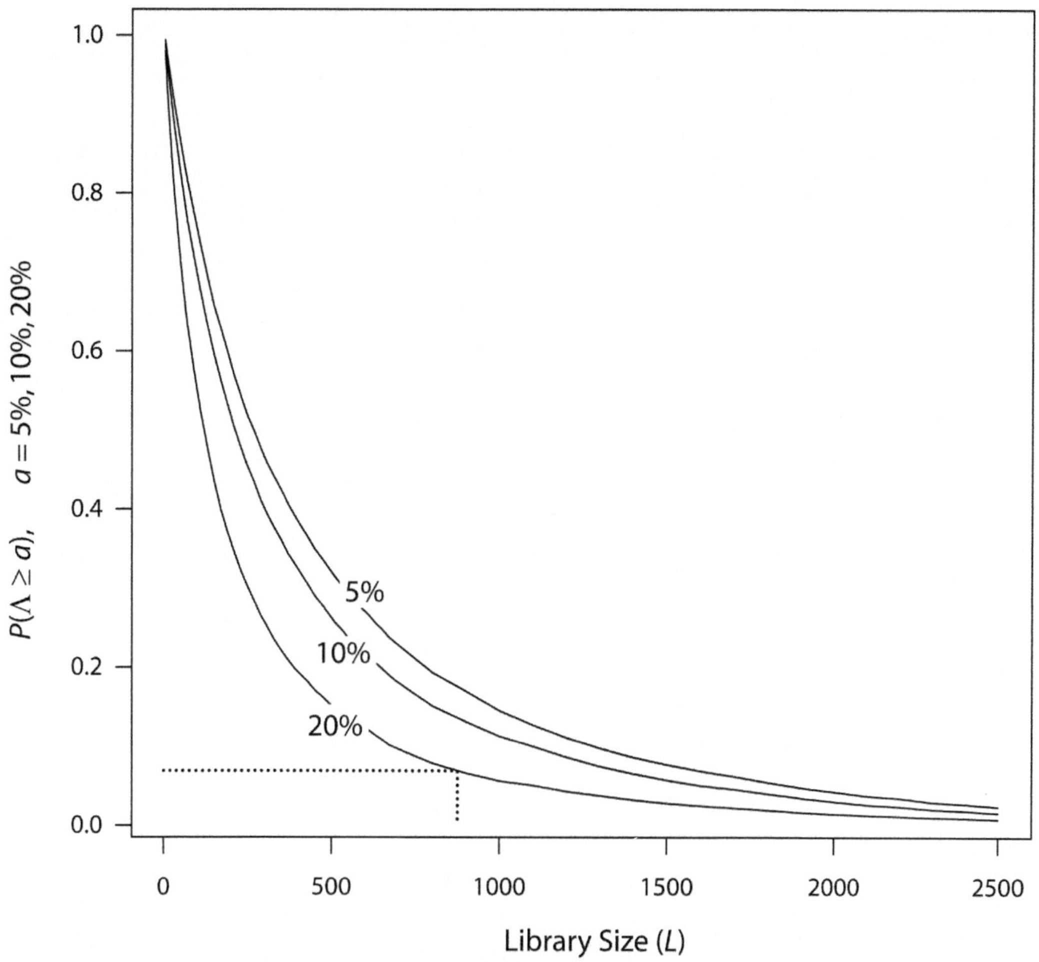


**Figure 5. Probability of losing more than 5%, 10%, 20% of fitness.** The probability that the fitness of the best variant discovered $E(F_{v*})$ will be at least 5%, 10%, and 20% lower than that of the best variant in variant space $E(F_{v**})$. Randomization, fitness model, parameters, and transformation as in Figure 1.

increases as $L$ increases, and the distribution reflected by the remaining columns becomes stochastically closer to zero.

## Discussion

It is difficult to establish universal, sharp guidelines for the optimal library size in a saturation mutagenesis experiment, as the requirements from the design process, the screening costs, and the fitness landscape of the specific protein at hand vary from one experiment to another. However, judging from the results of the extensive simulation study presented above, it seems that when randomizing two positions, a rough yet reasonable rule of thumb is to choose a library size no larger than a point about half way between $L_1^{0.95}$ and $L_2^{0.95}$, i.e., between the library size required to ensure a 0.95 probability of discovering the best variant, and that of discovering any of the top two variants. Using a similar methodology, one can establish similar rough guidelines for library size determination when randomizing $M \neq 2$ positions. We have done so (results not shown) for the cases $M=1$ and $M=3$: When randomizing a single position, we recommend a library no larger than about $L_1^{0.95}$ variants, and when randomizing three positions, about $L_2^{0.95}$ variants.

These recommendations are conservative, as they are based on the model that provides the largest library sizes among the three models considered (the model of Nov and Wein), and on the most challenging scenario from a wide range of model parameters (right panel of Figure 2).

In [29] it was suggested that the logarithmic transformation is perhaps "too concave" for converting the raw fitness measurements, and that to achieve a better fit to the model of Nov and Wein, a power transformation such as the Box–Cox transformation [41] with power $0 < \lambda \leq 1$ should be used instead (such transformations are used routinely in statistics to improve model fit, and $\lambda$ has no physicochemical interpretation). However, since a Box–Cox transformation with $0 < \lambda \leq 1$ is "less concave" than the logarithmic transformation, the resulting library sizes will be smaller; thus, assuming that a logarithmic transformation is required, as was done in this work, is again conservative.

The library sizes we recommend are *lower* than those actually chosen by some practitioners, sometimes by a very large margin. In two unrelated experiments, each involving randomization of a single NNS position, Boersma et al. [42] screened 1,250 variants and Maeda et al. [43] screened 360 variants, where ~80 variants seem to suffice. Similarly, Champion et al. [44] screened ~8,000 variants when randomizing two NNN positions, where ~2,000 seem to suffice.

One may use the approach described in this work to compare the efficiency of the various randomization schemes. We have

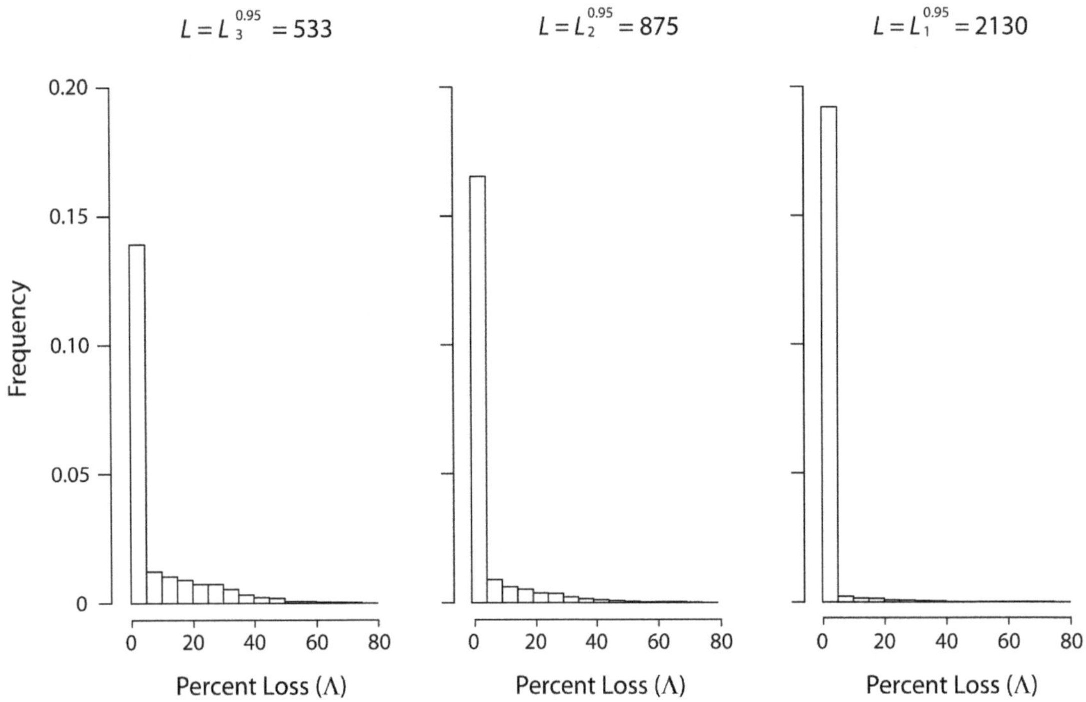


**Figure 6. Distribution of loss of fitness as a function of library size.** Distribution of the loss of fitness for three library sizes: $L_1^{0.95}$, $L_2^{0.95}$, and $L_3^{0.95}$. Randomization, fitness model, parameters, and transformation as in Figure 1.

done so, and the results are consistent with previously published findings [16,23]: the most efficient schemes are those that induce a 1/20 probability for each of the amino acids [17,18], next is NNK (or NNS, which is identical in this sense), then NNB, and the least efficient scheme is NNN.

Another approach to protein engineering, which often complements directed evolution, is rational design. Here, scientists use elaborate computational algorithms and detailed structural knowledge of the protein at hand, in order to introduce targeted mutations that are predicted to improve the protein's fitness. When taking a pure rational design approach, library size is minimal, as only the targeted mutants need to be generated via site-directed mutagenesis. Often, rational design and directed evolution are combined into the so-called semi-rational approach, whereby computational tools dictate which positions are best to be explored by saturation mutagenesis [45]. The library size analysis presented in this work is relevant to such experiments. Some rational design studies aim at designing highly stable, "ideal" protein structures; such structures may be used as scaffolds for the next generation of engineered proteins, over which saturation mutagenesis could be employed to focus the desired function [46].

## Conclusions

Our simulation results indicate that existing criteria and practices for determining the library size in saturation mutagenesis experiments are often too conservative: smaller libraries, sometimes by an order of magnitude, suffice to exhaust almost all the fitness that the fitness landscape has to offer. This observation is in line with other recent works, which advocate the use of "small but smart" libraries [9]. The reduction in library size carries clear benefits in laboratory expenses, time, and labor.

## Acknowledgments

I thank three referees whose comments significantly improved this paper.

## Author Contributions

Conceived and designed the experiments: YN. Performed the experiments: YN. Analyzed the data: YN. Wrote the paper: YN.

## References

1. Romero PA, Arnold FH (2009) Exploring protein fitness landscapes by directed evolution. Nature Reviews Molecular Cell Biology 10: 866–876.
2. Maynard Smith J (1970) Natural selection and the concept of a protein space. Nature 225: 563–564.
3. Dandekar T, Argos P (1992) Potential of genetic algorithms in protein folding and protein engineering simulations. Protein Engineering 5: 637–645.
4. Voigt CA, Kauffman S, Wang Z-G (2001) Rational evolutionary design: the theory of *in vitro* protein evolution. Advances in protein chemistry 55: 79–160.
5. Reetz MT (2013) The Importance of Additive and Non-Additive Mutational Effects in Protein Engineering. Angewandte Chemie 52: 2658–2666.
6. Reetz MT, Kahakeaw D, Lohmer R (2008) Addressing the numbers problem in directed evolution. ChemBioChem 9: 1797–1804.
7. Alcolombri U, Elias M, Tawfik DS (2011) Directed Evolution of Sulfotransferases and Paraoxonases by Ancestral Libraries. Journal of Molecular Biology 411: 837–853.
8. Herman A, Tawfik DS (2007) Incorporating Synthetic Oligonucleotides via Gene Reassembly (ISOR): a versatile tool for generating targeted libraries. Protein Engineering, Design and Selection 20: 219–226.
9. Goldsmith M, Tawfik DS (2013) Enzyme Engineering by Targeted Libraries. Methods in Enzymology 523: 257–283.
10. Prasad S, Bocola M, Reetz MT (2011) Revisiting the lipase from Pseudomonas aeruginosa: Directed evolution of substrate acceptance and enantioselectivity using iterative saturation mutagenesis. ChemPhysChem 12: 1550–1557.
11. Yeung YA, Leabman MK, Marvin JS, Qiu J, Adams CW, et al. (2009) Engineering human IgG1 affinity to human neonatal Fc receptor: Impact of

affinity improvement on pharmacokinetics in primates. The Journal of Immunology 182: 7663–7671.
12. Williams GJ, Goff RD, Zhang C, Thorson JS (2008) Optimizing glycosyltransferase specificity via "hot spot" saturation mutagenesis presents a catalyst for novobiocin glycorandomization. Chemistry and biology 15: 393–401.
13. Feingersch R, Shainsky J, Wood TK, Fishman A (2008) Protein engineering of toluene monooxygenases for synthesis of chiral sulfoxides. Applied and Environmental Microbiology 74: 1555–1566.
14. Shaner NC, Lin MZ, McKeown MR, Steinbach PA, Hazelwood KL, et al. (2008) Improving the photostability of bright monomeric orange and red fluorescent proteins. Nature Methods 5: 545–551.
15. Reetz M, Carballeira J (2007) Iterative saturation mutagenesis (ISM) for rapid directed evolution of functional enzymes. Nature protocols 2: 891–903.
16. Patrick WM, Firth AE (2005) Strategies and computational tools for improving randomized protein libraries. Biomolecular engineering 22: 105–112.
17. Hughes MD, Nagel DA, Santos AF, Sutherland AJ, Hine AV (2003) Removing the redundancy from randomised gene libraries. Journal of Molecular Biology 331: 973–979.
18. Tang L, Gao H, Zhu X, Wang X, Zhou M, et al. (2012) Construction of "small-intelligent" focused mutagenesis libraries using well-designed combinatorial degenerate primers. BioTechniques 52: 149–157.
19. Kille S, Acevedo-Rocha CG, Parra LP, Zhang ZG, Opperman DJ, et al. (2013) Reducing codon redundancy and screening effort of combinatorial protein libraries created by saturation mutagenesis. ACS Synthetic Biology 2(2) 83–92.
20. Patrick WM, Firth AE, Blackburn JM (2003) User-friendly algorithms for estimating completeness and diversity in randomized protein-encoding libraries. Protein Engineering 16: 451–457.
21. Bosley AD, Ostermeier M (2005) Mathematical expressions useful in the construction, description and evaluation of protein libraries. Biomolecular Engineering 22: 57–61.
22. Firth AE, Patrick WM (2008) GLUE-IT and PEDEL-AA: new programmes for analyzing protein diversity in randomized libraries. Nucleic Acids Research 36: W281–W285.
23. Nov Y (2012) When second best is good enough: Another probabilistic look at saturation mutagenesis. Applied and Environmental Microbiology 78: 258–262.
24. Kauffman SA, Weinberger ED (1989) The *NK* model of rugged fitness landscapes and its application to maturation of the immune response. Journal of Theoretical Biology 141: 211–245.
25. Aita T, Husimi Y (2000) Adaptive walks by the fittest among finite random mutants on a Mt. Fuji-type fitness landscape – II. Effects of small non-additivity. Journal of Mathematical Biology 41: 207–231.
26. Nov Y,Wein LM (2005) Modeling and analysis of protein design under resource constraints. Journal of Computational Biology 12: 247–282.
27. Fox RJ, Davis SC, Mundorff EC, Newman LM, Gavrilovic V, et al. (2008) Improving catalytic function by ProSAR-driven enzyme evolution. Nature Biotechnology 25: 338–344.
28. Barak Y, Nov Y, Ackerley DF, Matin A (2008) Enzyme improvement in the absence of structural knowledge: a novel statistical approach. The ISME Journal 2: 171–179.
29. Brouk M, Nov Y, Fishman A (2010) Improving biocatalyst performance by integrating statistical methods into protein engineering. Applied and Environmental Microbiology 76: 6397–6403.
30. Wu H, Beuerlein G, Nie Y, Smith H, Lee B, et al. (1998) Stepwise in vitro affinity maturation of Vitaxin, an $\alpha_v\beta_3$-specific humanized mAb. Proceedings of the National Academy of Science 95: 6037–6042.
31. Reetz M, Wu S (2008) Greatly reduced amino acid alphabets in directed evolution: making the right choice for saturation mutagenesis at homologous enzyme positions. Chemical Communications : 5499–5501.
32. Arnold FH (1996) Directed evolution: Creating biocatalysts for the future. Chemical Engineering Science 51: 5091–5102.
33. Sniegowski PD, Gerrish PJ, Lenski RE (1997) Evolution of high mutation rates in experimental populations of E. coli. Nature 387: 703–705.
34. Voigt CA, Kauffman S, Wang ZG (2001) Rational evolutionary design: the theory of in vitro protein evolution. Advances in Protein Chemistry 55: 79–160.
35. Drummond DA, Iverson BL, Georgiou G, Arnold FH (2005) Why high-error-rate random mutagenesis libraries are enriched in functional and improved proteins. Journal of Molecular Biology 350: 806–816.
36. Pavelka A, Chovancova E, Damborsky J (2009) HotSpot wizard: a web server for identification of hot spots in protein engineering. Nucleic Acids Research 37: W376–W383.
37. Skinner MM, Terwilliger TC (1996) Potential use of additivity of mutational effects in simplifying protein engineering. Proceedings of the National Academy of Sciences 93: 10753–10757.
38. Benos PV, Bulyk ML, Stormo GD (2002) Additivity in proteindna interactions: how good an approximation is it? Nucleic Acids Research 30: 4442–4451.
39. Carneiro M, Hartl DL (2010) Adaptive landscapes and protein evolution. Proceedings of the National Academy of Sciences 107: 1747–1751.
40. Slepian D (1962) The one-sided barrier problem for Gaussian noise. Bell System Technical Journal 41: 463–501.
41. Box GEP, Cox DR (1964) An analysis of transformations. Journal of the Royal Statistical Society Series B (Methodological) 26: 211–252.
42. Boersma YL, Pijning T, Bosma MS, van der Sloot AM, Godinho LF, et al. (2008) Loop grafting of bacillus subtilis lipase a: Inversion of enantioselectivity. Chemistry & Biology 15: 782–789.
43. Maeda T, Sanchez-Torres V, Wood T (2008) Protein engineering of hydrogenase 3 to enhance hydrogen production. Applied Microbiology and Biotechnology 79: 77–86.
44. Champion E, Moulis C, Morel S, Mulard LA, Monsan P, et al. (2010) A pH-based high-throughput screening of sucrose-utilizing transglucosidases for the development of enzymatic glucosylation tools. ChemCatChem 2: 969–975.
45. Lutz S (2010) Beyond directed evolution – semi-rational protein engineering and design. Current Opinion in Biotechnology 21: 734–743.
46. Koga N, Tatsumi-Koga R, Liu G, Xiao R, Acton TB, et al. (2012) Principles for designing ideal protein structures. Nature 491: 222–227.

# 21

# Mechanism of Protein Kinetic Stabilization by Engineered Disulfide Crosslinks

**Inmaculada Sanchez-Romero[1¤a], Antonio Ariza[2¤b], Keith S. Wilson[2], Michael Skjøt[3¤c], Jesper Vind[3], Leonardo De Maria[3], Lars K. Skov[3], Jose M. Sanchez-Ruiz[1]***

**1** Facultad de Ciencias, Departamento de Quimica Fisica, Universidad de Granada, Granada, Spain, **2** Structural Biology Laboratory, Department of Chemistry, University of York, Heslington, York, United Kingdom, **3** Novozymes A/S, Bagsværd, Denmark

## Abstract

The impact of disulfide bonds on protein stability goes beyond simple equilibrium thermodynamics effects associated with the conformational entropy of the unfolded state. Indeed, disulfide crosslinks may play a role in the prevention of dysfunctional association and strongly affect the rates of irreversible enzyme inactivation, highly relevant in biotechnological applications. While these kinetic-stability effects remain poorly understood, by analogy with proposed mechanisms for processes of protein aggregation and fibrillogenesis, we propose that they may be determined by the properties of sparsely-populated, partially-unfolded intermediates. Here we report the successful design, on the basis of high temperature molecular-dynamics simulations, of six thermodynamically and kinetically stabilized variants of phytase from *Citrobacter braakii* (a biotechnologically important enzyme) with one, two or three engineered disulfides. Activity measurements and 3D crystal structure determination demonstrate that the engineered crosslinks do not cause dramatic alterations in the native structure. The inactivation kinetics for all the variants displays a strongly non-Arrhenius temperature dependence, with the time-scale for the irreversible denaturation process reaching a minimum at a given temperature within the range of the denaturation transition. We show this striking feature to be a signature of a key role played by a partially unfolded, intermediate state/ensemble. Energetic and mutational analyses confirm that the intermediate is highly unfolded (akin to a proposed critical intermediate in the misfolding of the prion protein), a result that explains the observed kinetic stabilization. Our results provide a rationale for the kinetic-stability consequences of disulfide-crosslink engineering and an experimental methodology to arrive at energetic/structural descriptions of the sparsely populated and elusive intermediates that play key roles in irreversible protein denaturation.

**Editor:** Maria Gasset, Consejo Superior de Investigaciones Cientificas, Spain

**Funding:** This work was supported by grants BIO2009-09562, CSD2009-00088 from the Spanish Ministry of Science and Innovation, and FEDER Funds (JMS-R). The funders had no role in study design, data collection and analysis, decision to publish, or preparation of the manuscript.

**Competing Interests:** The study was supported by Novozymes A/S, the employer of JV, LDM and LKS. The Novozymes employees have financial interests, including stock options with the company. Novozymes is the producer of phytases for the animal feed market. Novozymes has patents on the Citrobacter braakii phytase and variants thereof. There are no further patents, products in development or marketed products to declare.

* E-mail: sanchezr@ugr.es

¤a Current address: Institute of Science and Technology Austria (IST Austria), Am Campus 1, Klosterneuburg, Austria
¤b Current address: Institute of Molecular and Cellular Biology, University of Leeds, Leeds, United Kingdom
¤c Current address: Novo Nordisk A/S, Måløv, Denmark

## Introduction

Natural disulfide bonds have been known for many years to contribute to native protein stabilization [1]. In the context of a two-state (native to unfolded) equilibrium denaturation process, disulfide-bond stabilization is often viewed as a consequence of the decrease in the conformational entropy of the unfolded state caused by the presence of the crosslink [2,3]. Hence, it is generally accepted that engineering disulfide bridges is an efficient way to stabilize proteins, provided that they are introduced at locations in which they do not distort or strain the native fold or perturb the active site [4,5,6,7,8,9,10,11].

However, there is a fundamental difference between a protein's thermodynamic and kinetic stability and clearly disulfide bridges may also have an important effect on the latter. Indeed, in a biotechnological setting, the rate of irreversible enzyme inactivation may well be affected by the presence of disulfide bridges and be more relevant than the unfolding free energy change [12,13]. In addition, disulfide bridges in aggregation prone regions can help prevent the dysfunctional association of proteins [14,15]. While thermodynamic stability is related to the free energy difference between the native and the unfolded states, in the context of transition-state theory, kinetic stability is determined by that between the native state and the transition states for the rate limiting steps in the irreversible denaturation pathway [16]. Furthermore, the mechanisms of irreversible denaturation can be exceedingly complex [17,18,19], with non-native intermediate species (states, ensembles) playing a fundamental role. Accordingly, while a correspondence between disulfide bridge effects on thermodynamic and kinetic stability cannot be postulated *a priori*, very few experimental studies have addressed the relative importance of these effects. For instance, the failure of designed disulfide bridges to stabilize some proteins has been related to the irreversible, kinetically-controlled nature of the denaturation process [20]. However, this interpretation remains unproven, given the low level of success often met in the rational design of stabilizing disulfide bridges [21,22].

We here use rational computation-based procedures to design stabilizing disulfide bridges in the phytase from *Citrobacter braakii*, a biotechnologically important enzyme currently used in RONOZYME HiPhos (DSM Nutritional Products, Basel, Switzerland). The design procedure, based on heated Molecular Dynamics simulations, was highly successful and allowed us to prepare several stabilized variants with 1, 2 or 3 engineered bridges. Differential scanning calorimetry combined with extensive thermal inactivation reveals the main features of the effect of the bridges on thermodynamic stability and their impact on kinetic stability. These variants thus provide an excellent comparison of the effects of engineered disulfide bridges on thermodynamic and kinetic stability. We report the crystal structure of one of the variants.

## Results and Discussions

### Computational Design of Disulfide Bridges

The design of the engineered disulfide bridges was based on molecular dynamics simulations performed at several temperatures on a homology model of the phytase from *C. braakii*, no 3D structure being available at that time (see Text S1 for details). Molecular dynamics simulations are well established for the study of protein folding and unfolding [23,24,25] and, indeed, the isotropic root mean square deviations (iRMSF) of the Cα carbons at the different temperatures suggest that the enzyme displays a clear unfolding behavior in the simulations performed at 500 K (see Figure 1) while at the lower temperatures the available thermal energy is only able to excite fluctuations of specific regions.

The design of the additional disulfide bridges targeted regions of the structure with a substantial tendency to unfold in the Molecular Dynamics simulations. The strategy was to introduce the disulfides in regions of the structure significantly displaced with respect to one another at 500 K, in comparison with the starting structure (Figure 2). Once these regions were identified, the positions to mutate were chosen by geometric considerations in the protein structure. This approach was based on the following hypotheses: i) the targeted regions are likely to be more flexible and may be expected to accommodate the engineered disulfides without straining or distorting the native structure; ii) the targeted regions could be unfolded in the kinetically relevant transition state and their stabilization by disulfide crosslinking may increase the free-energy barrier for irreversible denaturation and enhance kinetic stability. Since we aimed at enhancing both the thermodynamic and kinetic stabilities of the phytase without compromising function, we deemed it advisable to focus engineering on those regions with a substantial tendency to unfold which lined the rim of the substrate binding pocket or were in direct contact with them, involving key residues for the enzymatic activity.

The procedure described above led us to prepare three variants (see Text S1 for details) each with a single engineered disulfide bridge: S1 (N31C/T177C), S2 (G52C/A99C) and S3 (K141C/V199C). Among these three single disulfide variants, S2 proved to be the most stable (see below). Therefore, we deemed it appropriate also to study multiple-disulfide variants in which the bridge in S2 (G52C/A99C) was combined with the other bridges. The following variants with two or three of these bridges were therefore made: D1 (N31C/T177C+G52C/A99C), D2 (G52C/A99C+K141C/V199C) and T (N31C/T177C+G52C/A99C+K141C/V199C).

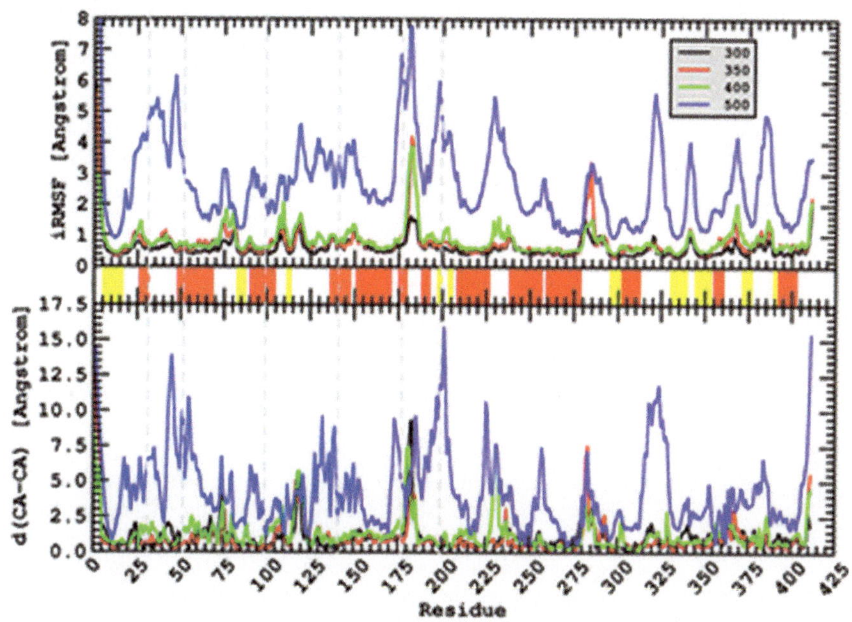


**Figure 1. Molecular Dynamics simulations performed on a homology model of the phytase from *C. braakii*.** Simulations were performed at several temperatures (shown in Kelvin in the inset of the upper panel). Upper panel: isotropic root mean square deviations (iRMSF) of the Cα at the different temperatures. Middle panel: secondary structure assignment with red regions representing α-helices and yellow regions representing β-sheets. Bottom panel: mean distances between Cα of the starting structure and Cα of the structures at different temperatures. These profiles suggest that at high temperature the enzyme displays an unfolding behavior while at low temperatures the available thermal energy is only able to excite fluctuations of specific regions. Dashed grey lines mark the position of the residues mutated to engineer disulfide bridges.

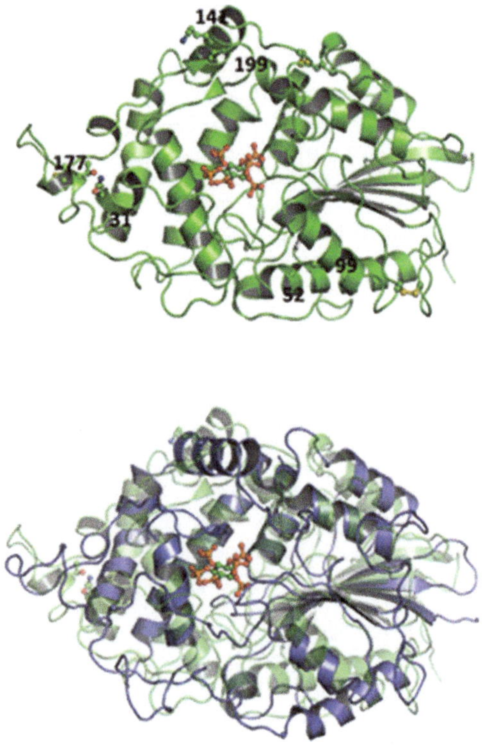


**Figure 2. Disulfide crosslink design.** Upper panel: structure of the homology model of the phytase from *C. braakii* used in Molecular Dynamics simulations showing the residues mutated to introduce disulfide bridges. Bottom panel: superposition of the MD structures at 500 K and low temperature (blue and green, respectively). Regions targeted for disulfide bridge engineering are those showing significant displacement between the two structures.

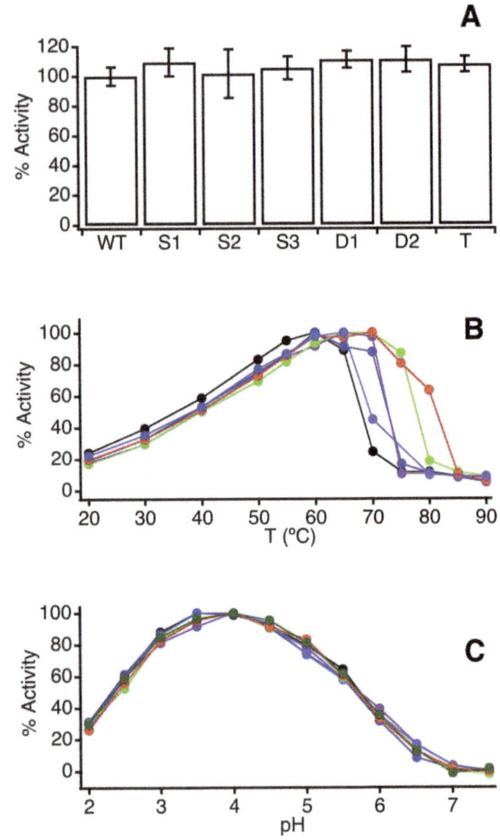


**Figure 3. Preliminary assessment of the activity and stability of the phytase variants with engineered disulfide bridges.** (A) Activity values for the wild-type and the variants studied at pH 4.5. (B) Profiles of activity vs. temperature at pH 4.5. The drop in activity at the higher temperatures indicates denaturation and provides a first estimate of the thermal stability. (C) Profiles of activity vs. pH for wild-type and the variants. In panels B and C, the maximum activity value of each profile is normalized to 100, while in panel A wild-type phytase is assigned a value of 100. Code color for the variants in panels B and C refers to the number of engineered bridges (black 0, blue 1, green 2, red 3) and is more clearly apparent in Figure 5A.

## Preliminary Assessment of the Activity and Stability of the Variants with Engineered Disulfide Bridges

All the variants showed similar activity levels in the standard phytase assay (Figure 3 A; see Text S1 for details) and essentially identical pH profiles (Figure 3 B). Furthermore, very similar temperature dependencies of activity are observed at sub-denaturational temperatures (Figure 3 C) while the high temperature activity drop associated with irreversible denaturation is shifted to higher temperatures as the number of disulfide bridges increases. These results suggest that, while the engineered crosslinks do enhance stability, they do not cause dramatic alterations in the native, functional structure of the enzyme, corroborated by the 3D-structure of one of the variants. Dynamic light scattering experiments support that irreversible denaturation in this system is linked to significant protein aggregation (Figure S1).

## 3D-structure of a Phytase Variant

The structure with a disulfide bridge engineered between residues 141 and 199 was refined to 2.3 Å spacing with two independent protein monomers in the asymmetric unit (see Text S1 for details). The fold (Figure 4) has two domains and is very similar to those of previous 3- and 6-phytases in the PDB from the bacteria *E. coli* [26], *Hafnia alvei* [27], *Klebsiella pneumoniae* [28] and *Yersinia kristensenii* [27] and the fungi *Aspergillus ficuum* [29], *A. fumigatus* [30], *A. niger* [31] and *Debaryomyces castellii* [32] and is typical of members of the histidine acid phosphatase superfamily. The major domain on the right is composed of residues 6–28, 47–135 and 260–410 and is colored blue and grey. At its backbone is a mainly parallel β-sheet, surrounded by α-helices and loops. The second domain, primarily α-helical, is composed of two insertions (residues 39–46 and 136–259) in the first domain and is seen on the left of the Figure in red/yellow. The active site is at the interface between the two domains, its position indicated in the Figure by superimposing the *myo*-inositol hexakissulfate complex of the *H. alvei* [27] enzyme (PDB 4aro, 4arv and 4aru). The engineered disulfide between residues 141 and 199 crosslinks an α-helix and a coiled loop of the α-helical domain and has well-defined electron density (Figure 4 B) indicating a highly successful introduction of this new element into the phytase fold.

The high degree of conservation of fold in the 3- and 6-Phytases of the HAP superfamily is illustrated by the superposition of the *C. braakii* structure on that of the *H. alvei* enzyme in Figure 4B. The rmsd in ~380 equivalent Cα positions is 1.18 Å. The four disulfide bridges can be seen to be totally conserved.

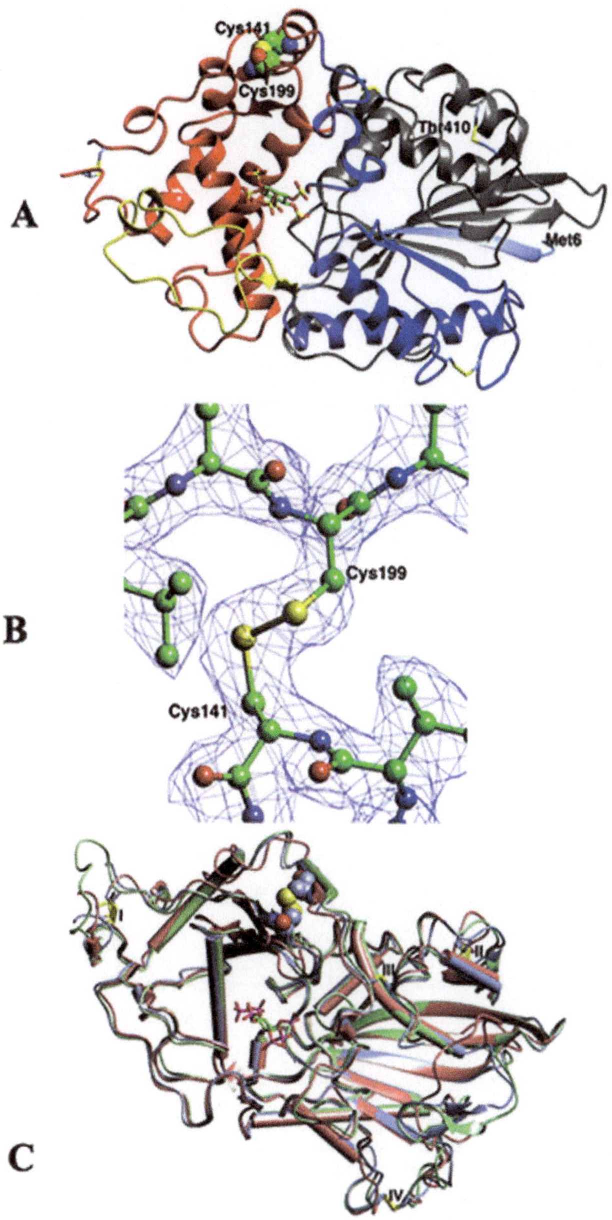


**Figure 4. Structural consequences of an engineered disulfide crosslink.** (A) The fold of Chain A of *C. braakii* phytase in ribbon format, with residues 6–18 in ice blue, 19–46 yellow, 47–135 blue, 136–259 red and 260–410 grey. The four conserved disulfide bridges are shown as cylinders, and the engineered ones as spheres. The phytate analogue myo-inositol hexakissulfate – shown as cylinders-has been modeled into the active site based on its position in its complex with *H. alvei* phytase (PDB 4aro). (B) The electron density in the final $2F_o$–$F_c$ synthesis contoured at the 1σ level around the engineered disulfide bridge between residues 141 and 199. (C) Superposition of the structures of the *E. coli* (green, PDB 1dkq) and *H. alvei* (coral, PDB 4ars) enzymes on *C. braakii* phytase (blue), using the SSM option [44] in CCP4mg. The structures are shown in worm and tube format. Figures 4 A–C were all made with CCP4mg [45].

## Determination of Termal Stability Through Differential Scanning Calorimetry

DSC transitions (Figure 5: see Text S1 for details) for the thermal denaturation of wild-type phytase and the engineered variants were only partially reversible and showed a small but significant scan-rate dependence (Figure 5E). This indicates that the thermally-induced transitions reflect a degree of irreversible denaturation and are probably kinetically distorted by the process responsible for the irreversibility. We have therefore refrained from carrying out a detailed equilibrium thermodynamic analysis of the transitions. However, some general and robust trends are evident. Firstly, the transition temperature increases monotonically with the number of engineered bridges (Figure 5B), which clearly validates the approach used for computational disulfide design. Indeed, the effects of the engineered disulfides on thermal stability appear to be roughly additive; for instance adding the $T_m$ enhancements (relative to wt) found for the single variants (2.2, 6.3 and 1.3°C for S1, S2 and S3, respectively), $\Delta T_m$ values of 8.5, 7.6 and 9.8°C are predicted for D1, D2 and T, while the experimental $T_m$ enhancements relative to wt for these variants are 9.8 (D1), 8.2 (D2) and 12.1°C (T). Secondly, the reversibility is higher for variants with two and three engineered bridges (Figure 5C). In addition, the degree of reversibility decreases with increasing protein concentration (Figure 5D), suggesting a role for aggregation in the irreversible denaturation, an interpretation supported by dynamic light scattering experiments (Figure S1).

## Thermal Inactivation Kinetics

Irreversible denaturation of wild-type phytase and the variants was characterized by thermal inactivation kinetics. Briefly, samples were kept at a given temperature, aliquots were withdrawn at several times, quickly cooled down to 0°C and assayed for phytase activity (see Text S1 for details). Unfolded phytase (as well as partially-unfolded states "reversibly-linked" to the native state) should be able to fold to the native, active protein upon cooling, so the observed fall in activity with time can be primarily attributed to irreversible denaturation.

Thermal inactivation profiles were determined for each variant (representative examples in Figure 6) at several temperatures and total protein concentrations. For any given temperature, kinetic profiles were found to depend on total protein concentration, again consistent with a role for aggregation in the irreversible denaturation. However, mechanisms of protein aggregation can be exceedingly complex, involving conformational re-arrangement, nucleation and growth steps, as well as processes of aggregate fragmentation, aggregate coalescence to yield larger aggregates and phase separation of insoluble aggregates [33,34]. Analysis of typically featureless thermal inactivation profiles, such as those in Figure 6, in terms of a detailed mechanism is clearly not possible. Our approach is more modest but more robust. We aim at a simple, approximate phenomenological description of the profiles so that a single metric for the time scale of the irreversible denaturation process can be derived. Since no lag phases were observed in the experimental profiles, we used a simple n-order chemical kinetics equation:

$$A(t) = A_0\left\{1 + C^{n+1}(n-1)k\right\}^{1/1-n} \qquad (1)$$

where $A(t)$ and $A_0$ stand for the activities at time $t$ and time zero, $C$ is the total protein concentration and $k$ is the $n$-order rate constant for the process. Fits of equation 1 to the thermal inactivation profiles were visually good (Figures 6 A and 6 B) and yielded $n$ values typically of about 3–4 (Figure 6 C), a result generally consistent with the proposed role of aggregation in phytase irreversible denaturation. The phenomenological adequacy of equation 1 within the 0.5–1 mg/mL range of total protein concentration is further supported by the agreement

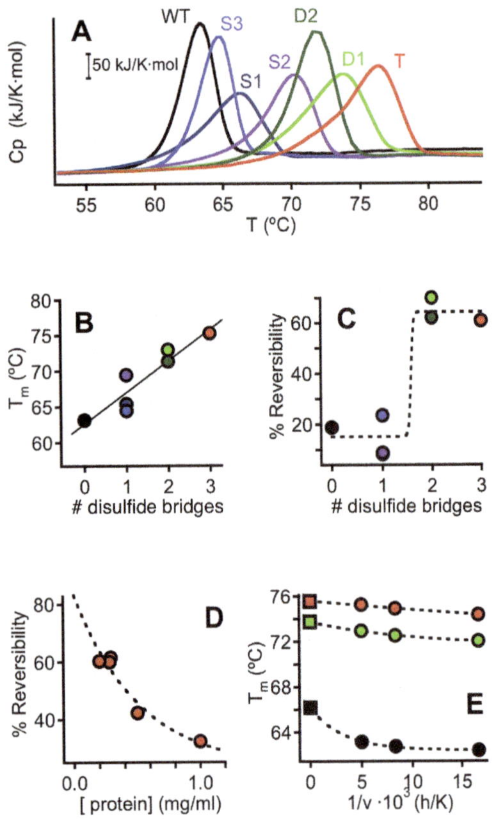


**Figure 5. Differential scanning calorimetry (DSC) of the thermal denaturation of wild-type phytase and the disulfide variants.** A) Representative DSC profiles at a protein concentration of 0.5 mg/mL and a scan rate of 200 degrees/hour. B) Plot of transition temperature (derived from the results in panel A) versus number of engineered bridges. C) Plot of degree of reversibility of the thermal denaturation process versus the number of engineered bridges. The degrees of reversibility shown were derived from DSC experiments performed with a protein concentration of 0.25 mg/mL, at a scanning rate of 200°C /hour. In all cases, the first scan was stopped at 100°C and a re-heating run was performed after letting the sample cool in the calorimetric cell. The degree of reversibility was calculated as the ratio between the maximum heat capacity values for the reheating and the first transitions after suitable chemical baseline correction. D) Degree of reversibility for the variant with three engineered bridges versus protein concentration. E) Scan-rate effect on the transition temperatures (circles) derived from DSC experiments. Extrapolation to 1/v=0 (i.e., to infinite scan-rate) should remove kinetic distortions associated to partial irreversibility (the squares at 1/v=0 actually represent the denaturation temperatures for equilibrium unfolding determined from model fitting to rate data). In panels B–E the colors refer to the different variant studied as specified in panel A.

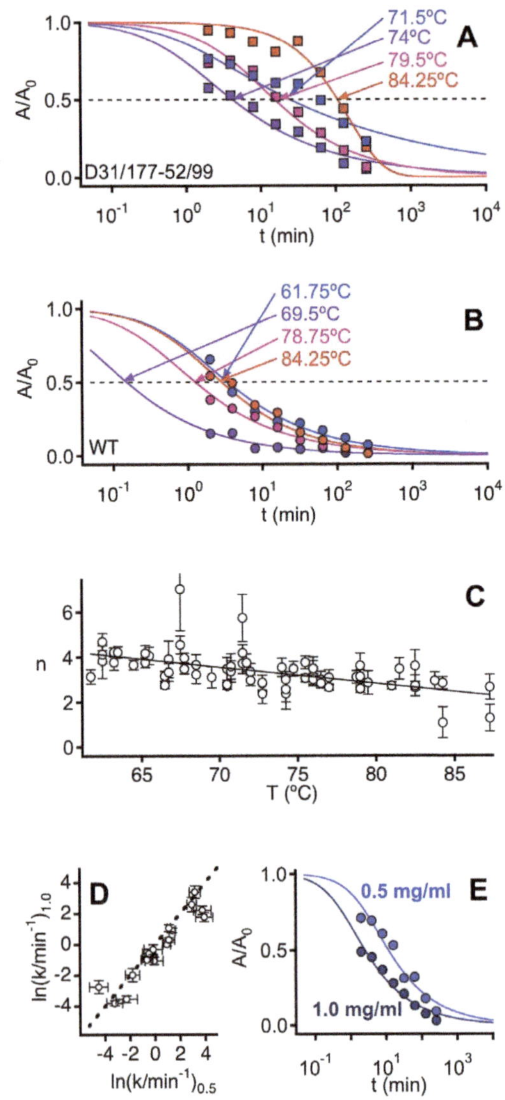


**Figure 6. Thermal inactivation kinetics data for wild-type phytase and variants.** (A) and (B) Illustrative plots of activity versus time for experiments performed at several temperatures with (A) a variant with two engineered bridges and (B) wild-type. Total protein concentration is 0.5 mg/mL. The continuous lines represent the best fits of equation 1 to the experimental data. The fastest kinetics are observed at an intermediate temperature (74°C in A and 69.5°C in (B). (C) Values of the reaction order derived from the fitting of equation 1 to inactivation profiles for wild-type phytase and all the variants. (D) Agreement between the rate constants obtained with total protein concentration of 0.5 and 1 mg/mL. (E) Representative example of the protein concentration effect on the rate of irreversible denaturation (variant D1, 76°C).

(Figures 6D,E) of the rate constants obtained at these two protein concentrations (equation 1 possibly breaks down at lower protein concentrations: see Figure S2).

This allows a suitable metric of the time scale for irreversible denaturation to be calculated as the time $\tau_{1/2}$ at which the activity falls to half the initial value:

$$\tau_{1/2} = \frac{(0.5)^{1-n} - 1}{C^{n-1}(n-1)k} \tag{2}$$

All our subsequent analyses are based upon the temperature and mutational effects on this time scale.

## Analysis of the Mutation and Temperature Dependencies of the Time Scale for Irreversible Denaturation

Figures 7A and 7B show plots of time scale for irreversible denaturation versus temperature for a total protein concentration of 0.5 mg/mL (similar plots are obtained at 1 mg/mL: see Figure S3). Figure 7 shows that, for a given temperature, irreversible

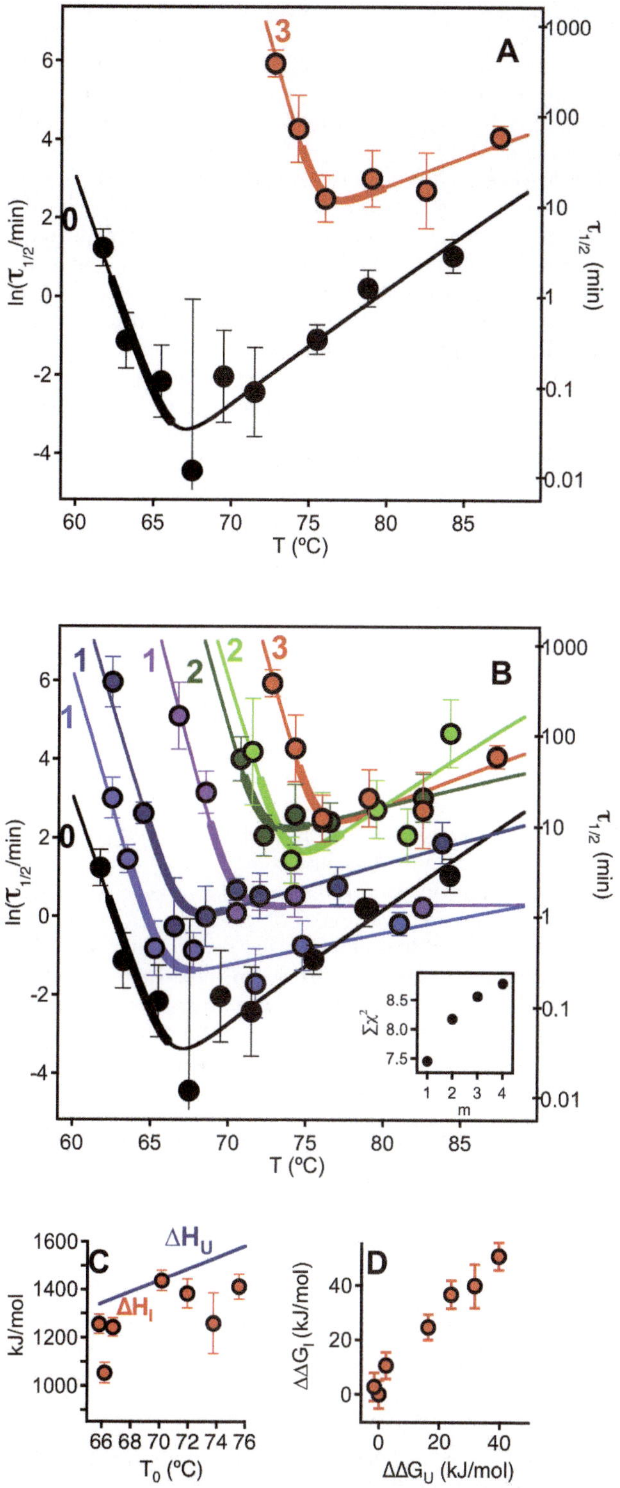


**Figure 7. Temperature dependence of the time scale ($\tau_{1/2}$) for the irreversible denaturation of wild type phytase and variants.** (A) and (B) Plots of $\ln(\tau_{1/2})$ versus temperature for the variants (for the sake of clarity, only wild type and the variant with three engineered bridges are included in panel A). The continuous lines represent the best fits of equation 3 and the Inset in panel B is a plot of a measure of goodness of fit ($\chi^2$) versus the reaction order m in equation 3 (see main text and Text S2 for details). Thicker lines indicate the temperature range of denaturation transition as seen by DSC (Figure 5A). The colors of the lines and data points refer to the variant, as specified in panel A of Figure 5, with the number of engineered bridges indicated. (C) Enthalpies of the unfolded and intermediate states (relative to the native state) plotted versus the corresponding $T_0$ value. (D) Correlation between the effect of engineered disulfide bridges on the free energy of the unfolded and intermediate states (relative to the native state) at a temperature of 70°C.

denaturation becomes slower (larger value of $\tau_{1/2}$) as the number of engineered disulfides is increased. However, the most surprising result is the unexpected temperature-dependence of the rate of irreversible denaturation: for essentially all variants the plot of $\ln(\tau_{1/2})$ versus temperature has an asymmetric-V shape with a minimum at a temperature that roughly agrees with the temperature range of the denaturation transition seen by DSC (see figure 5A and the thicker lines in the plots of Figure 7A). This pattern strongly suggests (Figure 8) that some species, different from the native state (N) and the fully unfolded state (U) are critical for the irreversible denaturation process and that such species represent a partially-folded intermediate state (I) that attains a kinetically relevant concentration within the denaturation transition range. Straightforward theoretical analysis of this model (see Text S2) leads to a simple equation describing the temperature dependence of the time scale for irreversible denaturation:

$$\ln\tau_{1/2} = A + \frac{m\Delta H_I}{R}\left(\frac{1}{T} - \frac{1}{T_0}\right) + m\ln\left\{1 + exp\left[-\frac{\Delta H_U}{R}\left(\frac{1}{T} - \frac{1}{T_0}\right)\right]\right\} \quad (3)$$

where $T_0$ is the equilibrium denaturation temperature (i.e., the temperature at which the equilibrium constant for the N to U conversion is unity), A is a constant (related to the value of the equilibrium constant for the N to I conversion at $T_0$), $m$ is the reaction order for species I in the phenomenological rate equation and $\Delta H_I$ and $\Delta H_U$ are, respectively, the enthalpies of I and U relative to the native state. Fits of equation 3 to the experimental $\tau_{1/2}$ versus $T$ data were excellent (Figure 7 and Figure S3; see Text S2 for additional details on the fitting process) and are further validated by the fact that the values obtained for $T_0$ (the equilibrium unfolding temperature) are consistent with the extrapolations to infinite scan-rate of the denaturation temperatures determined from the DSC transitions (Figure 5E).

It is important to note that our analysis of the model in Figure 8A assumes that the population of intermediate species is always much lower than the total protein concentration (see Text S2). While the species I is not assumed to be significantly populated, its critical role in kinetics is reflected in the experimental $\tau_{1/2}$ versus $T$ profiles and its energetics are readily derived from the fits of equation 3 to those profiles. Thus, the enthalpy of I relative to the native state ($\Delta H_I$) is directly determined as a fitting parameter in equation 3 and a suitable metric of the free energy of I relative to N ($\Delta G'_I$) can be easily calculated from the values of the fitting parameters $A$, $T_0$ and $\Delta H_I$ in a straightforward manner (see Text S2 for details).

The energetic description afforded by the $\Delta H_I$ and $\Delta G'_I$ values for the variants immediately indicates an extensively unfolded intermediate species, with the enthalpy of the intermediate being close to that of the unfolded state (both relative to the native state), Figure 7C. Furthermore, a plot of the free energy of the intermediate versus that of the unfolded state (both relative to the native state) including all variants studied is linear with a slope close to unity (Figure 7D). The plot of $\Delta G'_I$ versus $\Delta G_U$ is in fact a

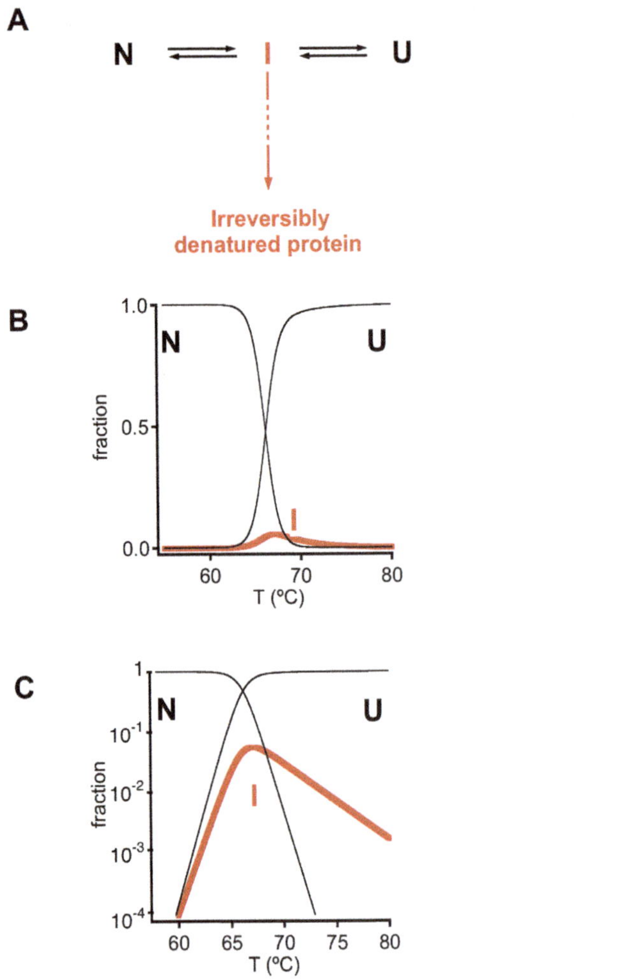


**Figure 8. Model used to derive equation 3 and describe the experimental thermal inactivation profiles for wild-type phytase and variants (Figures 7A and B).** (A) An intermediate state (or ensemble) is assumed to be critical for the irreversible denaturation process. (B) and (C) At equilibrium, the population of I is always low, although it reaches a maximum roughly within the temperature range of the transition. When using a logarithmic scale (panel C) the shape of the population of I versus temperature profiles matches that of the $\ln(\tau_{1/2})$ versus temperature plots of Figure 7, with the maximum of population of I corresponding to the minimum of $\tau_{1/2}$ (see **equation 3**). The profiles in panels B and C have been calculated using equations provided in Text S2.

comparison between the effect of the engineered disulfide bridges on the changes in free energy for the N→I and N→U processes and, therefore, it is equivalent to a mutational ϕ-value analysis [35] of the structure of the intermediate. Indeed, the plot suggests a ϕ value around unity, consistent with a substantially unfolded intermediate species.

## Methods

See Text S1.

## Conclusions

We have used *C. braakii* phytase to probe the effect of designed disulfide bridges on protein kinetic stability. The rational design procedure, based on high temperature molecular dynamics simulations, was highly successful, leading to variants with enhanced thermodynamic stability and, furthermore, with much slower inactivation kinetics. However, as expounded below, the congruence between the effect of the engineered disulfides in the thermodynamic and kinetic stabilities is actually a complex phenomenon.

The inactivation kinetics for the variants shows a striking, non-Arrhenius temperature dependence, with the time-scale for the irreversible denaturation process generally reaching a minimum at a given temperature within the range of the unfolding transitions. This pattern is a clear signature of the key role played by an intermediate state which, being partially unfolded, becomes maximally populated at intermediate temperatures. In fact, inspired by reasoning commonly found in the literature on protein aggregation [33], we can intuitively view the "unusual" temperature dependence reported as reflecting the conformational changes required to reach the critical intermediate species. Thus the native structure needs to undergo significant unfolding to produce the key intermediate and this unfolding process becomes favored as the temperature is increased. In contrast, the unfolded state needs to become structured to reach the intermediate and such folding is disfavoured at the higher temperatures. Accordingly, the rate of the irreversible denaturation decreases with temperature when the initial state of the process is the unfolded protein, while the rate of irreversible denaturation of the native state increases with temperature. The combination of these two opposing trends produces the "V-shaped" dependencies in plots of time-scale for irreversible denaturation *versus* temperature. The "V's" are clearly asymmetric with nearly flat high temperature branches, indicating that the intermediate is highly-unfolded retaining little structure. This interpretation is confirmed by the fittings of the experimental time-scale profiles on the basis of the model embodied in Equation 3 (Figures 7 and 8) and, in particular, by the observation that the derived enthalpy values for the intermediate species are close to those expected for the unfolded state (Figure 7C). This, of course, does not mean that the intermediate species is to be considered as roughly equivalent to the unfolded state. In fact, the intermediate species probably contains a structured region that, although necessarily small (on account of the substantially unfolded character of the intermediate), provides an efficient nucleus for the aggregation process. Indeed, since the order of reaction for the intermediate in the rate equation is approximately unity (see inset in Figure 7B) while the overall reaction order for the aggregation process is about 3–4 (Figure 6C), it appears plausible that the structured region in the intermediate is able to recruit protein molecules in other states (native and unfolded) for the aggregation process. Furthermore, the small structured region must contribute to a high free energy with respect to both the native and unfolded states to explain the low population of the intermediate and it is therefore probably non-native-like. Overall, it is clear that the intermediate species, although substantially unfolded, cannot be viewed as a "member" of the unfolded ensemble.

While many critical intermediates in protein irreversible denaturation processes (such as aggregation and fibrillogenesis) are typically described as "partially-folded", "molten-globule-like" or even "native-like" [36,37,38,39,40], a highly-unfolded intermediate state (with at most a small structured region) has been proposed to play a key role in the misfolding of the prion protein [41,42]. Since the key intermediate state in the thermal inactivation of phytase is also highly unfolded, we may expect mutation effects on its free energy value (measured with respect to the native state) to parallel the corresponding mutation effects on thermodynamic stability. The experimental data validate this expectation (Figure 7D).

Of course, the congruence found between the effect of the engineered disulfide bridges in the thermodynamic and kinetic stabilities is a direct consequence of the highly-unfolded character of the key intermediate in the particular case of phytase and cannot be taken as a general feature for all proteins. We can easily imagine protein systems for which the key intermediate is native-like (thermodynamic stability enhancements caused by engineered disulfide bridges will not lead to kinetic stabilization), partially unfolded (thermodynamic stability enhancements will be partially reflected in kinetic stability) or structurally polarized (only those thermodynamic stability enhancements associated to crosslinks in the region of the protein molecule that becomes unstructured in the intermediate will have consequences for kinetic stability). Nevertheless, the methodology we propose should discriminate between these alternatives. Furthermore, the usefulness of the approach goes beyond providing a general framework for the understanding of the effect of disulfide crosslinking on protein stability. While it is widely accepted that intermediate states/ensembles play key roles in many processes of protein aggregation and fibrillogenesis, their typically low population has made their characterization elusive. For instance, in the paradigmatic case of lysozyme amyloidoses, a recent characterization of the critical non-native states/ensembles relied upon sophisticated experimental methodologies combined with destabilizing solvent conditions [39,43]. The results reported here suggest that the temperature dependencies of suitable metrics of the rate of irreversible protein denaturation may reveal distinct signatures of key intermediate states in the process and that the corresponding energetic/structural descriptions can be derived from mutational analyses of these signatures.

## Supporting Information

**Figure S1 Dynamic light scattering experiments supporting aggregation upon irreversible denaturation of phytase.**

**Figure S2 Rate constants obtained with total protein concentration of 0.25 and 1 mg/mL.**

**Figure S3 Temperature dependence of the time scale ($\tau 1/2$) for the irreversible denaturation of wild type phytase and variants at a protein concentration of 1 mg/mL.**

**Text S1 Experimental section.**

**Text S2 Theoretical analysis of a thermal inactivation model involving a critical, lowly-populated intermediate species.**

## Acknowledgments

The authors thank the ESRF for access to beamline ID14-1 for synchrotron data collection and support by staff during visits. We also thank Mette B. Egede for her skillful technical assistance with the purification of the phytases and for performing the DLS experiments.

Atomic coordinates and structure factors for the reported crystal structure has been deposited with the Protein Data Bank as entry 3ZHC.

## Author Contributions

Conceived and designed the experiments: IS-R KSW LKS LDM JMS-R. Performed the experiments: IS-R AA MS JV. Analyzed the data: IS-R KSW LDM JMS-R. Wrote the paper: IS-R KSW LDM JMS-R.

## References

1. Anfinsen CB, Haber E, Sela M, White FH Jr (1961) The kinetics of formation of native ribonuclease during oxidation of the reduced polypeptide chain. Proc Natl Acad Sci U S A 47: 1309–1314.
2. Pace CN, Grimsley GR, Thomson JA, Barnett BJ (1988) Conformational stability and activity of ribonuclease T1 with zero, one, and two intact disulfide bonds. J Biol Chem 263: 11820–11825.
3. Zhang T, Bertelsen E, Alber T (1994) Entropic effects of disulphide bonds on protein stability. Nat Struct Biol 1: 434–438.
4. Villafranca JE, Howell EE, Voet DH, Strobel MS, Ogden RC, et al. (1983) Directed mutagenesis of dihydrofolate reductase. Science 222: 782–788.
5. Wells JA, Powers DB (1986) In vivo formation and stability of engineered disulfide bonds in subtilisin. J Biol Chem 261: 6564–6570.
6. Matsumura M, Becktel WJ, Levitt M, Matthews BW (1989) Stabilization of phage T4 lysozyme by engineered disulfide bonds. Proc Natl Acad Sci U S A 86: 6562–6566.
7. Betz SF (1993) Disulfide bonds and the stability of globular proteins. Protein Sci 2: 1551–1558.
8. Martensson LG, Karlsson M, Carlsson U (2002) Dramatic stabilization of the native state of human carbonic anhydrase II by an engineered disulfide bond. Biochemistry 41: 15867–15875.
9. Hagihara Y, Mine S, Uegaki K (2007) Stabilization of an immunoglobulin fold domain by an engineered disulfide bond at the buried hydrophobic region. J Biol Chem 282: 36489–36495.
10. Compton JR, Legler PM, Clingan BV, Olson MA, Millard CB (2011) Introduction of a disulfide bond leads to stabilization and crystallization of a ricin immunogen. Proteins 79: 1048–1060.
11. Wozniak-Knopp G, Stadlmann J, Ruker F (2012) Stabilisation of the Fc fragment of human IgG1 by engineered intradomain disulfide bonds. PloS one 7: e30083.
12. Wetzel R (1987) Harnessing disulfide bonds using protein engineering. Trends Biochem Sci 12: 478–482.
13. Ramakrishnan V, Srinivasan SP, Salem SM, Matthews SJ, Colon W, et al. (2012) Geofold: topology-based protein unfolding pathways capture the effects of engineered disulfides on kinetic stability. Proteins 80: 920–934.
14. Pechmann S, Levy ED, Tartaglia GG, Vendruscolo M (2009) Physicochemical principles that regulate the competition between functional and dysfunctional association of proteins. Proc Natl Acad Sci U S A 106: 10159–10164.
15. Grana-Montes R, de Groot NS, Castillo V, Sancho J, Velazquez-Campoy A, et al. (2012) Contribution of disulfide bonds to stability, folding, and amyloid fibril formation: the PI3-SH3 domain case. Antioxid Redox Signal 16: 1–15.
16. Sanchez-Ruiz JM (2010) Protein kinetic stability. Biophys Chem 148: 1–15.
17. Roberts CJ, Das TK, Sahin E (2011) Predicting solution aggregation rates for therapeutic proteins: approaches and challenges. Int J Pharm 418: 318–333.
18. Cohen SI, Vendruscolo M, Dobson CM, Knowles TP (2012) From macroscopic measurements to microscopic mechanisms of protein aggregation. J Mol Biol 421: 160–171.
19. Remmele RL Jr, Zhang-van Enk J, Dharmavaram V, Balaban D, Durst M, et al. (2005) Scan-rate-dependent melting transitions of interleukin-1 receptor (type II): elucidation of meaningful thermodynamic and kinetic parameters of aggregation acquired from DSC simulations. J Am Chem Soc 127: 8328–8339.
20. Mitchinson C, Wells JA (1989) Protein engineering of disulfide bonds in subtilisin BPN'. Biochemistry 28: 4807–4815.
21. Siadat OR, Lougarre A, Lamouroux L, Ladurantie C, Fournier D (2006) The effect of engineered disulfide bonds on the stability of Drosophila melanogaster acetylcholinesterase. BMC Biochem 7: 12.
22. Pecher P, Arnold U (2009) The effect of additional disulfide bonds on the stability and folding of ribonuclease A. Biophys Chem 141: 21–28.
23. Fersht AR, Daggett V (2002) Protein folding and unfolding at atomic resolution. Cell 108: 573–582.
24. Day R, Daggett V (2003) All-atom simulations of protein folding and unfolding. Adv Protein Chem 66: 373–403.
25. Daggett V, Fersht AR (2003) Is there a unifying mechanism for protein folding? Trends Biochem Sci 28: 18–25.
26. Lim D, Golovan S, Forsberg CW, Jia Z (2000) Crystal structures of Escherichia coli phytase and its complex with phytate. Nat Struct Biol 7: 108–113.
27. Ariza A, Moroz OV, Blagova EV, Turkenburg JP, Waterman J, et al. (2013) Degradation of Phytate by the 6-Phytase from *Hafnia alvei*: A Combined Structural and Solution Study. PLoS ONE 8(5): e65062. doi: 10.1371/journal.pone.0065062

28. Bohm K, Herter T, Muller JJ, Borriss R, Heinemann U (2010) Crystal structure of Klebsiella sp. ASR1 phytase suggests substrate binding to a preformed active site that meets the requirements of a plant rhizosphere enzyme. FEBS J 277: 1284–1296.
29. Kostrewa D, Gruninger-Leitch F, D'Arcy A, Broger C, Mitchell D, et al. (1997) Crystal structure of phytase from Aspergillus ficuum at 2.5 A resolution. Nat Struct Biol 4: 185–190.
30. Xiang T, Liu Q, Deacon AM, Koshy M, Kriksunov IA, et al. (2004) Crystal structure of a heat-resilient phytase from Aspergillus fumigatus, carrying a phosphorylated histidine. J Mol Biol 339: 437–445.
31. Oakley AJ (2010) The structure of Aspergillus niger phytase PhyA in complex with a phytate mimetic. Biochem Biophys Res Commun 397: 745–749.
32. Ragon M, Hoh F, Aumelas A, Chiche L, Moulin G, et al. (2009) Structure of Debaryomyces castellii CBS 2923 phytase. Acta Crystallogr Sect F Struct Biol Cryst Commun 65: 321–326.
33. Booth DR, Sunde M, Bellotti V, Robinson CV, Hutchinson WL, et al. (1997) Instability, unfolding and aggregation of human lysozyme variants underlying amyloid fibrillogenesis. Nature 385: 787–793.
34. Weiss WFt, Young TM, Roberts CJ (2009) Principles, approaches, and challenges for predicting protein aggregation rates and shelf life. J Pharm Sci 98: 1246–1277.
35. Fersht AR, Matouschek A, Serrano L (1992) The folding of an enzyme. I. Theory of protein engineering analysis of stability and pathway of protein folding. J Mol Biol 224: 771–782.
36. Uversky VN, Fink AL (2004) Conformational constraints for amyloid fibrillation: the importance of being unfolded. Biochim Biophys Acta 1698: 131–153.
37. Jahn TR, Radford SE (2008) Folding versus aggregation: polypeptide conformations on competing pathways. Arch Biochem Biophys 469: 100–117.
38. Jahn TR, Parker MJ, Homans SW, Radford SE (2006) Amyloid formation under physiological conditions proceeds via a native-like folding intermediate. Nat Struct Mol Biol 13: 195–201.
39. Dhulesia A, Cremades N, Kumita JR, Hsu ST, Mossuto MF, et al. (2010) Local cooperativity in an amyloidogenic state of human lysozyme observed at atomic resolution. J Am Chem Soc 132: 15580–15588.
40. Bemporad F, De Simone A, Chiti F, Dobson CM (2012) Characterizing intermolecular interactions that initiate native-like protein aggregation. Biophys J 102: 2595–2604.
41. Hosszu LL, Baxter NJ, Jackson GS, Power A, Clarke AR, et al. (1999) Structural mobility of the human prion protein probed by backbone hydrogen exchange. Nat Struct Biol 6: 740–743.
42. Yu H, Liu X, Neupane K, Gupta AN, Brigley AM, et al. (2012) Direct observation of multiple misfolding pathways in a single prion protein molecule. Proc Natl Acad Sci U S A 109: 5283–5288.
43. Buell AK, Dhulesia A, Mossuto MF, Cremades N, Kumita JR, et al. (2011) Population of nonnative states of lysozyme variants drives amyloid fibril formation. J Am Chem Soc 133: 7737–7743.
44. Krissinel E, Henrick K (2004) Secondary-structure matching (SSM), a new tool for fast protein structure alignment in three dimensions. Acta Crystallogr D Biol Crystallogr 60: 2256–2268.
45. McNicholas S, Potterton E, Wilson KS, Noble ME (2011) Presenting your structures: the CCP4mg molecular-graphics software. Acta Crystallogr D Biol Crystallogr 67: 386–394.

# 22

# A Thermostable β-Glucuronidase Obtained by Directed Evolution as a Reporter Gene in Transgenic Plants

**Ai-Sheng Xiong[1]*, Ri-He Peng[2], Jing Zhuang[3], Jian-Min Chen[4], Bin Zhang[2], Jian Zhang[3]*, Quan-Hong Yao[2]***

**1** State Key Laboratory of Crop Genetics and Germplasm Enhancement, College of Horticulture, Nanjing Agricultural University, Nanjing, China, **2** Biotechnology Research Institute, Shanghai Academy of Agricultural Sciences, Shanghai, China, **3** Alberta Innovates-Technology Futures, Vegreville, Alberta, Canada, **4** College of Bioscience and Biotechnology, Yangzhou University, Yangzhou, China

## Abstract

A β-glucuronidase variant, GUS-TR3337, that was obtained by directed evolution exhibited higher thermostability than the wild-type enzyme, GUS-WT. In this study, the utility of GUS-TR337 as an improved reporter was evaluated. The corresponding *gus-tr3337* and *gus-wt* genes were independently cloned in a plant expression vector and introduced into *Arabidopsis thaliana*. With 4-MUG as a substrate, plants containing the *gus-wt* gene showed no detectable β-glucuronidase activity after exposure to 60°C for 10 min, while those hosting the *gus-tr3337* gene retained 70% or 50% activity after exposure to 80°C for 10 min or 30 min, respectively. Similarly, *in vivo* β-glucuronidase activity could be demonstrated by using X-GLUC as a substrate in transgenic *Arabidopsis* plants hosting the *gus-tr3337* gene that were exposed to 80°C for up to 30 min. Thus, the thermostability of GUS-TR3337 can be exploited to distinguish between endogenous and transgenic β-glucuronidase activity, which is a welcome improvement in its use as a reporter.

**Editor:** Annalisa Pastore, National Institute for Medical Research, Medical Research Council, London, United Kingdom

**Funding:** The research is supported by the Priority Academic Program Development of Jiangsu Higher Education Institutions; National key Project of Transgenic Crops of China (2009ZX08002-011B); International Scientific and Technological Cooperation (2010DFA62320). The funders had no role in study design, data collection and analysis, decision to publish, or preparation of the manuscript.

**Competing Interests:** Jing Zhuang and Jian Zhang are employees of Alberta Innovates-Technology Futures. There are no patents, products in development or marketed products to declare.

* E-mail: xiongaisheng@njau.edu,cn (A-SX); Jian.Zhang@albertainnovates.ca (JZ); yaoquanhong@saas.sh.cn (Q-HY)

## Introduction

There are about fifty reporter genes that have proved useful in transgenic plant research, based on their efficiency, scientific applications and commercialization. Reporter genes serve as indicators to study transgenic events, for example by facilitating visual identification of successfully transformed entities in a large background of non-transformed material. Although many reporter genes have been proposed, only a few of them have been commercialized [1]. β-glucuronidase (GUS) [2,3], green fluorescent protein (GFP) [4,5], luciferase (LUC) [6,7,8] are three important reporters that have a proven track record in transgenic plant research.

The advantage of GFP and LUC as reporters includes the possibility of monitoring in live tissues and in real time [1,4,8]. These two reporters allow continuous monitoring of the gene activity even through the developmental stages of the transgenic plants [1,5,9]. However, there are several disadvantages in the use of these two reporter genes. For example, the half life of LUC is short (a few hours) in plants and the process of catalyzing the release of oxyluciferin is slow [6,7,10]. In case of GFP, there may be interference due to a high background of auto-fluorescence. Another major drawback in the use GFP as a reporter is that its detection requires relatively expensive instrumentation, namely the fluorescence microscope.

The bacterial enzyme GUS encoded by the *E. coli gusA* gene is to date the most widely used as a reliable transgenic reporter system in transgenic plant research [1]. GUS has some advantages as a reporter in transgenic plant research and commercialization. First, in genetically modified plants *E.coli* GUS can be regarded as safe for environment and consumers [11]. Second, in transgenic plants, overexpressed GUS is very stable and shows no toxicity in plant growth and development. In addition, detection of GUS does not require specialized equipment as it can easily be observed after staining with substrate such as 5-bromo-4-chloro-3-indolyl-β-D-glucuronic acid (X-GLUC). Thus, the GUS reporter system is a simple, reliable and cost-effective method to be used in transgenic studies [1,12]. Despite these benefits of the *gusA* gene, false-positives are difficult to eliminate in some plants due to endogenous GUS activity [13]. Increasing the thermal stability of GUS through directed evolution *in vitro* could potentially eliminate false-positives and dramatically increase its veracity as a reporter.

The methodology of *in vitro* directed evolution has been approved in the laboratory [14,15,16]. Directed evolution has provided significant advances in many fields, such as for biocatalysis with industrially important enzymes [17,18,19,20,21,22], plant improvement [23,24,25,26], or in vaccines and pharmaceuticals [27,28,29].

Directed evolution *in vitro* via DNA shuffling, is an important method for generating proteins with improved functions. Stemmer introduced the method of *in vitro* DNA shuffling for the formation of recombinant genes from a set of parental genes [30,31]. We have previously reported the directed evolution of a GUS variant,

GUS-TR3337, which is significantly more resistant to high temperatures than wild-type enzyme, GUS-WT [32,33]. Here, our objective was to test the utility of GUS-TR3337 as an improved reporter gene in a transgenic plant.

In order to study the activity of the thermostable GUS in plants, it was introduced into *Arabidopsis thaliana* by the floral dip method. The transgenic *Arabidopsis* plants containing the wild-type *gus-wt* gene (Genbank No. AF485783), lost most of the GUS activity within 5 min of exposure to 60°C. In contrast, *Arabidopsis* plants transformed with the *gus-tr3337* gene (GenBank No. DQ513152), retained the activity when heated upto 80°C for as long as 30 min. We confirm here that the mutant thermostable GUS gene, *gus-tr3337* is an improved reporter for transgenic research.

## Materials and Methods

### Chemicals, enzymes, plasmids and yeast strains

All reagents were purchased from Sigma Chemical Co. Ltd (St Louis, MO, USA) and Sinopharm Chemical Reagent Co. Ltd (Shanghai, China). Restriction enzymes were purchased from Promega (Madison, USA). High-fidelity DNA polymerase Pyrobest$^{TM}$ was purchased from Takara Co., Ltd (Dalian, China). DNA sequencing kit was purchased from Applied Systems Company (Foster City, CA, USA). QIAquick$^{TM}$ gel extraction kit was purchased from Qiagen (Stanford, CA, USA). X-GLUC and plasmid pBI121, containing the *gusA* gene, were purchased from Clontech (Palo Alto, CA, USA). *E. coli* strain DH5α used for amplification of plasmid DNA was purchased from Stratagene (La Jolla, CA, USA).

### Construction of plant expression vectors and plant transformation

The original *gus-wt* and mutant *gus-tr3337* genes were independently cloned in the binary vector pYF7716 that was constructed by our research team. This vector contains double-cauliflower mosaic virus 35S promoter, to direct transcription of the transgene as well as that of the intron-containing neomycin phosphotransferase II gene (*Km*). In addition, the vector carries the omega enhancer and the nopaline synthase terminator, that were fused to the tobacco SAR (scaffold attachment region) (Figure S1). The *gus-wt* and *gus-tr3337* constructs generated in this vector were introduced into *Agrobacterium tumefaciens* GV3101 by electroporation. *A. thaliana* cv. Columbia was transformed by the floral dip method as described previously [34]. The presence of the *gus* transgenes was confirmed via PCR amplification of the genes in *Arabidopsis* plants using the following specific primers pairs: GUS-WTF1 (5′-ATGTTACGTCCTGTAGAAACCCCAACC-3′) and GUS-WTR1 (5′-TCATTGTTTGCCTCCCTGCTGCGGTTT-TTC-3′), for the *gus-wt* gene, and GUS-TRF1 (5′-TCACGCCG-TACG TTATTGCCGGG-3′) and GUS-TRR1 (5′-TTGCT-ACCGCCAACGCGTAATATG-3′) for the *gus-tr3337* gene. The PCR conditions were: denaturation for 10 min at 94°C, followed by 30 cycles, each of 40 s at 94°C, 30 s at 58°C, 2 min at 72°C; and then extension for 10 min at 72°C. Standard DNA manipulation methods were used in plasmid construction and *E. coli* transformation [35].

### Histochemical analysis of GUS activity in transgenic *Arabidopsis* plants

For the analysis of expression of the *gus* transgenes, seeds of the trangenic *Arabidopsis* plants were sterilized with 1% (v/v) NaOCl solution and sown in pots containing a 1:1:1 mixture of peat moss/perlite/vermiculite. Pots were kept in a controlled-environment growth chamber at 22°C, programmed for a photoperiod of 16 h-day under cool white fluorescent light and 8 h-dark. High temperature treatments were performed on whole two-week old seedlings by exposure to temperatures of 70°C, 80°C, 30°C for different times (10 min, 20 min, 30, min). The heat treatment was given by placing a single seedling in a 0.6 mL capacity Eppendorf$^{TM}$ tube containing 150 μL ddH$_2$0, and then holding the tube at the desired heating regime in a PCR System PTC-200, using the lid heating function to avoid the water evaporation. Then, the water was removed. Based on the apparent optimal temperature of β-glucuronidase activity tested in transgenic *Arabidopsis* plants, the plant samples were incubated at 45°C for transgenic *gus-wt Arabidopsis* samples or 60°C for transgenic *gus-tr3337 Arabidopsis* samples, with 1 mg/mL X-GLUC in a GUS activity buffer (50 mM sodium phosphate, pH 7.0; 1 mM EDTA; 0.1% Triton X-100; 0.5 mM K3[Fe(CN)6]; 0.5 mM K4[Fe(CN)6]·3H2O). Histochemical staining for GUS activity was performed essentially as described by Jefferson et al. [2] and Sudan et al. [36] for most experiments. The stained samples were immersed in 70% ethanol to remove chlorophyll. The ethanol solution was changed three times at 1-h intervals.

### GUS enzyme kinetics by fluorometric assay in transgenic *Arabidopsis* plants

The GUS activity in plant samples was measured by a fluorometric assay using DyNA Quant 200$^{TM}$ Fluorometer (Hoefer Pharmacia Biotech, San Francisco, CA, USA) as described by Martin et al. [37] and Fior et al. [38] for most experiments. Standards were prepared with different concentrations of 4-methylumbelliferone sodium salt (4-MU, Sigma Chemical Co. Ltd., No. M1508): 1 μM, 1 mM and 100 mM, in 0.2 M sodium carbonate. 4-Methylumbelliferyl-β-D-glucuronide (4-MUG, Sigma Chemical Co. Ltd., No. M5664), 2 mM, was added to sample as substrate. GUS extraction buffer (50 mM sodium phosphate, pH 7.0; 10 mM β-mercaptoethanol; 10 mM EDTA and 0.1% Triton X-100) was used as reaction medium.

Extracts of whole three-week-old seedlings in GUS extraction buffer were used for the analysis of enzyme kinetics. Seedlings were frozen by immersion in liquid nitrogen and ground with a pestle and mortar in the presence of liquid nitrogen. GUS extraction buffer, 500 μL, was added and the plant material was vigorously ground to a finely pulverized powder. The extract thus obtained was centrifuged at 8,000× *g* for 1 min at 4°C. The supernatant was recovered and diluted one-hundred fold with GUS extraction buffer in order to decrease the effect of any potential indigenous GUS inhibitors [39]. The diluted extract, 10 μL was added to 4-MUG, 90 μL (2 mM), to initiate the reaction. After 1 h the reaction was stopped by the addition of 900 μL of 0.2 M $Na_2CO_3$, and fluorescence was measured using a DyNA Quant 200$^{TM}$ spectrofluorometer.

The optimal pH for the activity of enzymes GUS-WT and GUS-TR3337 from transgenic *Arabidopsis* plants was determined at 37°C at every half point between pH values 2.5 and 13.0. The activity of both enzymes from transgenic *Arabidopsis* plants was measured at 5-degree intervals between 25°C and 90°C to determine the optimum temperature at which maximum activity was retained. The thermostability of the two enzymes from transgenic *Arabidopsis* plants was determined by measuring the residual activity at 37°C after subjecting the samples to temperatures of 60°C, 70°C, 80°C and 90°C for different times (10 min, 20 min and 30 min), followed by chilling on ice. All reactions were performed in triplicate. Kinetic parameters were determined as described in previous reports [32,40,41].

## Results

### Over-expression of *gus-wt* and *gus-tr3337* in *Arabidopsis*

The cloned *gus-wt* or *gus-tr3337* genes under the control of the CaMV 35S promoter were independently introduced into *Arabidopsis* (Figure S1). The transgenic *Arabidopsis* plants thus recovered showed high levels of expression of the two genes (Figure S2). There were no noticeable differences in the morphology or growth among the above transgenic plants and the parental *Arabidopsis* plants (data not shown).

### Thermostability of the mutant GUS-TR3337 in transgenic *Arabidopsis* plant

High temperature treatments were performed with two-week-old whole seedlings to evaluate the thermostability of the mutant GUS-TR3337 in transgenic *Arabidopsis* plants. Seedlings of wild-type *Arabidopsis* could not been stained by X-GLUC when they were not or subjected to 60°C for 5 min (Figure 1). Seedlings containing the wild-type *gus-wt* gene or the mutant *gus-tr3337* gene exhibited stable GUS activity when they were not subjected to high temperature treatments (Figures 2A and 3A). The seedlings expressing the *gus-wt* gene were stained weakly by X-GLUC after being subjected to 60°C for 5 min or 10 min (Figures 2B and 2C) and the enzyme appeared to be inactivated by treatments for longer duration or at higher temperature (Figures 2D–F).

On the opposite, seedlings expressing the mutant *gus-tr3337* gene exhibited stable GUS activity when treated at 60°C and 70°C for 10 min, 20 min or 30 min (Figures 3C–E and F–H), and even at 80°C for 10 min, 20 min or 30 min (Figures 3I–K). However, treatment at 90°C for 10 min or longer appeared to inactivate the GUS-TR3337 enzyme (Figures 3H–I).

### Characterization of the GUS-TR3337 activity in transgenic *Arabidopsis* plants

Three-week-old transgenic *Arabidopsis* seedlings containing the mutant *gus-tr3337* gene or the wild-type *gus-wt* gene were used to characterize the GUS-TR3337 and GUS-WT enzymes in seedling extracts. Three representative seedlings each (P1, P2 and P3) expressing either of transgenes were tested using 4-MUG as substrate at 37°C. A single activity peak was found at pH 6.5 in all six transgenic *Arabidopsis* lines expressing *gus-tr3337* or *gus-wt* gene (Figure 4A).

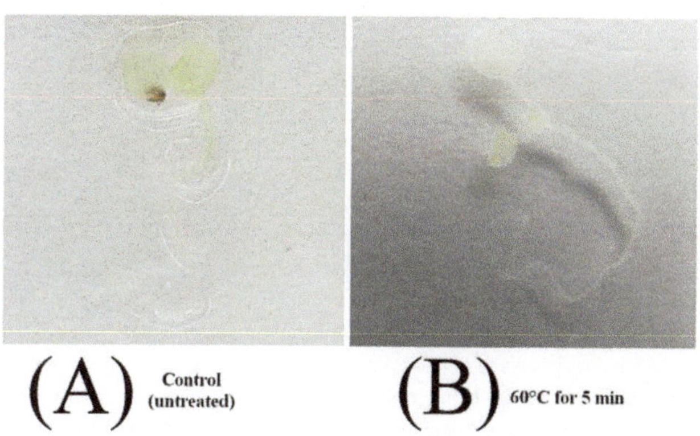


**Figure 1. Histochemical staining of GUS activity in wild-type *Arabidopsis* seedlings.** Seedlings were subjected to heat treatment and then stained at with X-GLUC 37°C overnight. A: Control (untreated); B: 60°C for 5 min.

The apparent optimal temperature of GUS activity tested at pH 7.0 in extracts of transgenic *Arabidopsis* plants over-expressing *gus-wt* gene was 45°C, while that of transgenic *Arabidopsis* plants over-expressing *gus-tr3337* gene was 60°C (Figure 4B). The GUS-WT activity in transgenic *Arabidopsis* lines at 45°C was about 30% higher than that at 35°C. The GUS-TR3337 activity in transgenic *Arabidopsis* lines at 60°C was >2 times of its activity at 35°C.

The GUS-WT activity in the extracts of transgenic *Arabidopsis* decreased significantly after being subjected to 60°C for 10 min, and was undetectable after treatment at 70°C or higher temperatures for 10 min or longer (Figure. 5). In contrast, GUS-TR3337 activity in the transgenic *Arabidopsis* plants was much more resistant to heat treatment. Nearly 90% and 70% of GUS activity was retained when plants were heated at 70°C for 10 min or 30 min, and about 70% and 50% was retained when heated at 80°C for 10 min or 30 min.

Similarly, when heated at 70°C for 10 min or 30 min, they retained nearly 90% and 70% of this activity, respectively; and when heated at 80°C for 10 min or 30 min, retained about 70% and 50%, of this activity, respectively. The GUS-TR3337 activity decreased significantly following at heat treatment 90°C for 10 min and was undetectable if subjected to 90°C or higher temperatures for 30 min or longer (Figure 5).

## Discussion

Various reporter genes encoding GFP, GAL, LUC, GUS, or oxalate oxidase (OxO) have been used in transgenic plant research or crop development, and have been assessed for the efficiency, scientific application and commercialization. Many of these genes have specific limitations or have not been sufficiently tested to merit their widespread use [1]. There are some reports on improved reporter genes that allow easy detection or a wide range of applications and that are generated by directing their evolution in the laboratory. Crameri and his colleagues used DNA shuffling to obtain GFP variants with 40 times increased fluorescence intensity [42]. A mutant GFP with stronger fluorescence intensity was selected by directed evolution in conjunction with a functional salvage screen [43].

The GUS enzyme encoded by *gusA* gene of *E. coli* is still to date the most widely used reporter gene in plants. Some limitations of using GUS as a reporter are the activity assay that is destructive to the plants and the false positive events due to endogenous GUS or GUS-like enzymes activity in some plants [44]. Hu et al. reported that there were intrinsic GUS-like activities in some seed plants [13]. Some novel GUS or GUS-like genes were also isolated and characterized from plants, such as *Scutellaria baicalensis Georgi* [45]. GUS-like enzymes were known to be present in microorganisms [2,3,46,47], animals [48,49] and plants [11,13,36,39,50], such as *Scutellaria baicalensis* [51], rye [52], and maize [53]. Sudan and his colleagues demonstrated that GUS is ubiquitously present in plants [36]. Over the past three decades, numerous reports have illustrated that there was GUS background activity in different plant species, including some model plants, such as *Arabidopsis thaliana*, *Oryza sativa*, *Nicotiana tabacum* and *Zea mays* [36,38,54,55,56].

Therefore, efforts have been focused on the suppression of the background activity [57,58,59]. As reported by Kosugi et al., the addition of methanol to 20% of the final volume during GUS staining, or during fluorimetric assays, eliminates the endogenous activities found in many plants [59]. The methanol appears to have no effect on the bacterial GUS encoded by the *gusA* from *E. coli*. However, methanol is not always effective in suppressing the endogenous GUS activities, and moreover, it is harmful to the experimentalists.

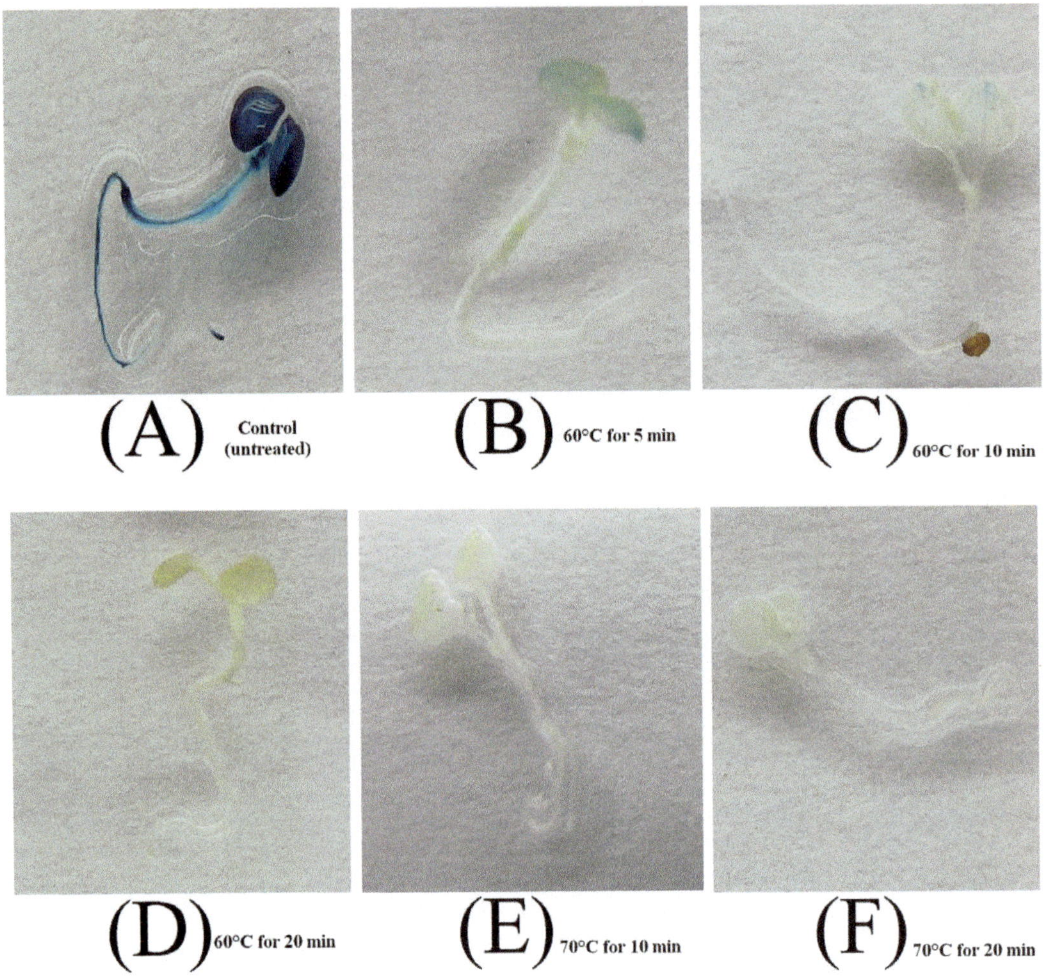


**Figure 2. Histochemical staining of GUS activity in transgenic *Arabidopsis* seedlings hosting the *gus-wt* gene.** Seedlings were subjected to various heat treatments and then stained at with X-GLUC 45°C overnight. (A) Control (untreated); (B) 60°C for 5 min; (C) 60°C for 10 min; (D) 60°C for 20 min; (E) 70°C for 10 min; (F) 70°C for 20 min.

The *E. coli* GUS system is a good model system for the study of directed evolution *in vitro*. Matsumura et al. have directed the evolution of a GUS variant that is significantly more resistant to both glutaraldehyde and formaldehyde than the wild-type enzyme [40]. This GUS variant could resolve the problem of loss of GUS activity in transgenic plants during tissue fixation by glutaraldehyde or formaldehyde. After four cycles of screening the best variant was more active than the wild-type enzyme, and retained function at 70°C, whereas the wild-type enzyme lost activity at 65°C [41]. DNA shuffling (sexual recombination) and a histochemical screen have also been used to direct the evolution of *E. coli* GUS variants with improved β-glucuronidase activity [60]. The same model evolutionary system was employed to test the efficiency of several other techniques [61]. One of the GUS variants was obtained by screening for increased p-nitrophenyl-β-d-xylopyranoside (pNP-xyl) activity [62]. Although there have been a few studies on directed evolution of GUS to obtain an improved reporter, none of them have been in transgenic plants.

We obtained a temperature resistant *gusA* variant *gus-tr3337* (GenBank No. DQ513151) by employing a high throughput system for directed evolution *in vitro* [32,33]. The mutant enzyme maintained its activity even when the nitrocellulose filter containing the variant colony was heated at 100°C for 30 min. Sequencing of the *gus-tr3337* revealed six mutations: Q493R, T509A, M532T, N550S, G559S and N566S that are present in GUS-TR3337, indicating that these were the key amino acids needed to confer its high thermostability. The purified GUS-TR3337 protein retained most of its activity when heated at 80°C for 10 min, while the wild-type GUS protein lost its activity completely after 10 min at 70°C. With pNPG as a substrate, the purified GUS-TR3337 protein retained 75% or 48% of its activity when heated at 80°C for 10 min or 30 min, respectively [33]. In the present study, with 4-MUG as substrate, transgenic *Arabidopsis* plants hosting the *gus-tr3337* gene retained about 70% and 50% of GUS activity when heated at 80°C for 10 min or 30 min, respectively. With pNPG as substrate, a single activity peak had been recorded at pH 6.5 at 37°C with the purified GUS-WT and GUS-TR3337 enzymes [33]. Similar results were obtained in the present study on GUS enzymes in extracts from transgenic *Arabidopsis* plants overexpressing the *gus-tr3337* gene and the *gus-wt* gene. The apparent optimal temperature for GUS-WT and GUS-TR3337 tested at pH 6.5 was 45°C and 65°C, respectively with pNPG as substrate; while temperature optima for the two enzymes expressed in *Arabidopsis* plants, as tested with 4-MUG as a substrate at pH 7.0 was 45°C and 60°C respectively. Thus, the apparent temperature optimum of GUS-TR3337 expressed in transgenic

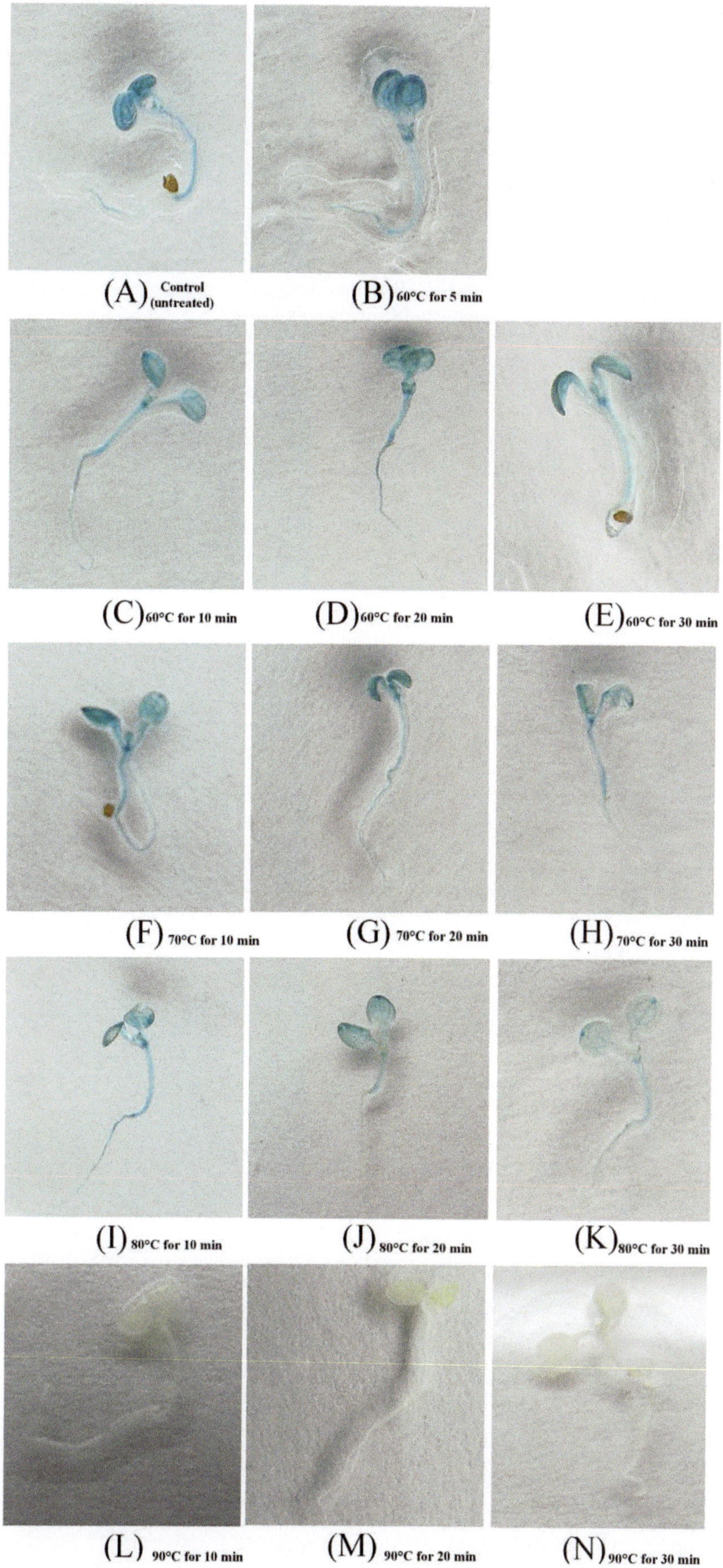
(A) Control (untreated)
(B) 60°C for 5 min
(C) 60°C for 10 min
(D) 60°C for 20 min
(E) 60°C for 30 min
(F) 70°C for 10 min
(G) 70°C for 20 min
(H) 70°C for 30 min
(I) 80°C for 10 min
(J) 80°C for 20 min
(K) 80°C for 30 min
(L) 90°C for 10 min
(M) 90°C for 20 min
(N) 90°C for 30 min

**Figure 3. Histochemical staining of GUS activity in transgenic *Arabidopsis* seedlings hosting the *gus-tr3337* gene.** Seedlings were subjected to various heat treatments and then stained at with X-GLUC 60°C overnight. (A) Control (untreated); (B) 60°C for 5 min; (C–E) 60°C for 10 min, 20 min or 30 min respectively; (F–H) 70°C for 10 min, 20 min or 30 min, respectively; (I–K) 80°C for 10 min, 20 min or 30 min, respectively; (L–N) 90°C for 10 min, 20 min or 30 min, respectively.

plants was 5°C lower than that of purified enzyme, under the conditions of the assays. We also repeated the characterization (determination of the optimal pH, temperature and of residual activity after heat treatment) of the purified enzymes of GUS-TR3337 and GUS-WT by comparing pNPG and MUG as substrates and found that the two substrates yield similar results (Figures S3 and S4).

Although the GUS system is the most popular transgenic reporter gene for plants [1], there were some limitations of the use GUS system as a reporter. Other potential problems with GUS, as discussed by Taylor [44], include its false positive, particularly in those plants with endogenous high endogenous GUS or GUS like activity that interferes with the histochemical detection of transgenic GUS. In the present study, transgenic *Arabidopsis* plants with the *gus-tr3337* gene (GenBank No. DQ513152) retained GUS enzyme activity in plant extracts as well *in vivo* even after being heated at 80°C for up to 30 min, while plants with the *gus-wt* gene lost both *in vitro* and *in vivo* enzyme activity almost completely after 10 min at 60°C. Based on the staining with 1 mg/mL X-GLUC the apparent temperature optimum for GUS-TR3337 activity was found to be 60°C. The transgenic *Arabidopsis* plant hosting the *gus-tr3337* gene could be easily staining with X-GLUC after heated at 80°C for 10 min to 30 min. The thermostability of GUS-TR3337 can be exploited to minimize the recovery of false positives. Some limitations of GUS as a reporter could be overcome by using this thermostable GUS gene (*gus-tr3337*). Staining with X-GLUC after

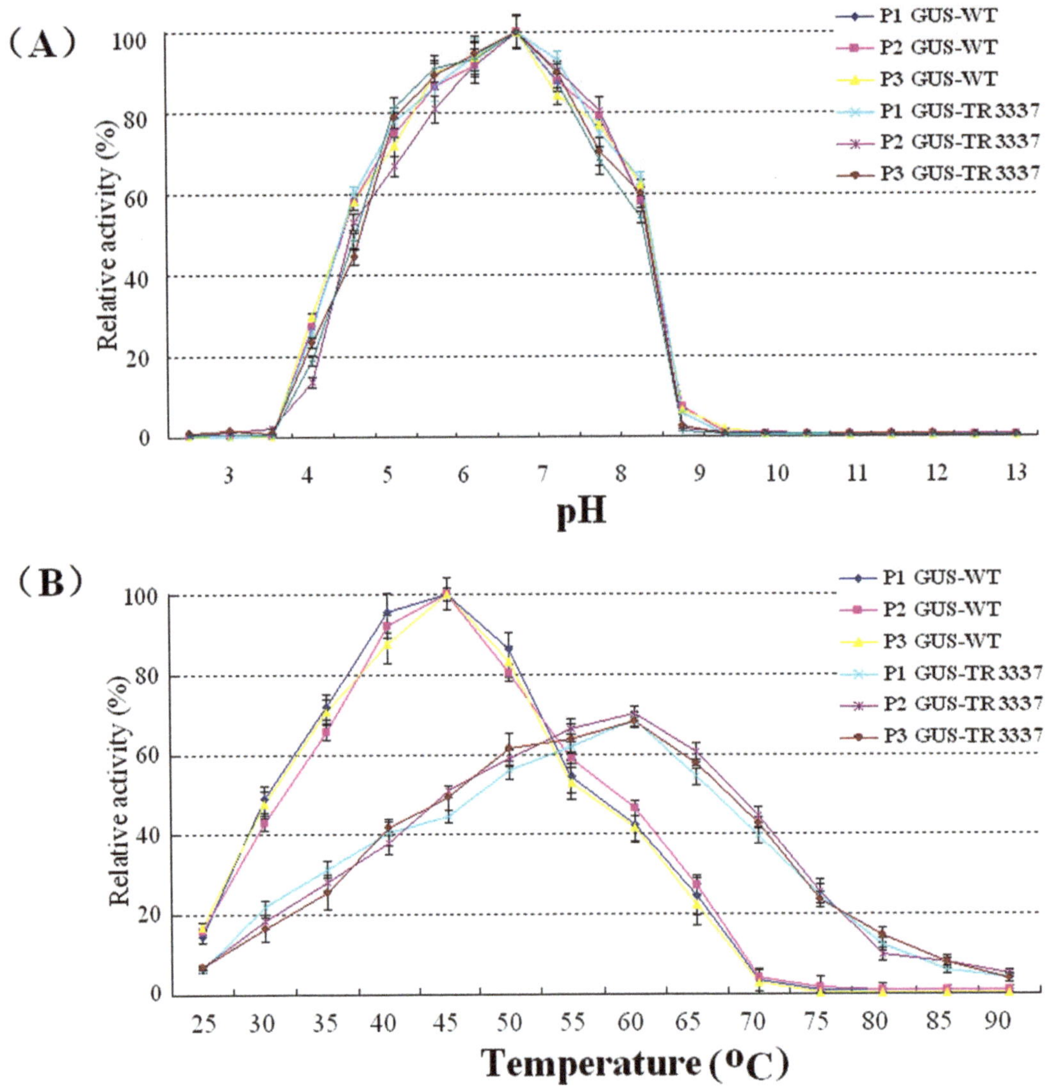


**Figure 4. pH- and temperature-dependence of GUS activity in transgenic *Arabidopsis* plants hosting the *gus-tr3337* gene or the *gus-wt* gene.** P1. P2. P3 are triplicates. (A) Enzymatic activity in seedling extracts was measured in buffers at the indicated pH; (B) Enzymatic activity in seedling extracts was measured in phosphate buffer at pH 7.0 at the indicated temperatures. Values are mean ± S.D. of triplicates.

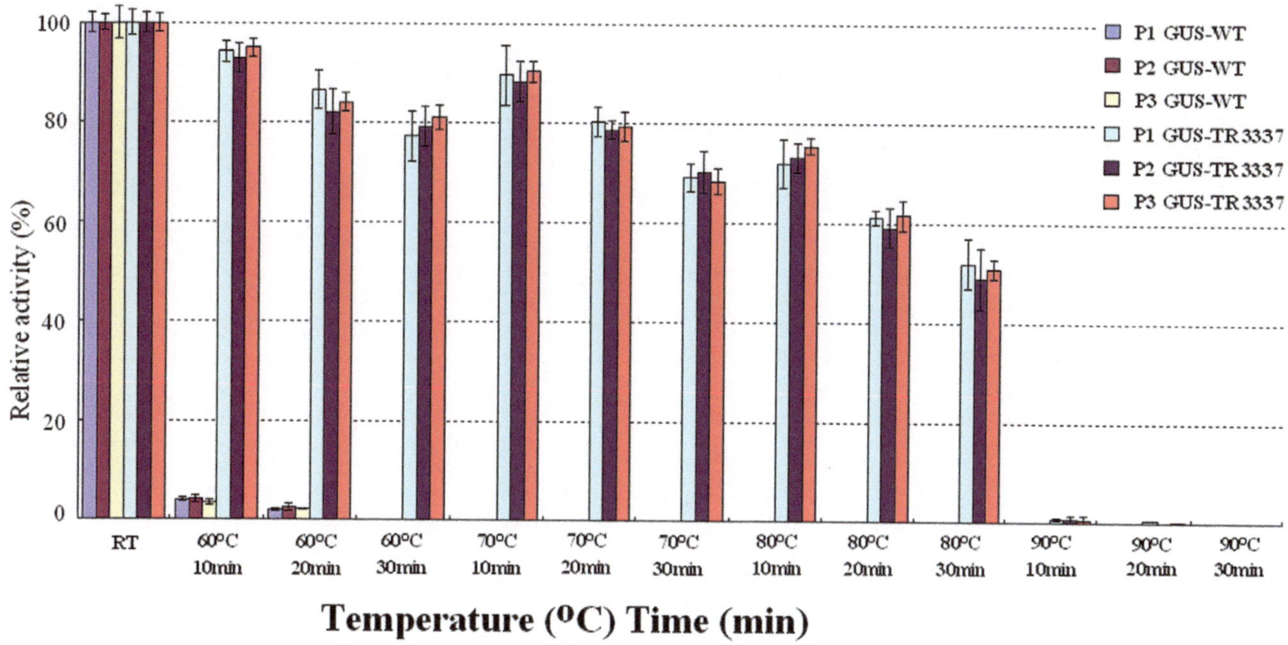


**Figure 5. Thermostability of GUS in transgenic *Arabidopsis* plants.** P1-, P2-, P3-GUS-TR3337: transgenic *Arabidopsis* plant lines hosting the mutant *gus-tr3337* gene; P1-, P2-, P3-GUS-WT: transgenic *Arabidopsis* plant lines hosting the wild-type *gus-wt* gene. Seedlings were subjected to heat treatments at the indicated temperature and duration and the residual GUS activity was measured in the plant extracts at 37°C. Values are mean ± S.D. of triplicates.

heat treatment at 70°C for 30 min or at 80° for 10–30 min will allow to distinguish the GUS activity associated to the transgenic gus-tr3337 expression from the endogenous one, responsible of false positives, in the plants where the endogenous Gus or GUS-like activity is not thermostable. In conclusion, the use of *gus-tr3337* transgene can enhance the precision, sensitivity, and reliability of *gus* gene as a reporter.

## Supporting Information

**Figure S1 Schematic diagram of the vectors used in this study.** The original (*gus-wt*) and mutant (*gus-tr3337*) genes was cloned into the binary vector pYF7716, at the *Bam* HI and *Sac* I restriction enzyme sites, under the control of double CaMV 35S promoter (D35S). For steady transmission of *gus-wt* and mutant *gus-tr3337*, two scaffold attachment regions (SAR) were fused upstream of the D35S promoter and downstream of the Nos-Terminator (Nos-T).

**Figure S2 Identification of $T_1$ generation of transgenic plants.** (A) Screening of $T_1$ generation seeds on kanamycin containing plates; (B) Screening of $T_1$ transgenic seedlings after staining for GUS activity.

**Figure S3 pH- and temperature-dependence of GUS activity in purified protein with 4-MUG as a substrate.** (A) Enzymatic activity was measured at the indicated pH; (B) Enzymatic activity was measured at the indicated temperatures. Values are mean ± S.D. of triplicates.

**Figure S4 Thermostability of GUS in purified protein with 4-MUG as a substrate.** Proteins were subjected to heat treatments at the indicated temperature and duration and the residual GUS activity was measured at 37°C. Values are mean ± S.D. of triplicates.

## Author Contributions

Conceived and designed the experiments: ASX QHY J. Zhang. Performed the experiments: ASX J. Zhuang BZ. Contributed reagents/materials/analysis tools: QHY RHP JMC. Wrote the paper: ASX. Revised the paper: ASX QHY. All authors read and approved the final manuscript.

## References

1. Miki B, McHugh S (2004) Selectable marker genes in transgenic plants: applications, alternatives and biosafety. J Biotechnol 107: 193–232.
2. Jefferson RA, Kavanaugh TA, Bevan MW (1987) GUS fusions: β-glucuronidase as a sensitive and versatile gene fusion marker for higher plants. EMBO J 6: 3901–3907.
3. Jefferson RA (1989) The GUS reporter gene system. Nature 342: 837–838.
4. Stewart CN (2001) The utility of green fluorescent protein in transgenic plants. Plant Cell Rep 20: 376–382.
5. Haseloff J, Siemering KR (2006) The uses of green fluorescent protein in plants. Methods Biochem Anal 47: 259–284.
6. Millar AJ, Short SR, Chua NH, Kay SA (1992) A novel circadian phenotype based on firefly luciferase expression in transgenic plants. Plant Cell 4: 1075–1087.
7. Millar AJ, Short SR, Hiratsuka K, Chua NH, Kay SA (1992) Firefly luciferase as a reporter of regulated gene expression in higher plants. Plant Mol Biol Rep 10: 324–333.
8. Hastings JW, Morin JG (2006) Photons for reporting molecular events: green fluorescent protein and four luciferase systems. Methods Biochem Anal 47: 15–38.
9. Verhees J, van der Krol AR, Vreugdenhil D, van der Plas LH (2002) Characterization of gene expression during potato tuber development in individuals and populations using the luciferase reporter system. Plant Mol Biol 50: 653–665.
10. van Leeuwen W, Hagendoorn MJM, Ruttink T, van Poecke R, van der Plas LHW, et al. (2000) The use of the luciferase reporter system for in planta gene expression studies. Plant Mol Biol Rep 18: 143–144.

11. Gilissen LJ, Metz PL, Stiekema WJ, Nap JP (1998) Biosafety of *E. coli* beta-glucuronidase (GUS) in plants. Transgenic Res 7: 157–163.
12. Mantis J, Tague RW (2000) Comparing the utility of β-glucuronidase and green fluorescent protein for detection of weak promoter activity in *Arabidopsis thaliana*. Plant Mol Biol Rep 18: 319–330.
13. Hu C, Chee PP, Robert H. Chesney, James H. Zhou JH, et al. (1990) Intrinsic GUS-like activities in seed plants. Plant Cell Rep 9: 1–5.
14. Dougherty MJ, Arnold FH (2009) Directed evolution: new parts and optimized function. Curr Opin Biotechnol 20: 486–491.
15. Shivange AV, Marienhagen J, Mundhada H, Schenk A, Schwaneberg U (2009) Advances in generating functional diversity for directed protein evolution. Curr Opin Chem Biol 13: 19–25.
16. Xiong AS, Yao QH, Peng RH, Cheng ZM (2010) Directed in vitro evolution of reporter genes based on semi-rational design and high throughput screening. *In Vitro* Mutagenesis Protocols. 3rd edition. Edited by Jeff Braman. Springer, Humana Press Inc. Totowa, NJ, USA.
17. Arnold FH, Volkov AA (1999) Directed evolution of biocatalysts. Curr Opin Chem Biol 3: 54–59.
18. Chirumamilla RR, Muralidhar R, Marchant R, Nigam P (2001) Improving the quality of industrially important enzymes by directed evolution. Mol Cell Biochem 224: 159–168.
19. Zhao H, Chockalingam K, Chen Z (2002) Directed evolution of enzymes and pathways for industrial biocatalysis. Curr Opin Biotechnol 13: 104–110.
20. Alexeeva M, Carr R, Turner NJ (2003) Directed evolution of enzymes: new biocatalysts for asymmetric synthesis. Org Biomol Chem 1: 4133–4137.
21. Otten LG, Quax WJ (2005) Directed evolution: selecting today's biocatalysts. Biomol Eng 22: 1–9.
22. Tian YS, Peng RH, Xu J, Zhao W, Gao F, et al. (2010) Mutations in two amino acids in phyI1s from *Aspergillus niger* 113 improve its phytase activity. World J Microbiol Biotechnol 26: 903–907.
23. Lassner M, Bedbrook J (2001) Directed molecular evolution in plant improvement. Curr Opin Plant Biol 4: 152–156.
24. Dixon DP, McEwen AG, Lapthorn AJ, Edwards R (2003) Forced evolution of a herbicide detoxifying glutathione transferase. J Biol Chem 278: 23930–23935.
25. Castle LA, Siehl DL, Gorton R, Patten PA, Chen YH, et al. (2004) Discovery and directed evolution of a glyphosate tolerance gene. Science 304: 1151–1154.
26. Zhou M, Xu H, Wei X, Ye Z, Wei L, et al. (2006) Identification of a glyphosate-resistant mutant of rice 5-enolpyruvylshikimate 3-phosphate synthase using a directed evolution strategy. Plant Physiol 40: 184–195.
27. Chartrain M, Salmon PM, Robinson DK, Buckland BC (2000) Metabolic engineering and directed evolution for the production of pharmaceuticals. Curr Opin Biotechnol 11: 209–214.
28. Locher CP, Paidhungat M, Whalen RG, Punnonen J (2005) DNA shuffling and screening strategies for improving vaccine efficacy. DNA Cell Biol 24: 256–263.
29. Whalen RG, Kaiwar R, Soong NW, Punnonen J (2001) DNA shuffling and vaccines. Curr Opin Mol Ther 3: 31–36.
30. Stemmer WP (1994) Rapid evolution of a protein in vitro by DNA shuffling. Nature 370: 389–391.
31. Stemmer WP (1994) DNA shuffling by random fragmentation and reassembly: in vitro recombination for molecular evolution. Proc Natl Acad Sci USA 91: 10747–10751.
32. Xiong AS, Peng RH, Liu JG, Zhuang J, Qiao YS, et al. (2007) High efficiency and throughput system in directed evolution in vitro of reporter gene. Appl Microbiol Biotechnol 74: 160–168.
33. Xiong AS, Peng RH, Cheng ZM, Li Y, Liu JG, et al. (2007) Concurrent mutations in six amino acids in beta-glucuronidase improve its thermostability. Protein Eng Des Sel 20: 319–325.
34. Zhang X, Henriques R, Lin SS (2006) Agrobacterium-mediated transformation of *Arabidopsis thaliana* using the floral dip method. Nat Protoc 1: 641–646.
35. Sambrook J, Russell DW (2001) Molecular Cloning; A laboratory Manual. (Cold Spring Harbor laboratory Press, cold Spring Harbor, New York, USA.
36. Sudan C, Prakash S, Bhomkar P, Jain S, Bhalla-Sarin N (2006) Ubiquitous presence of β-glucuronidase (GUS) in plants and its regulation in some model plants. Planta 224: 853–864.
37. Martin T, Wohner RV, Hummel S, Willmitzer L, Frommer WB (1992) The GUS reporter system as a tool to study plant gene expression. In GUS Protocols Galagher SR, ed. San Diego: Academic Press. pp 23–43.
38. Fior S, Vianelli A, Gerola PD (2009) A novel method for fluorometric continuous measurement of β-glucuronidase (GUS) activity using 4-methyl-umbelliferyl- β-D-glucuronide (MUG) as substrate. Plant Sci 176: 130–135.
39. Fior S, Gerola PD (2009) Impact of ubiquitous inhibitors on the GUS gene reporter system: evidence from the model plants Arabidopsis, tobacco and rice and correction methods for quantitative assays of transgenic and endogenous GUS. Plant Methods 5: 19.
40. Matsumura I, Wallingford JB, Surana NK, Vize PD, Ellington AD (1999) Directed evolution of the surface chemistry of the reporter enzyme beta-glucuronidase. Nat Biotechnol 17: 696–701.
41. Flores H, Ellington AD (2002) Increasing the thermal stability of an oligomeric protein, beta-glucuronidase. J Mol Biol 315: 325–337.
42. Crameri A, Whitehorn EA, Tate E, Stemmer WP (1996) Improved green fluorescent protein by molecular evolution using DNA shuffling. Nat Biotechnol 14: 315–319.
43. Nam SH, Oh KH, Kim GJ, Kim HS (2003) Functional tuning of a salvaged green fluorescent protein variant with a new sequence space by directed evolution. Protein Eng 16: 1099–1105.
44. Taylor CB (1997) Promoter fusion analysis: an insufficient measure of gene expression. Plant Cell 9: 273–275.
45. Sasaki K, Taura F, Shoyama Y, Morimoto S (2000) Molecular characterization of a novel beta-glucuronidase from *Scutellaria baicalensis georgi*. J Biol Chem 275: 27466–27472.
46. Levvy GA, Marsh CA (1959) Preparation and properties of β-glucuronidase. Adv Carbohydr Chem 14: 381–428.
47. Hawkesworth G, Draser BS, Hill MJ (1971) Intestinal bacteria and the hydrolysis of glycosidic bonds. J Med Microbiol 4: 451–459.
48. Langley SD, Wilson SD, Gros AS, Warner CK, Finnerty V (1983) A genetic variant of β-glucuronidase in Drosophila melanogaster. J Biol Chem 258: 7416–7424.
49. Sebastiano M, D'Alessio M, Bazzicalupo P (1986) β-glucuronidase mutants of the nematode *Caernohabditis elegans*. Genetics 112: 459–468.
50. Plegt L, Bino J (1989) β-glucuronidase activity during development of male gametophyte from transgenic and non-transgenic plants. Mol Gen Genet 216: 321–327.
51. Lewy GA (1954) Baicalinase, a plant β-glucuronidase. Biochem J 56: 462–469.
52. Anhalt S, Weissenbock G (1992) Subcellular localization of luteolin glucuronides and related enzymes in rye mesophyll. Planta 187: 83–88.
53. Muhitch MJ (1998) Characterization of pedicel β-glucuronidase activity in developing maize (*Zea mays*) kernels. Physiol Plant 104: 423–430.
54. Bekkaoui F, Pilon M, Laine E, Raju DSS, Crosby WL, et al. (1988) Transient gene expression in electroporated *Picea glauca* protoplasts. Plant Cell Rep 7: 481–484.
55. Martin T, Scmidt R, Altman T, Frommer WB (1990) Non-destructive assay system for detection of β-glucuronidase activity in higher plants. Plant Mol Biol Rep 10: 37–46.
56. Woznaik CA, Owens LD (1994) Native β-glucuronidase activity in sugarbeet (*Beta vulgaris*). Physiol Plant 90: 763–771.
57. Alwen A, Moreno RMB, Vicente O, Heberle-Bors E (1992) Plant endogenous activity: how to avoid interference with the use of *E.coli* β-glucuronidase as a reporter gene in transgenic plants. Transgenic Res 1: 63–70.
58. Hodal L, Bochardt A, Neilson JE, Mattson O, Okkels FT (1992) Detection, expression and specific eliminaton of endogenous β-glucuronidase activity in transgenic and non-transgenic plants. Plant Sci 87: 115–122.
59. Kosugi S, Ohashi Y, Nakajima K, Arai Y (1990) An improved assay for β-glucuronidase in transformed cells: Methanol almost completely suppresses a putative endogenous β-glucuronidase activity. Plant Sci 70: 133–140.
60. Matsumura I, Ellington AD (2001) *In vitro* evolution of betaglucuronidase into a beta-galactosidase proceeds through nonspecific intermediates. J Mol Biol 305: 331–339.
61. Rowe LA, Geddie ML, Alexander OB, Matsumura I (2003) A comparison of directed evolution approaches using the beta-glucuronidase model system. Mol Biol 332: 851–860.
62. Geddie ML, Matsumura I (2004) Rapid evolution of beta-glucuronidase specificity by saturation mutagenesis of an active site loop. J Biol Chem 279: 26462–8.

23

# Targeting G with TAL Effectors: A Comparison of Activities of TALENs Constructed with NN and NK Repeat Variable Di-Residues

**Michelle L. Christian[1], Zachary L. Demorest[1], Colby G. Starker[1], Mark J. Osborn[1], Michael D. Nyquist[1], Yong Zhang[1], Daniel F. Carlson[2], Philip Bradley[3], Adam J. Bogdanove[4], Daniel F. Voytas[1]***

**1** Department of Genetics, Cell Biology & Development and Center for Genome Engineering, University of Minnesota, Minneapolis, Minnesota, United States of America, **2** Department of Animal Science, University of Minnesota, Saint Paul, Minnesota, United States of America, **3** Division of Public Health Sciences, Fred Hutchinson Cancer Research Center, Seattle, Washington, United States of America, **4** Department of Plant Pathology, Iowa State University, Ames, Iowa, United States of America

**Abstract**

The DNA binding domain of Transcription Activator-Like (TAL) effectors can easily be engineered to have new DNA sequence specificities. Consequently, engineered TAL effector proteins have become important reagents for manipulating genomes *in vivo*. DNA binding by TAL effectors is mediated by arrays of 34 amino acid repeats. In each repeat, one of two amino acids (repeat variable di-residues, RVDs) contacts a base in the DNA target. RVDs with specificity for C, T and A have been described; however, among RVDs that target G, the RVD NN also binds A, and NK is rare among naturally occurring TAL effectors. Here we show that TAL effector nucleases (TALENs) made with NK to specify G have less activity than their NN-containing counterparts: fourteen of fifteen TALEN pairs made with NN showed more activity in a yeast recombination assay than otherwise identical TALENs made with NK. Activity was assayed for three of these TALEN pairs in human cells, and the results paralleled the yeast data. The *in vivo* data is explained by *in vitro* measurements of binding affinity demonstrating that NK-containing TAL effectors have less affinity for targets with G than their NN-containing counterparts. On targets for which G was substituted with A, higher G-specificity was observed for NK-containing TALENs. TALENs with different N- and C-terminal truncations were also tested on targets that differed in the length of the spacer between the two TALEN binding sites. TALENs with C-termini of either 63 or 231 amino acids after the repeat array cleaved targets across a broad range of spacer lengths – from 14 to 33 bp. TALENs with only 18 aa after the repeat array, however, showed a clear optimum for spacers of 13 to 16 bp. The data presented here provide useful guidelines for increasing the specificity and activity of engineered TAL effector proteins.

**Editor:** Shuang-yong Xu, New England Biolabs, Inc., United States of America

**Funding:** This work was supported by a grant from the National Science Foundation to DFV (DBI 0923827) and a grant from the National Institutes of Health to DFV and AJB (R01 GM098861). YZ received support from the Central Universities of China (ZYGX2009J084) and the National Natural Science Foundation of China (30900779). The funders had no role in study design, data collection and analysis, decision to publish, or preparation of the manuscript.

**Competing Interests:** MC, AJB and DFV are listed inventors on a patent application titled "TAL effector-mediated DNA modification" that is co-owned by Iowa State Univ. and the Univ. of Minnesota, and has been licensed to Cellectis, a European biotechnology company. There are no further patents, products in development or marketed products to declare.

* E-mail: voytas@umn.edu

## Introduction

The ability to target proteins to specific DNA sequences makes it possible to manipulate nucleic acids *in vivo*, for example, by creating artificial transcriptional regulators that alter gene expression or by engineering sequence-specific nucleases that modify genetic loci of interest. Work in the past few years has made it increasingly evident that the protein domain of choice for DNA targeting is the DNA binding domain derived from Transcription Activator-Like (TAL) effectors [1]. TAL effectors are proteins produced by bacterial plant pathogens of the genus *Xanthomonas*, and they are delivered to plant cells during infection where they activate the expression of select target genes and thereby make the plant more prone to bacterial colonization [2]. Targeting of TAL effectors to plant gene promoters is achieved by a simple and elegant mechanism of DNA binding [3,4]. This mechanism enables the pathogen to rapidly evolve TAL effectors with new DNA sequence specificities, and by extension, it enables scientists to engineer DNA binding domains that recognize novel sites for various genome engineering applications.

DNA binding by TAL effectors is mediated by the central domain of the protein, which comprises approximately 13–28 tandem repeats of a 34 amino acid motif [5]. Amino acid sequences of the repeats are largely invariant with the exception of two residues at positions 12 and 13, called the repeat variable di-residues (RVDs). Each repeat forms two alpha helices that are joined at one end by a short loop that contains the RVD: residue 12 reaches back to interact with one of the alpha helices to stabilize the loop, and residue 13 contacts a specific base within the major groove of DNA [3,4]. Among the most common RVDs in TAL effectors are HD, NG and NI, which specify the nucleotides C, T and A, respectively; these RVDs are only infrequently associated with other bases [5]. The mechanistic basis for this sequence specificity has been explained by the recently reported structure of two TAL effectors bound to DNA [3,4]. The most common RVD that specifies G, namely NN, also interacts well with A, and this interaction is achieved by hydrogen-bonding between the second asparagine in the RVD and the $N_7$ group on these purines [3,4,6].

The lack of specificity for G is one potential drawback in the use of engineered TAL effector proteins for DNA targeting. NK, an RVD only rarely found in nature, also interacts with G [7]. In vitro SELEX data suggested that NK RVDs may have higher specificity for G than NN [7]; however, DNA targeting to a locus in zebrafish was more effective when engineered TAL effector proteins were made using repeats with NN RVDs rather than NK [8]. Two recent studies also indicated that NK TAL effectors have less activity than NN-containing proteins, and further, the RVD NH was shown to provide G-specificity [9,10]. Clearly there is considerable interest in both understanding and increasing G-specificity of engineered TAL proteins.

Other efforts to adopt TAL effectors for genome engineering have focused on defining the minimal region of the protein required for DNA binding [7,8,11,12]. In the TAL effector PthXo1, for example, the repeat region is flanked by 288 and 295 amino acids at the N- and C-termini, respectively. To delimit the DNA binding domain, different research groups have made and tested various N- and C-terminal truncations. An N-terminal truncation at residue 152 and C-terminal truncations up to 18 amino acids after the repeat domain still allow for effective DNA binding [7,12]. The structure of TAL effectors bound to DNA is consistent with truncations of this extent still being able to bind DNA [3,4].

The length of the C-terminus after the TAL effector DNA binding domain has a direct impact on the activity of TAL effector nucleases (TALENs) – fusions of the TAL repeat arrays to the catalytic domain of the FokI endonuclease [11]. FokI functions as a dimer, and DNA cleavage is achieved using two TALENs that bind opposing DNA target sites separated by a spacer. The length of the spacer must allow for efficient FokI dimerization so that DNA cleavage is achieved. In all TALEN architectures tested to date, DNA cleavage is occurs across a broad range of spacer lengths [7,8,11,12], and this variability further motivates defining the minimal DNA binding domain so that TALENs can be engineered with maximal activity over a narrow range of spacer lengths.

Toward improving TAL protein engineering, we compared the activity of TALENs that use repeats with either NN or NK RVDs to target G. We conclude that whereas TALENs made with NK RVDs may have more specificity for G-containing targets, this specificity is offset by a considerable reduction in activity that is due to having less affinity for G than their NN-containing counterparts. We also compared the activity of TALENs with various N- and C-terminal deletions and found a novel architecture that cleaves optimally over a narrower spacer length range than any reported to date.

## Methods

### TALEN construction

TALEN target sites were identified using the TAL effector-Nucleotide Targeter (TALE-NT) program [11]. All targets chosen had a T at the -1 position. TALENs recognizing the target sites were constructed using our previously reported assembly method that is based on Golden Gate cloning [13]. In brief, Golden Gate cloning uses Type IIS restriction endonucleases (e.g. BsaI, Esp3I) to create unique 4 bp overhangs on DNA fragments so that they can be assembled in a precise, sequential order. Our library of plasmids encodes TAL effector repeats with five different RVDs (NI, HD, NN, NG and NK) that can be released by digestion with BsaI to create unique 4 bp ends. This enables up to 10 RVD-encoding plasmids to be ligated in the correct order in a single reaction. Sub-arrays of 10 TAL effector repeats can then be joined simultaneously in a second, similar reaction, resulting in a fully-assembled array (11–31 repeats) cloned upstream a FokI nuclease domain in a yeast expression vector.

The pTAL-BamHI expression vector contains 288 aa in the N-terminus of the repeat array and 231 aa in the C-terminus. Truncations of the pTAL-BamHI backbone were constructed by first PCR amplifying fragments encoding the desired length of N or C-termini, along with the sequences necessary to perform subsequent Golden Gate cloning reactions. These amplicons were then cloned into a linearized pTAL backbone lacking the TALE termini to yield the NΔ152/C+63 and NΔ152/C+18 truncated TALEN plasmids. These plasmids are compatible with the Golden Gate assembly platform.

### Yeast single-strand annealing assay

The yeast-based assay for testing TALEN function has previously been described [13,14]. Briefly, a yeast strain expressing an engineered TALEN(s) is mated with a yeast strain carrying a corresponding target plasmid. Cleavage of the target and subsequent recombination by single-strand annealing reconstitutes a functional *lacZ* gene and provides a quantitative readout of TALEN activity. β-galactosidase measurements for all data were normalized to the TALEN, SurB, which is highly active in both yeast and at its endogenous target (Y. Zhang and D. Voytas, unpublished). Statistical significance was assessed using a paired t-test. Differences at $p<0.05$ were considered to be statistically significant. For some datasets, a ZFN positive control derived from the Zif268 zinc finger array was also included.

### Mammalian single-strand annealing assay

Activity of TALENs was measured in HEK293 cells using a single-strand annealing assay that reconstitutes a functional luciferase reporter. The template plasmid for the assay (pSSA-1-3) was kindly provided by Dr. David Segal (UC Davis). PCR-based mutagenesis was used to insert TALEN cleavage sites in pSSA-1-3 between the two halves of the luciferase gene.

Twenty-four hours prior to transfection, HEK293 cells were seeded in quadruplicate in a 24-well dish at a density of 200,000 cells/well in DMEM media containing 10% serum. TALEN monomer plasmids at doses of 100 or 200 μg with 25 ng of the SSA reporter target were transfected into the cells using Lipofectamine 2000 (Invitrogen) according to the manufacturer's instructions. As a control for any basal activity generated by the SSA reporter, the reporter was transfected along with pUC19 plasmid DNA instead of the TALEN. In addition, all reactions included 4 ng of the pRL-TK Renilla luciferase plasmid (Promega) to normalize transfection. At twenty-four hours post transfection, the cells were lysed in 150 μl 1× passive lysis buffer (Promega). Lysate luminescence was measured using the Dual-Luciferase® Reporter Assay System (Promega) according to the manufacturer's instructions. The fold activity of a given TALEN was determined by normalizing the firefly luciferase values to Renilla luciferase, and that value was then divided by the value obtained with the SSA reporter alone.

### Expression and purification of recombinant TALE protein

The bacterial expression vector pGEX6P2-TALE was created by ligating a Golden Gate compatible fragment with the NΔ152/C+63 architecture into pGEX6P2 (GE Healthcare). RVD arrays for specific TALE proteins were then cloned in as described above. Expression constructs encoding TAL effector proteins 166, 167 and 312 were then transformed into Rosetta cells and selected on media containing carbenicillin and chloramphenicol. 200 mL cultures were grown to log phase at 37°C before induction for 3 hours with 1 mM IPTG. The cells were pelleted by centrifugation and lysed in GST lysis buffer (25 mM HEPES pH 7.4, 150 mM NaCl, 5 mM $MgCl_2$, 130 μM $CaCl_2$, 0.5% Triton X-100, 10% glycerol, 1 mM PMSF, 1 μg/mL Leupeptin, 100 nM Aprotinin, 1 μg/mL Pepstatin A). The lysates were treated with RNase A (20 μg/mL) and DNase I (10 U/mL), clarified by centrifugation (21,000× g, 10 minutes) and then loaded onto a column containing equilibrated Glutathione Sepharose (GE Healthcare). The columns were washed with GST lysis buffer and subsequently by cleavage buffer (50 mM Tris-HCl pH 8.0, 1 mM EDTA, 1 mM DTT, 10% glycerol). Elution of untagged purified TALE protein was performed by overnight incubation at 4°C with PreScission protease (GE Healthcare). Purified TALE proteins were separated by electrophoresis and stained with Coomassie to determine the purity of the samples (Fig. S1).

### Electrophoretic mobility shift assay (EMSA)

Double stranded DNA substrates were prepared by annealing fluorescently tagged complementary oligos. Sequences for substrates used were 5′- TGGACACGACTTGAGCTGTCG-TCTTCTGCACTCGTAGTGCTGTGATGA for 166, 5′-TGGACATGACTTGAGCTAGTCAGCACCAGGCATCGT-AGTGCTGTGCTGA for 167, 5′- TGGACACGAC TTGAGCTGGCGAAAGAGTCCACCACCATCGTAGTGC-TGTGCTGA for 312 and 5′-TGGACACGACTTGAGCTC-GACGCTCAGGCAACCGTAGTGCTGTGCTGA for the scrambled target. The purified proteins were diluted into binding buffer (10 mM HEPES pH 7.6, 10% glycerol, 100 mM KCl, 10 mM MgCl2, 100 μM EDTA, 500 μ M DTT) at varying concentrations with a fixed concentration of the labeled DNA substrate (20 nM). The reactions were incubated for 30 minutes at room temperature and then separated by electrophoresis on a 7% TBE-acrylamide gel. Detection of the labeled substrate was then performed on a fluorescent scanner (Storm 860, Molecular Dynamics).

### Modeling of NN and NK RVD-DNA interactions

To gain insight into potential determinants of affinity and specificity, NN and NK RVDs were modeled onto all RVD loop backbones in the available crystal structures [3,4]. Corresponding DNA target site positions were mutated sequentially to adenine and guanine, and the energy of the protein-DNA complex was optimized by Monte Carlo sampling of rotameric sidechain conformations together with gradient-based minimization, allowing small shifts to the protein and DNA backbones. Simulations were performed with the Rosetta software package [15].

## Results

### Activity of TALENs with NN versus NK RVDs

We compared the activities of engineered TALENs that specify the nucleotide G with either NN or NK RVDs. Using the Golden Gate assembly method for TALEN engineering [13], we constructed 15 TALEN pairs that incorporate either NN or separately NK RVDs to specify G. The total number of G's in the corresponding target DNA sequences ranged from 3 to 17, and no bias was imposed as to how the G's were distributed in the target. To test the activity of the TALENs, we used a yeast-based single-strand annealing (SSA) assay in which LacZ activity serves as an indicator of DNA cleavage by TALENs [11,14]. Briefly, the assay employs a target plasmid containing a *lacZ* reporter gene with an internal sequence duplication disrupted by a TALEN target sequence. Cleavage of the target by the corresponding TALENs results in reconstitution of a functional *lacZ* gene whose expression is measured by standard LacZ enzymatic assays. TALEN pairs containing NN RVDs showed 14–90% activity of the positive control TALEN pair, SurB, which was slightly more active than the potent zinc-finger nuclease (ZFN), Zif268 (Fig. 1). When NN was exchanged for the NK RVD, nuclease activity was reduced for 14 of 15 TALEN pairs, and for several NK-containing TALENs, no activity was detected. One exception was TALEN pair 248/249, which contained 6 NK RVDs in total (2 in TALEN 248 and 4 in 249). The activities of both NN- and NK-versions of this TALEN pair were comparable. Another exception was 143/144 in which the NK version had slightly higher activity than the NN version.

We sought to validate the trends observed in yeast by testing several TALEN pairs made with either NN or NK RVDs in human cells. As in yeast, we used a SSA assay in which a luciferase gene is interrupted by an internal sequence duplication and a TALEN recognition site. When the TALEN cleaves the target, the break is repaired, restoring the luciferase open reading frame and thereby allowing for comparison and quantification of TALEN activity. Three target sites were tested, one of which (143/144) was an exception in the yeast assay and showed higher activity for the NK-containing TALEN pair. For all three targets, the activity of TALENs with NN versus NK RVDs recapitulated in human cells what was observed in yeast, although in some cases the magnitude of the difference was considerably less (Fig. 2). For example, with TALEN pair 351/352, the NK version was 30-fold less active in yeast, whereas only 1.6-fold less activity was observed for the NK version in human cells.

### TAL effector arrays containing NK RVDs have lower affinities for DNA targets

To understand the basis for the lower activity of NK-containing TALENs *in vivo*, we performed a series of electrophoretic mobility shift assays (EMSAs) using purified TAL effectors 166, 167, and 312, which contain 3, 4, or 5 NN or NK RVDs, respectively. Each NN- or NK-containing array was expressed as a GST fusion protein, purified from *E. coli*, and used to perform EMSAs with cognate targets or a scrambled substrate. All NN-containing arrays bound their target significantly better than the scrambled substrate. For TAL effector 166, both the NN- and NK-containing proteins bound target substrates with similar affinities, as evidenced by the shift of the target probe in the presence of increasing concentration of protein (Fig. 3A). TAL effector 312NN, which contains five NN/NK repeats, showed the most dramatic difference between binding affinities when compared to TAL effector 312NK. A shift of the target probe can be seen for TAL effector 312NN at 40 nM of protein, while there is a lack of visible shift at the highest concentration (250 nM) of TAL effector 312NK. Notably, TAL effector 166NN and 312NN also bound the scrambled substrate with greater affinity than the NK versions. TAL effector 167NN bound its target substrate with higher affinity than the NK-containing array.

The data for all six TAL effector proteins were plotted as the fraction of bound substrate at each protein concentration from 0 nM to 250 nM (Fig. 3B). These data taken together indicate that

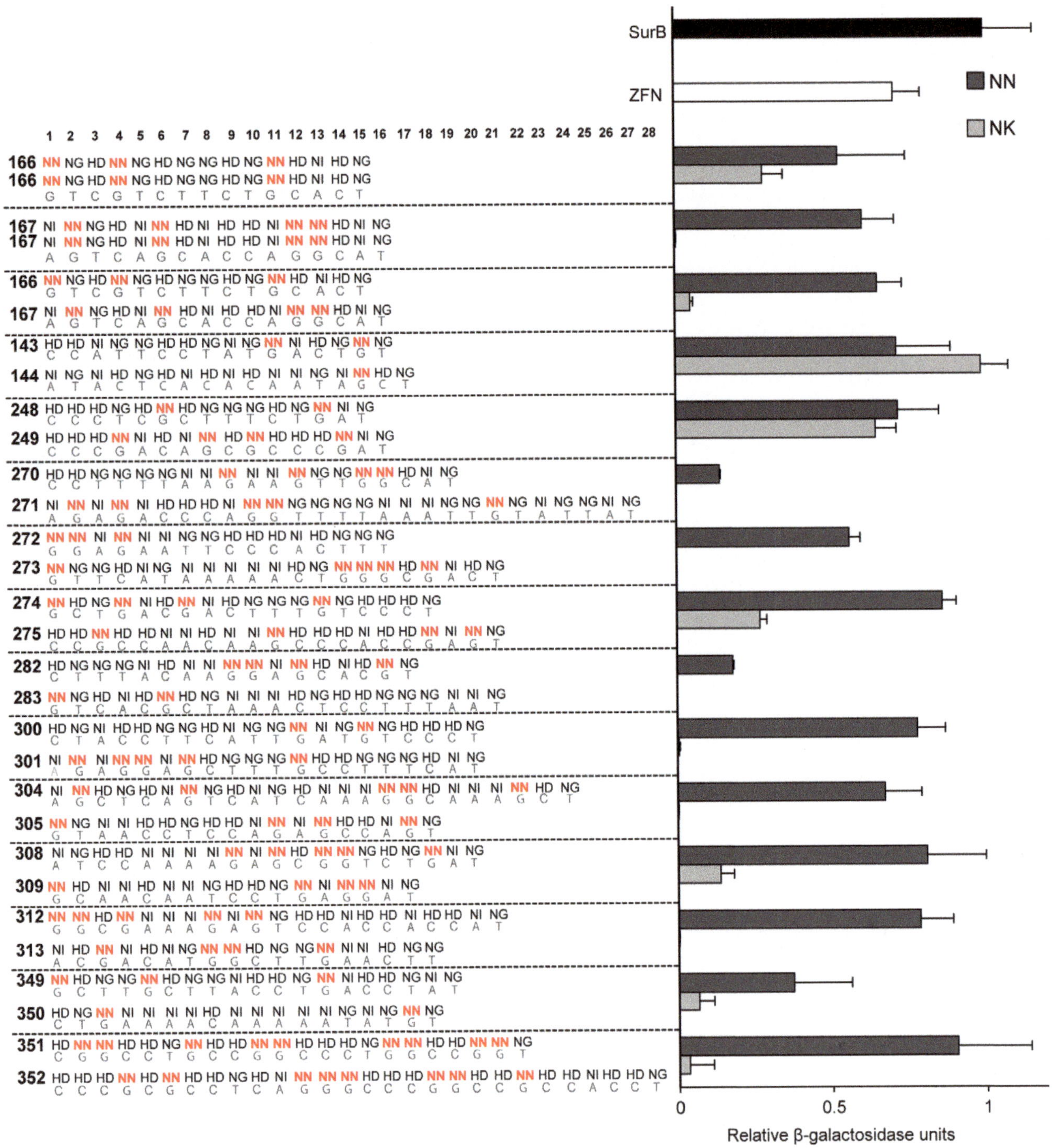


**Figure 1. Activity of TALENs containing NN or NK RVDs.** Plotted on the right are the relative activities of TALENs with either all NN (dark gray) or NK (light gray) RVDs for specifying G nucleotides. Activities are relative β-galactosidase units as determined in our yeast-based TALEN activity assay [11]. All values are normalized to a TALEN positive control, SurB (black). Also shown for reference is the activity obtained with the well-characterized ZFN, Zif268 (white). Left and right arrays comprising TALEN pairs are numbered (e.g. 166, 167) followed by RVD sequences of the various TALEN DNA binding domains with repeat 1 being nearest to the N-terminus of the protein. NN RVDs that were replaced with NK are in red. The DNA target sequences are shown below the RVDs. Values are expressed as the mean of triplicates with standard deviation (S.D.).

substituting NK for NN RVDs does contribute to overall lower binding affinities of the TAL effector proteins. Furthermore, the TAL effector protein binding affinities correlate well with the nuclease activities of the corresponding TALENs *in vivo*.

## Effects on TALEN activity of number and position of NK RVDs in repeat arrays

Having observed predominantly reduced nuclease activity for TALEN pairs with NK versus NN RVDs, we next evaluated the effect on activity of number and/or position of NK RVDs within

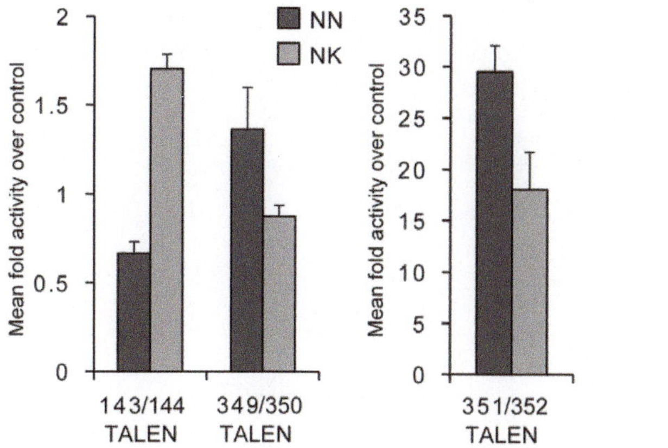


**Figure 2. Activity of TALENs containing NN or NK RVDs in human cells.** Three TALENs with either all NN (dark gray) or NK (light gray) RVDs were tested in a single-strand annealing (SSA) assay in HEK293 cells. Fold activation levels were determined by normalizing the levels of luciferase activity mediated by TALEN conversion of the SSA luciferase target to the basal level of luciferase exhibited by the SSA plasmid alone. Each treatment group contained a transfection/toxicity control consisting of the Renilla luciferase plasmid. Values are expressed as the mean of quadruplicates with S.D.

the array. The Golden Gate TALEN assembly method allows rearrangement of RVDs within existing TALEN pairs with relative ease [13], and so we altered the NN/NK composition of the 166 and 167 TALENs to create four variants of each (Fig. 4A). The 166 and 167 arrays and their variants were then tested as TALENs on homodimeric target sites (i.e. the target site for either the 166 or 167 array was duplicated in inverse orientation and separated by a 15 bp spacer). Two consistent observations were made: 1) as described above, TALENs with NK RVDs had the least activity compared to their NN-containing counterparts, and 2) the most significant impact on activity occurred when the NK RVDs were located within the N-terminal half of the array. This latter observation was particularly pronounced for NK substitutions in the 167 repeat array.

When the 166/167 array variants were tested together as heterodimeric TALENs the impact of the NK RVDs was diminished (Fig. 4B, G-containing targets). For example, when the 166 array variants were tested with a fully NN-containing 167 array, no difference in TALEN activity was observed. Further, the 167-N variant, which showed 5-fold reduction in activity when tested as a homodimer, showed only a modest decrease in activity when tested with the fully NN-containing 166 array. This suggests that NK substitutions that decrease TALEN activity can be ameliorated by pairing with a functional TALEN partner.

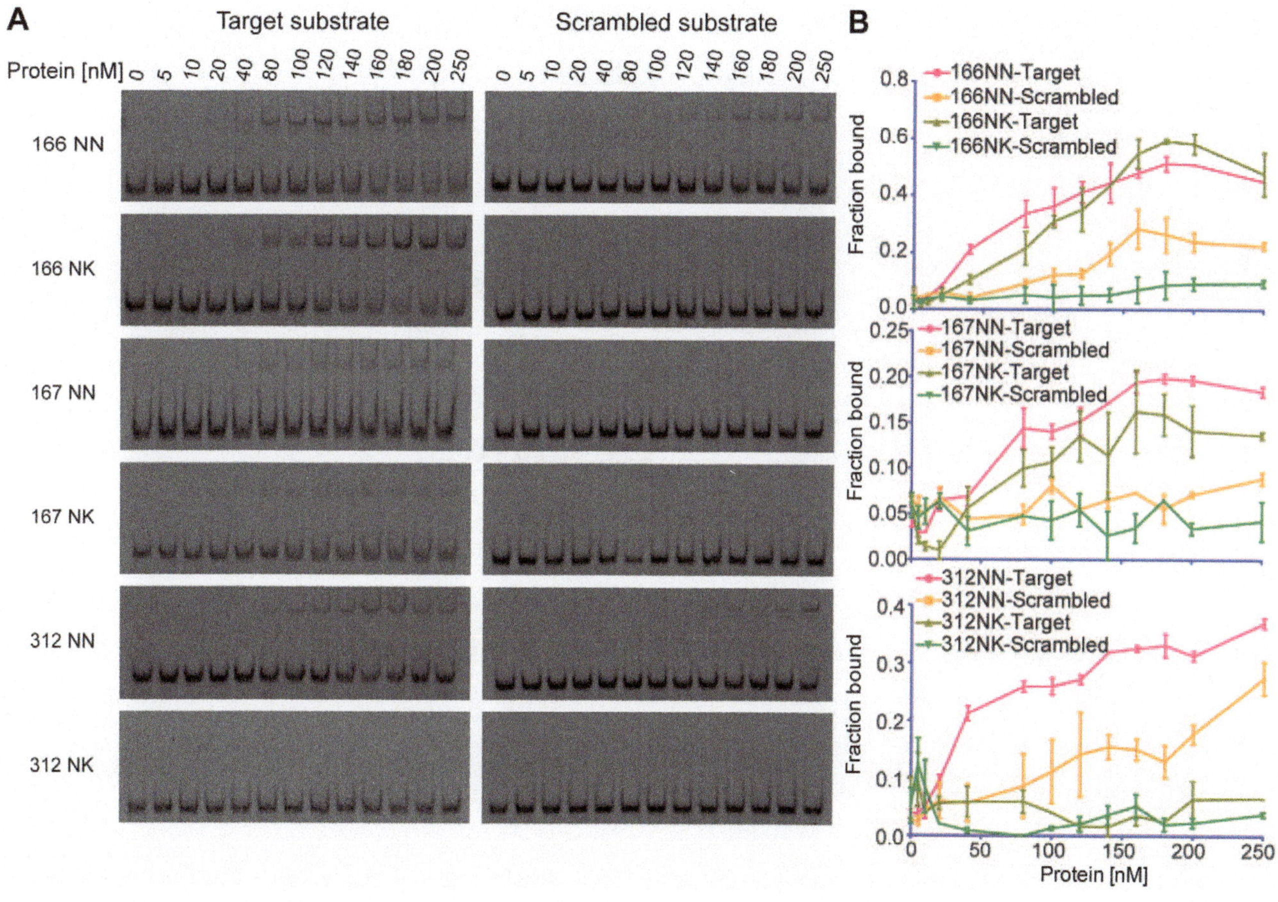


**Figure 3. Relative binding affinities of purified TAL effector proteins for DNA targets.** (**a**) Representative images of electrophoretic mobility shift assays. Fluorescently-labeled double-stranded oligonucleotide substrates corresponding to either TAL effector targets or a scrambled substrate were incubated with purified TAL effector proteins comprised of either NN or NK RVDs. (**b**) Graphs depicting results from densitometry performed on gel images. Shifted labeled substrate was detected using a fluorescent scanner and quantified by densitometry. The fraction of substrate bound by the TAL effector protein was plotted for each protein concentration. Data indicate the mean ± S.D. for 3 independent experiments.

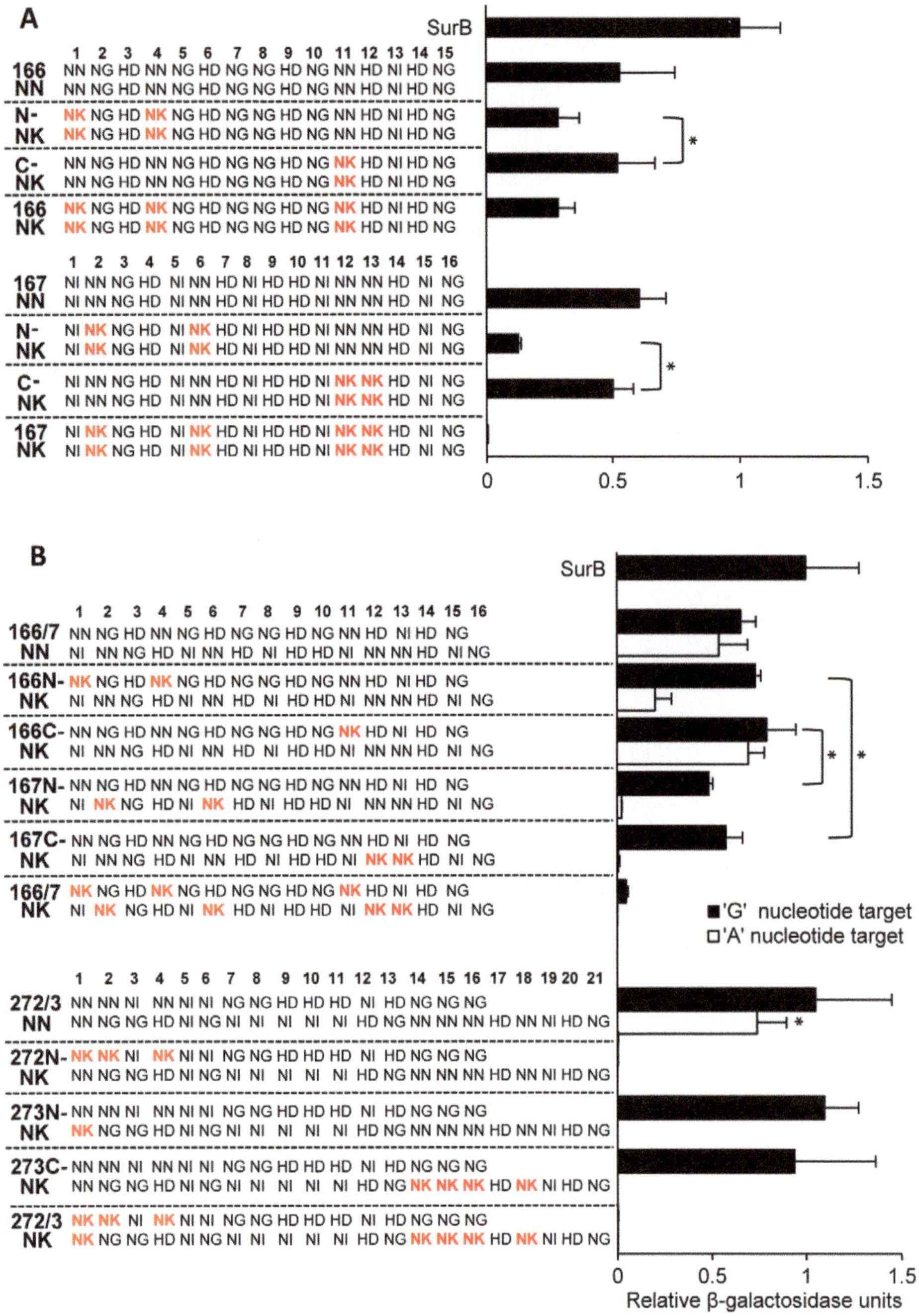


**Figure 4. Effects on TALEN activity of number and position of NK RVDs in repeat arrays.** (**a**) Activity of homodimeric TALENs. TALEN RVD sequences are shown on the left. Note that the left and right TALENs are identical except for the presence of NN versus NK RVDs. NK RVDs are shown in red for ease of visualization. The activity of the TALEN pairs is plotted on the right. (**b**) Activity of heterodimeric TALENs. TALEN arrays with NK substitutions were tested with TALEN array partners that contain only NN RVDs. Black bars denote activity of TALENs at target sequences with G nucleotides at positions corresponding to NN or NK RVDs. White bars denote TALEN activity with A nucleotides in place of G. All values are normalized to the SurB TALEN positive control. Values are expressed as the mean of quadruplicates with S.D. *, differences in relative β-galactosidase levels between the two indicated samples at $p<0.05$ were considered to be statistically significant.

NK substitutions were made in repeat arrays for a second pair of TALENs (272/273) (Fig. 4B), and the same general observations were made with respect to the impact of the substitutions on TALEN activity. When both the 272/273 TALEN pairs contained all NK RVDs to specify G, little or no activity was observed. Activity was also compromised when a fully NN-containing TALEN 273 was paired with a TALEN 272 variant with three NK substitutions at the N-terminus. In contrast, four NK substitutions at the C-terminus of a 273 variant had no effect on activity when this TALEN was paired with a fully NN-containing TALEN 272. It appears that the number of NK substitutions also impacts activity, as a single NK substitution at the N-terminus of TALEN 273 did not affect activity when paired with the NN-containing 272 TALEN. We conclude, therefore, that NK substitutions are more likely to compromise TALEN activity when located at the N-terminus of the protein and that the loss of activity is amplified when there are multiple NK substitutions.

### Testing TALENs on G- and A-containing targets

A previous report showed a specificity of NK RVDs for G nucleotides over A nucleotides [7]. In contrast, the NN RVD appears to recognize both G and A [3,4,7,16]. We therefore tested several of our TALEN variants against targets in which G's had been replaced with A's (Fig. 4B, Fig. S2). The fully NN-containing 166/167 and 272/273 TALEN pairs showed comparably high levels of activity on both G- and A-containing targets, whereas little or no activity was observed for the fully NK-containing arrays on both targets. All four 166/167 arrays that varied in the number and distribution of NK RVDs showed higher activity on G-containing targets, and the same was observed for 2 of 3 272/273 TALEN pairs. These data suggest that NK has higher specificity for G over A.

### Modeling of NN and NK RVD-DNA interactions

To assess whether our conclusions regarding the affinity and specificity of NN and NK RVDs for G and A have any structural basis, molecular modeling simulations of potential NN:G, NK:G, NN:A, and NK:A interactions were performed using the two solved TAL-DNA co-complexes as structural templates. As determined from the PthXo1 crystal structure, an NN RVD can donate a hydrogen bond to the $N_7$ atom of G or A and accept a hydrogen bond from the backbone nitrogen of residue 13 in the preceding repeat [3] (Fig. 5). When threaded onto the available RVD loop structures, the lysine at position 13 in an NK RVD is positioned to donate a hydrogen bond to the guanine $O_6$ atom of guanine, but is unable to participate in the two hydrogen bonds formed by the asparagine at that position in an NN RVD. The potential desolvation of the $N_7$ atom, together with the differential loss of sidechain entropy upon binding, may explain the lower affinity of the NK RVD for G relative to the NN:G interaction. In addition, the prediction that NK:G does form a hydrogen bond with guanine and not with adenine may provide the basis for its apparent G-specificity. It should be noted that the NK interactions are theoretical models based on existing TAL crystal structures; large-scale rearrangement of the RVD loop or interacting residues would likely change the set of potential interactions.

### Comparison of TALEN backbone architectures

The first TALEN architecture described contains 288 aa of the N-terminus of Tal1c upstream of the repeat arrays, and 231 aa of the C-terminus downstream of the repeats [11]. This architecture, termed the pTAL-BamHI backbone, has been shown to be active for many TALEN pairs [13]. A second TALEN architecture was described with enhanced nuclease activity, referred to here as the NΔ152/C+63-backbone [7]. The first 152 aa were deleted from the N-terminal portion of the TAL effector protein, upstream of the repeat array (Δ152). A second deletion to the C terminus leaves 63 aa immediately following the last repeat in the repeat array (+63). Other C-terminal truncations of TAL effector proteins have been described [12], and we constructed an additional TALEN backbone with only 18 aa following the half-repeat of the TAL effector array, designated NΔ152/C+63.

To evaluate the relationship between TALEN activity, spacer length and the length of the C-terminus, we cloned a single pair of TAL effector DNA-binding domains in the three architectures described above and tested their activities on targets with spacer lengths varying from 0 to 39 bases. We found that the TALEN pair with the pTAL-BamHI backbone yielded activity that was 25% or greater than the positive control on targets separated by 14–33 bp, with an optimal activity range on spacer lengths of 24–27 bp (Fig. 6). Similarly, the TALEN pair with the NΔ152/C+63-backbone had activity at least 25% of the positive control across spacer lengths ranging from 14–33 bp. In contrast, the NΔ152/C+18 TALEN displayed more focused activity, showing activity exceeding 25% of the positive control for spacer lengths of 13–16 bp. For spacer lengths of 13–15 bp, activities of the NΔ152/C+18 TALENs were 2–3 fold higher than the NΔ152/C+63 versions. We conclude that shorter C-terminal truncations make it possible to achieve optimal TALEN activity over a narrow range of spacer lengths.

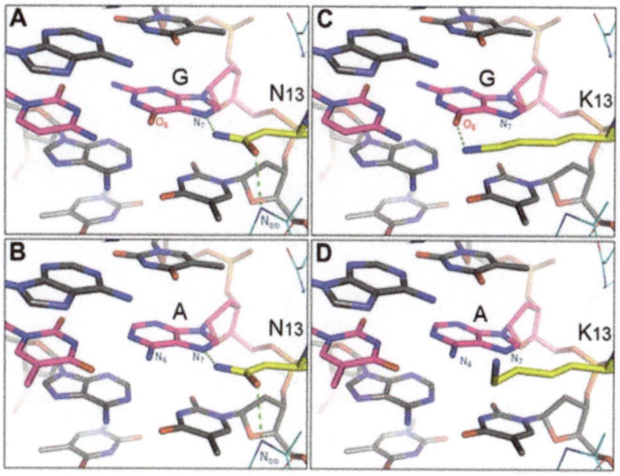


**Figure 5. Representative low-energy models of NN:G, NK:G, NN:A, and NK:A interactions.** (**a & b**) An NN RVD recognizing either G or A can donate a hydrogen bond to the purine $N_7$ atom and accept a hydrogen bond from the backbone nitrogen of residue 13 in the preceding repeat. (**c & d**) When threaded onto the available RVD loop structures, the lysine at position 13 (K13) in an NK RVD is positioned to donate a hydrogen bond to the guanine $O_6$ atom, but is unable to participate in the two hydrogen bonds formed by an asparagine (N13) at that position. The NK RVD is not predicted to form hydrogen bonds with adenine.

## Discussion

Our analysis of the activity of TALENs that target G with either NN or NK RVDs revealed three general conclusions: 1) NN-containing TALENs are more active than their NK-containing counterparts, which can be explained by their overall higher binding affinity for G-containing targets; 2) TALENs made with NK RVDs are more specific for G-containing targets, and 3) NK RVDs have the biggest impact on TALEN activity when they occur in the N-terminal half of the repeat array.

Our study presents, to the best of our knowledge, the first affinity data comparing the effect of RVDs specifying G to the overall affinity of the TAL effector protein. The EMSAs revealed that TAL effectors containing four or five NK RVD substitutions had lower binding affinities than their NN counterparts for the same target. Importantly, the data indicate that the difference in activity of the TALENs can be attributed to the affinity of the TAL effectors for their DNA targets. For example, the NK version of TALEN pair 166 retained over half of the activity of the NN version, which can be explained by the comparable binding affinities of the 166 NN/NK TAL effector arrays *in vitro*. Furthermore, TAL effector 166NN also bound its non-target

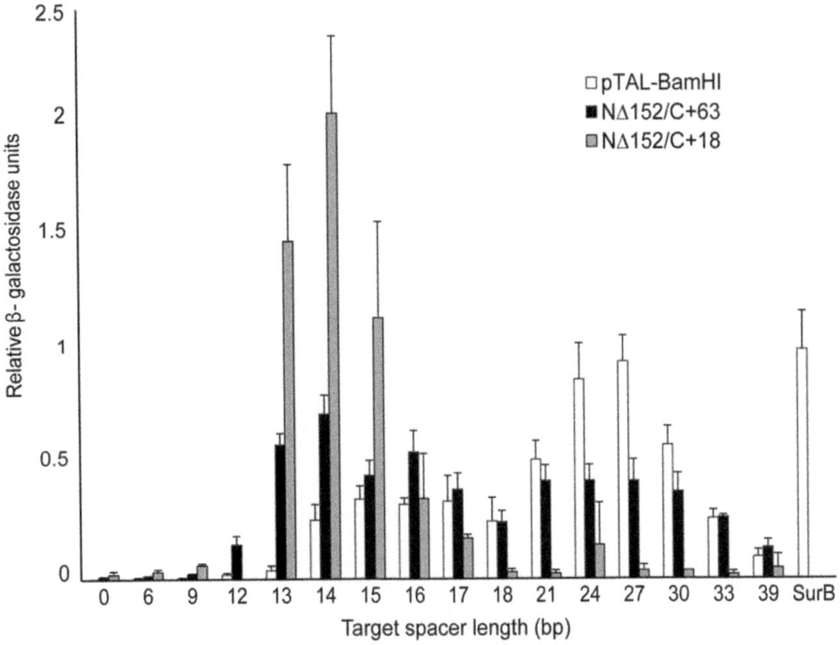


**Figure 6. TALEN backbone architectures and spacer length optima.** A single TALEN repeat array was placed in either the pTAL-BamHI backbone (white), the NΔ152/C+63 backbone (black) or the NΔ152/C+18 backbone (light gray) and tested against its corresponding palindromic target. The left and right TALENs (23.5 repeats) were separated by spacers of lengths indicated by the values shown along the x-axis. All values are normalized to SurB TALEN positive control. Values are expressed as the mean of quadruplicates with S.D.

substrate, indicating it has higher affinity for DNA in general, which could account for the increased nuclease activity observed for TALEN pair 166NN (Fig. 1). The nuclease activities of TALEN pairs 167 and 312 NN/NK also reflect the observed binding affinities of the TAL effector proteins, in that the NK-containing arrays showed both decreased nuclease activity and lower binding affinity.

Our conclusions are consistent with other studies that evaluated the activity of engineered TAL effector proteins with NN and/or NK RVDs. In one study, the activity of a single TALEN pair was evaluated that contained a mixture of both NK and NN RVDs [7]. These TALENs were active and able to introduce mutations in the intended target site at frequencies exceeding 3%. The TALENs making up this pair contained no NK RVDs within the first five repeats of either array, and of the 36 RVDs in both arrays, 5 were NKs (2 in one array and 3 in the other). The observed activity of this TALEN, therefore, is consistent with our finding that weak RVD/nucleotide associations are tolerated if they occur in the C-terminal half of the array. SELEX assays were also performed with four engineered TAL effector proteins to profile base specificity, and consistent with our data, NK was found to be more specific for G than NN.

A second study compared the activity of a TALEN pair targeting an endogenous gene in zebrafish [8]. Both NN and NK versions of the TALEN pair were evaluated, and the NK version was 6-fold less active in mutagenesis than its NN counterpart. Each array of this TALEN pair contained a single NK RVD: one was located in the first repeat of the left array and the other in the seventh repeat of the right. The observed difference in activity of the two TALEN pairs is consistent with our data showing that NK RVDs located in the N-terminal half of the array have the greatest impact on activity.

While this paper was in review, two additional studies appeared with data regarding the activity of TAL effectors targeting the nucleotide G. Both studies presented evidence consistent with our own observations, namely that TAL effectors with NK RVDs had compromised activity in a transcriptional activation assay compared to the corresponding NN-containing proteins [9,10]. This led to the conclusion that NK is a 'weak' RVD and does not contribute significantly to the affinity of the protein, but does retain specificity for G. The authors also suggested that NN is a 'strong' RVD and thus contributes more to the overall binding of the TALE. Our EMSA results are consistent with this conclusion, as illustrated by the 166NN TAL effector and TALEN data. Both studies also identified NH as an RVD that targets G with good specificity. It would be interesting to create NH-containing variants of the proteins tested here to directly measure their affinity and compare them to NN- and NK-containing counterparts.

It should also be noted that there is a difference in amino acid composition of the repeats used to make TALENs in this study as compared to the studies mentioned previously. TALENs assembled by our Golden-Gate method have a serine adjacent to the NN RVD (i.e. SNN), whereas previous studies use an NN preceded by asparagine (NNN). Although this difference is unlikely to contribute to the overall activity of the TALENs, the impact of this difference was not addressed experimentally in this study.

The recently reported crystal structures of TAL effectors bound to DNA provides some insight into the activity we observe for TALENs made with NN versus NK RVDs [3,4]. The N at position 13 in the NN RVD can form a hydrogen bond with $N_7$ of either guanine or adenine. For NK, the longer side chain of lysine likely cannot form this bond, and this would negatively impact the affinity of NK-containing repeat arrays for their targets. Additional structures of TAL effectors with NK RVDs should inform this hypothesis and also shed light on differences in how NN and NK RVDs achieve their base specificity.

The crystal structure of the PthXo1TAL effector, which contains 23.5 repeats, showed some disorder at the C-terminus with respect to its interaction with DNA [3]. Whether this disorder reflects a diminished role for the C-terminus in DNA binding remains to be determined; however, it is consistent with our finding that suboptimal RVDs at the C-terminus are better tolerated than those at the N-terminus. The activity of other engineered TAL effector proteins with so-called 'mismatches' between RVDs and their intended bases are also consistent with our findings. One study tested the impact of mismatches on transcriptional activation by engineered TAL effector transcriptional activators [17]. Mismatches near the N-terminus of the engineered TAL effectors typically had a greater negative impact on transcriptional activation than those near the C-terminus, and negative effects were more pronounced if two mismatches occurred in a row or there were more than two total mismatches in the array. We recognize that all mismatches are not equivalent due to differences in affinity and specificity of each RVD for its preferred base; further, TALEN cleavage or transcriptional activation by TAL effectors are indirect readouts of protein/DNA interactions. Direct biochemical measurements of TAL effector DNA binding affinity and specificity as reported here will provide further insight into how to optimize the engineering of these proteins for *in vivo* use.

In addition to issues pertaining to targeting the nucleotide G, TALEN specificity may also be compromised by the fact that a broad range of spacer lengths support FokI dimerization and target cleavage. For therapeutic purposes, it would be desirable to have a TALEN pair that only cleaves when both monomers are bound to target sites separated by a fixed spacer length. Other groups have accessed the effective range of spacers lengths over which TALENs can function, and their conclusions have been similar to ours, namely that truncations to the C-terminus narrows the spacer range over which TALENs are optimally active [7,11,12,18]. In this study we completed a side-by-side comparison of three different TAL effector architectures using the same TAL effector DNA binding domains. We observed that TALENs with 63 or 183 amino acids after the repeat array cleave spacer lengths ranging from 12–39 bp at 25% of the activity of our highly active positive control. This contrasts with zinc finger nucleases (ZFNs), which are most active when the range of spacer length is limited to a few base pairs [19]. Our data indicate that shortening the C-terminus to 18 residues after the repeat array considerably narrows the range of spacer lengths that support cleavage with the optimal range spanning only 3 bp. Further definition of the TAL effector DNA binding domain may additionally focus the range of spacer lengths that permit cleavage.

We evaluated two aspects of TAL effector protein engineering, namely the suitability of NN and NK RVDs for targeting the nucleotide G and the impact on TALEN activity of both the length of the spacer and the N- and C-termini on either side of the DNA binding domain. We used our yeast-based TALEN activity assay to quickly and quantitatively evaluate these design parameters. In the future, we hope to complement such *in vivo* studies with additional *in vitro* data on the binding activity of TAL effector proteins. We are optimistic that such an integrated approach will provide a deeper understanding of how these proteins interact with DNA and thereby establish best practices for TAL effector protein design.

## Supporting Information

**Figure S1** TALEN target sequences used in G- vs. A-target yeast SSA assays.

**Figure S2** Coomassie stained gel confirming the purity of the purified TALE proteins.

## Author Contributions

Conceived and designed the experiments: MLC ZLD. Performed the experiments: MLC ZLD CGS MJO MN PB. Analyzed the data: MLC ZLD CGS MJO DFV AJB PB. Contributed reagents/materials/analysis tools: YZ DFC PB. Wrote the paper: MLC ZLD DFV MJO AJB PB.

## References

1. Bogdanove AJ, Voytas DF (2011) TAL effectors: customizable proteins for DNA targeting. Science 333: 1843–1846.
2. Bogdanove AJ, Schornack S, Lahaye T (2010) TAL effectors: finding plant genes for disease and defense. Current Opinion in Plant Biology 13: 394–401.
3. Mak AN, Bradley P, Cernadas RA, Bogdanove AJ, Stoddard BL (2012) The crystal structure of TAL effector PthXo1 bound to its DNA target. Science 335: 716–719.
4. Deng D, Yan C, Pan X, Mahfouz M, Wang J, et al. (2012) Structural basis for sequence-specific recognition of DNA by TAL effectors. Science 335: 720–723.
5. Boch J, Bonas U (2010) *Xanthomonas* AvrBs3 family-type III effectors: discovery and function. Annual Review of Phytopathology 48: 419–436.
6. Mahfouz MM, Li L, Shamimuzzaman M, Wibowo A, Fang X, et al. (2011) De novo-engineered transcription activator-like effector (TALE) hybrid nuclease with novel DNA binding specificity creates double-strand breaks. Proc Natl Acad Sci U S A: in press, DOI: 10.1073/pnas.1019533108.
7. Miller JC, Tan S, Qiao G, Barlow KA, Wang J, et al. (2011) A TALE nuclease architecture for efficient genome editing. Nature Biotechnology 29: 143–148.
8. Huang P, Xiao A, Zhou M, Zhu Z, Lin S, et al. (2011) Heritable gene targeting in zebrafish using customized TALENs. Nature Biotechnology 29: 699–700.
9. Cong L, Zhou R, Kuo YC, Cunniff M, Zhang F (2012) Comprehensive interrogation of natural TALE DNA-binding modules and transcriptional repressor domains. Nature communications 3: 968.
10. Streubel J, Blucher C, Landgraf A, Boch J (2012) TAL effector RVD specificities and efficiencies. Nature Biotechnology 30: 593–595.
11. Christian M, Cermak T, Doyle EL, Schmidt C, Zhang F, et al. (2010) Targeting DNA double-strand breaks with TAL effector nucleases. Genetics 186: 757–761.
12. Mussolino C, Morbitzer R, Lutge F, Dannemann N, Lahaye T, et al. (2011) A novel TALE nuclease scaffold enables high genome editing activity in combination with low toxicity. Nucleic Acids Research 39: 9283–9293.
13. Cermak T, Doyle EL, Christian M, Wang L, Zhang Y, et al. (2011) Efficient design and assembly of custom TALEN and other TAL effector-based constructs for DNA targeting. Nucleic Acids Research 39: e82.
14. Townsend JA, Wright DA, Winfrey RJ, Fu F, Maeder ML, et al. (2009) High-frequency modification of plant genes using engineered zinc-finger nucleases. Nature 459: 442–445.
15. Leaver-Fay A, Tyka M, Lewis SM, Lange OF, Thompson J, et al. (2011) ROSETTA3: an object-oriented software suite for the simulation and design of macromolecules. Methods in enzymology 487: 545–574.
16. Boch J, Scholze H, Schornack S, Landgraf A, Hahn S, et al. (2009) Breaking the code of DNA binding specificity of TAL-type III effectors. Science 326: 1509–1512.
17. Zhang F, Cong L, Lodato S, Kosuri S, Church GM, et al. (2011) Efficient construction of sequence-specific TAL effectors for modulating mammalian transcription. Nature Biotechnology 29: 149–153.
18. Sun N, Liang J, Abil Z, Zhao H (2012) Optimized TAL effector nucleases (TALENs) for use in treatment of sickle cell disease. Molecular bioSystems 8: 1255–1263.
19. Bibikova M, Carroll D, Segal DJ, Trautman JK, Smith J, et al. (2001) Stimulation of homologous recombination through targeted cleavage by chimeric nucleases. Molecular and cellular biology 21: 289–297.

# PERMISSIONS

All chapters in this book were first published in PLOS ONE, by The Public Library of Science; hereby published with permission under the Creative Commons Attribution License or equivalent. Every chapter published in this book has been scrutinized by our experts. Their significance has been extensively debated. The topics covered herein carry significant findings which will fuel the growth of the discipline. They may even be implemented as practical applications or may be referred to as a beginning point for another development.

The contributors of this book come from diverse backgrounds, making this book a truly international effort. This book will bring forth new frontiers with its revolutionizing research information and detailed analysis of the nascent developments around the world.

We would like to thank all the contributing authors for lending their expertise to make the book truly unique. They have played a crucial role in the development of this book. Without their invaluable contributions this book wouldn't have been possible. They have made vital efforts to compile up to date information on the varied aspects of this subject to make this book a valuable addition to the collection of many professionals and students.

This book was conceptualized with the vision of imparting up-to-date information and advanced data in this field. To ensure the same, a matchless editorial board was set up. Every individual on the board went through rigorous rounds of assessment to prove their worth. After which they invested a large part of their time researching and compiling the most relevant data for our readers.

The editorial board has been involved in producing this book since its inception. They have spent rigorous hours researching and exploring the diverse topics which have resulted in the successful publishing of this book. They have passed on their knowledge of decades through this book. To expedite this challenging task, the publisher supported the team at every step. A small team of assistant editors was also appointed to further simplify the editing procedure and attain best results for the readers.

Apart from the editorial board, the designing team has also invested a significant amount of their time in understanding the subject and creating the most relevant covers. They scrutinized every image to scout for the most suitable representation of the subject and create an appropriate cover for the book.

The publishing team has been an ardent support to the editorial, designing and production team. Their endless efforts to recruit the best for this project, has resulted in the accomplishment of this book. They are a veteran in the field of academics and their pool of knowledge is as vast as their experience in printing. Their expertise and guidance has proved useful at every step. Their uncompromising quality standards have made this book an exceptional effort. Their encouragement from time to time has been an inspiration for everyone.

The publisher and the editorial board hope that this book will prove to be a valuable piece of knowledge for researchers, students, practitioners and scholars across the globe.

# LIST OF CONTRIBUTORS

**Ilana Berger, Chen Guttman and Dotan Amar**
Department of Life Sciences, Ben-Gurion University of the Negev, Be'er Sheva, Israel

**Raz Zarivach and Amir Aharoni**
National Institute for Biotechnology in the Negev, Ben-Gurion University of the Negev, Be'er Sheva, Israel

**Nagasundarapandian Soundrarajan, Sriram Sokalingam, Govindan Raghunathan, Sun-Gu Lee and Hyun-Jong Paik**
Department of Polymer Science and Chemical Engineering, Pusan National University, Busan, South Korea

**Nediljko Budisa**
Department of Chemistry, Technische Universitä t Berlin, Franklinstrasse, Berlin, Germany

**Tae Hyeon Yoo**
Department of Molecular Science and Technology, Ajou University, Suwon, South Korea
Department of Applied Chemistry and Biological Engineering, Ajou University, Suwon, South Korea

**Joseph R. Kramer and Ichiro Matsumura**
Emory University School of Medicine, Department of Biochemistry, Rollins Research Center, Atlanta, Georgia, United States of America

**John D. Steemson, Matthias Baake and Vickery L. Arcus**
Department of Biological Sciences, University of Waikato, Hamilton, New Zealand

**Jasna Rakonjac**
Institute of Molecular BioSciences, Massey University, Palmerston North, New Zealand

**Mark T. Liddament**
OBodies Limited, Hamilton, New Zealand

**Lisbeth Olsson and Jens Nielsen**
Department of Chemical and Biological Engineering, Chalmers University of Technology, Gothenburg, Sweden
Center for Microbial Biotechnology, Department of Systems Biology, Technical University of Denmark, Lyngby, Denmark

**José Manuel Otero**
Department of Chemical and Biological Engineering, Chalmers University of Technology, Gothenburg, Sweden
Center for Microbial Biotechnology, Department of Systems Biology, Technical University of Denmark, Lyngby, Denmark
Vaccine & Biologics Process Development, Bioprocess Research & Development, Merck Research Labs, West Point, Pennsylvania, United States of America

**Simon G. Poulsen**
Center for Microbial Biotechnology, Department of Systems Biology, Technical University of Denmark, Lyngby, Denmark

**Kiran R. Patil**
Center for Microbial Biotechnology, Department of Systems Biology, Technical University of Denmark, Lyngby, Denmark
Structural and Computational Biology Unit, European Molecular Biology Laboratory, Heidelberg, Germany

**Donatella Cimini**
Department of Experimental Medicine, Second University of Naples, Naples, Italy

**Vânia Brissos, Nádia Gonçalves and Lígia O. Martins**
Instituto de Tecnologia Química e Biológica, Universidade Nova de Lisboa, Oeiras, Portugal,

**Eduardo P. Melo**
Institute for Biotechnology and Bioengineering, Center for Molecular and Structural Biomedicine, Universidade do Algarve, Faro, Portugal

**Govindan Raghunathan, Nagasundarapandian Soundrarajan, Sriram Sokalingam and Sun-Gu Lee**
Department of Chemical Engineering, Pusan National University, Busan, South Korea,

**Hyungdon Yun**
School of Biotechnology, Yeungnam University, Gyeongsan, South Korea

**Rosa Lönneborg, Edina Varga and Peter Brzezinski**
Department of Biochemistry and Biophysics, Arrhenius Laboratories for Natural Sciences, Stockholm University, Stockholm, Sweden

**Swati Choudhary, Maureen B. Quin, Ethan T. Johnson and Claudia Schmidt-Dannert**
Department of Biochemistry, Molecular Biology and Biophysics, University of Minnesota, St. Paul, Minnesota, United States of America

**Mark A. Sanders**
University Imaging Centers, University of Minnesota, St. Paul, Minnesota, United States of America

**Luca Simonelli, Mattia Pedotti, Martina Beltramello, Elsa Livoti, Federica Sallusto and Luca Varani**
Institute for Research in Biomedicine, Bellinzona, Switzerland

**Antonio Lanzavecchia**
Institute for Research in Biomedicine, Bellinzona, Switzerland
Institute of Microbiology, Eidgenössische Technische Hochschule (ETH) Zürich, Zürich, Switzerland

**Luigi Calzolai**
Institute for Health and Consumer Protection, Joint Research Centre, Ispra, Italy

**Shan Zhang, Yongzhi He, Haiying Yu and Zhiyang Dong**
State Key Laboratory of Microbial Resources, Institute of Microbiology, Chinese Academy of Sciences, Beijing, P. R. China

**Gurkan Guntas, Manu Kanwar and Marc Ostermeier**
Department of Chemical and Biomolecular Engineering, Johns Hopkins University, Baltimore, Maryland, United States of America

**Sarah J. Moore and Heidi K. Norton**
Department of Bioengineering, Stanford University, Stanford, California, United States of America

**Cheuk Lun Leung**
Department of Chemical Engineering, Stanford University, Stanford, California, United States of America

**Jennifer R. Cochran**
Department of Bioengineering, Stanford University, Stanford, California, United States of America
Department of Chemical Engineering, Stanford University, Stanford, California, United States of America
Stanford Cancer Institute and Bio-X Program, Stanford, California, United States of America

**Soomin Yoon, Won Jun Yang and Junho Chung**
Cancer Research Institute, Seoul National University School of Medicine, Seoul, Republic of Korea
Department of Biochemistry and Molecular Biology, Seoul National University School of Medicine, Seoul, Republic of Korea

**Minsoo Kim**
Cancer Research Institute, Seoul National University School of Medicine, Seoul, Republic of Korea
Scripps Korea Antibody Institute, Chuncheon, Republic of Korea

**Yoon Sup Choi**
Cancer Research Institute, Seoul National University School of Medicine, Seoul, Republic of Korea
School of Interdisciplinary Bioscience and Bioengineering (I-Bio), Pohang University of Science and Technology (POSTECH), Pohang, Republic of Korea
KT Institute of Convergence Technology, Seocho-gu, Seoul, Korea

**Sung Ho Ryu**
School of Interdisciplinary Bioscience and Bioengineering (I-Bio), Pohang University of Science and Technology (POSTECH), Pohang, Republic of Korea
Division of Integrative Bioscience and Biotechnology, Pohang University of Science and Technology (POSTECH), Pohang, Republic of Korea

**Sanguk Kim**
School of Interdisciplinary Bioscience and Bioengineering (I-Bio), Pohang University of Science and Technology (POSTECH), Pohang, Republic of Korea
Division of IT Convergence Engineering, Pohang University of Science and Technology (POSTECH), Pohang, Republic of Korea

**Kyung-Lock Kim, Parkyong Song, Young-Eun Shin and Jung-eun Noh**
Division of Integrative Bioscience and Biotechnology, Pohang University of Science and Technology (POSTECH), Pohang, Republic of Korea

**Jiho Yoo and Hyun-soo Cho**
Department of Systems Biology, College of Life Science and Biotechnology, Yonsei University, Seoul, Republic of Korea

**Zenjiro Sampei, Tomoyuki Igawa, Tetsuhiro Soeda, Yukiko Okuyama-Nishida, Chifumi Moriyama, Tetsuya Wakabayashi, Eriko Tanaka, Atsushi Muto, Tetsuo Kojima, Takehisa Kitazawa, Kazutaka Yoshihashi, Aya Harada, Miho Funaki, Kenta Haraya, Tatsuhiko Tachibana, Sachiyo Suzuki, Keiko Esaki, Yoshiaki Nabuchi and Kunihiro Hattori**
Research Division, Chugai Pharmaceutical Co., Ltd., Gotemba, Shizuoka, Japan

**Carole Fruchart-Gaillard, Gilles Mourier, Laura Vera, Nicolas Gilles, Renée Ménez, Elodie Marcon, Enrico A. Stura and Denis Servent**
CEA, iBiTecS, Service d'Ingénierie Moléculaire des Protéines, Laboratoire de Toxinologie Moléculaire et Biotechnologies, Gif-sur-Yvette, France,

**Guillaume Blanchet**
CEA, iBiTecS, Service d'Ingénierie Moléculaire des Protéines, Laboratoire de Toxinologie Moléculaire et Biotechnologies, Gif-sur-Yvette, France,
Université Pierre et Marie Curie, Paris, France

**Wei Ke and Focco van den Akker**
Departments of Biochemistry, Case Western Reserve University, Cleveland, Ohio, United States of America,

**Abigail H. Laurent, Morgan D. Armstrong, Yuchao Chen, William E. Smith, Jing Liang, Chapman M. Wright and Marc Ostermeier**
Department of Chemical and Biomolecular Engineering, Johns Hopkins University, Baltimore, Maryland, United States of America

**Sandeep Chakraborty**
Department of Biological Sciences, Tata Institute of Fundamental Research, Mumbai, India

**Tomoyuki Igawa*, Atsuhiko Maeda, Kenta Haraya, Tatsuhiko Tachibana, Yuki Iwayanagi, Futa Mimoto, Yoshinobu Higuchi, Shinya Ishii, Shigero Tamba, Naoka Hironiwa, Kozue Nagano, Tetsuya Wakabayashi, Hiroyuki Tsunoda, Kunihiro Hattori**
Research Division, Chugai Pharmaceutical Co., Ltd., Gotemba, Shizuoka, Japan

**Yuval Nov**
Department of Statistics, University of Haifa, Haifa, Israel

**Jose M. Sanchez-Ruiz**
Facultad de Ciencias, Departamento de Quimica Fisica, Universidad de Granada, Granada, Spain

**Inmaculada Sanchez-Romero**
Facultad de Ciencias, Departamento de Quimica Fisica, Universidad de Granada, Granada, Spain,
Institute of Science and Technology Austria (IST Austria), Am Campus 1, Klosterneuburg, Austria

**Keith S. Wilson**
Structural Biology Laboratory, Department of Chemistry, University of York, Heslington, York, United Kingdom, 3 Novozymes A/S, Bagsværd, Denmark

**Antonio Ariza**
Structural Biology Laboratory, Department of Chemistry, University of York, Heslington, York, United Kingdom, Institute of Molecular and Cellular Biology, University of Leeds, Leeds, United Kingdom

**Jesper Vind, Leonardo De Maria and Lars K. Skov**
Novozymes A/S, Bagsværd, Denmark

**Michael Skjøt**
Novo Nordisk A/S, Måløv, Denmark

**Ai-Sheng Xiong**
State Key Laboratory of Crop Genetics and Germplasm Enhancement, College of Horticulture, Nanjing Agricultural University, Nanjing, China

**Ri-He Peng, Bin Zhang and Quan-Hong Yao**
Biotechnology Research Institute, Shanghai Academy of Agricultural Sciences, Shanghai, China,

**Jing Zhuang and Jian Zhang**
Alberta Innovates-Technology Futures, Vegreville, Alberta, Canada,

**Jian-Min Chen**
College of Bioscience and Biotechnology, Yangzhou University, Yangzhou, China

**Michelle L. Christian, Zachary L. Demorest, Colby G. Starker, Mark J. Osborn, Michael D. Nyquist, Yong Zhang and Daniel F. Voytas**
Department of Genetics, Cell Biology & Development and Center for Genome Engineering, University of Minnesota, Minneapolis, Minnesota, United States of America

**Daniel F. Carlson**
Department of Animal Science, University of Minnesota, Saint Paul, Minnesota, United States of America

**Philip Bradley**
Division of Public Health Sciences, Fred Hutchinson Cancer
Research Center, Seattle, Washington, United States of America

**Adam J. Bogdanove**
Department of Plant Pathology, Iowa State University, Ames, Iowa, United States of America

# Index

## R



## S



## T